Model-Based Design
for Embedded Systems

Computational Analysis, Synthesis, and Design of Dynamic Models Series

Series Editor

Pieter J. Mosterman

The MathWorks
Natick, Massachusetts

Model-Based Design for Embedded Systems

Gabriela Nicolescu
Pieter J. Mosterman

CRC Press
Taylor & Francis Group
Boca Raton   London   New York

CRC Press is an imprint of the
Taylor & Francis Group, an **informa** business

CRC Press
Taylor & Francis Group
6000 Broken Sound Parkway NW, Suite 300
Boca Raton, FL 33487-2742

First issued in paperback 2017

© 2010 by Taylor and Francis Group, LLC
CRC Press is an imprint of Taylor & Francis Group, an Informa business

No claim to original U.S. Government works

ISBN 13: 978-1-138-11472-2 (pbk)
ISBN 13: 978-1-4200-6784-2 (hbk)

Library of Congress Cataloging-in-Publication Data

Model-based design for embedded systems / Gabriela Nicolescu, Pieter J. Mosterman.
    p. cm. -- (Computational analysis, synthesis, and design of dynamic models series)
    Includes bibliographical references and index.
    ISBN 978-1-4200-6784-2 (hardcover : alk. paper)
    1. Embedded computer systems--Design and construction. I. Nicolescu, G. (Gabriela) II. Mosterman, Pieter J. III. Title. IV. Series.

TK7895.E42M62 2010
004.16--dc22
    2009036996

Visit the Taylor & Francis Web site at
http://www.taylorandfrancis.com

and the CRC Press Web site at
http://www.crcpress.com

Contents

Part II Design Tools and Methodology for Multiprocessor System-on-Chip

Part III Design Tools and Methodology for Multidomain Embedded Systems

Preface

The unparalleled flexibility of computation has been a key driver and feature bonanza in the development of a wide range of products across a broad and diverse spectrum of applications such as in the automotive aerospace, health care, consumer electronics, etc. Consequently, the embedded microprocessors that implement computational functionality have become a part of almost every facet of our world, thereby significantly improving the quality of our lives. The versatility of computational features invites and endorses a degree of imagination and creativity in design that has unlocked an almost insatiable demand for consistently increasing both the complexity of embedded systems and the performance of embedded computations. The quest to rise to these demands has resulted in computing architectures of a heterogeneous nature. These architectures often integrate several types of processors, analog and digital electronic components, as well as mechanical and optical components, all on a single chip. To efficiently design for such heterogeneity and to maximally exploit its capabilities have become one of the most prominent challenges that we are now faced with as a design automation community.

Model-Based Design is emerging as a solution to bridge the gap between computational capabilities that are available but that we are yet unable to exploit. Using a computational approach in the design itself allows raising the level of abstraction of the system specification at which novel and differentiating functionalities are captured. Automation can then assist in refining this specification to an implementation. For this to be successful, performance studies of potential implementations at a high level of abstraction are essential, combined with the necessity of traceability and parameterization throughout the refinement process.

This book provides a compilation of the work of internationally renowned authors on Model-Based Design. Each chapter contributes supreme results that have helped establish Model-Based Design and that continue to expand its barriers. The respective authors excel in their expertise on the automation of design refinement and how to relate properties throughout this refinement while enabling analytic and synthetic qualities. We are delighted and honored by their participation in the effort that led to this book, and we sincerely hope that the readers will find the indulgence of intellectual achievement as enjoyable and stimulating as we do.

In closing, we would like to express our genuine appreciation and gratitude for all the time and effort that each of the authors has put in. Our pleasant collaboration has certainly helped make the completion of this project as easy as possible. Of course, none of this would have been possible without the continuous support of the team at Taylor & Francis, especially our publisher, Nora Konopka, and the staff involved in the verification and production process: Amy Blalock, Ashley Gasque, and Catherine Giacari. Many thanks to each of you. A special word of thanks goes out to Jeanne Daunais for helping us with the extensive preparation of the final material.

Gabriela Nicolescu
Pieter J. Mosterman

MATLAB® is a registered trademark of The MathWorks, Inc. For product information, please contact:

The MathWorks, Inc.
3 Apple Hill Drive
Natick, MA 01760-2098 USA
Tel: 508 647 7000
Fax: 508-647-7001
E-mail: info@mathworks.com
Web: www.mathworks.com

See www.mathworks.com/trademarks<http://www.mathworks.com/trademarks> for a list of additional trademarks.

Introduction

Gabriela Nicolescu and Pieter J. Mosterman

The purpose of this book is to provide a comprehensive overview of the current state of Model-Based Design for embedded systems, the challenges involved, and the latest trends. To achieve this objective, the book offers a compilation of 21 outstanding contributions from industry and academia. The contributions are grouped into three main parts. Part I comprises the contributions that focus on a key dimension in the design of embedded systems: the performance analysis of real-time behavior based on computational models. Part II is composed of contributions proposing approaches that take into consideration the specific characteristics and design challenges of multiprocessor systems-on-chip (MPSoCs). Part III contains contributions in the field of system-level design of multidomain systems.

An "embedded system" is a system designed to perform a dedicated function, typically with tight real-time constraints, limited dimensions, and low cost and low-power requirements. It is a combination of computer hardware and software and additional mechanical, optical, or other parts that are typically used in the specific role of actuators, sensors, and transducers, in general. In some cases, embedded systems are part of a larger system or product, for example, an antilock braking system in a car. Examples of embedded systems are cell phones, digital cameras, GPS receivers, fax machines, printers, debit/credit card readers, heart rate monitors, blood gas monitors, etc. [Gan03].

The evolution of embedded systems parallels Moore's law, which states that the number of transistors on an integrated circuit doubles every 18 months. This technological progress enabled the integration of complex electronic systems on a single chip and the emergence of MPSoCs. An MPSoC is a system-on-chip that contains multiple interconnected instruction-set processors (CPUs). The typical MPSoC is a *heterogeneous* multiprocessor [Jer04]: it is composed of several different types of processing elements. Moreover, the memory architecture and the interconnection network may be heterogeneous as well. MPSoCs can be found in many products such as digital televisions, set-top boxes, telecommunication networks, cell phones, and video games.

In response to the challenges of further miniaturization, the International Technology Roadmap for Semiconductors (ITRS) emphasizes the More Than Moore's Law movement [ITR07]. This movement focuses on system integration rather than an increase in transistor density and leads to a functional diversification in integrated systems. This diversification allows for nondigital functionality such as radio-frequency (RF) circuitry; power control, optical, and/or mechanical components; sensors; and actuators to migrate from the system board level into the so-called system-in-package (SiP) level or system-on-chip (SoC) level implementation [TUM06]. These *multidomain heterogeneous systems* enable new applications and create new markets. System applications are in key fields such as transportation, mobility, security, health, energy, communication, education, and entertainment [ZHA06]. Some examples of applications of these systems are devices for nonintrusive surgery, sensors for harsh environments (e.g., chemically aggressive, extreme temperature, excessive vibration, and high shock), car surround sensors, precrash detection, energy autonomous systems, tire pressure monitoring, car-to-car communication and navigation, and ultrasonic devices (e.g., for distance measurement and three-dimensional imaging).

The heterogeneity of modern embedded systems is responsible for a complexity that is exceptionally challenging to their design. Moreover, these systems have particularly tight performance, time-to-market, and cost constraints. To meet these constraints, engineers must find solutions to efficiently design systems including complex electronic components that integrate several cores, RF circuitry, digital and analog hardware components, as well as mechanical and optical components. *Model-Based Design* addresses this issue by focusing on computational models as the core design artifact. The model enables a hierarchical design process where the entire system is first represented at an abstract level while model elaboration iteratively refines this design and includes details as necessary to implement the required functionality. Thus, different models that may be playing different roles are required for the main stages of the design: the specification, the test and validation, and the consecutive refinement. The ability to efficiently construct models combined with associated tools and systematic methodologies primes Model-Based Design for success by providing a complete solution that enables concurrent engineering, performance analysis, automatic test generation, building efficient specifications and execution models, code generation and optimization, and automatic refinement through different abstraction levels.

This book provides a comprehensive survey and overview of the benefits of Model-Based Design in the field of heterogeneous embedded systems. The selected contributions present successful approaches where models, tools, and methodologies result in important cost reduction and performance gain of heterogeneous embedded systems while decreasing their time-to-market.

Organization
This book is divided into three parts: Part I—Real-Time and Performance Analysis in Heterogeneous Embedded Systems, Part II—Design Tools and Methodology for Multiprocessor System-on-Chip, and Part III—Design

Tools and Methodology for Multidomain Embedded Systems. The following text presents an overview of each of the parts along with a brief introduction to the contents of each of the chapters.

Part I. Real-Time and Performance Analysis in Heterogeneous Embedded Systems

Part I highlights the importance of considering the real-time aspects of heterogeneous embedded systems along with analyses of their performance. This part comprises six chapters that focus on capturing the aspects of timing in models for embedded systems, and on defining tools that exploit these models in order to provide accurate performance prediction and analysis in the early stages of design. These aspects are illustrated by means of applications in the fields of signal and image processing, automotive, robotics, and wireless communications.

Chapter 1 provides a clear introduction to system-level performance prediction and analysis. It highlights its role in design and provides an overview of the two main approaches currently employed in this field: the analytical and the simulation-based approaches. The introduction to the performance prediction and analysis stage is realized by means of a concrete video-processing application scenario. Finally, this chapter describes a modular framework that enables the analysis of the flow of event streams through a network of computation and communication resources.

Chapter 2 discusses a hybrid approach that resolves performance analysis issues by combining the advantages of simulation-based and analytical approaches. A methodology is presented based on a cycle-accurate simulation approach for embedded software that also allows the integration of abstract SystemC models. The methodology is illustrated by an audio-processing application.

Chapter 3 provides a comprehensive overview of a generic and modular framework for formal performance analysis. After an introduction to hierarchical communications and MPSoC architectures and their implications on performance, this chapter presents a methodology to systematically investigate the sensitivity of a given system configuration and to explore the design space for optimal configurations. Finally, this chapter illustrates the timing bottlenecks in an illustrative heterogeneous automotive architecture, and shows how to improve the performance guided by sensitivity analysis and system exploration.

Chapter 4 proposes a modeling framework that may be instantiated to suit a variety of scheduling scenarios and can be easily extended. This chapter first introduces the formalism underlying the approach by means of an example. The framework that is used and the types of schedulability problems that can be analyzed using this framework are then presented. The framework is then applied to the analysis of an example system.

Chapter 5 presents the MOVeS analysis framework that can be used to provide schedulability analyses for multicore embedded systems. This framework is based on an embedded system model that consists of an application model, an execution platform model, and a system model, which is a particular mapping of the application onto the execution platform. The model is represented using timed automata. Finally, this chapter shows how the framework can be used to verify properties of an embedded system by means of a number of examples including that of a smart phone, showing the ability to handle systems of realistic size.

Chapter 6 introduces a MATLAB®/Simulink®-based simulation approach. It provides models of multitasking real-time kernels and networks that can be used in simulation models for network-embedded control systems. The application of this tool is illustrated by means of a simulation of mobile robots in a sensor network.

Part II. Design Tools and Methodology for Multiprocessor System-on-Chip

Part II addresses the Model-Based Design of MPSoCs. This part provides a comprehensive overview of current design practices, the MPSoC systems applications, as well as the theory behind the current and future tools and methodologies for MPSoC design. It consists of six chapters presenting solutions for the main challenges of MPSoC design. Tools and methodologies are proposed for modeling and programming complex applications for MPSoCs, mapping these applications manually and/or automatically onto parallel MPSoC platforms; defining programming models for abstracting the hardware/software interfaces; and exploiting novel, efficient platforms and developing unified methodologies for MPSoC platform-based designs. To introduce these concepts and to illustrate the efficiency of the proposed solutions, the chapters illustrate several case studies in the fields of multimedia, wireless communications, telecommunications, and control.

Chapter 7 starts with an overview of the market trends and the key role played by MPSoC systems in contemporary industrial practice. It introduces the programming models used for MPSoCs and the main characteristics of the MPSoC platforms. This chapter also presents the MultiFlex technology that supports the mapping of user-defined parallel applications, expressed in one or more programming models, onto an MPSoC platform. Finally, this chapter illustrates the application of the proposed technology to the design of a wireless system, a 3G WCDMA/FDD base-station.

Chapter 8 presents a novel methodology for embedded software design based on a parallel programming model, called common intermediate code (CIC). In a CIC, the function and data parallelisms of application tasks are specified independently of the target architecture and design constraints. Information on the target architecture and the design constraints is separately described in an architecture information file. Based on this information, the programmer maps tasks to processing components, either manually or automatically. The efficiency of the proposed methodology is illustrated using a multimedia application, the H.263 decoder.

Chapter 9 presents a definition of the programming models that abstract hardware/software interfaces in the case of heterogeneous MPSoCs. Then, a programming environment is proposed that identifies several programming models at different MPSoC abstraction levels. The proposed approach combines the Simulink environment for high-level programming and the SystemC design language for low-level programming. The proposed methodology is applied to a heterogeneous multiprocessor platform, to explore the communication architecture and to generate efficient executable code of the software stack for an H.264 video encoder application.

Chapter 10 discusses design principles and how a unified methodology together with a supporting software framework can be developed to improve the level of efficiency of the embedded electronics industry. This chapter first presents the design challenges for future systems and a manifesto espousing the benefits of a unified methodology. Then a methodology, a platform-based design, is summarized. The chapter proceeds to present Metropolis, a software framework supporting the methodology, and Metro II, a second-generation framework tailored to industrial test cases. It concludes with two test cases in diverse domains: semiconductor chips (a universal mobile telecommunication system multichip design) and energy-efficient buildings (an indoor air quality control system).

Chapter 11 presents reconfigurable heterogeneous and homogeneous multicore SoC platforms for streaming digital signal–processing (DSP) applications. Typical examples of streaming DSP applications are wireless baseband processing, multimedia processing, medical image processing, sensor processing (e.g., for remote surveillance cameras), and phased-array radars. This chapter first introduces streaming applications and multicore architectures, presents key design criteria for streaming applications, and concludes with a multidimensional classification of architectures for streaming applications. For each category, one or more sample architectures are presented.

Chapter 12 describes the use of partial reconfiguration capabilities of some field programmable gate array (FPGAs) to provide a platform that is similar to existing general-purpose FPGAs. Partial reconfiguration involves the reconfiguration of part of an FPGA (a reconfigurable region) while another part of the FPGA (a static region) remains active and operating. This chapter illustrates this approach by presenting a case study on the design of a software-defined radio platform.

Part III. Design Tools and Methodology for Multidomain Embedded Systems

Part III covers Model-Based Design for multidomain systems. Continuous-time and discrete-event models are at the core of Model-Based Design for these systems. This part of the book is composed of nine chapters and addresses the following challenges: validating and testing traditional formal models used for blending the continuous and discrete worlds, defining semantics for combining models specific to different domains, defining and exploiting new languages that embrace the heterogeneity of domains, unambiguous specification of semantics for domain-specific modeling languages (DSMLs), and developing new methodologies for Model-Based Design for that are able to take into account the heterogeneity in multidomain systems. Model-Based Design for illustrative heterogeneous systems such as optoelectromechanical and mixed-signal systems are discussed in detail.

Chapter 13 provides a comprehensive overview of modeling with timed and hybrid automata. These types of automata have been introduced in order to blend the discrete world of computers with the continuous physical world. This chapter presents the basics of timed and hybrid automata models and methods for exhaustive or partial verification, as well as testing for these models.

Chapter 14 captures the fundamental problems, methods, and techniques for specifying the semantics of DSMLs. The effective application of DSMLs for an embedded design requires developers to have an unambiguous specification of the semantics of modeling languages. This chapter explores two key aspects of this problem: the specifications of structural and behavioral semantics.

Chapter 15 emphasizes combining different modeling perspectives and provides a simple and elegant notion of parallel composition. This chapter first reviews the concepts of "component" and "contract" from a semantic point of view. Then, the extended state machine model is described. The syntax and the expressive power used for expressions in the transitions of the state-based model are reviewed, followed by the specialization of the model into different categories to support alternative perspectives.

Chapter 16 presents an approach to solve the problem of combining continuous-time and discrete-event execution models. This chapter focuses on the analysis of the two execution models and on the definition of models for simulation interfaces required for combining these models in a global continuous/discrete execution model. It proposes a generic methodology, independent of the simulation language, for the design of continuous/discrete cosimulation tools.

Chapter 17 provides an operational semantics that supports a combination of synchronous/reactive (SR) systems, discrete-event (DE) systems, and continuous-time (CT) dynamics. This chapter outlines a corresponding denotational semantics. Dialects of DE and CT are developed that generalize SR but provide complementary modeling and design capabilities.

Chapter 18 provides an overview of the analog, mixed signal (AMS) extensions for SystemC. With these extensions, SystemC becomes amenable to modeling HW/SW systems and—at the function and architecture levels—analog and mixed-signal subsystems. The intended uses include executable specification, architecture exploration, virtual prototyping, and integration validation. This chapter describes a methodology that efficiently exploits the AMS extensions together with newly introduced converter channels. The methodology is illustrated by applying it to a software-defined radio system.

Chapter 19 presents several aspects of heterogeneous design methods in the context of increasing diversification of integration technologies. This chapter first provides the rationale and analysis of the multitechnology need in terms of technological evolution and highlights the need for advances in this domain. It then presents Rune[II], a platform that addresses some of these needs. Finally, it illustrates the direct application of the proposed approach for optical link synthesis and technology performance characterization by analyzing optical link performance for two sets of photonic component parameters and three CMOS technology generations.

Chapter 20 concentrates on multidomain modeling and multirate simulation tools that are required to support mixed-technology system-level design. This chapter proposes the Chatoyant environment for simulating and analyzing optical microelectromechanical systems (MEMSs). By supporting a variety of multidomain components and signal modeling techniques at multiple levels of abstraction, Chatoyant has the ability to perform and analyze mixed-signal trade-offs, which makes it invaluable to multitechnology system designers.

Chapter 21 underscores the importance of the role of behavioral modeling in the design of multidomain systems. This chapter presents a case study where mixed-signal hardware description languages are used to specify and simulate systems composed of elements of a different nature. A VHDL-AMS-based approach is applied for the behavioral modeling of MEMS-based microinstrumentation.

References

[Gan03] J. Gannsle and M. Barr, *Embedded Systems Dictionary*, CMP Books, San Francisco, CA, 2003.

[ITR07] International Technology Roadmap for Semiconductors, ITRS 2007 Rapport.

[Jer04] A. Jerraya and W. Wolf, *Multiprocessors Systems-on-Chip*, Morgan Kaufmann, San Francisco, CA, 2004.

[TUM06] R. Tummala, Moore's law meets its match, *IEEE Spectrum*, 43(6), 44–49, June 2006 Issue.

[ZHA06] G. Q. Zhang, M. Graef, and F. van Roosmalen, Strategic research agenda of "More than Moore," in *Proceedings of EuroSime 2006*, Como, Italy, pp. 1–6, April 24–26, 2006.

Contributors

Karl-Erik Årzén
Department of Automatic Control
Lund Institute of Technology
Lund University
Lund, Sweden

Michael M. Bails
FedEx Ground
Pittsburgh, Pennsylvania

Felice Balarin
Cadence Berkeley Labs
Berkeley, California

Laura Barrachina-Saralegui
Institut de Microelectrònica de
  Barcelona
Centre Nacional de Microelectrònica
Barcelona, Spain

Olivier Benny
STMicroelectronics, Inc.
Ottawa, Ontario, Canada

Albert Benveniste
Institut de Recherche en
  Informatique et Systèmes
  Aléatoires
Institut National de Recherche
  en Informatique et en
  Automatique
Rennes, France

Jason M. Boles
Department of Computational
  Biology
University of Pittsburgh
Pittsburgh, Pennsylvania

Youcef Bouchebaba
STMicroelectronics, Inc.
Ottawa, Ontario, Canada

Hanifa Boucheneb
Department of Computer and
  Software Engineering
Ecole Polytechnique de Montreal
Montreal, Quebec, Canada

Aske W. Brekling
Department of Informatics and
  Mathematical Modelling
Technical University of Denmark
Lyngby, Denmark

Oliver Bringmann
Forschungszentrum Informatik
Karlsruhe, Germany

Benoît Caillaud
Institut de Recherche en
  Informatique et Systèmes
  Aléatoires
Institut National de Recherche
  en Informatique et en
  Automatique
Rennes, France

Anton Cervin
Department of Automatic Control
Lund Institute of Technology
Lund University
Lund, Sweden

Donald M. Chiarulli
Department of Computer Science
University of Pittsburgh
Pittsburgh, Pennsylvania

Massimiliano D'Angelo
PARADES GEIE
Rome, Italy

Markus Damm
Institute of Computer Technology
Vienna University of Technology
Vienna, Austria

Thao Dang
Verimag Laboratory
Centre National de la Recherche
  Scientifique
Grenoble, France

Abhijit Davare
Intel Corporation
Santa Clara, California

Alexandre David
Department of Computer Science
Center for Embedded Software
  Systems
Aalborg University
Aalborg, Denmark

Douglas Densmore
Department of Electrical
  Engineering and Computer Science
University of California at Berkeley
Berkeley, California

Samuel J. Dickerson
Department of Electrical and
  Computer Engineering
University of Pittsburgh
Pittsburgh, Pennsylvania

Rolf Ernst
Institute of Computer and
  Network Engineering
Technische Universität
  Braunschweig
Braunschweig, Germany

Carles Ferrer
Instituto de Microelectrònica de
  Barcelona
Centro Nacional de Microelectrónica
Universitat Autonòma de Barcelona
Barcelona, Spain

and

Department de Microelectrònica i
  Sistemes Electrònics
Universitat Autonòma de Barcelona
Barcelona, Spain

Vincent Gagne
STMicroelectronics, Inc.
Ottawa, Ontario, Canada

Luiza Gheorghe
Department of Computer and
  Software Engineering
Ecole polytechnique de Montreal
Montreal, Quebec, Canada

Christoph Grimm
Institute of Computer Technology
Vienna University of Technology
Vienna, Austria

Soonhoi Ha
School of Computer Science and
  Engineering
Seoul National University
Seoul, Republic of Korea

Jan Haase
Institute of Computer Technology
Vienna University of Technology
Vienna, Austria

Arne Hamann
Institute of Computer and
  Network Engineering
Technische Universität
  Braunschweig
Braunschweig, Germany

Michael R. Hansen
Department of Informatics and
  Mathematical Modelling
Technical University of Denmark
Lyngby, Denmark

Rafik Henia
Institute of Computer and
  Network Engineering
Technische Universität
  Braunschweig
Braunschweig, Germany

Jacob Illum
Department of Computer Science
Center for Embedded Software
  Systems
Aalborg University
Aalborg, Denmark

Ethan Jackson
Microsoft Research
Redmond, Washington

Jan W. M. Jacobs
OCE Technologies
Venlo, the Netherlands

Ahmed Jerraya
Atomic Energy Commission
Laboratory of the Electronics and
  Information Technology
MINATEC
Grenoble, France

André B. J. Kokkeler
Department of Electrical
  Engineering, Mathematics and
  Computer Science
University of Twente
Enschede, the Netherlands

Matthias Krause
Forschungszentrum Informatik
Karlsruhe, Germany

Timothy P. Kurzweg
Department of Electrical and
  Computer Engineering
Drexel University
Philadelphia, Pennsylvania

Michel Langevin
STMicroelectronics, Inc.
Ottawa, Ontario, Canada

Kim G. Larsen
Department of Computer Science
Center for Embedded Software
  Systems
Aalborg University
Aalborg, Denmark

Bruno Lavigueur
STMicroelectronics, Inc.
Ottawa, Ontario, Canada

Edward A. Lee
University of California at Berkeley
Berkeley, California

Steven P. Levitan
Department of Electrical and
  Computer Engineering
University of Pittsburgh
Pittsburgh, Pennsylvania

David Lo
STMicroelectronics, Inc.
Ottawa, Ontario, Canada

Bibiana Lorente-Alvarez
Department de Microelectrònica
Universitat Autonòma de Barcelona
Barcelona, Spain

Jan Madsen
Department of Informatics and
 Mathematical Modelling
Technical University of Denmark
Lyngby, Denmark

Jose A. Martinez
Cadence Design Systems, Inc.
San Jose, California

Michel Metzger
STMicroelectronics, Inc.
Ottawa, Ontario, Canada

Trevor Meyerowitz
Sun Microsystems
Menlo Park, California

Stephen Neuendorffer
Xilinx Research Labs
San Jose, California

Gabriela Nicolescu
Department of Computer and
 Software Engineering
Ecole Polytechnique de Montreal
Montreal, Quebec, Canada

Ian O'Connor
Lyon Institute of Nanotechnology
Ecole Centrale de Lyon
University of Lyon
Ecully, France

Roberto Passerone
Dipartimento di Ingegneia e
 Scienza dell' Informazione
University of Trento
Trento, Italy

and

PARADES S.c.a.r.l.
Rome, Italy

Pierre G. Paulin
STMicroelectronics, Inc.
Ottawa, Ontario, Canada

Simon Perathoner
Computer Engineering and
 Networks Laboratory
Swiss Federal Institute of
 Technology Zurich
Zurich, Switzerland

Chuck Pilkington
STMicroelectronics, Inc.
Ottawa, Ontario, Canada

Alessandro Pinto
United Technology Research
 Center
Berkeley, California

Katalin Popovici
TIMA Laboratory
Grenoble, France

and

The MathWorks, Inc.
Natick, Massachusetts

Joseph Porter
Institute for Software Integrated
 Systems
Vanderbilt University
Nashville, Tennessee

Razvan Racu
Institute of Computer and
 Network Engineering
Technische Universität
 Braunschweig
Braunschweig, Germany

Gerard K. Rauwerda
Recore Systems
Enschede, the Netherlands

David K. Reed
Keynote Systems
San Mateo, California

Wolfgang Rosenstiel
Forschungszentrum Informatik
Karlsruhe, Germany

Jonas Rox
Institute of Computer and
 Network Engineering
Technische Universität
 Braunschweig
Braunschweig, Germany

Alberto Sangiovanni-Vincentelli
Department of Electrical
 Engineering
 and Computer Science
University of California at Berkeley
Berkeley, California

and

Advanced Laboratory on Embedded
 Systems
Roma, Italy

Simon Schliecker
Institute of Computer and
 Network Engineering
Technische Universität
 Braunschweig
Braunschweig, Germany

Jürgen Schnerr
Forschungszentrum Informatik
Karlsruhe, Germany

Alena Simalatsar
Dipartimento di Ingegneria e
 Scienza dell' Informazione
University of Trento
Trento, Italy

Arne Skou
Department of Computer Science
Center for Embedded Software
 Systems
Aalborg University
Aalborg, Denmark

Gerard J. M. Smit
Department of Electrical
Engineering, Mathematics &
 Computer Science
University of Twente
Enschede, the Netherlands

Janos Sztipanovits
Institute for Software Integrated
 Systems
Vanderbilt University
Nashville, Tennessee

Ryan Thibodeaux
South West Research Institute
San Antonio, Texas

Lothar Thiele
Computer Engineering and
 Networks Laboratory
Swiss Federal Institute of
 Technology Zurich
Zurich, Switzerland

Stavros Tripakis
Verimag Laboratory
Centre National de la Recherche
 Scientifique
Grenoble, France

Alexander Viehl
Forschungszentrum Informatik
Karlsruhe, Germany

Yosinori Watanabe
Cadence Design Systems, Inc.
San Jose, California

Guang Yang
National Instruments
 Corporation
Austin, Texas

Haiyang Zheng
University of California at
 Berkeley
Berkeley, California

Qi Zhu
Intel Corporation
Santa Clara, California

Part I

Real-Time and Performance Analysis in Heterogeneous Embedded Systems

1

Performance Prediction of Distributed Platforms

Lothar Thiele and Simon Perathoner

CONTENTS

1.1 System-Level Performance Analysis

One of the major challenges in the design process of distributed embedded systems is to accurately predict performance characteristics of the final system implementation in early design stages. This analysis is generally referred to as the system-level performance analysis. In this section, we introduce the relevant properties of distributed embedded systems, we describe the role of the system-level performance analysis in the design process of such platforms, and we review different analysis approaches.

1.1.1 Distributed Embedded Platforms

Embedded systems are special-purpose computer systems that are integrated into products such as cars, telecommunication devices, consumer electronics, and medical equipment. In contrast to general-purpose computer systems, embedded systems are designed to perform few dedicated functions that are typically known at the time of design. In general, the knowledge about the specific application domain and the behavior of the system is exploited to develop customized and optimized system designs. Embedded systems must be efficient in terms of power consumption, size, and cost. In addition, they usually have to be fully predictable and highly dependable, as a malfunction or a breakdown of the device they may control is in general not acceptable.

The embedding into large products and the constraints imposed by the environment often require distributed implementations of embedded systems. In addition, the components of a distributed platform are typically heterogeneous, as they perform different functionalities and are adapted to the particular local environment. Also the interconnection networks are often not homogeneous, but may be composed of several interconnected subnetworks, each one with its own topology and communication protocol. The individual processing nodes are typically not synchronized. They operate in parallel and communicate via message passing. They make autonomous decisions concerning resource sharing and scheduling of tasks. Therefore, it is particularly difficult to maintain a global-state information of the system.

Many embedded systems are reactive systems that are in a continuous interaction with their environment through sensors and actuators. Thus, they often have to execute at a pace determined by their environment, which means that they have to meet real-time constraints. For these kinds of systems, the predictability in terms of execution time is as important as the result of the processing itself: a correct result arriving later (or even earlier) than expected is wrong.

Based on the characteristics described above, it becomes apparent that heterogeneous and distributed embedded real-time systems are inherently difficult to design and to analyze, particularly, as not only the availability and the correctness of the processed results, but also the timeliness of the computations are of major concern.

1.1.2 Role of Performance Analysis in the Design Process

Reliable predictions of performance characteristics of a system such as end-to-end delays of events, memory demands, and resource usages are required to support important design decisions. In particular, the designer of a complex embedded system typically has to cope with a large design space that is given by the numerous alternatives for partitioning, allocation, and binding in the system design. Thus, he or she often needs to evaluate the performance of many design options in order to optimize the trade-offs between several

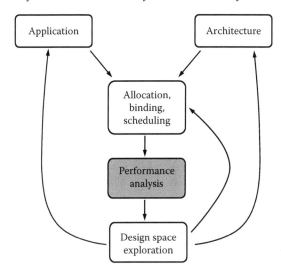

FIGURE 1.1
Performance analysis in the design space exploration cycle.

design objectives. In such a design space exploration, the performance analysis plays a crucial role, as can be seen in Figure 1.1.

Methods and tools for expedient and reliable performance analyses of system specifications at a high abstraction level are not only needed to drive the design space exploration but also for verification purposes. In particular, they permit to guarantee the functionality of a system in terms of real-time constraints before much time and resources are invested for its actual implementation.

1.1.3 Approaches to Performance Analysis

The need for accurate performance predictions in early design stages has driven research for many years. Most of the approaches for performance analysis proposed so far can be broadly divided into two classes: simulation-based methods and analytic techniques. There are also stochastic methods for performance analysis; however, we will not discuss them further in this context.

Simulation-based methods for performance estimation are widely used in industry. There are several commercial tools that support cycle-accurate cosimulation of complete HW/SW systems. Besides commercial tool suites, there also exist free simulation frameworks that can be applied for performance estimation, such as SystemC [9].

The main advantage of simulation-based performance estimation approaches is their large and customizable modeling scope, which permits to take into account various complex interactions and correlations in a system.

In addition, in many cases, the same simulation environment can be used for both function and performance verifications. However, most simulation-based performance estimation methods suffer from insufficient corner-case coverage. This means that they are typically not able to provide worst-case performance guarantees. Moreover, accurate simulations are often computationally expensive.

In other works [5,6], hybrid performance estimation methods have been presented that combine simulation and analytic techniques. While these approaches considerably shorten the simulation run-times, they still cannot guarantee full coverage of corner cases.

To determine guaranteed performance limits, analytic methods must be adopted. These methods provide hard performance bounds; however, they are typically not able to model complex interactions and state-dependent behaviors, which can result in pessimistic performance bounds.

Several models and methods for analytic performance verifications of distributed platforms have been presented so far. These approaches are based on essentially different abstraction concepts. The first idea was to extend well-known results of the classical scheduling theory to distributed systems. This implies the consideration of communication delays, which cannot be neglected in a distributed system. Such a combined analysis of processor and bus scheduling is often referred to as holistic scheduling analysis. Rather than a specific performance analysis method, holistic scheduling is a collection of techniques for the analysis of distributed platforms, each of which is tailored toward a particular combination of an event stream model, a resource-sharing policy, and communication arbitration (see [10,11,15] as examples). Several holistic analysis techniques are aggregated and implemented in the modeling and analysis suite for real-time applications (MAST) [3].*

In [12], a more general approach to extend the concepts of the classical scheduling theory to distributed systems was presented. In contrast to holistic approaches that extend the monoprocessor scheduling analysis to special classes of distributed systems, this compositional method applies existing analysis techniques in a modular manner: the single components of a distributed system are analyzed with classical algorithms, and the local results are propagated through the system by appropriate interfaces relying on a limited set of event stream models.

In this chapter, we will describe a different analytic and modular approach for performance prediction that does not rely on the classical scheduling theory. The method uses real-time calculus [13] (RTC), which extends the basic concepts of network calculus [7]. The corresponding modular performance analysis (MPA) framework [1] analyzes the flow of event streams through a network of computation and communication resources.

---

* Available as Open Source software at http://mast.unican.es

1.2 Application Scenario

In this section, we introduce the reader to the system-level performance analysis by means of a concrete application scenario from the area of video processing. Intentionally, this example is extremely simple in terms of the underlying hardware platform and the application model. On the other hand, it allows us to introduce the concepts that are necessary for a compositional performance analysis (see Section 1.4).

The example system that we consider is a digital set-top box for the decoding of video streams. The architecture of the system is depicted in Figure 1.2. The set-top box implements a picture-in-picture (PiP) application that decodes two concurrent MPEG-2 video streams and displays them on the same output device. The upper stream, V_{HR}, has a higher frame resolution and is displayed in full screen whereas the lower stream, V_{LR}, has a lower frame resolution and is displayed in a smaller window at the bottom left edge of the screen.

The MPEG-2 video decoding consists of the following tasks: variable length decoding (VLD), inverse quantization (IQ), inverse discrete cosine transformation (IDCT), and motion compensation (MC). In the considered set-top box, the decoding application is partitioned onto three processors: CPU_1, CPU_2, and CPU_3. The tasks VLD and IQ are mapped onto CPU_1 for the first video stream (process P_1) and onto CPU_2 for the second video stream (process P_3). The tasks IDCT and MC are mapped onto CPU_3 for both video streams (processes P_2 and P_4). A pre-emptive fixed priority scheduler is adopted for the sharing of CPU_3 between the two streams, with the upper stream having higher priority than the lower stream. This reflects the fact that the decoder gives a higher quality of service (QoS) to the stream with a higher frame resolution, V_{HR}.

As shown in the figure, the video streams arrive over a network and enter the system after some initial packet processing at the network interface. The inputs to P_1 and P_3 are compressed bitstreams and their outputs are partially decoded macroblocks, which serve as inputs to P_2 and P_4. The fully decoded video streams are then fed into two traffic-shaping components S_1 and S_2, respectively. This is necessary because the outputs of P_2 and P_4 are potentially bursty and need to be smoothed out in order to make sure that no packets are lost by the video interface, which cannot handle more than a certain packet rate per stream.

We assume that the arrival patterns of the two streams, V_{HR} and V_{LR}, from the network as well as the execution demands of the various tasks in the system are known. The performance characteristics that we want to analyze are the worst-case end-to-end delays for the two video streams from the input to the output of the set-top box. Moreover, we want to analyze the memory demand of the system in terms of worst-case packet buffer occupation for the various tasks.

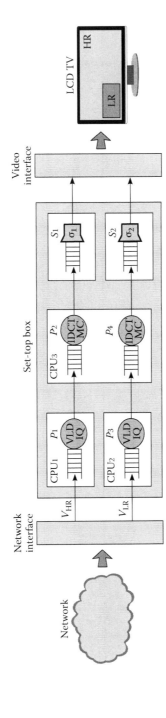

FIGURE 1.2
A PiP application decoding two MPEG-2 video streams on a multiprocessor architecture.

In Section 1.3, we at first will formally describe the above system in the concrete time domain. In principle, this formalization could directly be used in order to perform a simulation; in our case, it will be the basis for the MPA described in Section 1.4.

1.3 Representation in the Time Domain

As can be seen from the example described in Section 1.2, the basic model of computation consists of component networks that can be described as a set of components that are communicating via infinite FIFO (first-in first-out) buffers denoted as channels. Components receive streams of tokens via their input channels, operate on the arriving tokens, and produce output tokens that are sent to the output channels. We also assume that the components need resources in order to actually perform operations. Figure 1.3 represents the simple component network corresponding to the video decoding example.

Examples of components are tasks that are executed on computing resources or data communication via buses or interconnection networks. Therefore, the token streams that are present at the inputs or outputs of a component could be of different types; for example, they could represent simple events that trigger tasks in the corresponding computation component or they could represent data packets that need to be communicated.

1.3.1 Arrival and Service Functions

In order to describe this model in greater detail, at first we will describe streams in the concrete time domain. To this end, we define the concept of arrival functions: $R(s,t) \in \mathbb{R}^{\geq 0}$ denotes the amount of tokens that arrive in the time interval $[s,t)$ for all time instances, $s,t \in \mathbb{R}$, $s < t$, and $R(t,t) = 0$. Depending on the interpretation of a token stream, an arrival function may be integer valued, i.e., $R(s,t) \in \mathbb{Z}^{\geq 0}$. In other words, $R(s,t)$ "counts" the

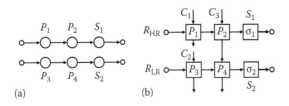

(a) (b)

FIGURE 1.3
Component networks corresponding to the video decoding example in Section 1.2: (a) without resource interaction, and (b) with resource interaction.

number of tokens in a time interval. Note that we are taking a very liberal definition of a token here: It just denotes the amount of data or events that arrive in a channel. Therefore, a token may represent bytes, events, or even demanded processing cycles.

In the component network semantics, tokens are stored in channels that connect inputs and outputs of components. Let us suppose that we had determined the arrival function $R'(s,t)$ corresponding to a component output (that writes tokens into a channel) and the arrival function $R(s,t)$ corresponding to a component input (that removes tokens from the channel); then we can easily determine the *buffer fill level*, $B(t)$, of this channel at some time t: $B(t) = B(s) + R'(s,t) - R(s,t)$.

As has been described above, one of the major elements of the model is that components can only advance in their operation if there are resources available. As resources are the first-class citizens of the performance analysis, we define the concept of service functions: $C(s,t) \in \mathbb{R}^{\geq 0}$ denotes the amount of available resources in the time interval $[s,t)$ for all time instances, $s,t \in \mathbb{R}$, $s < t$, and $C(t,t) = 0$. Depending on the type of the underlying resource, $C(s,t)$ may denote the accumulated time in which the resource is fully available for communication or computation, the amount of processing cycles, or the amount of information that can be communicated in $[s,t)$.

1.3.2 Simple and Greedy Components

Using the above concept of arrival functions, we can describe a set of very simple components that only perform data conversions and synchronization.

- *Tokenizer*: A tokenizer receives fractional tokens at the input that may correspond to a partially transmitted packet or a partially executed task. A discrete output token is only generated if the whole processing or communication of the predecessor component is finished. With the input and output arrival functions $R(s,t)$ and $R'(s,t)$, respectively, we obtain as a transfer function $R'(s,t) = \lfloor R(s,t) \rfloor$.
- *Scaler*: Sometimes, the units of arrival and service curves do not match. For example, the arrival function, R, describes a number of events and the service function, C, describes resource units. Therefore, we need to introduce the concept of scaling: $R'(s,t) = w \cdot R(s,t)$, with the positive scaling factor, w. For example, w may convert events into processor cycles (in case of computing) or into number of bytes (in case of communication). A much more detailed view on workloads and their modeling can be found in [8], for example, modeling time-varying resource usage or upper and lower bounds (worst-case and best-case resource demands).
- *AND and OR*: As a last simple example, let us suppose a component that only produces output tokens if there are tokens on all inputs (AND). Then the relation between the arrival functions at the inputs

$R_1(s,t)$ and $R_2(s,t)$, and output $R'(s,t)$ is $R'(s,t) = \min\{B_1(s) + R_1(s,t),$ $B_2(s) + R_2(s,t)\}$, where $B_1(s)$ and $B_2(s)$ denote the buffer levels in the input channels at time s. If the component produces an output token for every token at any input (OR), we find $R'(s,t) = R_1(s,t) + R_2(s,t)$.

The elementary components described above do not interact with the available resources at all. On the other hand, it would be highly desirable to express the fact that a component may need resources in order to operate on the available input tokens. A *greedy processing component* (GPC) takes an input arrival function, $R(s,t)$, and produces an output arrival function, $R'(s,t)$, by means of a service function, $C(s,t)$. It is defined by the input/output relation

$$R'(s,t) = \inf_{s \leq \lambda \leq t} \{R(s,\lambda) + C(\lambda,t) + B(s), C(s,t)\}$$

where $B(s)$ denotes the initial buffer level in the input channel. The remaining service function of the remaining resource is given by

$$C'(s,t) = C(s,t) - R'(s,t)$$

The above definition can be related to the intuitive notion of a greedy component as follows: The output between some time λ and t cannot be larger than $C(\lambda,t)$, and, therefore, $R'(s,t) \leq R'(s,\lambda) + C(\lambda,t)$, and also $R'(s,t) \leq C(s,t)$. As the component cannot output more than what was available at the input, we also have $R'(s,\lambda) \leq R(s,\lambda) + B(s)$, and, therefore, $R'(s,t) \leq \min\{R(s,\lambda) + C(\lambda,t) + B(s), C(s,t)\}$. Let us suppose that there is some last time λ^* before t when the buffer was empty. At λ^*, we clearly have $R'(s,\lambda^*) = R(s,\lambda^*) + B(s)$. In the interval from λ^* to t, the buffer is never empty and all available resources are used to produce output tokens: $R'(s,t) = R(s,\lambda^*) + B(s) + C(\lambda^*,t)$. If the buffer is never empty, we clearly have $R'(s,t) = C(s,t)$, as all available resources are used to produce output tokens. As a result, we obtain the mentioned input–output relation of a GPC.

Note that the above resource and timing semantics model almost all practically relevant processing and communication components (e.g., processors that operate on tasks and use queues to keep ready tasks, communication networks, and buses). As a result, we are not restricted to model the processing time with a fixed delay. The service function can be chosen to represent a resource that is available only in certain time intervals (e.g., time division multiple access [TDMA] scheduling), or which is the remaining service after a resource has performed other tasks (e.g., fixed priority scheduling). Note that a scaler can be used to perform the appropriate conversions between token and resource units. Figure 1.4 depicts the examples of concrete components we considered so far. Note that further models of computation can be described as well, for example, (greedy) *Shapers* that limit the amount of output tokens to a given shaping function, σ, according to $R'(s,t) \leq \sigma(t-s)$ (see Section 1.4 and also [19]).

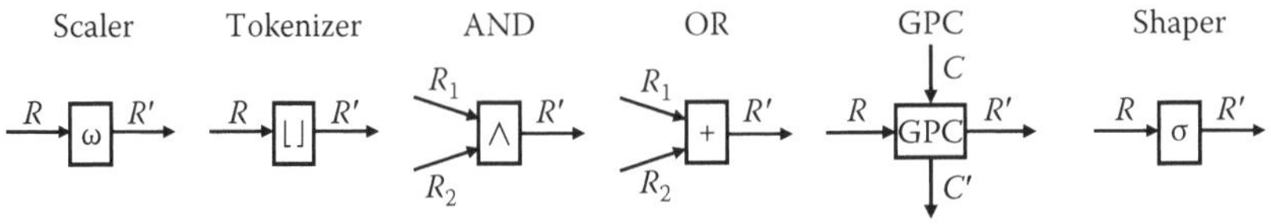

FIGURE 1.4
Examples of component types as described in Section 1.3.2.

1.3.3 Composition

The components shown in Figure 1.4 can now be combined to form a component network that not only describes the flow of tokens but also the interaction with the available resources. Figure 1.3b shows the component network that corresponds to the video decoding example. Here, the components, as introduced in Section 1.3.2, are used. Note that necessary scaler and tokenizer components are not shown for simplicity, but they are needed to relate the different units of tokens and resources, and to form tokens out of partially computed data.

For example, the input events described by the arrival function, R_{LR}, trigger the tasks in the process P_3, which runs on CPU_2 whose availability is described by the service function, C_2. The output drives the task in the process P_4, which runs on CPU_3 with a second priority. This is modeled by feeding the GPC component with the remaining resources from the process P_2.

We can conclude that the flow of event streams is modeled by connecting the "arrival" ports of the components and the scheduling policy is modeled by connecting their "service" ports. Other scheduling policies like the non-preemptive fixed priority, earliest deadline first, TDMA, general processor share, various servers, as well as any hierarchical composition of these policies can be modeled as well (see Section 1.4).

1.4 Modular Performance Analysis with Real-Time Calculus

In the previous section, we have presented the characterization of event and resource streams, and their transformation by elementary concrete processes. We denote these characterizations as *concrete*, as they represent components, event streams, and resource availabilities in the time domain and work on concrete stream instances only. However, event and resource streams can exhibit a large variability in their timing behavior because of nondeterminism and interference. The designer of a real-time system has to provide performance guarantees that cover all possible behaviors of a distributed system and its environment. In this section, we introduce the abstraction of the MPA with the RTC [1] (MPA-RTC) that provides the means to capture all possible interactions of event and resource streams in a system, and permits to derive safe bounds on best-case and worst-case behaviors.

This approach was first presented in [13] and has its roots in network calculus [7]. It permits to analyze the flow of event streams through a network of heterogeneous computation and communication resources in an embedded platform, and to derive hard bounds on its performance.

1.4.1 Variability Characterization

In the MPA, the timing characterization of event streams and of the resource availability is based on the abstractions of *arrival curves* and *service curves*, respectively. Both the models belong to the general class of variability characterization curves (VCCs), which allow to precisely quantify the best-case and worst-case variabilities of wide-sense-increasing functions [8]. For simplicity, in the rest of the chapter we will use the term VCC if we want to refer to either arrival or service curves.

In the MPA framework, an event stream is described by a tuple of arrival curves, $\alpha(\Delta) = [\alpha^l(\Delta), \alpha^u(\Delta)]$, where $\alpha^l : \mathbb{R}^{\geq 0} \mapsto \mathbb{R}^{\geq 0}$ denotes the lower arrival curve and $\alpha^u : \mathbb{R}^{\geq 0} \mapsto \mathbb{R}^{\geq 0}$ the upper arrival curve of the event stream. We say that a tuple of arrival curves, $\alpha(\Delta)$, conforms to an event stream described by the arrival function, $R(s, t)$, denoted as $\alpha \models R$ iff for all $t > s$ we have $\alpha^l(t - s) \leq R(s, t) \leq \alpha^u(t - s)$. In other words, there will be at least $\alpha^l(\Delta)$ events and at most $\alpha^u(\Delta)$ events in any time interval $[s, t)$ with $t - s = \Delta$.

In contrast to arrival functions, which describe one concrete trace of an event stream, a tuple of arrival curves represents all possible traces of a stream. Figure 1.5a shows an example tuple of arrival curves. Note that any event stream can be modeled by an appropriate pair of arrival curves, which means that this abstraction substantially expands the modeling power of standard event arrival patterns such as sporadic, periodic, or periodic with jitter.

Similarly, the availability of a resource is described by a tuple of service curves, $\beta(\Delta) = [\beta^l(\Delta), \beta^u(\Delta)]$, where $\beta^l : \mathbb{R}^{\geq 0} \mapsto \mathbb{R}^{\geq 0}$ denotes the lower service curve and $\beta^u : \mathbb{R}^{\geq 0} \mapsto \mathbb{R}^{\geq 0}$ the upper service curve. Again, we say that a tuple of service curves, $\beta(\Delta)$, conforms to an event stream described by the service function, $C(s, t)$, denoted as $\beta \models C$ iff for all $t > s$ we have $\beta^l(t - s) \leq C(s, t) \leq \beta^u(t - s)$. Figure 1.5b shows an example tuple of service curves.

Note that, as defined above, the arrival curves are expressed in terms of events while the service curves are expressed in terms of workload/ service units. However, the component model described in Section 1.4.2 requires the arrival and service curves to be expressed in the same unit. The transformation of event-based curves into resource-based curves and vice versa is done by means of so-called workload curves which are VCCs

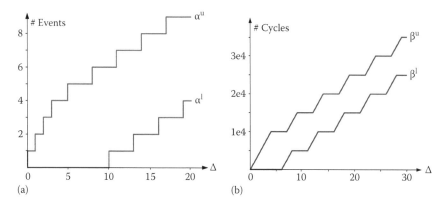

FIGURE 1.5
Examples of arrival and service curves.

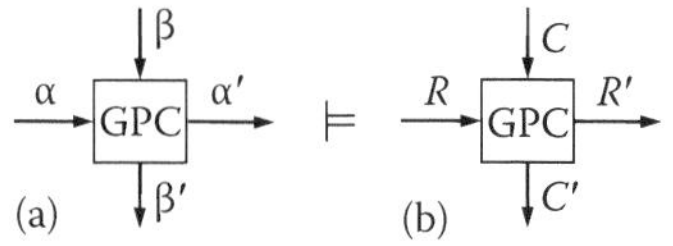

FIGURE 1.6
(a) Abstract and (b) concrete GPCs.

themselves. Basically, these curves define the minimum and maximum workloads imposed on a resource by a given number of consecutive events, i.e., they capture the variability in execution demands. More details about workload transformations can be found in [8]. In the simplest case of a constant workload w for all events, an event-based curve is transformed into a resource-based curve by simply scaling it by the factor w. This can be done by an appropriate scaler component, as described in Section 1.3.

1.4.2 Component Model

Distributed embedded systems typically consist of computation and communication elements that process incoming event streams and are mapped on several different hardware resources. We denote such event-processing units as components. For instance, in the system depicted in Figure 1.2, we can identify six components: the four tasks, P_1, P_2, P_3 and P_4, as well as the two shaper components, S_1 and S_2.

In the MPA framework, an *abstract component* is a model of the processing semantics of a concrete component, for instance, an application task or a concrete dedicated HW/SW unit. An abstract component models the execution of events by a computation or communication resource and can be seen as a transformer of abstract event and resource streams. As an example, Figure 1.6 shows an abstract and a concrete GPC.

Abstract components transform input VCCs into output VCCs, that is, they are characterized by a transfer function that relates input VCCs to output VCCs. We say that an abstract component conforms *to* a concrete component if the following holds: Given any set of input VCCs, let us choose an arbitrary trace of concrete component inputs (event and resource streams) that conforms to the input VCCs. Then, the resulting output streams must conform to the output VCCs as computed using the abstract transfer function. In other words, for any input that conforms to the corresponding input VCCs, the output must also conform to the corresponding output VCCs.

In the case of the GPC depicted in Figure 1.6, the transfer function Φ of the abstract component is specified by a set of functions that relate the incoming arrival and service curves to the outgoing arrival and service curves. In this case, we have $\Phi = [f_\alpha, f_\beta]$ with $\alpha' = f_\alpha(\alpha, \beta)$ and $\beta' = f_\beta(\alpha, \beta)$.

1.4.3   Component Examples

In the following, we describe the abstract components of the MPA framework that correspond to the concrete components introduced in Section 1.3: scaler, tokenizer, OR, AND, GPC, and shaper.

Using the above relation between concrete and abstract components, we can easily determine the transfer functions of the simple components, tokenizer, scaler, and OR, which are depicted in Figure 1.4.

- *Tokenizer*: The tokenizer outputs only integer tokens and is characterized by $R'(s,t) = \lfloor R(s,t) \rfloor$. Using the definition of arrival curves, we simply obtain as the abstract transfer function $\alpha'^u(\Delta) = \lceil \alpha^u(\Delta) \rceil$ and $\alpha'^l(\Delta) = \lfloor \alpha^l(\Delta) \rfloor$.
- *Scaler*: As $R'(s,t) = w \cdot R(s,t)$, we get $\alpha'^u(\Delta) = w \cdot \alpha^u(\Delta)$ and $\alpha'^l(\Delta) = w \cdot \alpha^l(\Delta)$.
- *OR*: The OR component produces an output for every token at any input: $R'(s,t) = R_1(s,t) + R_2(s,t)$. Therefore, we find $\alpha'^u(\Delta) = \alpha_1^u(\Delta) + \alpha_2^u(\Delta)$ and $\alpha'^l(\Delta) = \alpha_1^l(\Delta) + \alpha_2^l(\Delta)$.

The derivation of the AND component is more complex and its corresponding transfer functions can be found in [4,17].

As described in Section 1.3, a GPC models a task that is triggered by the events of the incoming event stream, which queue up in a FIFO buffer. The task processes the events in a greedy fashion while being restricted by the availability of resources. Such a behavior can be modeled with the following internal relations that are proven in [17]:*

---

*The deconvolutions in min-plus and max-plus algebra are defined as $(f \oslash g)(\Delta) = \sup_{\lambda \geq 0}\{f(\Delta + \lambda) - g(\lambda)\}$ and $(f \overline{\oslash} g)(\Delta) = \inf_{\lambda \geq 0}\{f(\Delta + \lambda) - g(\lambda)\}$, respectively. The convolution in min-plus algebra is defined as $(f \otimes g)(\Delta) = \inf_{0 \leq \lambda \leq \Delta}\{f(\Delta - \lambda) + g(\lambda)\}$.

$$\alpha'^{u}(\Delta) = \min\{(\alpha^{u} \otimes \beta^{u}) \oslash \beta^{l}, \beta^{u}\}$$

$$\alpha'^{l}(\Delta) = \min\{(\alpha^{l} \oslash \beta^{u}) \otimes \beta^{l}, \beta^{l}\}$$

$$\beta'^{u}(\Delta) = \max\{\inf_{\Delta \leq \lambda}\{\beta^{u}(\lambda) - \alpha^{l}(\lambda)\}, 0\}$$

$$\beta'^{l}(\Delta) = \sup_{0 \leq \lambda \leq \Delta}\{\beta^{l}(\lambda) - \alpha^{u}(\lambda)\}$$

In the example system of Figure 1.2, the processing semantics of the tasks P_1, P_2, P_3, and P_4 can be modeled with abstract GPCs.

Finally, let us consider a component that is used for event stream shaping. A *greedy shaper component* (GSC) with a shaping curve σ delays events of an input event stream such that the output event stream has σ as an upper arrival curve. Additionally, a greedy shaper guarantees that no events are delayed longer than necessary. Typically, greedy shapers are used to reshape bursty event streams and to reduce global buffer requirements. If the abstract input event stream of a GSC with the shaping curve, σ, is represented by the tuple of arrival curves, $[\alpha^{l}, \alpha^{u}]$, then the output of the GSC can be modeled as an abstract event stream with arrival curves:

$$\alpha^{u'}_{\text{GSC}} = \alpha^{u} \otimes \sigma \qquad \alpha^{l'}_{\text{GSC}} = \alpha^{l} \otimes (\sigma \overline{\oslash} \sigma)$$

Note that a greedy shaper does not need any computation or communication resources. Thus, the transfer function of an abstract GSC considers only the ingoing and the outgoing event stream, as well as the shaping curve, σ. More details about greedy shapers in the context of MPA can be found in [19].

In the example system of Figure 1.2, the semantics of the shapers, S_1 and S_2, can be modeled with abstract GSCs.

1.4.4 System Performance Model

In order to analyze the performance of a distributed embedded platform, it is necessary to build a system performance model. This model has to represent the hardware architecture of the platform. In particular, it has to reflect the mapping of tasks to computation or communication resources and the scheduling policies adopted by these resources.

To obtain a performance model of a system, we first have to model the event streams that trigger the system, the computation and communication resources that are available, and the processing components. Then, we have to interconnect the arrival and service inputs and outputs of all these elements so that the architecture of the system is correctly represented.

Figure 1.7 depicts the MPA performance model for the example system described in Figure 1.2. Note that the outgoing abstract service stream of GPC$_2$ is used as the ingoing abstract service stream for GPC$_4$, i.e., GPC$_4$ gets only the resources that are left by GPC$_2$. This represents the fact that the two tasks share the same processor and are scheduled according to a

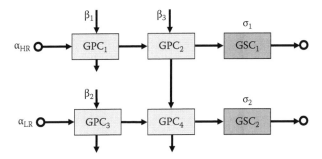

FIGURE 1.7
Performance model for the example system in Figure 1.2.

pre-emptive fixed priority scheduling policy with GPC_2 having a higher priority than GPC_4.

In general, scheduling policies for shared resources can be modeled by the way the abstract resources β are distributed among the different abstract tasks. For some scheduling policies, such as earliest deadline first (EDF) [16], TDMA [20], nonpreemptive fixed priority scheduling [4], various kinds of servers [16], or any hierarchical composition of these elementary policies, abstract components with appropriate transfer functions have been introduced. Figure 1.8 shows some examples of how to model different scheduling policies within the MPA framework.

1.4.5 Performance Analysis

The performance model provides the basis for the performance analysis of a system. Several performance characteristics such as worst-case end-to-end delays of events or buffer requirements can be determined analytically within the MPA framework.

The performance of each abstract component can be determined as a function of the ingoing arrival and service curves by the formulas of the RTC. For instance, the maximum delay, d_{max}, experienced by an event of an event stream with arrival curves, $[\alpha^l, \alpha^u]$, that is processed by a GPC on a resource with service curves, $[\beta^l, \beta^u]$, is bounded by

$$d_{max} \leq \sup_{\lambda \geq 0} \left\{ \inf\{\tau \geq 0 : \alpha^u(\lambda) \leq \beta^l(\lambda + \tau)\} \right\} \overset{\text{def}}{=} \text{Del}(\alpha^u, \beta^l)$$

The maximum buffer space, b_{max}, that is required to buffer an event stream with arrival curves, $[\alpha^l, \alpha^u]$, that is processed by a GPC on a resource with service curves, $[\beta^l, \beta^u]$, is bounded by

$$b_{max} \leq \sup_{\lambda \geq 0} \{\alpha^u(\lambda) - \beta^l(\lambda)\} \overset{\text{def}}{=} \text{Buf}(\alpha^u, \beta^l)$$

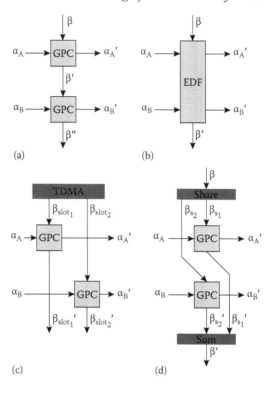

FIGURE 1.8
Modeling scheduling policies in the MPA framework: (a) preemptive fixed priority, (b) EDF, (c) TDMA, and (d) generalized processor sharing.

Figure 1.9 shows the graphical interpretation of the maximum delay experienced by an event at a GPC and the maximum buffer requirement of the GPC: d_{max} corresponds to the maximum horizontal distance between α^u and β^l, and b_{max} corresponds to the maximum vertical distance between α^u and β^l.

In order to compute the end-to-end delay of an event stream over several consecutive GPCs, one can simply add the single delays at the various components. Besides this strictly modular approach, one can also use a holistic delay analysis that takes into consideration that in a chain of task the worst-case burst cannot appear simultaneously in all tasks. (This phenomenon is described as "pay burst only once" [7].) For such a task chain the total delay can be tightened to

$$d_{max} \leq \text{Del}(\alpha^u, \beta_1^l \otimes \beta_2^l \otimes \ldots \otimes \beta_n^l)$$

For an abstract GSC, the maximum delay and the maximum backlog are bounded by

$$d_{max} = \text{Del}(\alpha^u, \sigma) \qquad b_{max} = \text{Buf}(\alpha^u, \sigma)$$

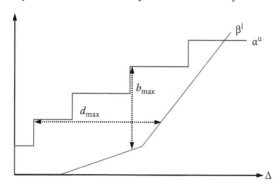

FIGURE 1.9
Graphical interpretation of d_{max} and b_{max}.

Let us come back to the example of Figure 1.2. By applying the above reasoning, the worst-case end-to-end delay for the packets of the two video streams can be analytically bounded by

$$d_{HR} \leq \text{Del}(\alpha_{HR}^u, \beta_1^l \otimes \beta_3^l \otimes \sigma_1) \qquad d_{LR} \leq \text{Del}(\alpha_{LR}^u, \beta_2^l \otimes \beta_3^{\prime l} \otimes \sigma_2)$$

1.4.6 Compact Representation of VCCs

The performance analysis method presented above relies on computations on arrival and service curves. While the RTC provides compact mathematical representations for the different operations on curves, their computation in practice is typically more involved. The main issue is that the VCCs are defined for the infinite range of positive real numbers. However, any computation on these curves requires a finite representation.

To overcome this problem, we introduce a compact representation for special classes of VCCs. In particular, we consider piecewise linear VCCs that are finite, periodic, or mixed.

- Finite piecewise linear VCCs consist of a finite set of linear segments.
- Periodic piecewise linear VCCs consist of a finite set of linear segments that are repeated periodically with a constant offset between consecutive repetitions.
- Mixed piecewise linear VCCs consist of a finite set of linear segments that are followed by a second finite set of linear segments that are repeated periodically, again with a constant offset between consecutive repetitions.

Figure 1.10a through c shows examples of these three classes of curves.

Many practically relevant arrival and service curves are piecewise linear. For example, if a stream consists of a discrete token, the corresponding

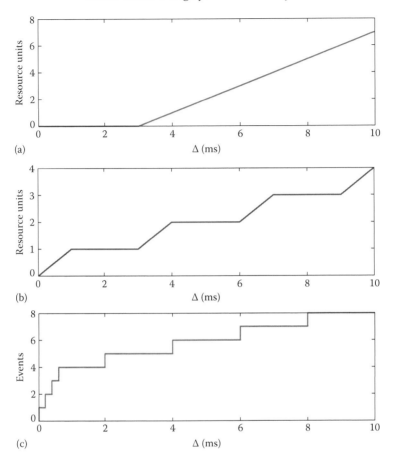

FIGURE 1.10
(a) A finite piecewise linear VCC, (b) a periodic piecewise linear VCC, and
(c) a mixed piecewise linear VCC.

arrival curve is an integer and can be represented as a piecewise constant
function. For the service curves, one could use the same reasoning, as the basic
resource units (number of clock cycles, number of bytes, etc.) are typically
also atomic. However, these units are often too fine-grained for a practical
analysis and hence it is preferable to use a continuous model. In most practical
applications, the fluid resource availability is piecewise constant over time,
that is, practically relevant service curves are also piecewise linear.

Here, we want to note that there are also piecewise linear VCCs that are
not covered by the three classes of curves that we have defined above. In
particular, we have excluded irregular VCCs, that is, VCCs with an infinite
number of linear segments that do not eventually show periodicity.
However, most practically relevant timing specifications for event streams
and availability specifications for resources can be captured by either finite,

periodic, or mixed piecewise linear VCCs. In addition, note that VCCs only describe bounds on token or resource streams, and, therefore, one can always safely approximate an irregular VCC to a mixed piecewise VCC.

In the following, we describe how these three classes of curves can be represented by means of a compact data structure. First, we note that a single linear segment of a curve can be represented by a triple $\langle x, y, s \rangle$ with $x \in \mathbb{R}_{\geq 0}$ and $y, s \in \mathbb{R}$ that specifies a straight line in the Cartesian coordinate system, which starts at the point (x, y) and has a slope s. Further, a piecewise linear VCC can be represented as a (finite or infinite) sequence $\langle \langle x_1, y_1, s_1 \rangle, \langle x_2, y_2, s_2 \rangle, \ldots \rangle$ of such triples with $x_i < x_{i+1}$ for all i. To obtain a curve defined by such a sequence, the single linear segments are simply extended with their slopes until the x-coordinate of the starting point of the next segment is reached.

The key property of the three classes of VCCs defined above is that these VCCs can be represented with a finite number of segments, which is fundamental for practical computations: Let ρ be a lower or an upper VCC belonging to a set of finite, periodic, or mixed VCCs. Then ρ can be represented with a tuple

$$\nu_\rho = \left\{ \Sigma_A, \Sigma_P, p_x, p_y, x_{p0}, y_{p0} \right\}$$

where

Σ_A is a sequence of linear segments describing a possibly existing irregular initial part of ρ

Σ_P is a sequence of linear segments describing a possibly existing regularly repeated part of ρ

If Σ_P is not an empty sequence, then the regular part of ρ is defined by the period p_x and the vertical offset p_y between two consecutive repetitions of Σ_P, and the first occurrence of the regular sequence Σ_P starts at (x_{p0}, y_{p0}). In this compact representation, we call Σ_A the aperiodic curve part and Σ_P the periodic curve part.

In the compact representation, a finite piecewise linear VCC has $\Sigma_P = \{\}$, that is, it consists of only the aperiodic part, Σ_A, with $x_{A,1} = 0$. A periodic piecewise linear VCC can be described with $\Sigma_A = \{\}$, $x_{P,1} = 0$, and $x_{p0} = 0$, that is, it has no aperiodic part. And finally, a mixed piecewise linear VCC is characterized by $x_{A,1} = 0$, $x_{P,1} = 0$, and $x_{p0} > 0$.

As an example, consider the regular mixed piecewise linear VCC depicted in Figure 1.10c. Its compact representation according to the definition above is given by the tuple

$$\nu_C = \left\{ \langle \langle 0, 1, 0 \rangle, \langle 0.2, 2, 0 \rangle, \langle 0.4, 3, 0 \rangle, \langle 0.6, 4, 0 \rangle \rangle, \langle \langle 0, 0, 0 \rangle \rangle, 2, 1, 2, 5 \right\}$$

The described compact representation of VCCs is used as a basis for practical computations in the RTC framework. All the curve operators adopted in the RTC (minimum, maximum, convolutions, deconvolutions, etc.) are closed on the set of mixed piecewise linear VCCs. This means that the result of the operators, when applied to finite, periodic, or mixed piecewise linear

VCCs, is again a mixed piecewise linear VCC. Further details about the compact representation of VCCs and, in particular, on the computation of the operators can be found in [17].

1.5   RTC Toolbox

The framework for the MPA with the RTC that we have described in this chapter has been implemented in the RTC Toolbox for MATLAB® [21], which is available at http://www.mpa.ethz.ch/Rtctoolbox.

The RTC Toolbox is a powerful instrument for system-level performance analysis of distributed embedded platforms. At its core, the toolbox provides a MATLAB type for the compact representation of VCCs (see details in Section 1.4) and an implementation of a set of the RTC curve operations. Built around this core, the RTC Toolbox provides libraries to perform the MPA, and to visualize VCCs and the related data.

Figure 1.11 shows the underlying software architecture of the toolbox. The RTC toolbox internally consists of a kernel that is implemented in Java, and a set of MATLAB libraries that connect the Java kernel to the MATLAB command line interface. The kernel consists of classes for the compact representation of VCCs and classes that implement the RTC operators. These two principal components are supported by classes that provide various utilities. On top of these classes, the Java kernel provides APIs that provide methods to create compact VCCs, compute the RTC operations, and access parts of the utilities.

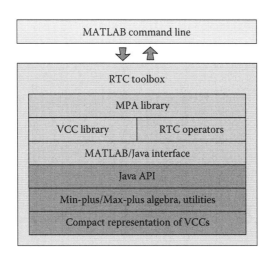

FIGURE 1.11
Software architecture of the RTC toolbox.

The Java kernel is accessed from MATLAB via the MATLAB Java Interface. However, this access is completely hidden from the user who only uses the MATLAB functions provided by the RTC libraries. The MATLAB libraries of the RTC Toolbox provide functions to create VCCs, plot VCCs, and apply operators of the RTC on VCCs. From the point of view of the user, the VCCs are MATLAB data types, even if internally they are represented as Java objects. Similarly, the MATLAB functions for the RTC operators are wrapper functions for the corresponding methods that are implemented in the Java kernel.

On top of the VCC and the RTC libraries, there is the MPA library. It provides a set of functions that facilitate the use of the RTC Toolbox for the MPA. In particular, it contains functions to create commonly used arrival and service curves, as well as functions to conveniently compute the outputs of the various abstract components of the MPA framework.

1.6   Extensions

In the previous sections, we have introduced the basics of the MPA approach based on the RTC. Recently, several extensions have been developed to refine the analysis method.

In [4], the existing methods for analyzing heterogeneous multiprocessor systems are extended to nonpreemptive scheduling policies. In this work, more complex task-activation schemes are investigated as well. In particular, components with multiple inputs and AND- or OR-activation semantics are introduced.

The MPA approach also supports the modeling and analysis of systems with dynamic scheduling policies. In [16], a component for the modeling of the EDF scheduling is presented. This work also extends the ability of the MPA framework to model and analyze hierarchical scheduling policies by introducing appropriate server components. The TDMA policies have been modeled using the MPA as well [20].

In Section 1.4, we have briefly described the GSC. More details about traffic shaping in the context of multiprocessor embedded systems and the embedding of the GSC component into the MPA framework can be found in [19].

In many embedded systems, the events of an event stream can have various types and impose different workloads on the systems depending on their types. Abstract stream models for the characterization of streams with different event types are introduced in [18]. In order to get more accurate analysis results, these models permit to capture and exploit the knowledge about correlations and dependencies between different event types in a stream.

Further, in distributed embedded platforms, there often exist correlations in the workloads imposed by events of a given type on different system components. In [22], a model is introduced to capture and characterize such workload correlations in the framework of the MPA. This work shows that the exploitation of workload correlations can lead to considerably improved analysis results.

The theory of real-time interfaces is introduced in [14]. It connects the principles of the RTC and the interface-based embedded system design [2]. The real-time interfaces represent a powerful extension of the MPA framework. They permit an abstraction of the component behavior into interfaces. This means that a system designer does not need to understand the details of a component's implementation, but only needs to know its interface in order to ensure that the component will work properly in the system. Before the introduction of the real-time interfaces, the MPA method was limited to the a posteriori analysis of component-based real-time system designs. With the real-time interfaces, it is possible to compose systems that are correct by construction.

1.7   Concluding Remarks

In this chapter, we have introduced the reader to the system-level performance prediction of distributed embedded platforms in the early design stages. We have defined the problem and given a brief overview of approaches to performance analysis.

Starting from a simple application scenario, we have presented a formal system description method in the time domain. We have described its usefulness for the simulation of concrete system executions, but at the same time we have pointed out that the method is inappropriate for worst-case analysis, as in general it cannot guarantee the coverage of corner cases.

Driven by the need to provide hard performance bounds for distributed embedded platforms, we have generalized the formalism to an abstraction in the time interval domain based on the VCCs and the RTC. We have presented the essential models underlying the resulting framework for the MPA and we have demonstrated its application. Finally, we have described a compact representation of the VCCs that enables an efficient computation of RTC curve operations in practice, and we have presented the RTC Toolbox for MATLAB, the implementation of the MPA analysis framework.

Acknowledgments

The authors would like to thank Ernesto Wandeler for contributing to some part of this chapter and Nikolay Stoimenov for helpful comments on an earlier version.

References

1. S. Chakraborty, S. Künzli, and L. Thiele. A general framework for analysing system properties in platform-based embedded system designs. In *Design Automation and Test in Europe (DATE)*, pp. 190–195, Munich, Germany, March 2003. IEEE Press.

2. L. de Alfaro and T. A. Henzinger. Interface theories for component-based design. In *EMSOFT '01: Proceedings of the First International Workshop on Embedded Software*, pp. 148–165, London, U.K., 2001. Springer-Verlag.

3. M. G. Harbour, J. J. Gutiérrez García, J. C. Palencia Gutiérrez, and J. M. Drake Moyano. Mast: Modeling and analysis suite for real time applications. In *Proceedings of 13th Euromicro Conference on Real-Time Systems*, pp. 125–134, Delft, the Netherlands, 2001. IEEE Computer Society.

4. W. Haid and L. Thiele. Complex task activation schemes in system level performance analysis. In *5th International Conference on Hardware/Software Codesign and System Synthesis (CODES+ISSS'07)*, pp. 173–178, Salzburg, Austria, October 2007.

5. S. Künzli, F. Poletti, L. Benini, and L. Thiele. Combining simulation and formal methods for system-level performance analysis. In *Design Automation and Test in Europe (DATE)*, pp. 236–241, Munich, Germany, 2006. IEEE Computer Society.

6. K. Lahiri, A. Raghunathan, and S. Dey. System-level performance analysis for designing on-chip communication architectures. *IEEE Transactions on CAD of Integrated Circuits and Systems*, 20(6):768–783, 2001.

7. J.-Y. Le Boudec and P. Thiran. *Network Calculus: A Theory of Deterministic Queuing Systems for the Internet*. Springer-Verlag, New York, Inc., 2001.

8. A. Maxiaguine, S. Künzli, and L. Thiele. Workload characterization model for tasks with variable execution demand. In *Design Automation and Test in Europe (DATE)*, pp. 1040–1045, Paris, France, February 2004. IEEE Computer Society.

9. The Open SystemC Initiative (OSCI). http://www.systemc.org.

10. J. C. Palencia Gutiérrez and M. G. Harbour. Schedulability analysis for tasks with static and dynamic offsets. In *Proceedings of the 19th Real-Time Systems Symposium*, Madrid, Spain, 1998. IEEE Computer Society.

11. T. Pop, P. Eles, and Z. Peng. Holistic scheduling and analysis of mixed time/event-triggered distributed embedded systems. In *CODES '02:*

Proceedings of the Tenth International Symposium on Hardware/Software Codesign, pp. 187–192, New York, 2002. ACM.

12. K. Richter, M. Jersak, and R. Ernst. A formal approach to mpsoc performance verification. *IEEE Computer*, 36(4):60–67, 2003.

13. L. Thiele, S. Chakraborty, and M. Naedele. Real-time calculus for scheduling hard real-time systems. In *Proceedings Symposium on Circuits and Systems*, volume 4, pp. 101–104, Geneva, Switzerland, 2000.

14. L. Thiele, E. Wandeler, and N. Stoimenov. Real-time interfaces for composing real-time systems. In *International Conference on Embedded Software EMSOFT 06*, pp. 34–43, Seoul, Korea, 2006.

15. K. Tindell and J. Clark. Holistic schedulability analysis for distributed hard real-time systems. *Microprocess. Microprogram.*, 40(2–3):117–134, 1994.

16. E. Wandeler and L. Thiele. Interface-based design of real-time systems with hierarchical scheduling. In *12th IEEE Real-Time and Embedded Technology and Applications Symposium (RTAS)*, pp. 243–252, San Jose, CA, April 2006.

17. E. Wandeler. Modular performance analysis and interface-based design for embedded realtime systems. PhD thesis, ETH Zürich, 2006.

18. E. Wandeler, A. Maxiaguine, and L. Thiele. Quantitative characterization of event streams in analysis of hard real-time applications. *Real-Time Systems*, 29(2):205–225, March 2005.

19. E. Wandeler, A. Maxiaguine, and L. Thiele. Performance analysis of greedy shapers in real-time systems. In *Design, Automation and Test in Europe (DATE)*, pp. 444–449, Munich, Germany, March 2006.

20. E. Wandeler and L. Thiele. Optimal TDMA time slot and cycle length allocation. In *Asia and South Pacific Desing Automation Conference (ASP-DAC)*, pp. 479–484, Yokohama, Japan, January 2006.

21. E. Wandeler and L. Thiele. Real-Time Calculus (RTC) Toolbox. http://www.mpa.ethz.ch/Rtctoolbox, 2006.

22. E. Wandeler and L. Thiele. Workload correlations in multi-processor hard real-time systems. *Journal of Computer and System Sciences*, 73(2):207–224, March 2007.

2

SystemC-Based Performance Analysis of Embedded Systems

Jürgen Schnerr, Oliver Bringmann, Matthias Krause, Alexander Viehl, and Wolfgang Rosentiel

CONTENTS

This chapter presents a methodology for SystemC-based performance analysis of embedded systems. This methodology is based on a cycle-accurate simulation approach for the embedded software that also allows the integration of abstract SystemC models. Compared to existing simulation-based approaches, a hybrid method is presented that resolves performance issues by combining the advantages of simulation-based and analytical approaches. In the first step, cycle-accurate static execution time analysis is applied at each basic block of a cross-compiled binary program using static processor models. After that, the determined timing information is back-annotated into SystemC for a fast simulation of all effects that cannot be resolved statically. This allows the consideration of data dependencies during runtime, and the incorporation of branch prediction and cache models by efficient source-code instrumentation. The major benefit of our approach is that the generated code can be executed very efficiently on the simulation host with approximately 90% of the speed of the untimed software without any code instrumentation.

2.1   Introduction

In the future, new system functionality will be realized less by the sum of single components, but more by cooperation, interconnection, and distribution of these components, thereby leading to distributed embedded systems. Furthermore, new applications and innovations arise more and more from a distribution of functionality as well as from a combination of previously independent functions. Therefore, in the future, this distribution will play an important part in the increase of the product value.

The system responsibility of the supplier is also currently increasing. This is because the supplier is not only responsible for the designed subsystem, but additionally for the integration of the subsystem in the context of the entire system. This integration is becoming more complex: today, requirements of single components are validated; in future, the requirements validation of the entire system has to be achieved with regard to the designed component.

What this means is that changes in the product area will lead to a paradigm shift in the design. Even in the design stage, the impact of a component on an entire system has to be considered. A comprehensive modeling of distributed systems, and an early analysis and simulation of the system integration have to be considered.

Therefore, a methodical design process of distributed embedded systems has to be established, taking into account the timing behavior of the embedded software very early in the design process. This methodical design process can be implemented by using a comprehensive modeling of distributed systems and by using a platform-independent development of the application software (UML [6], MATLAB®/Simulink® [24], and C++).

What is also important is the early inclusion of the intended target platform in the model-based system design (UML), the mapping of function blocks on platform components, and the use of virtual prototypes for the abstract modeling of the target architecture.

An early evaluation of the target platform means that the application software can be evaluated while considering the target platform. Hence, an optimization of the target platform under consideration of the application software, performance requirements, power dissipation, and reliability can take place.

An early analysis of the system integration is provided by an early verification and exposure of integration faults using virtual prototypes. After that, a seamless transition to the physical prototype can take place.

2.2 Performance Analysis of Distributed Embedded Systems

The main question of performance analysis of distributed embedded systems is: What is the global timing behavior of a system and how can it be determined? The central issue is that computation has no timing behavior as long as the target platform is not known because the target platform has a major effect on timing.

The specification, however, can contain global performance requirements. The fulfillment of these requirements depends on local timing behaviors of system parts. A solution for determining local timing properties is an early inclusion of the target architecture.

Several analytical and simulative approaches for performance analysis have previously been proposed. In this chapter, a hybrid approach for performance analysis will be presented.

2.2.1 Analytical Approaches

Analytical approaches perform a formal analysis of pessimistic corner cases based on a system model. Corner cases are hard bounds of the temporal system behavior. The approaches can be divided into two categories: black-box approaches and white-box approaches. Furthermore, both approaches can be categorized depending on the level of system abstraction and with regard to the model of computation that is employed.

Black-box approaches consider functional system components as black boxes and abstract from their internal behavior.

Black-box abstraction commonly uses a task model [33] with abstract task activation and event streams representing activation patterns [34] at the task level. Using event stream propagation, fixed points are calculated. For this, no modification of the event streams is necessary. Examples for black-box approaches are the real-time calculus (see Chapter 1 or [44]), the system-level composition by event stream propagation as it is used in SymTA/S (see Chapter 3 or [11]), the MAST framework [9], and the framework proposed by Pop et al. [31].

White-box approaches include an abstract control-flow representation of each process within the system model. Then, a global performance and communication analysis considering (data-dependent) control structures of all processes can take place. For this analysis, an extraction of the control flow from the application software or from UML models [47] is required. Then, the environment can be modeled using event models or processes. Examples for white-box approaches are the communication dependency analysis [41], the control-flow-based extraction of hierarchical event streams [1], and timed automata [27].

Analytical approaches that only rely on best-case and worst-case timing estimates are very often too pessimistic, hence risk estimation for concrete scenarios is difficult to carry out. Different probabilistic analytic approaches attempt to tackle this issue by considering probabilities of timing quantities in white-box system analysis.

Timed Petri nets [49] are able to represent the internal behavior of a system. Although there exist stochastic extensions by generalized stochastic Petri nets (GSPN) [23], these do not consider execution times of the actual system components. Furthermore, synchronization by communication and the specification of communication protocols have to be modeled explicitly and cannot be extracted from executable functional implementations of a design.

System-level performance and power estimation based on stochastic automata networks (SAN) are introduced in [22]. The system including probabilities of execution times is modeled explicitly in SAN. The actual execution behavior of the components related to timing and control flow of a functional implementation is not considered. Stochastic automata [3] extend the model of communicating I/O automata [42] by general probability distributions for verifying performance requirements of systems. The system and timing probabilities have to be modeled explicitly and no bottom-up evaluation of a functional system implementation is given.

2.2.2 Simulative Approaches

Simulative approaches perform a simulation of the entire communication infrastructure and the processing elements. If necessary, this simulation includes a hardware IP.

Depending on the underlying model of computation, a network simulator such as the OPNET [28], Simulink, or SystemC [14] can be employed to simulate a network between communicating C/C++ processes. Timing annotation of such a network simulation is possible, but the exact timing behavior of the software is missing. To obtain this timing behavior, it is necessary to simulate the software execution on the target processor. For this simulation, the binary code for the target platform component is required.

This binary code can run on an instruction set simulator (ISS). An ISS is an abstract model for executing instructions at the binary level and can be implemented either as an interpreter or as a binary code translator. It does not consider modeling of the bus behavior. The binary code translation can be realized in two different ways: either as a static or as a dynamic compilation, also called the just-in-time (JIT) compilation [26]. An ISS is used in several commercial solutions, like the CoWare Processor Designer [5], CoMET from VaST Systems Technology [45], or Synopsys Virtual Platforms [43].

Furthermore, the binary code can be executed using a processor model that captures the complete processor (functional units, pipelines, caches, register, counter, I/Os, etc.). Such a model can have several levels of accuracy. For example, it can be a transaction-level model or a register transfer model. Since our approach uses transaction-level modeling (TLM), we will describe the different levels of abstraction of TLM models in more detail in Section 2.3.

In addition to simulating the processor, peripheral components and custom hardware have to be simulated as well, either by a co-simulation with HDL (hardware description language) simulators or by using SystemC.

An abstract processor model with an integrated RTOS (real-time operating system) model using task scheduling was presented in [35]. Additionally, a processor model using neural networks for execution-cycle estimation was presented in [30]. A transaction-level approach for the performance evaluation of SoC (System-on-Chip) architectures was presented in [48]. This approach is trace-based, and, therefore, cannot guarantee a sufficient path coverage of control-flow-dominated applications.

Furthermore, the integration of a so-called cycle-approximate retargetable processor model for software performance estimation at the transaction level was presented in [13]. The major drawback of this approach is that microarchitecture-dependent properties are measured on the target platform and are included probabilistically during execution. The comparable low deviation from on-board measurements of only 8% results from the fact that the reference measurements used the same examples and input data that the models were built from. It is likely that data-dependent effects will lead to larger accuracy errors.

2.2.3 Hybrid Approaches

Hybrid approaches combine the advantages of analytical and simulative approaches. A hybrid approach for combining simulation and formal analysis for tightening bounds of system-level performance analysis was presented in [20]. The objectives are to determine timing characteristics of nonformally specified components by simulation and to integrate simulation results into a framework for formal performance analysis. In comparison to the approach shown in [20], we focus on a fast timing simulation of the embedded software. The results determined using our approach may be included in system-level performance methodologies with the benefit of high accuracy and large time savings in the simulation stage.

Analytic performance risk quantification based on profiled execution times is presented in [46]. The model is derived from physical implementations. Although it is able to represent the temporal behavior of communication, computation, and synchronization, data-dependent timing effects cannot be detected reliably.

A hybrid model for the fast simulation that allows switching between native code execution and ISS-based simulation was presented in [17]. Another approach using a hybrid model was shown in [38] and [36]. This approach is based on the translation of an object code into an annotated binary code for the target processor. For the cycle-accurate execution of the annotated code on this processor, a special hardware is needed.

2.3 Transaction-Level Modeling

The TLM is a high-level approach to model systems where computation and communication between system modules are separated for each module of the proposed target architecture. Components that are described at different levels of abstraction can be integrated and exchanged in one common system model using standardized interfaces. Furthermore, an exploration and a refinement of components and their implementation in the global architecture can be performed.

Transaction-level models address the problem of designing increasingly complex systems by raising the level of design abstraction above the register transfer level (RTL). The Open SystemC Initiative (OSCI) Transaction-Level Working Group has defined different levels of abstraction. Of these abstraction levels, transaction-level models apply at the levels between the Algorithmic Level (AL) and the RTL. These levels are introduced in [2] and also are briefly presented here.

- Algorithmic Level (AL): Purely behavioral, no architectural detail whatsoever.
- Untimed (UT) Modeling: Notion of simulation time is not required, each process runs up to the next explicit synchronization point before yielding.
- Loosely Timed (LT) Modeling: The simulation time is used, but processes are temporally decoupled from the simulation time. Each process keeps a tally of the time it consumes, and may yield because it reaches an explicit synchronization point or because it has consumed its time quantum.
- Approximately Timed (AT) Modeling: Processes run in lockstep with the SystemC simulation time. Delays of process interactions are annotated by using timeouts (wait) or timed event notifications.
- Register Transfer Level (RTL): Has the description of the register and combination logic.

2.3.1 Accuracy and Speed Trade-Off during Refinement Process

The proposed approach allows for an early incorporation of the effects of the underlying target platform into the embedded software design. Platform architectures are not limited to single-core processors with simple communication architectures. The approach also applies to multi-core architectures and distributed embedded systems with complex network architectures, for instance, networks of interconnected electronic control units (ECUs) in the automotive domain. This flexibility requires a seamless refinement flow for the embedded software beginning at the platform-independent software down to the platform-specific target software. By stepwise refinement of the system model, a design at lower levels of abstraction, where the simulation is more accurate at the expense of increasing the simulation time, can be obtained. Two different refinement strategies have to be distinguished: computation refinement and communication refinement. Computation refinement is especially applicable for single-processor embedded systems without a special focus on communication aspects. In this case, the complexity of executing a cross-compiled binary code may be acceptable. But with an increasing number of processing units and network complexity (e.g., hierarchical automotive networks consisting of FlexRay, CAN, LIN, and MOST buses), the simulation speed for analyzing the timing influences of the embedded software on the distributed system becomes unacceptable. This issue is addressed by a highly scalable performance simulation approach for networked embedded systems because the integration of the ISSs with a high simulation time into each processing element becomes obsolete. A decreasing simulation time is specifically enabled by keeping computation at a high level of abstraction whereas communication is refined to a lower level or vice versa. During the refinement flow, different levels of abstraction are traversed. This strategy is supported by the TLM in SystemC. More detailed information about the modeling and refinement of SystemC simulation models within the scope of the automotive embedded software and AUTOSAR [10] is presented in [19].

2.3.1.1 Communication Refinement

As shown in Figure 2.1, there exists a communication scheme at the UT level that is called point-to-point communication. The point-to-point communication can be timed or untimed. A timed representation means that an abstract timing behavior is provided by use of wait(T) statements, which are allowed to be introduced within the point-to-point communication. However, only certain cases can be considered during simulation. The consideration of all cases possibly results in an infinite or at least in an unacceptable simulation time. This is a general problem of simulation, and only a formal analysis can solve this problem to cover each corner case of the system behavior. Such a method is also introduced in [39] and [40].

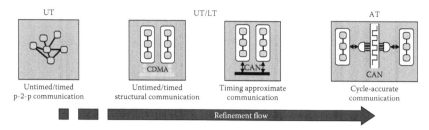

FIGURE 2.1
The communication refinement flow. (From Krause, M. et al., *Des. Automat. Embed. Syst.*, 10, 237, 2005. With permission.)

The refinement from untimed modeling to loosely timed modeling introduces abstract or dedicated buses respectively. The ports and interfaces of the untimed modeling remain and only the channel implementation is replaced. Figure 2.1 illustrates the communication refinement process for a CAN bus.

Refinement from the TLM to the RTL description means replacing transactions by signals. This refinement technique is described in [8] in detail.

2.3.1.2 Computation Refinement of Software Applications

Considering computation, the design is transformed to a structural representation by specifying the desired target architecture. Using untimed modeling, processes are still simulated as parallel processes by the SystemC simulation kernel. The most important impact to a software realization is the implemented scheduling of threads that are assigned to the same processing elements.

The refinement from an unstructured to a structured execution order is done by introducing a scheduler model to the system description, or, for more detailed modeling, an abstract RTOS model. However, this requires the specification of preemption points. Together with such preemption points, the timing information of the runtime is annotated. This chapter presents an approach on how to obtain and integrate the accurate timing information. Figure 2.2 illustrates the computation refinement process. Detailed information about refinement is presented in [18].

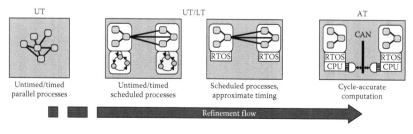

FIGURE 2.2
The computation refinement flow. (From Krause, M. et al., *Des. Automat. Embed. Syst.*, 10, 238, 2005. With permission.)

2.4 Proposed Hybrid Approach for Accurate Software Timing Simulation

In this section, a hybrid approach for the performance simulation of the embedded software [37] will be presented. Hybrid approaches consist of a combination of analytic and simulative approaches with the objective of gaining simulation speed while maintaining sufficient accuracy.

The integratability in a global refinement flow for the software down to the cycle-approximate level is given by the automated generation of the TLM interfaces.

The static worst-case/best-case execution time (WCET/BCET) analysis abstracts the influence of data dependencies on the software execution time. Because of this, the BCET/WCET analysis delivers very good results of the entire basic blocks, but it is too pessimistic across the basic block boundaries. Furthermore, the effects of a concurrent cache usage of different applications on multi-core architectures lead to even wider bounds. An analytic solution for this issue is still unknown. The objective of the presented approach is the reduction of pessimism that is contained in the WCET/BCET boundaries.

Simulative techniques that consider an application with concrete input data and the target architecture can be used to determine the timing behavior of the software on the underlying architecture. The proposed approach tries to prevent repeated time-consuming interpretation and repeated timing determination of all executed binary code instructions on the target architecture.

The hybrid approach provided in this chapter applies back-annotation of the WCET/BCET values. These values are determined statically at the basic block level using the binary code that was generated from the C source code. Additionally, the timing impact of data-dependent architectural properties such as branch prediction is also considered effectively. The tool that implements the proposed methodology generates the SystemC code. This code can be compiled for any host machine to be used for a target platform-independent simulation.

Communication calls in the automatically created SystemC models are encapsulated in the TLM [7] communication primitives. In this way, a clean and standardized ability to integrate the timed embedded software in virtual SystemC prototypes is provided.

One major advantage of the presented methodology is in the area of multi-core processors with shared caches. Whereas static analysis has no knowledge of concurrent cache usage of different applications and the impact on execution time, the presented methodology is able to handle these issues. How this is done will be described in more detail in Section 2.4.6.

Another possibility would be a translation of the binary code into the annotated SystemC code. One of the main advantages of such an approach is that no source code is needed, as the binary code is used for determining cycle counts and for generating the SystemC code. Another advantage is that a translation of the binary code into the SystemC code generates a fast code compared to an interpreting ISS, as no decoding of instructions is needed and the generated SystemC code can be easily used within a SystemC simulation environment. However, this approach has some major disadvantages. One main drawback is that the same problems that have to be solved in the static compilation (binary translation) have to be solved here (e.g., addresses of calculated branch targets have to be determined). Another disadvantage is that the automatically generated code is not very easily read by humans.

2.4.1   Back-Annotation of WCET/BCET Values

In this section, we will describe our approach in more detail. Figure 2.6 shows an overview of the approach.

First, the C source code has to be taken and translated using an ordinary C (cross)-compiler into the binary code for the embedded processor (source processor). After that, our back-annotation tool reads the object file and a description of the used source processor. This description contains both a description of the architecture and a description of the instruction set of the processor.

Figure 2.4 shows an example for the description of the architecture. It contains information about the resources of the processor (Figure 2.4a). This information is used for the modeling of the pipeline. Furthermore, it contains a description of the properties of the instruction (Figure 2.4b) and data caches (Figure 2.4c). Furthermore, such a description can contain information about the branch prediction of the processor.

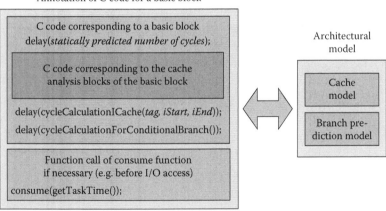

FIGURE 2.3
Back-annotation of WCET/BCET values. (From Schnerr, J. et al., High-performance timing simulation of embedded software, in: *Proceedings of the 45th Design Automation Conference (DAC)*, Anaheim, CA, pp. 290–295, June 2008. Copyright: ACM. Used with permission.)

```
<architecture>
  <resource>FI</resource>
  <resource>DI</resource>                                          (a)
  <resource>EX</resource>
  <resource>WB</resource>
  <icache>
    <associativity>2</associativity>
    <cachelinesize>8</cachelinesize>                               (b)
    <cachesize>4096</cachesize>
    <replacement>lru</replacement>
  </icache>
  <dcache>
    <associativity>2</associativity>
    <cachelinesize>8</cachelinesize>
    <cachesize>4096</cachesize>                                    (c)
    <replacement>lru</replacement>
    <writebackpolicy>write-back</writebackpolicy>
  </dcache>
</architecture>
```

FIGURE 2.4
Example for a description of the architecture.

Figure 2.5 shows an example for the description of the instruction set. This description contains information about the structure of the bit image of the instruction code (Figure 2.5c). It also contains information to determine the timing behavior of instructions and the timing behavior of instructions that are executed in context with other instructions (Figure 2.5d). Furthermore, for debugging and documentation purposes more information about the instruction can be given (Figure 2.5a and b).

Using this description, the object code is decoded and translated into an intermediate representation consisting of a list of objects. Each of these objects represents one intermediate instruction.

In the next step, the basic blocks of this program are determined using the intermediate representation consisting of a list of objects. As a result, using this list, a list of basic blocks is built.

After that, the execution time is statically calculated for each basic block with respect to the provided pipeline model of the proposed source processor. This calculation step is described in more detail in Section 2.4.3.

Subsequently, the back-annotation correspondences between the C source code and the binary code are identified. Then, the back-annotation process takes place. This is done by automated code instrumentation for cycle generation and dynamic cycle correction. The structure and functionality of this code are described in Section 2.4.2.

Not every impact of the processor architecture on the number of cycles can be predicted statically. Therefore, if dynamic, data-dependent effects (e.g., branch prediction and caches) have to be taken into account, an

```
<processor>
<defr>a 4</defr>
<defr>b 4</defr>
<defr>c 4</defr>
<defr>d 4</defr>
<def>n 2</def>
   ⋮

<!-- 0x06000001 addsc.a Ac, Ab, Da, n (RRS) -->
<instruction>
 <syntax>
 addsc.a A<par>c</par>, A<par>b</par>, D<par>a</par>, <par>n</par>      (a)
 </syntax>

 <description>
 Left-shift the contents of data register Da by the amount specified
 by n, where n can be 0, 1, 2, or 3. Add that value to the contents    (b)
 of address register Ab and put the result in address register Ac.
 </description>

 <image>
 <par>c</par>0110000000<par>n</par><par>b</par><par>a</par>00000001    (c)
 </image>

 <uses>FI 1</uses>
 <uses>DI 1</uses>                                                      (d)
 <uses>EX 1</uses>
 <uses>WB 1</uses>
 </instruction>
   ⋮

</processor>
```

FIGURE 2.5

Example for a description of an instruction.

additional code needs to be added. Further details concerning this code are described in Section 2.4.5.

During back-annotation, the C program is transformed into a cycle-accurate SystemC program that can be compiled to be executed on the processor of the simulation host (target processor).

One advantage of this approach is a fast execution of the annotated code as the C source code does not need major changes for back-annotation. Moreover, the generated SystemC code can be easily used within a SystemC simulation environment. The difficulty in using this approach is to find the corresponding parts of the binary code in the C source code if the compiler optimizes or changes the structure of the binary code too much. If this happens, recompilation techniques [4] have to be used to find the correspondences.

2.4.2 Annotation of SystemC Code

On the left-hand side of Figure 2.3, there is the necessary annotation of a piece of the C code that corresponds to a basic block. The right-hand side of

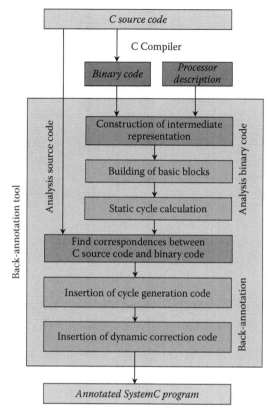

FIGURE 2.6

General principle for a basic block annotation. (Copyright: ACM. Used with permission.)

this figure shows the cache model and the branch prediction model that are used during runtime.

As described in further detail in Section 2.4.7, a function `delay` is used for accumulating the execution time of an annotated basic block during simulation. At the beginning of the annotated basic block code, the annotation tool adds a call of the delay function that contains the statically determined number of cycles this basic block would use on the source processor as a parameter. How this number is calculated is described in more detail in Section 2.4.3. In modern processor architectures, the impact of the processor architecture on the number of executed cycles cannot be completely predicted statically. Especially the branch prediction and the caches of a processor have a significant impact on the number of used cycles. Therefore, the statically determined number of cycles has to be corrected dynamically. The partitioning of the basic block for the calculation of additional cycles of instruction cache misses, as shown in Figure 2.3, is explained in Section 2.4.5.

If there is a conditional branch at the end of a basic block, branch prediction has to be considered and possible correction cycles have to be added. This is described in more detail in Section 2.4.5.

As shown in Figure 2.3, the back-annotation tool adds a call to the `consume` function that performs cycle generation at the end of each basic block code. If necessary, this instruction generates the number of cycles this basic block would need on the source processor. How this `consume` function works is described in Section 2.4.7.

In order to guarantee both—as fast as possible the execution of the code as well as the highest possible accuracy—it is possible to choose different accuracy levels of the generated code that parameterize the annotation tool. The first and the fastest one is a purely static prediction. The second one additionally includes the modeling of the branch prediction. And the third one takes also the dynamic inclusion of instruction caches into account.

The cycle calculation in these different levels will be discussed in more detail in the following sections.

2.4.3   Static Cycle Calculation of a Basic Block

In modern architectures, pipeline effects, superscalarity, and caches have an important impact on the execution time. Because of this, a calculation of the execution time of a basic block by summing the execution or latency times of the single instructions of this block is very inaccurate.

Therefore, the incorporation of a pipeline model per basic block becomes necessary [21]. This model helps statically predict pipeline effects and the effects of superscalarity. For the generation of this model, informations about the instruction set and the pipelines of the used processor are needed. These informations is contained in the processor description that is used by the annotation tool. With regard to this, the tool uses a modeling of the pipeline to determine which instructions of the basic block will be executed in parallel on a superscalar processor and which combinations of instructions in the basic block will cause pipeline stalls. Details of this will be described in the next section.

With the information gained by basic block modeling, a prediction is carried out. This prediction determines the number of cycles the basic block would have needed on the source processor.

Section 2.4.5 will show how this kind of prediction is improved during runtime, and how a cache model is included.

2.4.4   Modeling of Pipeline for a Basic Block

As previously mentioned, the processor description contains informations of the resources the processor has and of the resources a certain instruction uses. These informations about the resources are used to build a resource usage model that specifies microarchitecture details of the used processor.

For this model, it is assumed that all units in the processor such as functional units, pipeline stages, registers, and ports form a set of resources. These resources can be allocated or released by every instruction that is executed. This means that the resource usage model is based on the assumption that every time when an instruction is executed, this instruction allocates a set of resources and carries out an action. When the execution proceeds, the allocated resources and the carried-out actions change.

If two instructions wait for the same resource, then this is resolved by allocating the resource to the instruction that entered the pipeline earlier. This model is powerful enough to describe pipelines, superscalar execution, and other microarchitectures.

2.4.4.1 Modeling with the Help of Reservation Tables

The timing information of every program construct can be described with a reservation table. Originally, reservation tables were proposed to describe and analyze the activities in a pipeline [32]. Traditionally, reservation tables were used to detect conflicts for the scheduling of instructions [25]. In a reservation table, the vertical dimension represents the pipeline stages and the horizontal dimension represents the time. Figure 2.7 shows an example of a basic block and the corresponding reservation table. In the figure, every entry in the reservation table shows that the corresponding pipeline stage is used in the particular time slot. The entry consists of the number of the instruction that uses the resource. The timing interdependencies between the instructions of a basic block are analyzed using the composition of their basic block.

In the reservation table, not only conflicts that occur because of the different pipeline stages, but also data dependencies between the instructions can be considered.

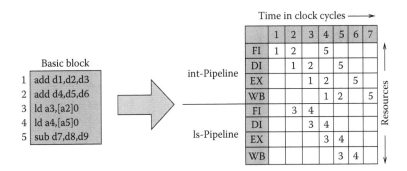

FIGURE 2.7
Example of a reservation table for a basic block.

2.4.4.1.1   Structural Hazards

In the following, a modeling of the instructions in a pipeline using reservation tables will be described [12,32]. To determine at which time after the start of an instruction the execution of a new instruction can start without causing a collision, these reservation tables have to be analyzed. One possibility to determine if two instructions can be started in the distance of K time units is to overlap the reservation with itself using an offset of K time units. If a used resource is overlapped by another, then there will be a collision in this segment and K is a forbidden latency. Otherwise, no collision will occur and K is an allowed latency.

2.4.4.1.2   Data Hazards

The time delay caused by data hazards is modeled in the same way as the delay caused by structural hazards. As the result of the pipelining of an instruction sequence should be the same as the result of sequentially executed instructions, register accesses should be in the same order as they are in the program. This restriction is comparable with the usage of pipeline stages in the order they are in the program, and, therefore, it can be modeled by an extension of the reservation table.

2.4.4.1.3   Control Hazards

Some processors (like the MIPS R3000 [12]) use delayed branches to avoid the waiting cycle that otherwise would occur because of the control hazard. This can be modeled by adding a delay slot to the basic block with the branch instruction. Such a modeling is possible, because the instruction in the delay slot is executed regardless of the result of the branch instruction.

2.4.4.2   Calculation of Pipeline Overlapping

In order to be able to model the impact of architectural components such as pipelines, the state of these components has to be known when the basic block is entered. If the state is known, then it is possible to find out the gain that results from the use of this component.

   If it is known that in the control-flow graph of the program, node e_i is the predecessor of node e_j, and the pipeline state after the execution of node e_i is also known, then the information about this state can be used to calculate the execution time of node e_j. This means the gain resulting from the fact that node e_i is executed before node e_j can be calculated.

   The gain will be calculated for every pair of succeeding basic blocks using the pipeline overlapping. This pipeline overlapping is determined using reservation tables [29]. Appending a reservation table of a basic block to a reservation table of another basic block works the same way as appending an instruction to this reservation table. Therefore, it is sufficient to consider only the first and the last columns. The maximum number of columns that have to be considered does not have to be larger than the maximum number of cycles for which a single instruction can stay in the pipeline [21].

2.4.5   Dynamic Correction of Cycle Prediction

As previously described, the actual cycle count a processor needs for executing a sequence of instructions cannot be predicted correctly in all cases. This is the case if, for example, a conditional branch at the end of a basic block produces a pipeline flush, or if additional delays occur because of cache misses in instruction caches. The combination of static analysis and dynamic execution provides a well-suited solution for this problem, since statically unpredictable effects of branch and cache behaviors can be determined during execution. This is done by inserting appropriate function calls into the translated basic blocks. These calls interact with the architectural model in order to determine the additional number of cycles caused by mispredicted branch and cache behaviors. At the end of each basic block, the generation of previously calculated cycles (static cycles plus correction cycles) can occur (Figure 2.3).

2.4.5.1   Branch Prediction

Conditional branches have different cycle times depending on four different cases resulting from the combination of predicted and mispredicted branches, as well as taken and non-taken branches. A correctly predicted branch needs less cycles for execution than a mispredicted one. Furthermore, additional cycles can be needed if a correctly predicted branch is taken, as the branch target has to be calculated and loaded in the program counter. This problem is solved by implementing a model of the branch prediction and by a comparison of the predicted branch behavior with the executed branch behavior. If dynamic branch prediction is used, a model of the underlying state machine is implemented and its results are compared with the executed branch behavior. The cycle count of each possible case is calculated and added to the cumulative cycle count before the next basic block is entered.

2.4.5.2   Instruction Cache

Figure 2.3 shows that for the simulation of the instruction cache, every basic block of the translated program has to be divided into several cache analysis blocks. This has to be done until the tag changes or the basic block ends. After that, a function call to the cache handling model is added. This code uses a cache model to find out possible cache hits or misses.

The cache simulation will be explained in more detail in the next few paragraphs. This explanation will start with a description of the cache model.

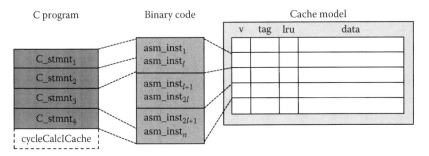

FIGURE 2.8
Correspondence C—assembler—cache line. (Copyright: ACM. Used with permission.)

2.4.5.3   Cache Model

The cache model, as it can be seen on the right-hand side of Figure 2.8, contains data space that is used for the administration of the cache. In this space, the valid bit, the cache tag, and the *least recently used* (*lru*) information (containing the replacement strategy) for each cache set during runtime is saved.

The number of cache tags and the according amount of valid bits that are needed depend on the associativity of the cache (e.g., for a two-way set associative cache, two sets of tags and valid bits are needed).

2.4.5.4   Cache Analysis Blocks

In the middle of Figure 2.8, the C source code that corresponds to a basic block is divided in several smaller blocks, the so-called cache analysis blocks. These blocks are needed for the consideration of the effects of instruction caches. Each one of these blocks contains the part of a basic block that fits into a single cache line.

As every machine language instruction in such a cache analysis block has the same tag and the same cache index, the addresses of the instructions can be used to determine how a basic block has to be divided into cache analysis blocks. This is because each address consists of the tag information and the cache index.

The cache index information (*iStart* to *iEnd* in Figure 2.3) is used to determine at which cache position the instruction with this address is cached. The tag information is used to determine which address was cached, as there can be multiple addresses with the same cache index. Therefore, a changed cache tag can be easily determined during the traversal of the binary code with respect to the cache parameters. The block offset information is not needed for the cache simulation, as no real caching of data takes place.

After the tag has been changed or at the end of a basic block, a function call that handles the simulated cache and the calculation of the additional cycles of cache misses are added to this block. More details about this function are described in the next section.

```
int cycleCalculationICache( tag, iStart, iEnd )
{
  for index = iStart to iEnd
  {
    if tag is found in index and valid bit is set then
    { // cache hit
      renew lru information
      return 0
    }
    else
    { // cache miss
      use lru information to determine tag to overwrite
      write new tag
      set valid bit of written tag
      renew lru information
      return additional cycles needed for cache miss
    }
  }
}
```

Listing 2.1
Function for cache cycle correction.

2.4.5.5 Cycle Calculation Code

As previously mentioned, each cache analysis block is characterized by a combination of tag and cache-set index informations. At the end of each basic block, a call to a function is included. During runtime, this function should determine whether the different cache analysis blocks that the basic block consists of are in the simulated cache or not. This way, cache misses are detected.

The function is shown in Listing 2.1. It has the tag and the range of cache-set indices (*iStart* to *iEnd*) as parameters.

To find out if there is a cache hit or a cache miss, the function checks whether the tag of each cache analysis block can be found in the specified set and whether the valid bit for the found tag is set.

If the tag can be found and the valid bit is set, the block is already cached (cache hit) and no additional cycles are needed. Only the *lru* information has to be renewed.

In all other cases, the *lru* information has to be used to determine which tag has to be overwritten. After that, the new tag has to be written instead of the found old one, and the valid bit for this tag has to be set. The *lru* information has to be renewed as well. In the final step, the additional cycles are returned and added to the cycle correction counter.

2.4.6 Consideration of Task Switches

In modern embedded systems, software performance simulation has to handle task switching and multiple interrupts. Cooperative task scheduling can already be handled by the previously mentioned approach since the presented cache model is able to cope with nonpreemptive task switches. Interrupts, and cooperative and nonpreemptive task scheduling can be handled similarly because the task preemption is usually implemented by using software interrupts. Therefore, the incorporation of interrupts is discussed in the following.

Software interrupts had to be included in the SystemC model. This has been achieved by the automatic insertion of dedicated preemption points after cycle calculation. This approach provides an integration of different user-defined task scheduling policies, and a task switch generates a software interrupt. Since the cycle calculation is completed before a task switch is executed and a global cache and branch prediction model is used, no other changes are necessary. A minor deviation of the cycle count for certain processes can occur because of the actual task switch that is carried out with a small delay caused by the projection of the task preemption at the binary-code level to the C/C++ source-code level. But, nevertheless, the cumulative cycle count is still correct. The accuracy can be increased by the insertion of the cycle calculation code after each C/C++ statement.

If the additional delay caused by the context switch itself has to be included, the (binary) code of the context switch routine can be treated like any other code.

2.4.7 Preemption of Software Tasks

For the modeling of unconditional time delays, there is the function `wait(sc_time)` in SystemC. The call of `wait(`Δt`)` by a SystemC thread at the simulation time t suspends the calling thread until the simulation time $t + \Delta t$ is reached, and after that it continues its execution with the proceeding instruction. The time that Δt needs is independent of the number of other active tasks at that time in the system. Therefore, the `wait` function is suitable for the delay of hardware functionality, as this is inherently parallel. In contrast, software tasks can only be executed if they are allocated to a corresponding execution unit. This means that the execution of a software task will be suspended as soon as the execution unit is withdrawn by the operating system. In order to model the software timing behavior, two functions have to be used. The first function is the `delay(int)` function, as shown in Listing 2.2. As previously mentioned, this function is used for a fine granular addition of time. The second one is the `consume(sc_time)` function that does a coarse-grained consumption of time of the accumulated delays. This function is an extension of the function `wait(sc_time)` with an appropriate condition as needed. Listing 2.3 shows such a `consume(sc_time)` function.

```
int taskTime;
const sc_time t_PERIOD (timePeriod, SC_NS);

void delay(int c)
{
  taskTime+=c;
}

sc_time getTaskTime()
{
  return taskTime*t_PERIOD;
}

void resetTaskTime()
{
  taskTime=0;
}
```

Listing 2.2
The delay function.

If a software task calls the consume function with a time value, T, as a parameter, it decrements the time only if the calling software task is in the state RUNNING. If the execution unit is withdrawn by the RTOS scheduler by a change of the execution state, the decrementation of the time in the consume function will be suspended. By changing the state to RUNNING by the scheduler, the software task can allocate an execution unit again, leading to a continuation of the decrementation of the time that was suspended before.

2.5 Experimental Results

In order to test the execution speed and the accuracy of the translated code, a few examples were compiled using a C compiler into an object code for the Infineon TriCore processor [15]. This object code was also used to generate an annotated SystemC code from the C code, as described in Section 2.4.1. As a reference, the execution speed and the cycle count of the TriCore code have been measured on a TriCore TC10GP evaluation board and on a TriCore ISS [16].

The examples consist of two filters (fir and ellip) and two programs that are part of audio-decoding routines (dpcm and subband).

```
void consume(sc_time T)
{
  while(T > SC_ZERO_TIME || state != _state)
  {
   if (signals.empty())
   {
    sc_time time = sc_time_stamp();
    wait(T, signal_event);
    if (state == _state)
      T−= sc_time_stamp() − time;
   }
  }
}
```

Listing 2.3
The consume function.

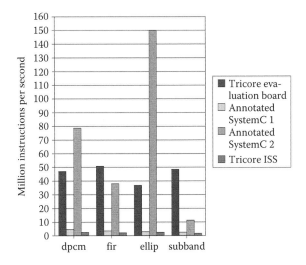

FIGURE 2.9
Comparison of speed. (Copyright: ACM. Used with permission.)

Figure 2.9 shows the comparison of the execution speed of the generated code with the execution speed of the TriCore evaluation board and the ISS. The execution speed in this figure is represented by million instructions of the TriCore Processor per second. The Athlon 64 processor running the SystemC code and the ISS had a clock rate of 2.4 GHz. The TriCore processor of the evaluation board ran at 48 MHz.

Using the annotated SystemC code, two different types of annotations have been used: the first one generates the cycles after the execution of each basic block, the second one adds cycles to a cycle counter after each basic block. The cycles are only generated when it is necessary (e.g., when communication with the hardware takes place). This is much more efficient and is depicted in Figure 2.9.

The execution speed of the TriCore processor ranges from 36.8 to 50.8 million instructions per second, whereas the execution speed of the annotated SystemC that models with immediate cycle generation ranges from 3.5 to 5.7 millions of simulated TriCore instructions per second. This means that the execution speed of the SystemC model is only about ten times slower than the speed of a real processor. The execution speed of the annotated SystemC code with on-demand cycle generation ranges from 11.2 to 149.9 million Tri-Core instructions per second.

In order to compare the SystemC execution speed with the execution speed of a conventional ISS, the same examples were run using the Tri-Core ISS. The result was an execution speed ranging from 1.5 to 2.4 million instructions per second. This means our approach delivers an execution speed increase of up to 91%.

A comparison of the number of simulated cycles of the generated SystemC code using branch prediction and cache simulation with the number of executed cycles of the TriCore evaluation board is shown in Figure 2.10. The deviation of the cycle counts of the translated programs (with branch

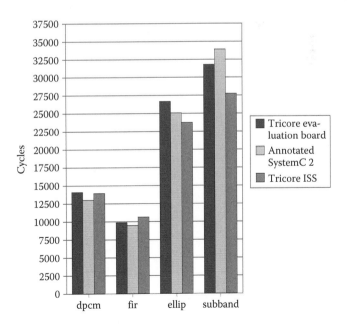

FIGURE 2.10

Comparison of cycle accuracy. (Copyright: ACM. Used with permission.)

prediction and caches included) compared to the measured cycle count from the evaluation board ranges between 4% for the program `fir` to 7% for the program `dpcm`. This is in the same range as it is using conventional ISS.

2.6   Outlook

As clock frequencies cannot be increased as linearly as the number of cores, modern processor architectures can exploit multiple cores to satisfy increasing computational demands. The different cores can share architectural resources such as data caches to speed up the access to common data. Therefore, access conflicts and coherency protocols have a potential impact on the runtimes of tasks executing on the cores.

The incorporation of multiple cores is directly supported by our SystemC approach. Parallel tasks can easily be assigned to different cores, and the code instrumentation by cycle information can be carried out independently. However, shared caches can have a significant impact on the number of executed cycles. This can be solved by the inclusion of a shared cache model that executes global cache coherence protocols, such as the MESI protocol. A clock calculation after each C/C++ statement is strongly recommended here to increase the accuracy.

2.7   Conclusions

This chapter presented a methodology for the SystemC-based performance analysis of embedded systems. To obtain a high accuracy with an acceptable runtime, a hybrid approach for a high-performance timing simulation of the embedded software was given. The approach shown was implemented in an automated design flow. The methodology is based on the generation of the SystemC code out of the original C code and the back-annotation of the statically determined cycle information into the generated code. Additionally, the impact of data dependencies on the software runtime is analytically handled during simulation. Promising experimental results from the application of the implemented design flow were presented. These results show a high execution performance of the timed embedded software model as well as good accuracy. Furthermore, the created SystemC models representing the timed embedded software could be easily integrated into virtual SystemC prototypes because of the generated TLM interfaces.

References

1. K. Albers, F. Bodmann, and F. Slomka. Hierarchical event streams and event dependency graphs: A new computational model for embedded real-time systems. In *Proceedings of the 18th Euromicro Conference on Real-Time Systems (ECRTS)*, Dresden, Germany, pp. 97–106, 2006.

2. J. Aynsley. *OSCI TLM2 User Manual*. Open SystemC Initiative (OSCI), November 2007.

3. J. Bryans, H. Bowman, and J. Derrick. Model checking stochastic automata. *ACM Transactions on Computational Logic (TOCL)*, 4(4):452–492, 2003.

4. C. Cifuentes. Reverse compilation techniques. PhD thesis, Queensland University of Technology Brisbane, Australia, November 19, 1994.

5. CoWare Inc. CoWare Processor Designer. http://www.coware.com/PDF/products/ProcessorDesigner.pdf.

6. L. B. de Brisolara, Marcio F. da S. Oliveira, R. Redin, L. C. Lamb, L. Carro, and F. R. Wagner. Using UML as front-end for heterogeneous software code generation strategies. In *Proceedings of the Design, Automation and Test in Europe (DATE) Conference*, Munich, Germany, pp. 504–509, 2008.

7. A. Donlin. Transaction level modeling: Flows and use models. In *Proceedings of the 2nd IEEE/ACM/IFIP International Conference on Hardware/Software Codesign and System Synthesis (CODES+ISSS)*, San Jose, CA, pp. 75–80, 2004.

8. T. Grötker, S. Liao, G. Martin, and S. Swan. *System Design with SystemC*. Kluwer, Dordrecht, the Netherlands, 2002.

9. M. González Harbour, J. J. Gutiérrez García, J. C. Palencia Gutiérrez, and J. M. Drake Moyano. MAST: Modeling and analysis suite for real time applications. In *Proceedings of the 13th Euromicro Conference on Real-Time Systems (ECRTS)*, Delft, the Netherlands, pp. 125–134, 2001.

10. H. Heinecke. Automotive open system architecture – An industry-wide initiative to manage the complexity of emerging automotive E/E architectures. In *Convergence International Congress & Exposition On Transportation Electronics*, Detroit, MI, 2004.

11. R. Henia, A. Hamann, M. Jersak, R. Racu, K. Richter, and R. Ernst. System level performance analysis—the SymTA/S approach. *IEE Proceedings Computers and Digital Techniques*, 152(2):148–166, March 2005.

12. Y. Hur, Y. H. Bae, S.-S. Lim, S.-K. Kim, B.-D. Rhee, S. L. Min, C. Y. Park, H. Shin, and C.-S. Kim. Worst case timing analysis of RISC processors: R3000/R3010 case study. In *Proceedings of the IEEE Real-Time Systems Symposium (RTSS)*, Pisa, Italy, pp. 308–319, 1995.

13. Y. Hwang, S. Abdi, and D. Gajski. Cycle-approximate retargetable performance estimation at the transaction level. In *Proceedings of the Design, Automation and Test in Europe (DATE) Conference*, Munich, Germany, pp. 3–8, 2008.

14. IEEE Computer Society. *IEEE Standard SystemC Language Reference Manual*, March 2006.

15. Infineon Technologies AG. *TC10GP Unified 32-bit Microcontroller-DSP—User's Manual*, 2000.

16. Infineon Technologies Corp. *TriCore^{TM} 32-bit Unified Processor Core—Volume 1: v1.3 Core Architecture*, 2005.

17. S. Kraemer, L. Gao, J. Weinstock, R. Leupers, G. Ascheid, and H. Meyr. HySim: A fast simulation framework for embedded software development. In *Proceedings of the 5th IEEE/ACM International Conference on Hardware/Software Codesign and System Synthesis (CODES+ISSS)*, Salzburg, Austria, pp. 75–80, 2007.

18. M. Krause, O. Bringmann, and W. Rosenstiel. Target software generation: An approach for automatic mapping of SystemC specifications onto real-time operating systems. *Design Automation for Embedded Systems*, 10(4):229–251, December 2005.

19. M. Krause, O. Bringmann, and W. Rosenstiel. *Hardware-dependent Software: Principles and Practice*, Chapter 10 Verification of AUTOSAR Software by SystemC-based virtual prototyping. pp. 261–293, Springer, Netherlands, 2009.

20. S. Künzli, F. Poletti, L. Benini, and L. Thiele. Combining simulation and formal methods for system-level performance analysis. In *Proceedings of the Design, Automation and Test in Europe (DATE) Conference*, Munich, Germany, pp. 236–241, 2006.

21. S.-S. Lim, Y. H. Bae, G. T. Jang, B.-D. Rhee, S. L. Min, C. Y. Park, H. Shin, K. Park, S.-M. Moon, and C. S. Kim. An accurate worst case timing analysis for RISC processors. *IEEE Transactions on Software Engineering*, 21(7):593–604, 1995.

22. R. Marculescu and A. Nandi. Probabilistic application modeling for system-level performance analysis. In *Proceedings of the Conference on Design, Automation and Test in Europe (DATE)*, Munich, Germany, pp. 572–579, 2001.

23. M. Ajmone Marsan, G. Conte, and G. Balbo. A class of generalized stochastic petri nets for the performance evaluation of multiprocessor systems. *ACM Transactions on Computer Systems*, 2(2):93–122, 1984.

24. The MathWorks, Inc. Real-Time Workshop® Embedded Coder 5, September 2007.

25. Steven S. Muchnick. *Advanced Compiler Design and Implementation*. Morgan Kaufmann Publishers, San Francisco, CA, 1997.

26. A. Nohl, G. Braun, O. Schliebusch, R. Leupers, H. Meyr, and A. Hoffmann. A universal technique for fast and flexible instruction-set architecture simulation. In *Proceedings of the 39th Design Automation Conference (DAC)*, New York, pp. 22–27, 2002.

27. C. Norström, A. Wall, and W. Yi. Timed automata as task models for event-driven systems. In *Proceedings of the Sixth International Conference on Real-Time Computing Systems and Applications (RTCSA)*, Hong Kong, China, pp. 182–189, 1999.

28. OPNET Technologies, Inc. http://www.opnet.com.

29. G. Ottosson and M. Sjödin. Worst-case execution time analysis for modern hardware architectures. In *Proceedings of the ACM SIGPLAN 1997 Workshop on Languages, Compilers, and Tools for Real-Time Systems (LCT-RTS '97)*, Las Vegas, NV, pp. 47–55, 1997.

30. M. Oyamada, F. R. Wagner, M. Bonaciu, W. O. Cesário, and A. A. Jerraya. Software performance estimation in MPSoC design. In *Proceedings of the 12th Asia and South Pacific Design Automation Conference (ASP-DAC)*, Yokohama, Japan, pp. 38–43, 2007.

31. P. Pop, P. Eles, Z. Peng, and T. Pop. Analysis and optimization of distributed real-time embedded systems. In *Proceedings of the 41st Design Automation Conference (DAC)*, San Diego, CA, pp. 593–625, 2004.

32. C. V. Ramamoorthy and H. F. Li. Pipeline architecture. *ACM Computing Surveys*, 9(1):61–102, 1977.

33. K. Richter, M. Jersak, and R. Ernst. A formal approach to MpSoC performance verification. *Computer*, 36(4):60–67, 2003.

34. K. Richter, D. Ziegenbein, M. Jersak, and R. Ernst. Model composition for scheduling analysis in platform design. In *Proceedings of the 39th Design Automation Conference (DAC)*, New Orleans, LA, pp. 287–292, 2002.

35. G. Schirner, A. Gerstlauer, and R. Dömer. Abstract, multifaceted modeling of embedded processors for system level design. In *Proceedings of the 12th Asia and South Pacific Design Automation Conference (ASP-DAC)*, Yokohama, Japan, pp. 384–389, 2007.

36. J. Schnerr, O. Bringmann, and W. Rosenstiel. Cycle accurate binary translation for simulation acceleration in rapid prototyping of SoCs. In *Proceedings of the Design, Automation and Test in Europe (DATE) Conference*, Munich, Germany, pp. 792–797, 2005.

37. J. Schnerr, O. Bringmann, A. Viehl, and W. Rosenstiel. High-performance timing simulation of embedded software. In *Proceedings of the 45th Design Automation Conference (DAC)*, Anaheim, CA, pp. 290–295, June 2008.

38. J. Schnerr, G. Haug, and W. Rosenstiel. Instruction set emulation for rapid prototyping of SoCs. In *Proceedings of the Design, Automation and Test in Europe (DATE) Conference*, Munich, Germany, pp. 562–567, 2003.

39. A. Siebenborn, O. Bringmann, and W. Rosenstiel. Communication analysis for network-on-chip design. In *International Conference on Parallel Computing in Electrical Engineering (PARELEC)*, Dresden, Germany, pp. 315–320, 2004.

40. A. Siebenborn, O. Bringmann, and W. Rosenstiel. Communication analysis for system-on-chip Design. In *Proceedings of the Design, Automation and Test in Europe (DATE) Conference*, Paris, France, pp. 648–655, 2004.

41. A. Siebenborn, A. Viehl, O. Bringmann, and W. Rosenstiel. Control-flow aware communication and conflict analysis of parallel processes. In *Proceedings of the 12th Asia and South Pacific Design Automation Conference (ASP-DAC)*, Yokohama, Japan, pp. 32–37, 2007.

42. E. W. Stark and S. A. Smolka. Compositional analysis of expected delays in networks of probalistic I/O Automata. In *IEEE Symposium on Logic in Computer Science*, Indianapolis, IN, pp. 466–477, 1998.

43. Synopsys, Inc. Synopsys Virtual Platforms. http://www.synopsys.com/products/designware/virtual_platforms.html.

44. L. Thiele, S. Chakraborty, and M. Naedele. Real-time calculus for scheduling hard real-time systems. In *IEEE International Symposium on Circuits and Systems (ISCAS)*, Geneva, Switzerland, volume 4, pp. 101–104, 2000.

45. VaST Systems Technology. CoMET®. http://www.vastsystems.com/docs/CoMET_mar2007.pdf.

46. A. Viehl, M. Schwarz, O. Bringmann, and W. Rosenstiel. Probabilistic performance risk analysis at system-level. In *Proceedings of the 5th IEEE/ACM International Conference on Hardware/Software Codesign and System Synthesis (CODES+ISSS)*, Salzburg, Austria, pp. 185–190, 2007.

47. A. Viehl, T. Schönwald, O. Bringmann, and W. Rosenstiel. Formal performance analysis and simulation of UML/SysML Models for ESL Design. In *Proceedings of the Design, Automation and Test in Europe (DATE) Conference*, Munich, Germany, pp. 242–247, 2006.

48. T. Wild, A. Herkersdorf, and G.-Y. Lee. TAPES – Trace-based architecture performance evaluation with systemC. *Design Automation for Embedded Systems*, 10(2–3):157–179, September 2005.

49. A. Yakovlev, L. Gomes, and L. Lavagno, editors. *Hardware Design and Petri Nets*. Kluwer Academic Publishers, Dordrecht, the Netherlands, March 2000.

3

Formal Performance Analysis for Real-Time Heterogeneous Embedded Systems

Simon Schliecker, Jonas Rox, Rafik Henia, Razvan Racu, Arne Hamann, and Rolf Ernst

CONTENTS

3.1 Introduction

Formal approaches to system performance modeling have always been used in the design of real-time systems. With increasing system complexity, there is a growing demand for the use of more sophisticated formal methods in a wider range of systems to improve system predictability, and determine system robustness to changes, enhancements, and design pitfalls. This demand can be addressed by the significant progress in the last couple of years in performance modeling and analysis on all levels of abstraction.

New modular models and methods now allow the analysis of large-scale, heterogeneous systems, providing reliable data on transitional load situations, end-to-end timing, memory usage, and packet losses. A compositional performance analysis allows to decompose the system into the analysis of individual components and their interaction, providing a versatile method to approach real-world architectures. Early industrial adopters are already using such formal methods for the early evaluation and exploration of a design, as well as for a formally complete performance verification toward the end of the design cycle—neither of which could be achieved solely with simulation-based approaches.

The formal methods, as presented in this chapter, are based on abstract load and execution data, and are thus applicable even before executable hardware or software models are available. Such data can even be estimates derived from previous product generations, similar implementations, or simply engineering competence allowing for first evaluations of the application and the architecture. This already allows tuning an architecture for maximum robustness against changes in system execution and communication load, reducing the risk of late and expensive redesigns. During the design process, these models can be iteratively refined, eventually leading to a verifiable performance model of the final implementation.

The multitude of diverse programming and architectural design paradigms, often used together in the same system, call for formal methods that can be easily extended to consider the corresponding timing effects. For example, formal performance analysis methods are also becoming increasingly important in the domain of tightly integrated multiprocessor system-on-chips (MPSoCs). Although such components promise to deliver higher performance at a reduced production cost and power consumption, they introduce a new level of integration complexity. Like in distributed embedded systems, multiprocessing comes at the cost of higher timing complexity of interdependent computation, communication, and data storage operations.

Also, many embedded systems (distributed or integrated) feature communication layers that introduce a hierarchical timing structure into the communication. This is addressed in this chapter with a formal representation and accurate modeling of the timing effects induced during transmission.

Finally, today's embedded systems deliver a multitude of different software functions, each of which can be particularly important in a specific situation (e.g., in automotives: an electronic stability program (ESP) and a parking assistance). A hardware platform designed to execute all of these functions at the same time will be expensive and effectively overdimensioned given that the scenarios are often mutually exclusive. Thus, in order to supply the desired functions at a competitive cost, systems are only dimensioned for subsets of the supplied functions, so-called scenarios, which are investigated individually. This, however, poses new pitfalls when dimensioning distributed systems under real-time constraints. It becomes mandatory to also consider the scenario-transition phase to prevent timing failures.

This chapter presents an overview of a general, modular, and formal performance analysis framework, which has successfully accommodated many extensions. First, we present its basic procedure in Section 3.2. Several extensions are provided in the subsequent sections to address specific properties of real systems: Section 3.3 visits multi-core architectures and their implications on performance; hierarchical communication as is common in automotive networks is addressed in Section 3.4; the dynamic behavior of switching between different application scenarios during runtime is investigated in Section 3.5. Furthermore, we present a methodology to systematically investigate the sensitivity of a given system configuration and to explore the design space for optimal configurations in Sections 3.6 and 3.7. In an experimental section (Section 3.8), we investigate timing bottlenecks in an example heterogeneous automotive architecture, and show how to improve the performance guided by sensitivity analysis and system exploration.

3.2 Formal Multiprocessor Performance Analysis

In past years, compositional performance analysis approaches [6,14,16] have received an increasing attention in the real-time systems community. Compositional performance analyses exhibit great flexibility and scalability for timing and performance analyses of complex, distributed embedded real-time
systems. Their basic idea is to integrate local performance analysis techniques, for example, scheduling analysis techniques known from real-time research, into system-level analyses. This composition is achieved by connecting the component's inputs and outputs by stream representations of their communication behaviors using event models. This procedure is illustrated in Sections 3.2.1 through 3.2.4.

3.2.1 Application Model

An embedded system consists of hardware and software components inter-acting with each other to realize a set of functionalities. The traditional approach to formal performance analysis is performed bottom-up. First, the behavior of the individual functions needs to be investigated in detail to gather all relevant data, such as the execution time. This information can then be used to derive the behavior within individual components, accounting for local scheduling interference. Finally, the system-level timing is derived on the basis of the lower-level results.

For an efficient system-level performance verification, embedded systems are modeled with the highest possible level of abstraction. The smallest unit modeling performance characteristics at the application level is called a task. Furthermore, to distinguish computation and communication, tasks are cat-egorized into computational and communication tasks. The hardware plat-form is modeled by computational and communication resources, which are referred to as CPUs and buses, respectively. Tasks are mapped on resources in order to be executed. To resolve conflicting requests, each resource is asso-ciated with a scheduler.

Tasks are activated and executed due to activating events that can be gen-erated in a multitude of ways, including timer expiration, and task chaining according to inter-task dependencies. Each task is assumed to have one input first-in first-out (FIFO) buffer. In the basic task model, a task reads its acti-vating data solely from its input FIFO and writes data into the input FIFOs of dependent tasks. This basic model of a task is depicted in Figure 3.1a. Var-ious extensions of this model also exist. For example, if the task may be sus-pended during its execution, this can be modeled with the requesting-task model presented in Section 3.3. Also, the direct task activation model has been extended to more complex activation conditions and semantics [10].

3.2.2 Event Streams

The timing properties of the arrival of workload, i.e., activating events, at the task inputs are described with an activation model. Instead of considering each activation individually, as simulation does, formal performance anal-ysis abstracts from individual activating events to event streams. Generally,

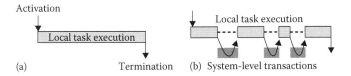

FIGURE 3.1
Task execution model.

event streams can be described using the upper and lower event-arrival functions, η^+ and η^-, as follows.

Definition 3.1 (Upper Event-Arrival Function, η^+) *The upper event-arrival function, $\eta^+(\Delta t)$, specifies the maximum number of events that may occur in the event stream during any time interval of size Δt.*

Definition 3.2 (Lower Event-Arrival Function, η^-) *The lower event-arrival function, $\eta^-(\Delta t)$, specifies the minimum number of events that may occur in the event stream during any time interval of size Δt.*

Correspondingly, an event model can also be specified using the functions $\delta^-(n)$ and $\delta^+(n)$ that represent the minimum and maximum distances between any n events in the stream. This representation is more useful for latency considerations, while the η-functions better express the resource loads. Each can be derived from the other (as they are "pseudo-inverse," as defined in [5]). Different parameterized event models have been developed to efficiently describe the timings of events in the system [6,14].

One popular and computationally efficient abstraction for representing event streams is provided by so-called standard event models [33], as visualized in Figure 3.2. Standard event models capture the key properties of event streams using three parameters: the activation period, P; the activation jitter, J; and the minimum distance, d. *Periodic* event models have one parameter P stating that each event arrives periodically at exactly every P time units. This simple model can be extended with the notion of jitter, leading to periodic with jitter event models, which are described by two parameters, namely, P and J. Events generally occur periodically, yet they can jitter around their exact position within a jitter interval of size J. If the jitter value is larger than the period, then two or more events can occur simultaneously, leading to bursts. To describe *bursty* event models, periodic with jitter event models can be extended with the parameter d^{min} capturing the minimum distance between the occurrences of any two events.

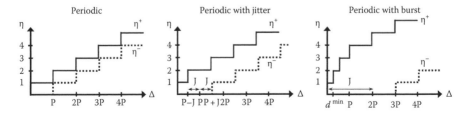

FIGURE 3.2
Standard event models.

3.2.3   Local Component Analysis

Based on the underlying resource-sharing strategy, as well as stream repre-sentations of the incoming workload modeled through the activating event models, local component analyses systematically derive worst-case scenarios to calculate worst-case (sometimes also best-case) task response times (BCRT and WCRT), that is, the time between task activation and task completion, for all tasks sharing the same component (i.e., the processor). Thereby, local component analyses guarantee that all observable response times fall into the calculated [best-case, worst-case] interval. These analyses are therefore considered conservative.

Note that different approaches use different models of computation to perform local component analyses. For instance, SymTA/S [14,43] is based on the algebraic solution of response time formulas using the sliding-window technique proposed by, for example, Lehoczky [23], whereas the real-time calculus utilizes arrival curves and service curves to characterize the workload and processing capabilities of components, and determines their real-time behavior [6]. These concepts are based on the network cal-culus. For details please refer to [5].

Additionally, local component analyses determine the communication behaviors at the outputs of the analyzed tasks by considering the effects of scheduling. The basic model assumes that tasks produce output events at the end of each execution. Like the input timing behavior, the output event tim-ing behavior can also be captured by event models. The output event models can then be derived for every task, based on the local response time analysis.

For instance, standard event models used by SymTA/S allow the specifi-cation of very simple rules to obtain output event models during the local component analysis. Note that in the simplest case (i.e., if tasks produce exactly one output event for each activating event) the output event model period equals the activation period. A discussion on how output event model periods are determined for more complex semantics (when considering rate transitions) can be found in [19]. The output event model jitter, J_{out}, is calcu-lated by adding the difference between maximum and minimum response times, $R^{max} - R^{min}$, the response time jitter, to the activating event model jitter, J_{in} [33]:

$$\mathcal{J}_{out} = \mathcal{J}_{in} + (R^{max} - R^{min}) \tag{3.1}$$

The output event model calculation can also be performed for general event models that are specified solely with the upper and lower event-arrival functions. This method will be applied in Section 3.4 to hierarchical event models (HEMs). Recently, a more exact output jitter calculation algorithm was proposed for the local component analysis based on standard event models [15] and general event models [43]. The approaches exploit the fact that the response time of a task activation is correlated with the timings of preceding events—the task activation arriving with worst-case jitter does not necessarily experience the worst-case response time.

3.2.4 Compositional System-Level Analysis Loop

On the highest level of the timing hierarchy, the compositional system-level analysis [6,14] derives the system's timing properties from the lower-level results. For this, the local component analysis (as explained in Section 3.2.3) is alternated with the output event model propagation. The basic idea is visualized on the right-hand side of Figure 3.3. (The shared-resource analysis depicted on the left-hand side will be explained in Section 3.3.) In each global iteration of the compositional system-level analysis, input event model assumptions are used to perform local scheduling analyses for all components. From this, their response times and output event models are derived as described above. Afterward, the calculated output event models are propagated to the connected components, where they are used as activating input event models for the subsequent global iteration. Obviously, this iterative analysis represents a fix-point problem. If all calculated output event models remain unmodified after an iteration, the convergence is reached and the last calculated task response times are valid [20,34].

To successfully apply the compositional system-level analysis, the input event models of all components need to be known or must be computable by the local component analysis. Obviously, for systems containing feedback between two or more components, this is not the case, and, thus, the system-level analysis cannot be performed without additional measures. The concrete strategies to overcome this issue depend on the component types and their input event models. One possibility is the so-called starting point generation of SymTA/S [33].

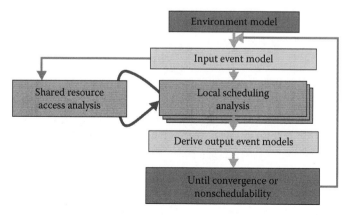

FIGURE 3.3
MPSoC performance analysis loop.

3.3 From Distributed Systems to MPSoCs

The described procedure appropriately covers the behaviors of hardware and software tasks that consume all relevant data upon activation and produce output data into a single FIFO. This represents the prevailing design practice in many real-time operating systems [24] and parallel programming concepts [21]. However, it is also common—particularly in MPSoCs—to access shared resources such as a memory during the execution of a task. The diverse interactions and correlations between integrated system components then pose fundamental challenges to the timing predictions. Figure 3.4 shows an example dual-core system in which three tasks access the same shared memory during execution. In this section, the scope of the above approach is extended to cover such behaviors.

The model of the task is for this purpose extended to include local execution as well as memory transactions during the execution [38]. While the classical task model is represented as an execution time interval (Figure 3.1a), a so-called requesting task performs transactions during its execution, as depicted in Figure 3.1b. The depicted task requires three chunks of data from an external resource. It issues a request and may only continue execution after the transaction is transmitted over the bus, processed on the remote component, and transmitted back to the requesting source. Thus, whenever a transaction has been issued, but is not finished, the task is not ready. The accesses to the shared resource may be logical shared resources (as in [27]), but for the scope of this chapter, we assume that the accesses go to a shared memory. Such memory accesses may be explicit data-fetch operations or implicit cache misses.

The timing of such memory accesses, especially cache misses, is extremely difficult to accurately predict. Therefore, an analysis cannot predict the timing of each individual transaction with an acceptable effort. Instead, a shared-resource access analysis algorithm will be utilized in Section 3.3.3 that subsumes all transactions of a task execution and the interference by

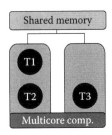

FIGURE 3.4
Multicore component with three requesting tasks that access the same shared memory during execution.

other system activities. Even though this presumes a highly unfavorable and unlikely coincidence of events, this approach is much less conservative than the consideration of individual transactions.

The memory is considered to be a separate component, and an analysis must be available for it to predict the accumulated latency of a set of memory requests. For this analysis to work, the event models for the amount of requests issued from the various processors are required. The outer analysis loop in the procedure of Figure 3.3, as described in Section 3.2, provides these event models for task activations throughout the system. These task activating event models allow the derivation of bounds on the task's number of requests to the shared-resource. These bounds can be used by the shared-resource analysis to derive the transaction latencies. The processor's scheduling analysis finally needs to account for the delays experienced during the task execution by integrating the transaction latencies. This intermediate analysis is shown on the left hand side of Figure 3.3. As it is based on the current task activating event model assumptions of the outer analysis, the shared-resource analysis possibly needs to be repeated when the event models are refined. In order to embed the analysis of requesting tasks into the compositional analysis framework described in Section 3.2, three major building blocks are required:

1. Deriving the number of transactions issued by a task and all tasks on a processor
2. Deriving the latency experienced by a set of transactions on the shared resource
3. Integrating the transaction latency into the tasks' worst-case response times

These three steps will be carried out in the following. We begin with the local investigation of deriving the amount of initiated transactions (Section 3.3.1) and the extended worst-case response time analysis (Section 3.3.2). Finally, we turn to the system-level problem of deriving the transaction latency (Section 3.3.3).

3.3.1 Deriving Output Event Models

For each individual task-activation, the amount of issued requests can be bound by closely investigating the task's internal control flow. For example, a task may explicitly fetch data each time it executes a for-loop that is repeated several times. By multiplying the maximum number of loop iterations with the amount of fetched data, a bound on the memory accesses can be derived. Focused on the worst-case execution time problem, previous research has provided various methods to find the longest path through such a program description with the help of integer linear programming (see [49]).

Implicit data fetches such as cache misses are more complicated to capture, as they only occur during runtime and cannot be directly identified in the application binary. Depending on the current cache state and the execution history, cache misses may occur at different points in time. However, formal methods are able to identify for each basic block the maximum number of cache misses that may occur during the execution [46]. The control flow graph can be annotated with this information, making the longest path analyses feasible again.

Depending on the actual system configuration, the upper bound on the number of transactions per task execution may not be sufficiently accurate. In a formal model, this could translate into an assumed burst of requests that may not occur in practice. This can be addressed with a more detailed analysis of the task control flow, as is done in [1,39], which provides bounds on the minimum distances between any n requests of an activation of that task. This pattern will then repeat with each task activation.

This procedure allows to conservatively derive the shared resource request bound functions $\tilde{\eta}_{\tau}^{+}(w)$ and $\tilde{\eta}_{\tau}^{-}(w)$ that represent the transaction traffic that each task τ in the system can produce within a given time window of size w. Requesting tasks that share the same processor may be executed in alternation, resulting in a combined request traffic for the complete processor. This again can be expressed as an event model. For example, a straightforward approach is to approximate the processor's request event model (in a given time window) with the aggregation of the request event models of each individual task executing on that processor. Obviously, this is an overestimation, as the tasks will not be executed at the same time, but rather the scheduler will assign the processor exclusively. The resulting requests will be separated by the intermediate executions, which can be captured in the joint shared resource request bound by a piecewise assembly from the elementary streams [39].

3.3.2 Response Time Analysis in the Presence of Shared Memory Accesses

Memory access delays may be treated differently by various processor implementations. Many processors, and some of the most commonly used, allow tasks to perform coprocessor or memory accesses by offering a multi-cycle operation that stalls the entire processor until the transaction has been processed by the system [44]. In other cases, a set of hardware threads may allow to perform a quick context switch to another thread that is ready, effectively keeping the processor utilized (e.g., [17]). While this behavior usually has a beneficial effect on the average throughput of a system, multithreading requires caution in priority-based systems with reactive or control applications. In this case, the worst-case response time of even high-priority tasks may actually increase [38].

The integration of dynamic memory access delays into the real-time analysis will in the following be performed for a processor with priority-based preemptive scheduling that is stalled during memory accesses. In such a system, a task's worst-case response time is determined by the task's worst-case execution time plus the maximum amount of time the task can be kept from executing because of preemptions by higher-priority tasks and blocking by lower-priority tasks. A task that performs memory accesses is additionally delayed when waiting for the arrival of requested data. Furthermore, preemption times are increased, as the remote memory accesses also cause high-priority tasks to execute longer.

A possible runtime schedule is depicted in Figure 3.5. In the case where both tasks execute in the local memory (Scenario 3.5a), the low-priority task is kept from executing by three invocations of the high-priority tasks. Local memory accesses are not explicitly shown, as they can be considered to be part of the execution time. When both tasks access the same remote memory (Scenario 3.5b), the finishing time of the lower-priority task increases, because it itself fetches data from the remote memory, and also because of the prolonged preemptions by the higher-priority task (as its request also stalls the processor). The execution of the low-priority task in the example is now stretched such that it suffers from an additional preemption of the other task. Finally, Scenario 3.5c shows the effect of a task on another core CPUb that is also accessing the same shared memory, in this case, periodically. Whenever the memory is also used by a task on CPUb, CPUa is stalled for a longer time, again increasing the task response times, possibly leading to the violation of a given deadline. As the busy wait adds to the execution

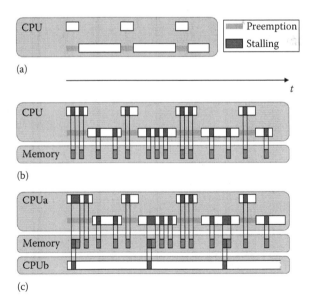

FIGURE 3.5

Tasks on different processors accessing a shared memory. (a and b) Single processor case and (c) conflicts from another CPU.

time of a task, the total processor load increases—possibly making the overall system unschedulable.

On the basis of these observations, a response time equation can be derived for the example scheduler. The response time represents the sum of the following:

- The core execution times of all tasks mapped to the processor, and their activation event models.
- The increased blocking time due to the resources being stalled during memory accesses. (This is not shown in the example.)
- The aggregate delay caused by the memory accesses that is a function of the memory accesses of a specific task and its higher-priority tasks. This is investigated in Section 3.3.3.

Variations of such a response time analysis have been presented for single- and multithreaded static-priority preemptive scheduling [38], as well as for round-robin scheduling [41]. Other scheduling policies for which classical real-time analysis is available can be straight-forwardly extended to include memory delays by including a term that represents the aggregate busy time due to memory accesses.

3.3.3 Deriving Aggregate Busy Time

Deriving the timing of many memory accesses has recently become an important topic in real-time research. Previously, the worst-case timing of individual events was the main concern. Technically, a sufficient solution to find the delay that a set of many events may experience, is to derive the single worst-case load scenario and assume it for every access. However, not every memory request will experience a worst-case system state, such as worst-case time wheel positions in the time division multiple access (TDMA) schedules, or transient overloads in priority-based components. For example, the task on CPUb in Figure 3.5 will periodically access the shared memory, and, as a consequence, disturb the accesses by the two tasks on CPUa. A "worst-case memory access" will experience this delay, but of all accesses from CPUb, this happens maximally three times in this example. Thus, accounting this interference for every single memory access leads to very unsatisfactory results—which has previously prevented the use of conservative methods in this context.

The key idea is instead to consider all requests that are processed during the lifetime of a task jointly. We therefore introduce the worst-case *accumulated busy time*, defined as the total amount of time, during which at least one request is issued but is not finished. Multiple requests in a certain amount of time can in total only be delayed by a certain amount of interference, which is expressed by the aggregate busy time.

This aggregate busy time can be efficiently calculated (e.g., for a shared bus): a set of requests is issued from different processors that may interfere with each other. The exact individual request times are unknown and their actual latency is highly dynamic. Extracting detailed timing information (e.g., when a specific cache miss occurs) is virtually impossible and considering such details in a conservative analysis yields exponential complexity. Consequently, we disregard such details and focus on bounding the aggregate busy time. Given a certain level of dynamism in the system, this consideration will not result in excessive overestimations. Interestingly, even in multithreaded multicore architectures, the conservatism is moderate, summing up to less than a total of 25% of the overestimated response time, as shown in practical experiments [42].

Without bus access prioritization, it has to be assumed that it is possible for every transaction issued by any processor during the lifetime of a task activation i that it will disturb the transactions issued by i. Usually, the interference is then given by the transactions issued by the other concurrently active tasks on the other processors, as well as the tasks on the same processor as their requests are treated on a first-come-first-served basis. The interested readers are referred to [40] for more details on the calculation of aggregate memory access latencies.

If a memory controller is utilized, this can be very efficiently considered. For example, all requests from a certain processor may be prioritized over those of another. Then, the imposed interference by all lower-priority requests equals zero. Additionally, a small blocking factor of one elementary memory access time is required, in order to model the time before a transaction may be aborted for the benefit of a higher-priority request.

The compositional analysis approach of Section 3.2, used together with the methods of Section 3.3, now delivers a complete framework for the performance analysis of heterogeneous multiprocessor systems with shared memories. The following section turns to detailed modeling of interprocessor communication with the help of HEMs.

3.4 Hierarchical Communication

As explained in Section 3.2, traditional compositional analysis models bus communication by a simple communication task that is directly activated by the sending task, and which directly activates the receiving task. Figure 3.6 shows a simple example system that uses this model for communication, where each output event of the sending tasks, T_a and T_b, triggers the transmission of one message over the bus.

However, the modern communication stacks employed in today's embedded control units (ECUs), for example, in the automotive domain, make this abstraction inadequate. Depending on the configuration of the communication layer, the output events (called signals here) may or may

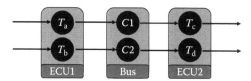

FIGURE 3.6
Traditional model.

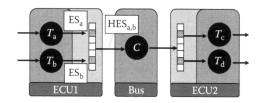

FIGURE 3.7
Communication via ComLayer.

not directly trigger the transmissions of messages (called frames here). For instance, AUTOSAR [2] defines a detailed API for the communication stack, including several frame transmission modes (direct, periodic, mixed, or none) and signal transfer properties (triggered or pending) with key influences on communication timings. Hence, the transmission timings of messages over the bus do not have to be directly connected to the output behaviors of the sending tasks anymore, but they may even be completely independent of the task's output behavior (e.g., sending several output signals in one message).

In the example shown in Figure 3.7, the tasks T_a and T_b produce output signals that are transmitted over the bus to the tasks T_c and T_d. The sending tasks write their output data into registers provided by the communication layer, which is responsible for packing the data into messages, called frames here, and triggering the transmission of these frames according to the signal types and transmission modes. On the receiving side, the frames are unpacked, which means that the contained signals are again written into different registers for the corresponding receiving task. Using flat event models, the timings of signal arrivals can only be bound with a large overestimation.

To adequately consider such effects of modern communication stacks in the system analysis, two elements must be determined:

1. The activation timings of the frames
2. The timings of signals transmitted within these frames arriving at the receiving side

To cope with both the challenges, we introduce hierarchical event streams (HESs) modeled by a HEM, which determines the activating function of the frame and also captures the timings of the signals assigned to that frame, and, most importantly, defines how the effects on the frame timings influence the timings of the transmitted signals. The latter allows to unpack the signals on the receiving side, giving tighter bounds for the activations of those tasks receiving the signals.

The general idea is that a HES has one outer representation in the form of an event stream ES_{outer}, and each combined event stream has one inner representation, also in the form of an event stream ES'_i, where i denotes the task to which the event stream corresponds. The relation between the outer event stream and the inner event stream depends on the hierarchical stream constructor (HSC) that combined the event streams. Each of the involved event streams is defined by functions $\delta^-(n)$ and $\delta^+(n)$ (see Section 3.2.2), returning the minimum and the maximum distance, respectively, between n consecutive events.

Figure 3.8 illustrates the structure of the HES at the input of the channel C of the example shown in Figure 3.7. The HSC combines the output streams of the tasks T_a and T_b, resulting in the hierarchical input stream of the communication task C. According to the properties and the configuration of the communication layer that is modeled by the HSC, the inner and outer event streams of the HES are calculated. Each event of the outer event stream, ES_{outer}, represents the sending of one message by the communication layer. The events of a specific inner event stream, ES'_a and ES'_b, model the timings of only those messages that contain data from the corresponding sending tasks. The detailed calculations of the inner and outer event streams, considering the different signal properties and frame transmission modes, are presented in [37].

For the local scheduling analysis of the bus, only the outer event stream is relevant. As a result, the best-case response time, R^{min}, and the worst-case

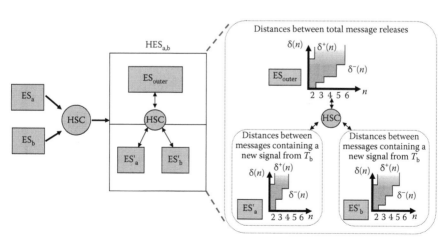

FIGURE 3.8
Structure of the hierarchical input stream of C.

response time, R^{max}, are obtained. Based on the outer event stream, ES_{outer}, of the hierarchical input stream, we obtain the outer event stream, ES'_{outer}, of the hierarchical output stream by using the following equations:

$$\delta'^{-}_{outer}(n) = \max\{\delta^{-}_{outer}(n) - J^{resp}, \delta'^{-}_{outer}(n-1) + d^{min}\} \qquad (3.2)$$

$$\delta'^{+}_{outer}(n) = \max\{\delta^{+}_{outer}(n) + J^{resp}, \delta'^{+}_{outer}(n-1) + d^{min}\} \qquad (3.3)$$

In fact, Equations 3.2 and 3.3 are generalizations of the output model calculation presented in Equation 3.1. As can be seen, actually two changes have been made to the message timing. First, the minimum (maximum) distance between a given number of events decreases (increases) by no more than the response time jitter, $J^{resp} = R^{max} - R^{min}$. Second, two consecutive events at the output of the channel are separated by at least a minimum distance, $d^{min} = R^{min}$. The resulting event stream, modeled by $\delta'^{-}_{outer}(n)$ and $\delta'^{+}_{outer}(n)$, becomes the outer stream of the output model.

To obtain the inner event streams, ES''_i, of the hierarchical output stream, we adapt the inner event streams, ES'_i, of the hierarchical input stream according to the changes applied to the outer stream. For the adaptation, we consider the two changes mentioned above separately. First, consider that the minimum distance between n messages decreases by J^{resp}. Then, the minimum distance between k messages that contain the data of a specific task decreases by J^{resp}. Second, we must consider that two consecutive messages become separated by a minimum distance d^{min}. Figure 3.9a illustrates a sequence of events consisting of two different event types, a and b. Assume that this event sequence models the message timing, where the events labeled by a lowercase a correspond to the messages containing data from task T_a, and the events labeled by a lowercase b correspond to the messages containing data from task T_b. Figure 3.9b shows how this event sequence changes when a minimum distance d^{min} between two consecutive events is considered. As indicated, the distance between the last two events of type b further decreases because of the minimum distance. Likewise, the maximum distance increases because of the minimum distance, d^{min}, as can be seen for the first and the second of the events of type b. Based on the minimum distance, d^{min}, the maximum possible decrease (increase), D^{max}, in the

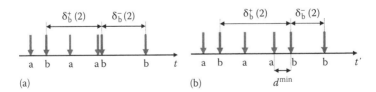

(a)   (b)

FIGURE 3.9
(a) The event sequence before applying the minimum distance and (b) the event sequence after considering the minimum distance d^{min}.

minimum (maximum) distance between events that can occur because of the minimum distance can be calculated. Note that, in the case of large bursts, $D^{\max}$ can be significantly larger than $d^{\min}$, since an event can be delayed by its predecessor event, which itself is delayed by its predecessor and so on. More details can be found in [37].

In general, considering the response time jitter, J^{resp}, and the minimum distance, $d^{\min}$, the inner stream of the hierarchical output stream, modeling messages that contain data from the task T_i, can be modeled by

$$\delta_i''^-(n) = \max\{\delta_i'^-(n) - J^{\text{resp}} - D^{\max}, \delta_i'^-(n-1) + d^{\min}\},$$
$$\delta_i''^+(n) = \delta_i'^+(n) + J^{\text{resp}} + D^{\max}$$

To determine the activation timings of the receiving tasks, T_c and T_d, we now have not only the arrival times of messages, but also the timings of exactly those messages that contain new data from a certain sending task, given by the corresponding inner stream. Assuming that the task T_c is only activated every time a new signal from the task T_a arrives, then the inner event stream ES_a'' of the hierarchical output stream of the communication task C can directly be used as an input stream of the task T_c.

It is also possible to have event streams with multiple hierarchical layers, for example, when modeling several layers of communication stacks or communications over networks interconnected by gateways, where several packets may be combined into some higher-level communication structure. This can be captured by our HEM by having an inner event stream of a HES that is the outer event stream of another HEM. For more details on multilevel hierarchies, refer to [36].

3.5 Scenario-Aware Analysis

Because of the increasing complexity of modern applications, hard real-time systems are often required to run different scenarios (also called operating modes) over time. For example, an automotive platform may exclusively execute either an ESC or a parking-assistant application. While the investigation of each static scenario can be achieved with classical real-time performance analysis, timing failures during the transition phase can only be uncovered with new methods, which consider the transient overload situation during the transition phase in which both scenarios can impress load artifacts on the system.

Each scenario is characterized by a specific behavior and is associated with a specific set of tasks. A scenario change (SC) from one scenario to another is triggered by a scenario change request (SCR) which may be caused either by the need to change the system functionality over time or by a system transition to a specific internal state requiring an SC. Depending on the task behavior across an SC, three types of tasks are defined:

- *Unchanged task*: An unchanged task belongs to both task sets of the initial (old) and the new scenario. It remains unchanged and continues executing normally after the SCR.
- *Completed task*: A completed task only belongs to the old scenario task set. However, to preserve data-consistency, completed task jobs activated before the SC are allowed to complete their execution after the SCR. Then the task terminates.
- *Added task*: An added task only belongs to the new scenario task set. It is initially activated after the SCR. Each added task is assigned an offset value, ϕ, that denotes its earliest activation time after the SCR.

During an SC, executions of completed, unchanged, and added tasks may interfere with one another, leading to a transient overload on the resource. Since the timing requirements in the system have to be met at any time during the system execution, it is necessary to verify if task deadlines could be missed because of an SC.

Methods analyzing the timing behavior across an SC under static-priority preemptive scheduling already exist [32,45,47]. However, they are limited to independent tasks mapped on single resources. Under such an assumption, the worst-case response time for an SC for a given task under analysis is proved to be obtained within the busy window during which the SCR occurs, called the transition busy window. These approaches can however not be applied to distributed systems because of the so-called echo effect. The echo effect is explained in the following section using the system example in Figure 3.11.

3.5.1   Echo Effect

The system used in the experiments of Section 3.8 (depicted in Figure 3.11) represents a hypothetical automotive system consisting of two IP components, four ECUs, and one multicore ECU connected via a CAN bus. The system is assumed to run two mutually exclusive applications: an ESP application (*Sens1, Sens2* → *eval1, eval2*) and a parking-assistant application (*Sens3* → *SigOut*). A detailed system description can be found in Section 3.8.

Let us focus on what happens on the CAN bus when the ESP application is deactivated (Scenario 1) and the parking-assistant application becomes active (Scenario 2). Depending on which application a communication task belongs to, we can determine the following task types on the bus when an SC occurs from Scenario 1 to Scenario 2: $C1$ and $C5$ are unchanged communication tasks, $C3$ and $C4$ are added communication tasks, and $C2$ is a completed communication task. Furthermore, we assume the following priority ordering on the bus: $C1 > C2 > C3 > C4 > C5$.

When an SC occurs from Scenario 1 to Scenario 2, the added communication task C3 is activated by events sent by the task *mon3*. However, C3 may have to wait until the prior completed communication task C2 finishes executing before being deactivated. This may lead to a burst of events waiting at the input of C3 that in turn may lead to a burst of events produced at its output. This burst of events is then propagated through the task *ctrl3* on ECU4 to the input of C4. In between, this burst of events may have been amplified because of scheduling effects on ECU4 (the task *ctrl3* might have to wait until *calc* finishes executing). Until this burst of events arrives at C4's input—which is a consequence of the SC on the bus—the transition busy window might already be finished on the bus. The effect of the transient overload because of the SC on the bus may therefore not be limited to the transition busy window but be recurrent. We call this recurrent effect the echo effect. As a consequence of the echo effect, for the worst-case response time calculation across the SC of the low-priority unchanged communication task C5, it is not sufficient to consider only its activations within the transition busy window. Rather, the activations within the successive busy windows need to be considered.

3.5.2   Compositional Scenario-Aware Analysis

The previous example illustrates how difficult it is to predict the effect of the recurrent transient overload after an SC in a distributed system. As a consequence of this unpredictability, it turns to be very difficult to describe the event timings at task outputs and therefore to describe the event timings at the inputs of the connected tasks, needed for the response time calculation across the SC. To overcome this problem, we need to describe the event timing at each task output, in a way that covers all its possible timing behaviors, even those resulting from the echo effect that might occur after an SC.

This calculation is performed by extending the compositional methodology presented in Section 3.2 as follows. As usual, all external event models at the system inputs are propagated along the system paths until an initial activating event model is available at each task input. Then, global system analysis is performed in the following way. In the first phase, two task response time calculations are performed on each resource. First, for each task we calculate its worst-case response time during the transition busy window. This calculation is described in detail in [13]. Additionally, for each unchanged or added task, using the classical analysis techniques we calculate its worst-case response times assuming the exclusive execution of the new scenario. Then, for each task, a response time interval is built into which all its observable response times may fall (i.e., the maximum of its response time during the transition busy window and its response time assuming the exclusive execution of the new scenario). The tasks' best-case response times are given by their minimum execution times in all scenarios.

Having determined a response time interval across the SC for each task, the second phase of the global system analysis is performed as usual, describing the traffic timing behavior at task outputs, by using event models. Afterward, the calculated output event models are propagated to the connected components, where they are used as activating event models for the subsequent global iteration. If after an iteration all calculated output event models remain unmodified, convergence is reached. As the propagated event models contain all potential event timings—during and after the transition—the calculated task response times are considered valid.

3.6 Sensitivity Analysis

As a result of an intensive HW/SW component reuse in the design of embedded systems, there is a need for analysis methods that, besides validating the performance of a system configuration, are able to predict the evolution of the system performance in the context of modifications of component properties.

The system properties represent intrinsic system characteristics determined by the configuration of the system components and the system's interaction with the environment. These include the execution/communication time intervals of the computational/communication tasks, the timing parameters of the task activation models, or the speed factor of the HW resources. These properties are used to build the performance model of the system. Based on this, the system quality is evaluated using a set of performance metrics such as response times, path latencies, event timings at components' outputs, and buffer sizes. These are used to validate the set of performance constraints, determined by local and global deadlines, jitter and buffering constraints, and so on.

The sensitivity analysis of real-time systems investigates the effects on the system quality (e.g., end-to-end delays, buffer sizes, and energy consumption) when subjected to system property variations. It will in the following be used to cover two complementary aspects of real-time system designs: performance characterization and the evaluation of the performance slack.

3.6.1 Performance Characterization

This aspect investigates the behavior of the performance metrics when applying modifications of different system properties. Using the mathematical properties of the functions describing the performance metrics, one can show the dependency between the values of the system properties specified in the input model and the values of the performance metrics. This is especially important for system properties leading to a discontinuous behavior of the performance metrics. The system designer can efficiently use this information to apply the required modifications without critical consequences on the system's performance.

Of special interest is the characterization of system properties whose variation leads to a nonmonotonic behavior of the performance metrics, referred to as timing anomalies. Such anomalies mostly occur because of inter-task functional dependencies, which are directly translated into timing dependencies in the corresponding performance model. The analyses of timing anomalies become relevant in the later design phases, when the estimated property values turn into concrete, fixed values. It is important for the system designer to know the source of such anomalies and which of the property values correspond to nonmonotonic system performance behavior. Since it divides the nonmonotonic—performance unpredictable—design configuration space into monotonic—performance predictable—subspaces, timing anomaly analysis is an important element of the design space exploration process.

3.6.2   Performance Slack

In addition to performance characterization, sensitivity analysis determines the bound between feasible and infeasible property values. This bound is called the sensitivity front. The maximum amount by which the initial value of a system property can be modified without jeopardizing the system feasibility is referred to as performance slack.

In general, to preserve the predictability, the modification of the design data is performed in isolation. This means, the system designer assumes the variation of a single system characteristic at a time, for example, the implementation of a software component. In general, in terms of performance, this corresponds to the variation of a single system property, for example, the worst-case execution time of the modified task. Such modifications are subject to one-dimensional sensitivity analysis. When the modification of a single system property has only local effects on the performance metrics, the computation of the performance slack is quite simple. Several formal methods were previously proposed [3,48]. However, in some other situations, the variation of the initial value affects several local and global performance metrics. Therefore, in order to compute the performance slack, sensitivity analysis has to be assisted by an appropriate system-level performance analysis model.

In many design scenarios though, changing the initial design data cannot be performed in isolation, such that a required design modification involves the simultaneous variation of several system characteristics, and thus system properties. For example, changes in the execution time of one task may coincide with a changed communication load of another task. Such modifications are the subject of multidimensional sensitivity analysis. Since the system performance metrics are represented as functions of several system properties and the dependency between these properties is generally unknown, the sensitivity front is more difficult to determine. In general, the complexity of the sensitivity analysis exponentially increases with the number of variable properties in the design space.

Based on the design strategy, two scenarios for using the performance slack are identified:

- *System robustness optimization*: Based on the slack values, the designer defines a set of robustness metrics to cover different possible design scenarios. In order to maximize the system robustness at a given cost level, the defined robustness metrics are used as optimization objectives by automatic design space exploration and optimization tools [12]. The scope is to obtain system configurations with less sensitivities to later design changes. More details are given in Section 3.7.
- *System dimensioning*: To reduce the global system cost, the system designer can decide to use the performance slack for efficient system dimensioning. In this case, instead of looking for system configurations that can accommodate later changes, the performance slack is used to optimize the system cost by selecting cheaper variants for processors, communication resources, or memories. A sufficient slack may even suggest the integration of the entire application on alternative platforms, reducing the number of hardware components [30]. Note that lower cost implies lower hardware costs on one side, and lower power consumption and smaller size on the other.

The sensitivity analysis approach has been tailored differently in order to achieve the previous design goals. Thus, to perform robustness optimization, the sensitivity analysis was integrated into a global optimization framework. For the evaluation of the robustness metrics it is not necessary to accurately determine the sensitivity front. Instead, using stochastic analysis, the sensitivity front can be approximated using two bounds for the sensitivity front: the lower bound determines the minimum guaranteed robustness (MGR), while the upper bound determines the maximum possible robustness (MPR). The benefit of using a stochastic analysis instead of an exact analysis is the nonexponential complexity with respect to the number of dimensions, which makes it suitable for a large number of variable system properties. Details about the implementation are given in [12].

For the second design scenario, the exact sensitivity front is required in order to perform the modifications of the system properties. Compared to previous formal sensitivity analysis algorithms, the proposed approach uses a binary search technique, ensuring complete transparency with respect to the application structure, the system architecture, and scheduling algorithms. In addition, the compatibility with the system-level performance analysis engine allows analysis of systems with global constraints. The one-dimensional sensitivity analysis algorithms are also used to bound the search space investigated by the stochastic sensitivity analysis approach.

The advantage of the exact analysis, when compared to the stochastic analysis, is the ability to handle nonmonotonic search spaces. A detailed description of the sensitivity analysis algorithms for different system properties can be found in [31].

3.7 Robustness Optimization

In the field of embedded system design, robustness is usually associated with reliability and resilience. Therefore, many approaches to fault tolerance against transient and permanent faults with different assumptions and for different system architectures can be found in the literature (e.g., in [9,22,28]). These approaches increase the system robustness against effects of external interferences (radiation, heat, etc.) or partial system failure, and are, therefore, crucial for safety critical systems.

In this chapter, a different notion of robustness for embedded systems is introduced and discussed: robustness to variations of system properties. Informally, a system is called robust if it can sustain system property modifications without severe consequences on system performance and integrity. In contrast to fault tolerance requiring the implementation of specific methods such as replication or reexecution mechanisms [18] to ensure robustness against faults, robustness to property variations is a meta problem that does not directly arise from the expected and specified functional system behaviors. It rather represents an intrinsic system property that depends on the system organization (architecture, application mapping, etc.) and its configuration (scheduling, etc.).

Accounting for property variations early during design is key, since even small modifications in systems with complex performance dependencies can have drastic nonintuitive impacts on the overall system behavior, and might lead to severe performance degradation effects [29]. Since performance evaluation and exploration do not cover these effects, it is clear that the use of these methods alone is insufficient to systematically control system performance along the design flow and during system lifetime. Therefore, explicit robustness evaluation and optimization techniques that build on top of performance evaluation and exploration are needed. They enable the designer to introduce robustness at critical positions in the design, and thus help to avoid critical performance pitfalls.

3.7.1 Use-Cases for Design Robustness

In the following, we discuss situations and scenarios where robustness of hardware and run-time system performance against property variations is expected and is crucial to efficiently design complex embedded systems.

Unknown quality of performance data: First, robustness is desirable to account for data quality issues in early design phases, where data that are required for performance analysis (e.g., task execution times and data rates) are often estimated or based on measurements. As a result of the unknown input data quality, also the expressiveness and accuracy of performance analysis results are unknown. Since even small deviations from estimated property values can have severe consequences on the final system performance, it is obvious that robustness against property variations leverages the applicability of formal analysis techniques during design. Clearly, design risks can be considerably reduced by identifying performance critical data and systematically optimizing the system for robustness.

Maintainability and extensibility: Secondly, robustness is important to ensure system maintainability and extensibility. Since major system changes in reaction to property variations are not usually possible during late design phases or after deployment, it is important to choose system architectures and configurations that offer sufficient robustness for future modifications and extensions, as early on as possible. For instance, the huge number of feature combinations in modern embedded systems has led to the problem of product and software variants. Using robustness optimization techniques, systems can be designed, at the outset, to accommodate additional features and changes. Other situations where robustness can increase system maintainability and extensibility include late feature requests, product and software updates (e.g., new firmware), bug fixes, and environmental changes.

Reusability and modularity: Finally, robustness is crucial to ensure component reusability and modularity. Even though these issues can be solved on the functional level by applying middleware concepts, they are still problematic from the performance verification point of view. The reason is that system performance is not composable, prohibiting the straightforward combination of individually correct components in a cut-and-paste manner to whole systems. In this context, robustness to property variations can facilitate the reuse of components across product generations and families, and simplify platform porting.

3.7.2 Evaluating Design Robustness

Sensitivity analysis has already been successfully used for the evaluation and optimization of specific system robustness aspects. In [26], for instance, the authors present a sensitivity analysis technique calculating maximum input rates that can be processed by stream-processing architectures without violating on-chip buffer constraints. The authors propose to integrate this technique into automated design space exploration to find architectures with optimal stream-processing capabilities, which exhibits a high robustness against input rate increases.

In this chapter, sensitivity analysis is systematically utilized for general robustness evaluation and optimization purposes. More precisely, instead of consuming the available slack for system dimensioning, and thus cost minimization, the slack is distributed so that the system's capability of supporting property variations is maximized. Using sensitivity analysis as a basis for robustness evaluation and optimization has two important advantages compared to previous approaches.

1. State-of-the-art modular sensitivity analysis techniques capture complex global effects of local system property variations. This ensures the applicability of the proposed robustness evaluation and optimization techniques to realistic performance models, and increases the expressiveness of the results.
2. Rather than providing the system behavior for some isolated discrete design points [4,7], sensitivity analysis characterizes continuous design subspaces with identical system states. It thus covers all possible system-property variation scenarios.

3.7.3 Robustness Metrics

In order to optimize robustness, we need, on the one hand, expressive robustness metrics and, on the other hand, efficient optimization techniques. In general, robustness metrics shall cover different design scenarios.

3.7.3.1 Static Design Robustness

The first considered design scenario assumes that system parameters are fixed early during design and cannot be modified later (e.g., at late design stages or after deployment) to compensate for system property modifications. This scenario is called static design robustness (SDR).

The SDR metric expresses the robustness of parameter configurations with respect to the simultaneous modifications of several given system properties. Since the exact extent of system property variations can generally not be anticipated, it is desirable that the system supports as many as possible modification scenarios. This shall be transparently expressed by the SDR metric: the more different modification scenarios represent feasible system states for a specific parameter configuration, the higher the corresponding SDR value. Note that the SDR optimization yields a single parameter configuration possessing the highest robustness potential for the considered system properties.

3.7.3.2 Dynamic Design Robustness

The SDR metric assumes static systems with fixed parameter configurations. However, the system may react to excessive system property variations with dynamic counteractions, such as parameter reconfigurations, which potentially increases the system's robustness. When such potential designer or system counteractions are included in the robustness evaluation, this view is expressed with the dynamic design robustness (DDR). The DDR metric expresses the robustness of given systems with respect to the simultaneous modifications of several system properties that can be achieved through reconfiguration. Consequently, it is relevant for the design scenario where parameters can be (dynamically) modified during design or after deployment. Obviously, the DDR metric depends on the set of possible parameter configurations, $\mathcal{C}$ ("counteractions"), that can be adopted through reconfiguration. For instance, it may be possible to react to a property variation by adaptation of scheduling parameters (e.g., adaptive scheduling strategies [25] and network management techniques [8]) or application remapping.

Application scenarios for the DDR metric include the evaluation of dynamic systems and, more generally, the assessment of the design risks connected to specific components. More precisely, already during early design, the DDR metric can be used to determine bounds for property values of specific components ensuring their correct functioning in the global context. This information efficiently facilitates feasibility and requirements analysis and greatly assists the designer in pointing out critical components requiring special focus during specification and implementation. Another use-case concerns reconfigurable systems. The DDR metric can be used to maximize the dynamic robustness headroom for crucial components. Obviously, by choosing a system architecture offering a high DDR for crucial system parts early, the designer can significantly increase system stability and maintainability.

Note that the DDR optimization yields multiple parameter configurations, each possessing partially disjoint robustness properties. For instance, one parameter configuration might exhibit high robustness for some system properties, whereas different parameter configurations might offer more robustness for other system properties.

Figure 3.10a and b visualize the conceptual difference between the notions of the SDR and the DDR by means of a simple example. Figure 3.10a shows the feasible region of two properties p_1 and p_2, i.e., the region containing all feasible property value combinations, of a given parameter configuration. This corresponds to the static robustness, where a single parameter configuration with high robustness needs to be chosen. Figure 3.10b visualizes the dynamic robustness. In the considered case, there exist two additional parameter configurations in the underlying reconfiguration space with interesting robustness properties. Both new parameter configurations contain feasible regions that are not covered by the first parameter configuration. The union of all three feasible regions corresponds to the dynamic robustness.

3.8   Experiments

In this section, the formal methods presented in this chapter are applied to the example system illustrated in Figure 3.11. The entire system consists

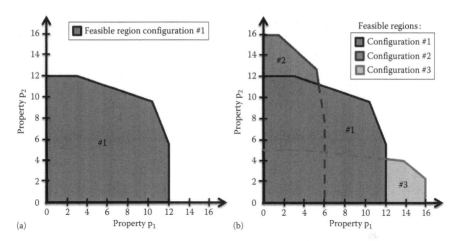

FIGURE 3.10

Conceptual difference between the SDR and the DDR for two considered system properties subject to maximization.

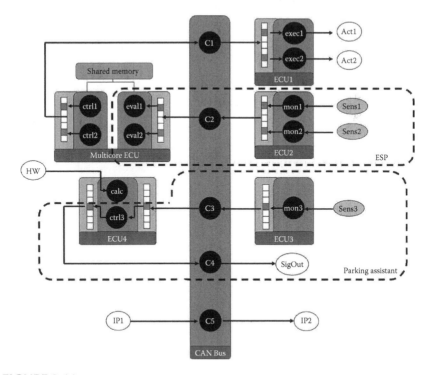

FIGURE 3.11

A hypothetical example system.

of four ECUs and one multicore ECU that are connected via a CAN bus. Additionally, there are two IP components that also communicate over the CAN bus. We assume that two applications from the automotive domain are running on this platform. The Sensors 1 and 2 collect the ESP-related data, which are preprocessed on ECU2. These data are then sent to the multicore ECU where the data are evaluated and appropriate control data are periodically generated based on the evaluated data. These data are then sent to ECU1 where the commands are processed. Sensor 3 collects data relevant for a parking-assistant application. The collected data are preprocessed on ECU3 and sent to ECU4, where the data are further processed before they are passed as an audio signal to the driver.

In the following, we will assume that these two applications are running mutually exclusive. For example, as soon as the driver shifts into the reverse gear, the parking-assistant application (Scenario 2) becomes active, and the ESP (Scenario 1) is deactivated.

The tasks on ECU1 are scheduled according to a round-robin scheduling policy, while all other ECUs implement a static-priority preemptive scheduling policy. Core execution and communication times, and the scheduling parameters (priority and time slot size) of all tasks in the system are specified in Table 3.1. Additionally, for tasks on the multicore ECU, the memory access time is explicitly given for each task, to allow considering the contention on the shared memory. (On single-core ECUs, the memory access time is contained in the core execution time.) For the communication, we assume that the transmission mode of the communication layers is direct and that

TABLE 3.1

Core Execution/Communication Time and Memory Access Time Per Task

HW	Task Name	Exec./Comm. Time (in ms)	Memory Access Time	Scheduling Parameter
Multicore ECU	ctrl1	[10:22]	[0:2]	Prio: High
	ctrl2	[20:20]	[0:1]	Prio: Low
	eval1	[12:14]	[0:4]	Prio: High
	eval2	[26:26]	[0:6]	Prio: Low
ECU1	exec1	[20:20]	—	Time Slot size: 5
	exec2	[30:30]	—	Time Slot size: 10
ECU2	mon1	[10:15]	—	Prio: High
	mon2	[12:18]	—	Prio: Low
ECU3	mon3	[20:20]	—	Prio: High
ECU4	calc	[1:1]	—	Prio: High
	ctrl3	[20:20]	—	Prio: Low
CAN Bus	C1	[6:6]	—	Prio: Highest
	C2	[5:5]	—	Prio: High
	C3	[10:10]	—	Prio: Med
	C4	[10:10]	—	Prio: Low
	C5	[7:7]	—	Prio: Lowest

all sending tasks produce triggering signals. This implies that whenever the sending tasks produce an output value, the transmission of a message is triggered. We suppose that Sensor 1 produces a new signal every 75 ms, Sensor 2 every 80 ms, and Sensor 3 every 75 ms. The hardware component activating the task *calc* on ECU4 performs this calculation every 2 ms and the control tasks ctrl1 and ctrl2 are activated every 100 ms and 120 ms, respectively. The system is subject to two end-to-end latency constraints. The latency of the parking-assistant application (Sens3 $\rightarrow$ SigOut) may not exceed 150 ms, whereas the communication between the two IP components (IP1 $\rightarrow$ IP2) must not last longer than 35 ms.

3.8.1 Analyzing Scenario 1

Initially, assume that in previous product generations the parking-assistant application used a dedicated communication bus, and, thus, only the ESP application initially needs to be contained in the model.

In this setup, when accounting for the communication of the ESP application and the communication between the two IP components, the bus load is only 47.21% and the maximum latency for the communication between IP1 and IP2 is 29 ms. Our analysis yields that all response times and end-to-end latencies remain within the given constraints.

By using the HEMs, we obtain very accurate input event models for the receiving tasks (exec1 and exec2). For example, Figure 3.12 illustrates the maximum number of message arrivals vs. signal arrivals for ECU1. The upper curve (marked by circles) represents the maximum number of

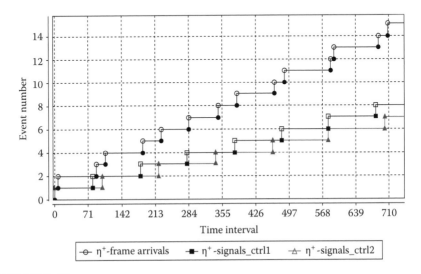

FIGURE 3.12
Message arrivals at ECU1 vs. signal arrivals.

messages that can arrive at ECU1. Using flat event models, we could only assume that every message contains a new signal for both receiving tasks, which results in a load of 91.58% of ECU1. With the HEM, we also obtain the maximum number of messages that contain a signal that was sent by task ctrl1 (marked by squares), and the maximum number of messages containing a signal from ctrl2 (marked by triangles). If we now use the timings of signal arrivals as activation timings of the receiving tasks, we obtain a much smaller load of only 45% for ECU1.

Hence, the system is not only schedulable, but it also appears that the bus with less than 50% utilization still has sufficient reserves to accommodate the additional communication of the parking-assistant application. Especially, since the time the parking assistant is enabled, the ESP communication is disabled.

3.8.2   Analyzing Scenario 2

In Scenario 2, Sensors 1 and 2 are disabled, and therefore tasks mon1 and mon2 are never activated. Consequently, they will not send data to the tasks eval1 and eval2. The control tasks ctrl1 and ctrl2 are still executed and send their data to the execution tasks running on ECU1. Their local response times will slightly decrease, as there will now be no competition for the shared memory from the second core. On the CAN bus we have the two additional communication tasks C3 and C4, representing the communication of the parking-assistant application. When we analyze this system, we obtain a maximum latency of 22 ms for the path IP1 → IP2 and 131 ms for the path Sens3 → SigOut. Therefore, the system is also schedulable when only the parking-assistant application is running.

3.8.3   Considering Scenario Change

Having analyzed the two scenarios in isolation from each other, we neglected the (recurrent) transient overload that may occur during the SC. This may lead to optimistic analysis results. Thus, the SC analysis is needed to verify the timing constraints across the SC. In the first experiment, we perform an SC analysis assuming an "all scenarios in one" execution, that is, all tasks belonging to both scenario task sets are assumed to be able to execute simultaneously. We obtain a maximum latency of 59 ms for the path IP1 → IP2 and 151 ms for the parking-assistant application path (path Sens3 → SigOut). So, the system is not schedulable, since neither constraint is met. In the second experiment, we use the compositional scenario-aware analysis presented in Section 3.5.2 for the timing verification across the SC. We calculate a maximum latency of 39 ms for the path IP1 → IP2 and 131 ms for the parking-assistant application path. Thus, we notice that there is an improvement in the calculated maximum latencies of the constrained application paths. However, the path IP1 → IP2 slightly exceeds its constraint.

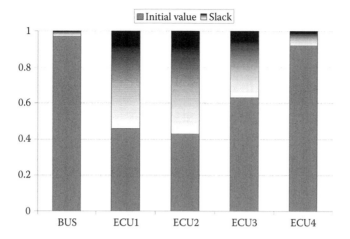

FIGURE 3.13
One-dimensional slack of the resource speeds.

3.8.4 Optimizing Design

As the design is not feasible in its current configuration, we need to optimize the critical path IP1 → IP2 latency. For this, we can explore the priority configuration of the communication tasks on the CAN bus. This can be performed automatically on the basis of genetic algorithms (refer to [11] for details). A feasible configuration is obtained for the following priority order: $C1 > C2 > C5 > C3 > C4$. The obtained maximum path IP1 → IP2 latency is equal to 29. Even though the maximum latency of the parking-assistant application increased from 131 to 138, this is still less than the imposed constraint.

3.8.5 System Dimensioning

According to Section 3.6, the performance slack of the system components can be efficiently used in order to select hardware components that are optimal with respect to cost. The diagram presented in Figure 3.13 shows the minimum speed of the CAN bus and the single-core ECUs. The presented values are relative to the resource speed values in the initial configuration. These values were individually obtained for each resource, which means that the speed of only one resource was changed at any one time.

3.9 Conclusion

This chapter has given an overview of state-of-the-art compositional performance analysis techniques for distributed systems and MPSoCs. Furthermore, we have highlighted specific timing implications that require attention when addressing the MPSoC setups, hierarchical communication networks, and SCs. To leverage the capabilities of the overall approach, sensitivity analysis and robustness optimization techniques were implemented that work without executable code and that are based on robustness metrics.

By means of a simple example, we have demonstrated that modeling and formal performance analysis are adequate for the verifying, optimizing, and dimensioning heterogeneous multiprocessor systems. Many of the techniques presented here are already used in industrial practice [35].

References

1. K. Albers, F. Bodmann, and F. Slomka. Hierarchical event streams and event dependency graphs: A new computational model for embedded real-time systems. *Proceedings of the 18th Euromicro Conference on Real-Time Systems*, Dresden, Germany, pp. 97–106, 2006.

2. AUTOSAR. AUTOSAR Specification of Communication V. 2.0.1, AUTOSAR Partnership, 2006. http://www.autosar.org.

3. P. Balbastre, I. Ripoll, and A. Crespo. Optimal deadline assignment for periodic real-time tasks in dynamic priority systems. In *18th Euromicro Conference on Real-Time Systems*, Dresden, Germany, 2006.

4. I. Bate and P. Emberson. Incorporating scenarios and heuristics to improve flexibility in real-time embedded systems. In *Proceedings of the IEEE Real-Time and Embedded Technology and Applications Symposium (RTAS)*, San Jose, CA, April 2006.

5. J. L. Boudec and P. Thiran. *Network Calculus: A Theory of Deterministic Queuing Systems for the Internet*. Springer, Berlin, 2001.

6. S. Chakraborty, S. Künzli, and L. Thiele. A general framework for analysing system properties in platform-based embedded system designs. In *Proceedings of the IEEE/ACM Design, Automation and Test in Europe Conference (DATE)*, Munich, Germany, 2003.

7. P. Emberson and I. Bate. Minimising task migration and priority changes in mode transitions. In *Proceedings of the IEEE Real-Time and Embedded Technology and Applications Symposium (RTAS)*, Seatlle, WA, April 2007.

8. J. Filipiak. *Real Time Network Management*. North-Holland, Amsterdam, the Netherlands, 1991.

9. O. Gonzalez, H. Shrikumar, J. Stankovic, and K. Ramamritham. Adaptive fault tolerance and graceful degradation under dynamic hard

real-time scheduling. In *Proceedings of the IEEE International Real-Time Systems Symposium (RTSS)*, San Francisco, CA, December 1997.

10. W. Haid and L. Thiele. Complex task activation schemes in system level performance analysis. In *Proceedings of the IEEE/ACM International Conference on HW/SW Codesign and System Synthesis (CODES-ISSS)*, Salzburg, Austria, September 2007.

11. A. Hamann, M. Jersak, K. Richter, and R. Ernst. Design space exploration and system optimization with SymTA/S-symbolic timing analysis for systems. In *Proceedings 25th International Real-Time Systems Symposium (RTSS04)*, Lisbon, Portugal, December 2004.

12. A. Hamann, R. Racu, and R. Ernst. A formal approach to robustness maximization of complex heterogeneous embedded systems. In *Proceedings of the IEEE/ACM International Conference on HW/SW Codesign and System Synthesis (CODES-ISSS)*, Seoul, South Korea, October 2006.

13. R. Henia and R. Ernst. Scenario aware analysis for complex event models and distributed systems. In *Proceedings of the Real-Time Systems Symposium*, Jucson, AZ, 2007.

14. R. Henia, A. Hamann, M. Jersak, R. Racu, K. Richter, and R. Ernst. System level performance analysis—the SymTA/S approach. *IEE Proceedings Computers and Digital Techniques*, 152(2):148–166, March 2005.

15. R. Henia, R. Racu, and R. Ernst. Improved output jitter calculation for compositional performance analysis of distributed systems. *Parallel and Distributed Processing Symposium, 2007. IPDPS 2007. IEEE International*, Long Beach, CA, pp. 1–8, 2007.

16. T. Henzinger and S. Matic. An interface algebra for real-time components. In *Proceedings of the IEEE Real-Time and Embedded Technology and Applications Symposium (RTAS)*, San Jose, CA, April 2006.

17. I. IXP2400. IXP2800 Network Processors.

18. V. Izosimov, P. Pop, P. Eles, and Z. Peng. Design optimization of time- and cost-constrained fault-tolerant distributed embedded systems. In *Proceedings of the IEEE/ACM Design, Automation and Test in Europe Conference (DATE)*, Munich, Germany, March 2005.

19. M. Jersak. Compositional performance analysis for complex embedded applications. PhD thesis, Technical University of Braunschweig, Braunschweig, Germany, 2004.

20. B. Jonsson, S. Perathoner, L. Thiele, and W. Yi. Cyclic dependencies in modular performance analysis. In *ACM & IEEE International Conference*

on Embedded Software (EMSOFT), Atlanta, GA, October 2008. ACM Press.

21. E. Lee, S. Neuendorffer, and M. Wirthlin. Actor-oriented design of embedded hardware and software systems. *Journal of Circuits Systems and Computers*, 12(3):231–260, 2003.

22. P. Lee, T. Anderson, J. Laprie, A. Avizienis, and H. Kopetz. *Fault Tolerance: Principles and Practice*. Springer Verlag, Secaucus, NJ, 1990.

23. J. Lehoczky. Fixed priority scheduling of periodic task sets with arbitrary deadlines. In *Proceedings of the IEEE Real-Time Systems Symposium (RTSS)*, Lake Buena Vista, FL, 1990.

24. J. Lemieux. *Programming in the OSEK/VDX Environment*. CMP Books, Lawrence, KS, 2001.

25. C. Lu, J. Stankovic, S. Son, and G. Tao. Feedback control real-time scheduling: Framework, modeling, and algorithms. *Real-Time Systems Journal*, 23(1–2):85–126, 2002.

26. A. Maxiaguine, S. Künzli, S. Chakraborty, and L. Thiele. Rate analysis for streaming applications with on-chip buffer constraints. In *Proceedings of the IEEE/ACM Asia and South Pacific Design Automation Conference (ASP-DAC)*, Yokohama, Japan, pp. 131–136, January 2004.

27. M. Negrean, S. Schliecker, and R. Ernst. Response-time analysis of arbitrarily activated tasks in multiprocessor systems with shared resources. In *Proceedings of Design, Automation and Test in Europe (DATE 2009)*, Nice, France, April 2009.

28. K. Poulsen, P. Pop, V. Izosimov, and P. Eles. Scheduling and voltage scaling for energy/reliability trade-offs in fault-tolerant time-triggered embedded systems. In *Proceedings of the IEEE/ACM International Conference on HW/SW Codesign and System Synthesis (CODES-ISSS)*, Salzburg, Austria, October 2007.

29. R. Racu and R. Ernst. Scheduling anomaly detection and optimization for distributed systems with preemptive task-sets. In *12th IEEE Real-Time and Embedded Technology and Applications Symposium (RTAS)*, San Jose, CA, April 2006.

30. R. Racu, A. Hamann, and R. Ernst. Automotive system optimization using sensitivity analysis. In *International Embedded Systems Symposium (IESS), Embedded System Design: Topics, Techniques and Trends*, Irvine, CA, pp. 57–70, June 2007. Springer.

31. R. Racu, A. Hamann, and R. Ernst. Sensitivity analysis of complex embedded real-time systems. *Real-Time Systems Journal*, 39(1–3):31–72, 2008.

32. J. Real and A. Crespo. Mode change protocols for real-time systems: A survey and a new proposal. *Real-Time System*, 26(2):161–197, 2004.

33. K. Richter, D. Ziegenbein, M. Jersak, and R. Ernst. Model composition for scheduling analysis in platform design. In *Proceedings of the 39th Design Automation Conference (DAC 2002)*, New Orleans, LA, June 2002.

34. K. Richter. Compositional performance analysis. PhD thesis, Technical University of Braunschweig, Braunschweig, Germany, 2004.

35. K. Richter. New kid on the block: Scheduling analysis improves quality and reliability of ecus and busses. *Embedded World Conference*, Nuremberg, Germany, 2008.

36. J. Rox and R. Ernst. Construction and deconstruction of hierarchical event streams with multiple hierarchical layers. In *Proceedings of the Euromicro Conference on Real-Time Systems (ECRTS 2008)*, Prague, Czech Republic, July 2008.

37. J. Rox and R. Ernst. Modeling event stream hierarchies with hierarchical event models. In *Proceedings of the Design, Automation and Test in Europe (DATE 2008)*, Munich, Germany, March 2008.

38. S. Schliecker, M. Ivers, and R. Ernst. Integrated analysis of communicating tasks in MPSoCs. *Proceedings of the 4th International Conference on Hardware/Software Codesign and System Synthesis*, Seoul, Korea, pp. 288–293, 2006.

39. S. Schliecker, M. Ivers, and R. Ernst. Memory access patterns for the analysis of MPSoCs. *2006 IEEE North-East Workshop on Circuits and Systems*, Gatineau, Quebec, Canada, pp. 249–252, 2006.

40. S. Schliecker, M. Ivers, J. Staschulat, and R. Ernst. A framework for the busy time calculation of multiple correlated events. *6th International Workshop on WCET Analysis*, Dresden, Germany, July 2006.

41. S. Schliecker, M. Negrean, and R. Ernst. Reliable performance analysis of a multicore multithreaded system-on-chip (with appendix). Technical report, Technische Universität Braunschweig, Braunschweig, Germany, 2008.

42. S. Schliecker, M. Negrean, G. Nicolescu, P. Paulin, and R. Ernst. Reliable performance analysis of a multicore multithreaded system-on-chip. In *Proceedings of the 6th IEEE/ACM/IFIP International Conference on Hardware/Software Codesign and System Synthesis*, pp. 161–166. ACM, New York, 2008.

43. S. Schliecker, J. Rox, M. Ivers, and R. Ernst. Providing accurate event models for the analysis of heterogeneous multiprocessor systems. In *Proceedings of the 6th IEEE/ACM/IFIP International Conference on*

Hardware/Software Codesign and System Synthesis, pp. 185–190. ACM, New York, 2008.

44. S. Segars. The ARM9 family-high performance microprocessors for embedded applications. *Proceedings of the International Conference on Computer Design: VLSI in Computers and Processors, 1998. ICCD'98.*, Austin, TX, pp. 230–235, 1998.

45. L. Sha, R. Rajkumar, J. Lehoczky, and K. Ramamritham. Mode change protocols for priority-driven preemptive scheduling. Technical Report UM-CS-1989-060, 31, 1989.

46. J. Staschulat and R. Ernst. Worst case timing analysis of input dependent data cache behavior. *Euromicro Conference on Real-Time Systems*, Dresden, Germany, 2006.

47. K. W. Tindell, A. Burns, and A. J. Wellings. Mode changes in priority pre-emptively scheduled systems. In *IEEE Real-Time Systems Symposium*, Phoenix, AZ, pp. 100–109, 1992.

48. S. Vestal. Fixed-priority sensitivity analysis for linear compute time models. *IEEE Transactions on Software Engineering*, 20(4):308–317, April 1994.

49. R. Wilhelm, J. Engblom, A. Ermedahl, N. Holsti, S. Thesing, D. Whalley, G. Bernat, C. Ferdinand, R. Heckmann, T. Mitra, F. Mueller, I. Puaut, P. Puschner, J. Staschulat, and P. Stenström, The worst-case execution-time problem—overview of methods and survey of tools, *Transactions on Embedded Computing Systems*, 7(3):1–53, 2008.

4

Model-Based Framework for Schedulability Analysis Using UPPAAL *4.1*

Alexandre David, Jacob Illum, Kim G. Larsen, and Arne Skou

CONTENTS

4.1 Introduction

Embedded systems involve the monitoring and control of complex physical processes using applications running on dedicated execution platforms in a resource-constrained manner in terms of, for example, memory, processing power, bandwidth, energy consumption, and timing behavior.

Viewing the application as a collection of interdependent tasks, various "scheduling principles" may be applied to coordinate the execution of tasks in order to ensure orderly and efficient usage of resources. Based on the physical process to be controlled, timing deadlines may be required for the individual tasks as well as the overall system. The challenge of "schedulability analysis" is now concerned with guaranteeing that the applied scheduling principle(s) ensure that the timing deadlines are met.

For single-processor systems, industrial applied schedulability analysis tools include TimeWiz from TimeSys Corporation [10] and RapidRMA from TriPacific [11], based on rate monotonic analysis. More recently, Sym-TA/S has emerged as an efficient tool for system-level performance and timing analysis based on formal scheduling analysis techniques and symbolic simulation [26]. These tools benefit from the great success of real-time scheduling theories: results that were developed in the 1970s and the 1980s, and are now well established. However, these theories and tools have become seriously challenged by the rapid increase in the use of multi-cores and multiprocessor systems-on-chips (MPSoCs).

To overcome the limitation to single-processor architectures, applications of simulation have been pursued, including—in the case of MPSoCs—the ARTS framework (based on SystemC) [22,23], the Daedaleus simulation tool [25], and the Design-Trotter [24].

Though extremely useful for early design exploration by providing very adequate performance estimates, for example, memory usage, energy consumption, and options for parallelizations, the use of simulation makes the schedulability analysis provided by these tools unreliable; though no deadline violation may be revealed after (even extensive) simulation, there is no guarantee that this will never occur in the future. For systems with hard real-time requirements, this is not satisfactory.

During recent years, the use of real-time model checking has become an attractive and maturing approach to schedulability analysis providing absolute guarantees: if after model checking no violations of deadlines have been found, then it is guaranteed that no violations will occur during execution. In this approach, the (multiprocessor) execution platform, the tasks, the interdependencies between tasks, their execution times, and mapping to the platform are modeled as timed automata [3], allowing efficient tools such as Uppaal [28] to "verify" schedulability using model checking.

The tool TIMES [4] has been pioneering this approach, providing a rather expressive task-model called time-triggered architecture (TTA) allowing for complex task-arrival patterns, and using the verification engine of Uppaal to verify schedulability. However, so far the tool only supports single-processor scheduling and limited dependencies between tasks. Other schedulability frameworks using timed automata as a modeling formalism and Uppaal as a backend are given in [8,13,14,17,27]. Also, related to schedulability analysis, a number of real-time operating systems (RTOS) have been formalized and analyzed using UPPAAL [16,20].

The MOVES analysis framework [19], presented in Chapter 5 of this book, is closely related to this chapter. Whereas the chapter on MOVES reports on the ability to apply UPPAAL to verify properties and schedulability of embedded systems through a number of (realistic size) examples, we provide in this chapter a detailed—and compared with [5], alternative—account on how to model multiprocessor-scheduling scenarios most efficiently, by making full use of the modeling formalism of UPPAAL. This chapter offers an UPPAAL modeling framework [15]) that may be instantiated to suit a variety of scheduling scenarios, and which can be easily extended. In particular, the framework includes

- A rich collection of attributes for tasks, including the offset, best- and worst-case execution times, minimum and maximum interarrival times, deadlines, and task priorities
- Task dependencies
- Assignment of resources, for example, processors or busses, to tasks
- Scheduling policies, including first-in first-out (FIFO), earliest deadline first (EDF), and fixed priority scheduling (FPS)
- Possible preemption of resources

The combination of task dependencies, execution time uncertainties, and preemption makes schedulability of the above framework undecidable [21]. However, the recent support for stopwatch automata [9] in UPPAAL leads to an efficient approximate analysis that has proved adequate for several concrete instances, as demonstrated in [19].

The outline of the remaining chapter is as follows: In Section 4.2, we show the formalism of UPPAAL by the use of an example. In Section 4.3, we give an introduction to the types of schedulability problems that can be analyzed using the framework presented in Section 4.4. Following the framework, in Section 4.5, we show how to instantiate the framework for a number of different schedulability problems by way of an example system. Finally, we conclude the chapter in Section 4.6.

4.2 UPPAAL and Its Formalism

In this section, we provide an introductory description of the UPPAAL modeling language.

4.2.1 Modeling Language

The tool UPPAAL is designed for design, simulation, and verification of real-time systems that can be modeled as networks of timed automata [2],

extended with integer variables and richer user-defined data types. A timed
automaton is a finite-state machine, extended with clock variables. The tool
uses a dense time-model of time, so clock variables evaluate to real numbers.
All the clocks progress synchronously.

We use, in this section, the train-gate example (distributed with the tool).
It is a railway system that controls, access to a bridge. The bridge has only
one track, and a gate controller ensures that at most one approaching train
is granted access to this track. Stopping and restarting a train takes time.
Figure 4.1 shows the model of a train in the editor of UPPAAL. When a train
is approaching (Appr), it must be stopped before 10 time units, otherwise, it
is too late and the train must cross the bridge (Cross). When it is stopped
(Stop), it must be restarted (Start) before crossing the bridge.

The timing constraints associated with the locations are "invariants."
They give a bound on how long these locations can be active: A train can
stay in Appr at most 20 time units and then must leave this location. Edges
between locations have guards (x<=10) to constrain when they can be taken,
synchronizations (stop[id]?) for communication, and updates (x=0) to
reset the clock x. Automata communicate with each other by means of channels. Here, we have an array of channels and every train has its own id. The
gate automaton selects a train and synchronizes with it with stop[id]!. In
UPPAAL, it is possible to declare arrays of clocks or any other type. Channels
can be declared to be "urgent" to prevent delays if a synchronization is possible, or "broadcast" to achieve broadcast synchronization instead of hand-
shake. Listing 4.1 shows the global declaration of the model with the channel

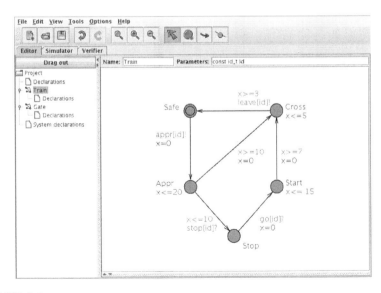

FIGURE 4.1
View of the train template in the editor.

Listing 4.1   Global declarations for the train-gate model.

```
1   const int  N = 6;                    // Number of trains
2   typedef int [0, N−1] id_t;
3   chan appr[N], stop[N], leave[N];
4   urgent chan go[N];
```

declarations. A constant is declared to size the model to the desired number of trains. The train model here is in fact a "template" for trains. Trains
are instantiated with a given id. In this case, having in the system declaration system Gate, Train; will instantiate an automaton for the Gate
controller and all the possible trains ranging over their missing argument
types. This is the "auto-instantiation" feature. It is possible to give specific
argument values too. The type id_t is a user-defined type declared in Listing 4.1. UPPAAL supports more complex user-defined types such as structures.
It is possible to combine arrays, structures, bounded integers, channels, and
clocks.

Figure 4.2 shows the simulator with the Gate automaton. On the message sequence chart, the different synchronizations between the automata
are shown. The automaton has two main locations where the bridge is free
or occupied. When a train is approaching it synchronizes with the gate, and
with a function call it is queued by the gate. If more trains are approaching,
then they are queued and stopped. Queuing followed by stopping is atomic,

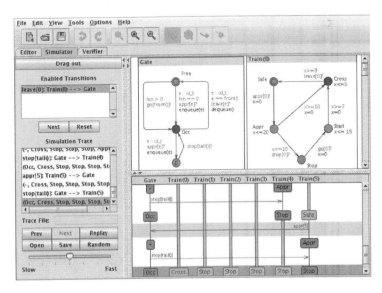

FIGURE 4.2
View of a simulation of the train-gate model showing the gate and one train.

Listing 4.2 Local declarations of the Gate template.

```
1   id_t  list [N+1];
2   int [0,N] len;
3
4   void enqueue(id_t element)  // Put an element at the end of the queue
5   {
6       list [len++] = element;
7   }
8
9   void dequeue()                 // Remove the front element of the queue
10  {
11      int  i  = 0;
12      len  -= 1;
13      while ( i  < len)
14      {
15          list [i]  =  list [i  + 1];
16          i++;
17      }
18      list [i]  = 0;
19  }
20
21  id_t  front ()                 // Returns the  front  element of the queue
22  {
23      return  list [0];
24  }
25
26  id_t  tail ()                  // Returns the  last  element of the queue
27  {
28      return  list [len  − 1];
29  }
```

which is modeled by marking the location "committed." Such a location forbids interleaving with other automata when it is active. A location can be marked "urgent" to mean that time cannot be delayed while it is active. After this, the gate dequeues a train and leaves it to cross the bridge. After that it will try to dequeue more trains and restart them with go[front()]!. Here, a function call is used to return the front of the queue. The Gate "picks" a train with the "select" statement, e:id_t. This allows the modeler to scale the model with the number of trains while still keeping the automaton compact. Listing 4.2 gives the complete local declarations of the Gate automaton. UPPAAL supports a C-like syntax that allows us to implement a queue here. One of the locations of the gate is marked "C."

Another feature of the language is the "scalar" type to define scalar sets. When these sets are used, the model checker takes advantage of their symmetries. Different variants of UPPAAL exist in other specific problem domains.

UPPAAL-TIGA [6,18] is based on timed game automata and is targeted toward code synthesis. UPPAAL-CORA [7,29] is designed for cost-optimal reachability analysis. UPPAAL-PRO [12,28] extends timed automata with probabilities.

Further features of UPPAAL include meta-variables, which can be used to store values such as regular variables; however, meta-variables are not included in the state inclusion check when doing model checking. That is, two states are considered identical if all but the meta-variables agree. Meta-variables are declared using the "meta" keyword (meta int i).

Finally, UPPAAL (as of version 4.1) supports stopwatches. Stopwatches are like clock variables; however, the progress of stopwatches can be set to either zero or one in automata locations, which is specified as an invariant ($x' == 0$ for clock x). When analyzing models using stopwatches, UPPAAL computes a finite overapproximation of the state space and is, thus, guaranteed to terminate even though the model-checking problem for stopwatch automata is, in general, undecidable. Checking properties such as avoidance of deadlocks can be meaningful for stopwatch automata, since if the overapproximation does not have a deadlock then neither will the real system. In Section 4.4, stopwatches are used to model preemptive schedulability problems using UPPAAL.

4.2.2 Specification Language

The specification language of UPPAAL is a subset of the timed computation tree logic (TCTL) [1]. The following properties are supported: (1) A[] ϕ, (2) E<> ϕ, (3) A<> ϕ, (4) E[] ϕ, and (5) $\phi -\!-\!> \psi$, where ϕ and ψ are state predicates. The safety property (1) specifies that ϕ must be satisfied for all states. The reachability property (2) specifies that there exists a path on which a state satisfies ϕ. The reachability property (3) specifies that for all paths there must be a state that satisfies ϕ. The liveness property (4) specifies that there is a path on which all states satisfy ϕ. The "leads-to" property (5) specifies that whenever a state satisfying ϕ is reached, then for all subsequent paths a state satisfying ψ is reached. In addition, UPPAAL can check for deadlocks with the property A[] not deadlock.

4.3 Schedulability Problems

At the core of any schedulability problem are the notions of tasks and resources. Tasks are jobs that require the usage of resources for a given duration after which tasks are considered done/completed. The added constraints to this basic setup is what defines a specific schedulability problem. In this section, we define a range of classical schedulability problems.

4.3.1   Tasks

A schedulability problem always consists of a finite set of tasks, which we consistently will refer to as $T = t_1, t_2, ..., t_n$. Each task has a number of attributes that we refer to by the following functions:

- INITIAL_OFFSET: $T \rightarrow \mathbb{N}$
  Time offset for initial release of task.
- BCET: $T \rightarrow \mathbb{N}_{\geq 0}$
  Best case execution time of task.
- WCET: $T \rightarrow \mathbb{N}_{\geq 0}$
  Worst case execution time of task.
- MIN_PERIOD: $T \rightarrow \mathbb{N}$
  Minimum time between task releases.
- MAX_PERIOD: $T \rightarrow \mathbb{N}$
  Maximum time between task releases.
- OFFSET: The time offset into every period, before the task is released.
- DEADLINE: $T \rightarrow \mathbb{N}_{\geq 0}$
  The number of time units within which a task must finish execution after its release. Often, the deadline coincides with the period.
- PRIORITY: Task priority.

These attributes are subject to the obvious constraints that BCET$(t) \leq$ WCET $(t) \leq$ DEADLINE$(t) \leq$ MIN_PERIOD$(t) \leq$ MAX_PERIOD(t). The periods are ignored if the task is nonperiodic.

The interpretation of these attributes is that a given task t_i cannot execute for the first OFFSET (t_i) time units, and should hereafter execute exactly once in every period of PERIOD (t_i) time units. Each such execution has a duration in the interval [BCET (t_i),WCET (t_i)]. The reason why tasks have a duration interval instead of a specific duration is that tasks are often complex operations that need to be executed, and the specific computation of a task depends on conditionals, loops, etc. and can vary between invocations. Furthermore, for multiprocessor scheduling, considering only worst-case execution times is insufficient as deadline violations can result from certain tasks exhibiting the best-case behavior.

We say that a task t is "ready" (to execute) at time τ iff

1. $\tau \geq$ INITIAL_OFFSET(t).
2. t has not executed in the given period dictated by τ.
3. All other constraints on t are satisfied. (Refer Section 4.3.2 for a discussion on task constraints.)

4.3.2   Task Dependencies

Task execution is often not just constrained by periods, but also by inter-dependencies among tasks, for example, because one task requires data that is computed by other tasks. Such dependencies among a set of tasks,

$T = t_1, t_2, ..., t_n$, are modeled as a directed acyclic graph (V, E), where tasks are nodes (i.e., $V = T$) and dependencies are directed edges between nodes. That is, an edge $(t_i, t_j) \in E$ from task t_i to task t_j indicates that task t_j cannot begin execution until task t_i has completed execution.

4.3.3   Resources

Resources are the elements that execute tasks. Each resource uses a scheduler to determine which task gets executed on a given resource at any point in time. Resources are limited by allowing the execution of only a single task at any given time.

Tasks are assigned a priori to resources. For a set of resources, $R = r_1, ..., r_k$, and a set of tasks, $T = t_1, ..., t_n$, we capture with the function ASSIGN : $T \rightarrow R$.

In a real-time system, resources function as different types of processors, communication busses, etc. Combined with task graphs we can use tasks and resources to emulate complex systems with such task interdependencies on different processors. For example, if we want to model two tasks t_i and t_j with dependency $t_i \rightarrow t_j$, but the tasks are executed on different processors and t_j needs the results of t_i to be communicated across a data bus, we introduce an auxiliary task t_{ic} that requires the bus resource and update the dependencies to $t_i \rightarrow t_{ic} \rightarrow t_j$. We illustrate this concept in Figure 4.8.

4.3.3.1   Scheduling Policies

In order for a resource to determine which task to execute and which tasks to hold, the resource applies a certain scheduling policy implemented in a scheduler. Scheduling strategies vary greatly in complexity depending on the constraints of the schedulability problem. In this section, we discuss a subset of scheduling policies for which we have included models in our scheduling framework.

- **First-In First-Out** Ready tasks are added to a queue in the order they become ready.
- **Earliest Deadline First** Ready tasks are added to a sorted list and executed in the order of the earliest deadline.
- **Fixed Priority Scheduling** Each task is given an extra attribute, PRIORITY, and ready tasks are executed according to the highest PRIORITY.

Schedulers operate in such a manner that resources are never idle while there are ready tasks assigned to them. That is, as soon a task has finished execution a new task is set for execution.

4.3.3.2   Preemption

Resources come in two shapes: preemptive and nonpreemptive. A nonpreemptive resource means that once a task has been assigned to execute on a given resource, that task will run until completion before another task is assigned to the resource. Preemption means that a task assigned to a resource can be temporarily halted from execution if the scheduler decides to assign another task to the resource. We say that the first task has been preempted. A preempted task can later resume execution for the remaining part of its duration.

Preemption allows for a greater responsiveness to tasks that are close to missing their deadline, but that flexibility is on behalf of the increased complexity of the schedulability analysis. The framework we define in the following section will include a model for schedulability analysis with preemption.

4.3.4 Schedulability

Now, we define what it means for a system to be schedulable. A system of tasks with constraints and resources with scheduling policies is said to be schedulable if no execution satisfying the constraints of the system violates a deadline.

4.4 Framework Model in Uppaal

In this section, we will describe our Uppaal framework for analyzing the scheduling problems defined in Section 4.3. The framework is constructed such that a model of a particular scheduling problem consists of three different timed automata templates: a generic task template, a generic resource template, and a scheduling policy model for each applied policy. We will describe the templates in this order.

4.4.1 Modeling Idea

In order to best explain the framework models, we will provide an abstract scheduling model that will serve as a base for the framework models. The abstract scheduling model is based on the basic scheduling model defined in [7].

The model depicted in Figure 4.3 naturally divides the scheduling problem into tasks (Figure 4.3a) and resources (Figure 4.3b). Each task and resource has a unique identifier (id). Initially, tasks are Waiting, and when a task is ready to execute, this is signaled to the resource to which the task is assigned, using the channel "ready," indexed with the appropriate resource id (i.e., the variable resource). This moves the task to Ready where it remains until either the deadline has passed, in which case the task moves to Error, or it receives a signal that the execution is complete via the channel "finished" indexed with the task id, in which case the task moves to Done.

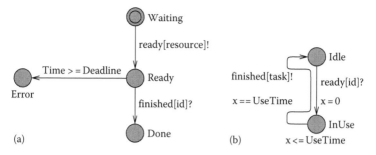

FIGURE 4.3

Abstract task and resource models. (a) Abstract task automaton and (b) abstract resource automaton.

Resources have two locations Idle and InUse indicating the state of the resource. Resources move from Idle to InUse upon a "ready" signal from a task, and return to Idle after the appropriate execution time. Resources signal that the task has finished execution using "finish," indexed with the appropriate task.

With this model, schedulability can be verified with the following CTL query:

A[] forall(i : task_id) not Task(i).Error

That is, is it always the case that on all execution paths no task will ever be in the Error location?

This is the base of the framework model introduced in the following sections. The added complexity of these models is because of the handling of preemption, periods, and different scheduling policies. Before introducing the models, we will introduce some of the basic data structures used in the code.

4.4.2 Data Structures

For each scheduling problem with tasks $T = t_0, ..., t_n$ and resources $R = r_0, ..., r_k$, we define the following data types for convenience:

t_id: Task ids ranging from 0 to n
r_id: Resource ids ranging from 0 to k
time_t: Integer value between zero and the largest period among all tasks

Having established the above data types we can move to more complex data types such as the data structure representing a task, which is called task_t and is depicted in Listing 4.3. In other words, the data structure of the task holds all task attributes defined in Section 4.3.1. Note that the "priority" is given by pri as 'priority' is a reserved keyword in UPPAAL.

Listing 4.3 Task structure.

```
 1   typedef struct {
 2      time_t   initial_offset ;
 3      time_t   min_period;
 4      time_t   max_period;
 5      time_t   offset ;
 6      time_t   deadline;
 7      time_t   bcet;
 8      time_t   wcet;
 9      r_id     resource;
10      int      pri ;
11   } task_t ;
```

Listing 4.4 Resource buffer.

```
 1   typedef struct {
 2      int [0, Tasks]  length;
 3      t_id  element[Tasks];
 4   } buffer_t ;
```

To specify a set of tasks, we create a global array called task of type task_t with one entry per task. The index of the array is the unique task identifier. See Section 4.5 for an example of task instantiation.

The final data structure is buffer_t which, is the central data structure of the resource template. Each resource has a buffer that keeps track of the tasks ready to execute on a given resource and is sorted according to the respective scheduling policy. The buffer element is defined in Listing 4.4. Resource buffers are held in a global array called buffer with one index per resource.

The above data structures serve as the foundation of the template models defined in the following sections.

4.4.3 Task Template

The task template serves as a model for both periodic and nonperiodic tasks. The type of scheduling problem at hand is specified using the global Boolean parameter Periodic. This variable is tested in the task template to guarantee that tasks observe correct periodic or nonperiodic behaviors.

The task template, depicted in Figure 4.4, takes a single parameter, namely, the task id, which is used to index the task array. The basic structure of a task consists of five locations named,

- Initial (initial): The task is waiting for the initial offset time to elapse.
- Waiting: The task is waiting for certain conditions in order to be ready to execute. This location is actually split into two locations representing whether the task is waiting for the period offset to have elapsed or

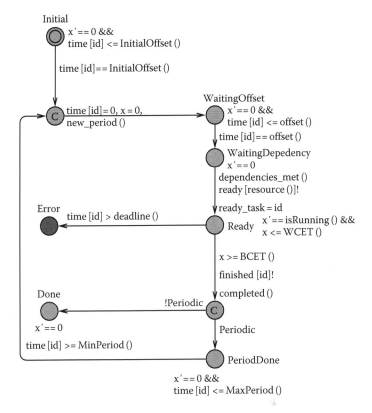

FIGURE 4.4
Task template—takes argument id of type t_id.

some other user-defined requirement. We return to the latter later in this section.

- Ready: The task is either executing or waiting to execute in the buffer of the respective resource.
- Error: The task did not manage to complete execution before the deadline.
- Done: The task has successfully completed execution within its deadline. This location is split into two locations. Which one of the two Done locations is used depends on whether the scheduling problem is periodic or not. In case the problem is nonperiodic, the done location is final; otherwise, the done location is a holding location waiting for the next period.

For every task attribute, we define a local function to access that attribute in the global task array. That is, the function resource() returns task[id].resource, BCET() returns task[id].bcet, etc.

Listing 4.5 Function to determine clock rate of the x clock of a given task.

```
1  int [0,1] isRunning() { return ( buffer [resource ()]. element[0] == id? 1 : 0);}
```

Each task uses two clock variables called time and x, where time represents the time since the beginning of the current period, and x represents how long the task has executed in the current period. The variable x is a local variable whereas time is global, and, thus, indexed by the task id. As we are using stopwatch automata, note that the progress rate of x is set to zero in all but the Ready location. In Ready, the rate determines whether the task is currently executing on the given resource. This is checked using the local function isRunning() defined in Listing 4.5.

On the other hand, time is always running and is reset at every period, so that when time exceeds the deadline, the task can move to Error.

Upon entering Ready, the task updates the variable ready_task with the task id to indicate which task has signaled to the given resource. The resource utilizes this id when inserting the task in the queue for execution. We return to this in Section 4.4.4.

To allow for problem-specific requirements, such as individual task-constraints, the task template includes the following three functions: new_period, dependencies_met, and completed. These functions can be used for a variety of problem-specific purposes, the most obvious of which is the task graph dependency definition. How to model task graphs is illustrated as follows: new_period is executed on the edges leading to Waiting and used for updating data structures, indicating that the task is beginning a new period; dependencies_met is tested in the guard leading from WaitingDependency to Ready; and completed is executed on the edge exiting Ready toward either Done location.

4.4.3.1 Modeling Task Graphs

To model task graphs using the customizable functions described above, we first need a global data structure to hold the task graph itself. In Listing 4.6, the task graph is modeled using a square Boolean dependency matrix where entry (i,j) dictates whether task i depends on task j. In the sample, we have four tasks, $t_0, ..., t_3$, where task 0 depends on task 1, which depends on task 2, which depends on task 3.

Furthermore, we have a Boolean array variable complete that determines whether a given task has finished execution. It is the responsibility of every function to reset the corresponding entry of this array in every period. In Listing 4.6, this is handled in the local task function, new_period. The value of complete is set to true when a task finishes execution in the completed function call. Finally, tasks cannot enter the Ready location from WaitingDependency until the function dependencies_met evaluates to true. This function simply iterates the corresponding row of the task graph

Listing 4.6   Modelling task graphs.

```
1   //Global declaration
2   const bool TaskGraph[Tasks][Tasks] = {
3     {0,1,0,0},
4     {0,0,1,0},
5     {0,0,0,1},
6     {0,0,0,0}
7   };
8   bool complete[Tasks];
9
10  //Task local declaration
11  void new_period() {
12    complete[id] = false ;
13  }
14
15  bool dependencies_met() {
16    return forall (j : t_id)  TaskGraph[id][j] imply complete[j];
17  }
18
19  void completed() {
20    complete[id] = true ;
21  }
```

matrix and asserts that if the task depends on another task, the entry of the complete array for that task must be true.

With these steps, we have successfully modeled a task graph using our framework.

Note that extra care must be taken about specifying task periods when using task dependencies, as the meaning of a dependency can become unclear if the task periods are out of sync. The above handling of task graphs reset the complete variable on every new period, which is safe if dependent tasks have the same period, but not necessarily so if the tasks do not. Thus, a good rule of thumb is to only specify task dependencies between tasks that have identical periods and no nondeterminism on periods.

4.4.4   Resource Template

The resource template described in this section is identical for both preemptive and nonpreemptive schedulers. The resource model is depicted in Figure 4.5, and the main difference between this model and the idea described in Section 4.4.1 is that the resource template does not include a clock. All timings are handled solely by the tasks.

A resource takes three input parameters that are the resource id (id), a Boolean preempt to indicate whether the resource is preemptive, and a scheduling policy of integer type policy_t.

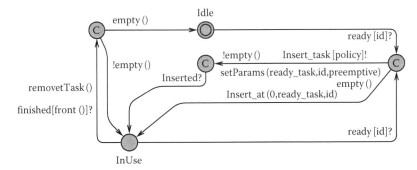

FIGURE 4.5
Resource—takes argument's id of type r_id, preempt of type bool, and policy
of type policy_t.

Listing 4.7   Insert a task in the buffer of a resource at a given position.

```
1   void  insert_at ( int [0, Tasks]  pos,  t_id  tid ,  r_id  rid) {
2       int  i ;
3       for ( i  =  buffer [ rid ]. length;  i  >  pos;  i−−) {
4           buffer [ rid ]. element[i]  =  buffer [ rid ]. element[i−1];
5       }
6       buffer [ rid ]. element[pos] = tid ;
7       buffer [ rid ]. length++;
8   }
```

The template has the two main locations Idle and InUse to indicate
the status of the resource. Idle resources are waiting for tasks to signal that
they are ready via the "ready" channel of the resource. Upon receiving such a
signal from either Idle or InUse, the resource will place the task at the front
of its buffer if the buffer is empty. Emptiness is tested with the local function
empty that reads the length of the respective buffer. Inserting a task in the
buffer is done via the global insert_at function defined in Listing 4.7, and
takes a position, a task id, and a resource id as an argument. The inserting
procedure moves all tasks in the buffer from the insert position one position
back (lines 3–5), and then inserts the task at the desired position, and, finally,
updates the length of the buffer.

In case there is at least one element in the resource buffer already, the
scheduling principle determines where to place the ready task in the buffer.
Each scheduling principle is a separate model that is activated by synchronization over the channel "insert_task" indexed which the scheduling policy.
The scheduling policy needs information about the task, the resource, and
whether the resource is preemptive, in order to make the decision of where
to place the incoming task in the buffer. This information is transferred via

Listing 4.8   Remove the front task from the resource buffer of the given resource.

```
1  void removeTask() {
2      int  i  = 0;
3      buffer [id]. length−−;
4      do {
5          buffer [id]. element[i]  = buffer [id]. element[i+1];
6          i++;
7      } while (i < buffer [id]. length );
8      buffer [id]. element[buffer [id]. length]  = 0;
9  }
```

the setParams function that copies this information to meta-variables, which can be read by the scheduler. We return to this in Section 4.4.5.

The resource remains in the committed location waiting for the signal "inserted" indicating that the task has been inserted in the resource buffer according to the scheduling policy. The waiting location is committed to guarantee that all task-insertion processes should take place as atomic operations. If a scheduling policy does not adhere to this constraint, the system deadlocks.

Upon inserting the task in the buffer according to the scheduling policy, the resource moves to the InUse location. The resource remains there until either the running task signals the completion of execution through "finished" or another task signals that it is ready. In the latter case, the resource repeats the insertion process described above. In the former case, the resource removes the task from its buffer with the local function remove-Task (see Listing 4.8). Depending on whether the resource buffer is empty, the resource returns to either Idle or InUse.

The above resource template can be used to model different types of resources and supports an easy manner to implement specialized scheduling policies with minimal overhead. In the following section, we describe how to model three types of scheduling policies in our framework.

4.4.5   Scheduling Policies

In this section, we describe models for three types of scheduling policies: FIFO, FPS, and EDF. Common to all scheduling principles in our framework is that the scheduling policy is only called if there is at least one element in the resource buffer already. This precondition is guaranteed by the resource template by not calling the scheduling policy when the resource buffer is empty and instead inserting the ready task at the front of the buffer.

The parameters transferred to the scheduling policies via the setParams function are stored in a struct meta-variable called param. This way, the

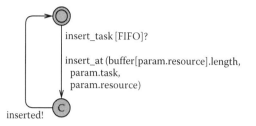

FIGURE 4.6
The FIFO scheduling policy.

parameters are accessible to the scheduling policies without increasing the state space, as meta-variables are ignored for state comparison. The parameters are availble to the policy via param.task, param.prempt, and param.resource, respectively.

4.4.5.1 First-In First-Out (FIFO)

FIFO is the simplest scheduling policy as tasks are simply put into the resource buffer in the order in which they arrive. This implies that FIFO scheduling disregards whether the resource is preemptive or not. The model for the FIFO policy is depicted in Figure 4.6, where insertion of the task is handled on the edge synchronizing on "insert_task."

4.4.5.2 Fixed Priority

The fixed priority scheduling policy (FPS) model is similar to that of FIFO in structure, except that the task insertion is handled by the function call insert_task_in_buffer (Listing 4.9) instead of a call insert_task. The function iterates (lines 6–8) through the resource buffer and compares the priority of the incoming task with the tasks in the buffer, and sets a local variable place such that the incoming task has a lower priority than all tasks in front of it and a higher priority than all tasks behind it. In case the incoming task has the lowest priority of all tasks in the buffer, the iteration is terminated by reaching the end of the buffer and inserts the task here. Obviously, this method requires the buffer to be sorted. However, as the method is used for all insertions, except the first, the buffer remains sorted. When the place of insertion has been established, the task is inserted using the insert_at function call (line 9).

Note in Listing 4.9 that preemption is handled in line 4 where the first assumed place of the incoming task is either the first buffer entry, in case of preemption, or the second index, in case of no preemption. This code utilizes the precondition that the buffer has at least one element when the function is called.

Listing 4.9   Function for task insertion according to FPS.

```
1   void  insert_task_in_buffer ()  {
2       t_id  t  = param.task;
3       r_id  r  = param.resource;
4       int  place = (param.preempt ? 0 :  1);
5       int  i ;
6       while(place<buffer[r].length  &&
7             task[buffer[r].element[place]].pri>=task[t].pri ){
8         place++;
9       }
10      insert_at (place, t, r );
11  }
```

4.4.5.3   Earliest Deadline First

The two scheduling policies above do not, in principle, need separate models, as task insertion can be handled by a simple function call. However, because of the UPPAAL modeling language constraints, scheduling policies such as the EDF cannot be handled in our framework as a simple function call, but need to implement the insertion as a model. The problem lies in that to determine how far a task i is from its deadline, we need to look at the difference between task[i].deadline and time[i]. UPPAAL does not allow such a comparison in function calls as it requires operations on the clock zone data structure. However, UPPAAL does allow such comparisons in the guard expression, and, thus, in order to find the position of an incoming task in a resource buffer we need to iterate through the buffer as model transitions comparing deadlines using guards.

In order to compare whether an incoming task i has an earlier deadline than a buffered task j, we need to check the following:

$$\text{task[i].deadline-time[i]} < \text{task[j].deadline - time[j]}$$
$$\Leftrightarrow \text{time[i] -task[i].deadline} > \text{time[j] - task[j].deadline}$$

Note that UPPAAL only allows subtraction of constants from clock values and not the reverse, hence the rewriting of the expression.

For convenience, we introduce a number of local variables and functions for the EDF model, as defined in Listing 4.10. These are variables to hold the parameters that are transferred from the resource, a function readParameters to set the values of the local variables, and a function resetVars to reset the variables after task insertion to minimize the search space.

Figure 4.7 depicts the model for the EDF task-insertion. Initially, the model reads the parameters and sets the initial assumed place variable of insertion according to whether or note the calling resource is preemptive (same as for the FPS). Now, the deadline of the incoming task is compared to the deadline of the task at place. If place is either past the last element in the buffer or the incoming task has an earlier deadline than the task at place, the

Listing 4.10   Local variables and function for the EDF scheduling policy.

```
1   int [0, Tasks] place;
2   t_id tid;
3   r_id rid;
4   bool preempt;
5   void readParameters() {
6       tid = param.task; rid = param.resource; preempt = param.preempt;
7   }
8   void resetVars () { place = tid = rid = 0; }
```

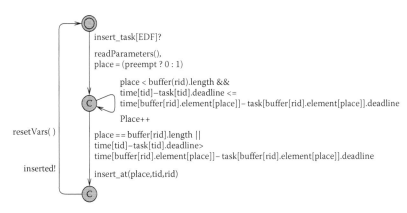

FIGURE 4.7
The EDF scheduling policy.

incoming task is inserted at this place in the buffer with a call to the global function insert_at. Otherwise, place is incremented and the deadline check is performed again. Note that this is nothing more than a model implementation of a regular while-loop as used for the FPS.

This concludes our model framework for schedulability analysis, and in the following section we proceed by showing how to instantiate the framework and formulate schedulability queries.

4.5   Framework Instantiation

In this section, we focus on instantiating our scheduling framework for standard scheduling problems. Instantiation requires data entry in the global declaration as well as in the system definition. The data needed in the global declaration is shown in Listing 4.11.

Listing 4.11   Local variables and function for the EDF scheduling policy.

```
1  const bool Periodic = .. ;    // Periodic Scheduling?
2  const int Tasks = .. ;        // Number of tasks
3  const int Procs = .. ;        // Number of resources
4  const int MaxTime = .. ;      // Maximum time constant
5  const task_t task[Tasks] = {  ...  };
```

Listing 4.12   Local variables and function for the EDF scheduling policy.

```
1  P1 = Resource(0, true/false ,  EDF/FPS/FIFO);
2  P2 = Resource(1, true/false ,  EDF/FPS/FIFO);
3  ...
4  system Task, P1, P2, ... , Policy_EDF, Policy_FPS, Policy_FIFO;
```

For a particular scheduling problem, the user needs to specify whether it is a periodic scheduling problem, how many tasks and resources there are, the maximum time constant for all tasks,* and the task attributes.

When defining the system, the user needs to specify the properties of the different resources, that is, preemption and scheduling policy.

In Listing 4.12, we show a sample system declaration with a number of resources $(P_1, P_2, \ldots)$. When declaring the actual system with the **system** keyword, all defined resources should be included with a single copy of each scheduling policy used. Note that all tasks are included with the single Task instantiation. This is because the parameter spaces of tasks are bound by the t_id type, so if no parameter is used, UPPAAL instantiates a copy of a task for every possible id.

4.5.1   Schedulability Query

For any given scheduling problem, the schedulability of checking whether all tasks always meet their respective deadlines can be stated as in Section 4.4.1:

$$A[] \ \text{forall} \ (\ i : t\_id \ ) \ \text{not} \ Task(i).Error$$

In other words, "Does it hold for all paths that no task is ever in the Error location?" Note that UPPAAL using stopwatches creates an over-approximation of the state space, meaning that if the above query is satisfiable, it is guaranteed that the system is schedulable under all circumstances. However, if the query is not satisfiable, this means that a counterexample has been established in the overapproximation. The system may still be schedulable however, since the counterexample is not necessarily a feasible run of the original system.

---

*Used for UPPAAL state footprint reduction.

4.5.2   Example Framework Instantiation

In this section, we provide a sample periodic schedulability problem and illustrate how to instantiate our framework for this problem. Part of the schedulability problem is depicted in Figure 4.8a. It is a simple system with four tasks, $T = t_0, ..., t_3$, and two resources, P_0, and P_1 and with task dependencies given as a task graph. As indicated by dashed lines in Figure 4.8, tasks t_0 and t_1 are assigned to P_0 and tasks t_2 and t_3 are assigned to P_1.

Note that task t_3 depends on task t_0, but the tasks are assigned to different resources. This is a common scheduling situation, and, usually, requires the communication of data between resources using a transport medium such as a data bus. The communication of data takes time, but does not block either of the resources. In our framework, such a scenario can be modeled as illustrated in Figure 4.8, introducing a new task t_4 that requires a new resource, *Bus*. The task t_4 is inserted into the task graph such that t_4 depends on t_0 and t_3 depends on t_4. We assume for the reasons explained in Section 4.4.3.1 that all tasks have fixed and identical periods. Thus, the attributes of t_4 should be set such that

- INITIAL_OFFSET(t_4) = INITIAL_OFFSET(t_0).
- BCET(t_4): Specific to the bus.
- WCET(t_4): Specific to the bus.
- MIN_PERIOD(t_4) = MIN_PERIOD(t_0). Given our assumption about identical periods, the minimum and maximum periods are equal and the same for all tasks.
- MAX_PERIOD(t_4) = MIN_PERIOD(t_4).
- OFFSET(t_4) = 0. Offset not needed as the task release is determined by the completion of t_0.
- DEADLINE(t_4) = MIN_PERIOD(t_4). Could be tightened with respect to the deadline of t_3 and XCET(t_4).
- PRIORITY(t_4) = 0. Irrelevant, as the bus is a FIFO resource.

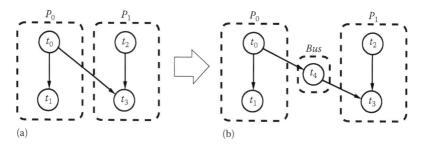

(a)   (b)

FIGURE 4.8
Sample schedulability problem. (a) Task graph with interdependent tasks across resources and (b) how to solve this problem by introducing a new task and a bus resource.

Listing 4.13   Modelling task graphs.

```
1   //Global Declaration
2   const bool Periodic = true;
3   const int MaxTime = 20;
4   const int Tasks = 5;              // Number of tasks
5   const int Procs = 3;             // Number of resources
6
7   const task_t task[Tasks] = {
8     {0,20,20,0,20,4,7,0,1},
9     {0,20,20,1,20,8,12,0,1},
10    {0,20,20,0,20,10,12,1,1},
11    {0,20,20,1,20,6,7,1,1},
12    {0,20,20,1,20,5,5,2,1}
13   };
14
15   const bool Depend[Tasks][Tasks] = {  // Task graph
16    {0,0,0,0,0},
17    {1,0,0,0,0},
18    {0,0,0,0,0},
19    {0,0,1,0,1},
20    {1,0,0,0,0}
21   };
22
23   //System Declaration
24   P0  = Resource(0,true,FPS);
25   P1  = Resource(1,true,FPS);
26   Bus = Resource(2,false,FIFO);
27
28   system Task, P0, P1, Bus, Policy_FPS, Policy_FIFO;
```

The "bus" resource will be modeled as a nonpreemptive FIFO resource to mimic the behavior of a data bus. Assuming some specific best-case and worst-case execution times, periods, deadlines, etc., the schedulability problem described above can be instantiated as shown in Listing 4.13.

Note in Listing 4.13 that the three tasks that depend on other tasks have been given an offset of 1. This is for convenience, as it is the responsibility of individual tasks to reset their own entry in complete. Upon the start of a new period without the offset, task t_1 could signal ready to the resource while reading an old value of complete for task t_0 before t_0 resets the value. Another way to handle this when all dependent tasks have the same fixed periods would be to let any task reset all values to complete when starting a new period. However, there is no general good way of handling this problem and it is left to the modeler to prevent unwanted situations.

4.6 Conclusion

We have provided a framework that allows the modeling and analysis of a variety of schedulability scenarios. In particular, our framework supports multi-processor systems, rich task-models with timing uncertainties in arrival and execution times, possible dependencies, a range of scheduling policies, and possible preemption of resources. The support of an approximate analysis of stopwatch automata in UPPAAL 4.1 is key to the successful schedulability analysis.

Furthermore, the uncertainty on the periods used in our framework could be generalized to more general task-arrivals where a separate process determines the arrival of tasks. Such situations can be modeled using the structure of our framework by letting the starting of periods be dictated through channel synchronization with the model controlling arrival times. Even with such liberty, the overapproximation is still finite and the termination is guaranteed.

The scheduling framework provided in this chapter is structured such that an adaptation can be made to accommodate other scheduling polices and inter-task constraints. The former can be achieved by adding another policy model similarly to the three built-in policies, FIFO, the FPS, and the EDF. The latter is achieved through the use of the function calls, new_period, dependencies_met, and completed.

Acknowledgment

The authors would like to thank Marius Mikučionis for providing the format for listing UPPAAL code.

References

1. R. Alur, C. Courcoubetis, and D. Dill. Model-checking for real-time systems. In *Proceedings of the Fifth IEEE Symposium on Logic in Computer Science (LICS'90)*, pp. 414–425, Philadelphia, PA, 1990. IEEE Computer Society Press, 1990.

2. R. Alur and D. Dill. Automata for modeling real-time systems. In *Proceedings of the 17th International Colloquium on Automata, Languages and Programming (ICALP'90)*, Warwick University, Couentry, U.K., 1990. *Lecture Notes in Computer Science*, 443:322–335. Springer, 1990.

3. R. Alur and D. Dill. A theory of timed automata. *Theoretical Computer Science (TCS)*, 126(2):183–235, 1994.

4. T. Amnell, E. Fersman, L. Mokrushin, P. Pettersson, and W. Yi. Times— a tool for modelling and implementation of embedded systems. In J.-P. Katoen and P. Stevens (editors), *TACAS*, Grenoble, France, 2002. *Lecture Notes in Computer Science*, 2280:460–464. Springer, 2002.

5. J. Madsen, A. Brekling, and M.R. Hansen. Models and formal verification of multiprocessor system-on-chips. *The Journal of Logic and Algebraic Programming*, 77(1):1–19, 2008.

6. G. Behrmann, A. Cougnard, A. David, E. Fleury, D. Larsen, K.G. Larsen, and D. Lime. Uppaal tiga: Time for playing games! In *Proceedings of Computer Aided Verification (CAV'07)*, Berlin, Germany, July 2007, *Lecture Notes in Computer Science*, 4590:121–125. Springer, 2007.

7. G. Behrmann, K.G. Larsen, and J.I. Rasmussen. Optimal scheduling using priced timed automata. *ACM SIGMETRICS Performance Evaluation Review*, 32(4):34–40, 2005.

8. T. Bœgholm, H. Kragh-Hansen, P. Olsen, B. Thomsen, and K.G. Larsen. Model-based schedulability analysis of safety critical hard real-time java programs. In *JTRES '08: Proceedings of the Sixth International Workshop on Java Technologies for Real-Time and Embedded Systems*, pp. 106–114, New York, 2008. ACM, 2008.

9. F. Cassez and K.G. Larsen. The impressive power of stopwatches. In C. Palamidesi (editor), *11th International Conference on Concurrency Theory, (CONCUR'2000)*, *University Park, PA, July 2000, Lecture Notes in Computer Science*, 1877:138–152. Springer-Verlag, 2000.

10. Timesys Corporation. Pittsburgh, PA, http://www.timesys.com.

11. Timesys Corporation. Pittsburgh, PA, http://www.tripac.com.

12. R.J. Engdahl and A.M. Haugstad. Efficient model checking for probabilistic timed automata. Master thesis, Aalborg University, Aalborg, Denmark, 2008.

13. E. Fersman, L. Mokrushin, P. Pettersson, and W. Yi. Schedulability analysis of fixed-priority systems using timed automata. *Theoretical Computer Science*, 354(2):301–317, 2006.

14. E. Fersman, P. Pettersson, and W. Yi. Timed automata with asynchronous processes: Schedulability and decidability. In *Proceedings of TACAS 2002*, pp. 67–82, Grenoble, France, Springer-Verlag, 2002.

15. UPPAAL Scheduling Framework
    `http://www.uppaal.com/SchedulingFramework`, January 2009.

16. K. Godary, I. Augé-Blum, and A. Mignotte. Sdl and timed petri nets versus uppaal for the validation of embedded architecture in automotive. In *Forum on Specification and Design Language (FDL'04)*, Lille, France, September 2004.

17. N. Guna, Z. Gu, Q. Deng, S. Gao, and G. Yu. Exact schedulability analysis for static-priority global multiprocessor scheduling using model-checking. In *Software Technologies for Embedded and Ubiquitous Systems*, Santorini Island, Greece, *Lecture Notes in Computer Science*, pp. 263–272. Springer, Berlin, 2007.

18. Uppaal Tiga Homepage. `http://www.cs.aau.dk/~adavid/tiga`, 2006.

19. A. Brekling, J. Madsen, and M.R. Hansen. A modelling and analysis framework for embedded systems. In *Model-Based Design for Embedded Systems*, G. Nicolescu and P.J. Mosterman (editors), Taylor & Francis, Boca Raton, FL, 2009.

20. J. Krakora and Z. Hanzalek. Timed automata approach to CAN verification. *INCOM*, 2004.

21. P. Krcál and W. Yi. Decidable and undecidable problems in schedulability analysis using timed automata. In K. Jensen and A. Podelski (editors), *TACAS*, Barcelona, Spain, 2004. *Lecture Notes in Computer Science*, 2988: 236–250. Springer, 2004.

22. J. Madsen, K. Virk, and M.J. Gonzalez. A systemC-based abstract real-time operating system model for multiprocessor system-on-chip. In *Multiprocessor System-on-Chip*. Morgan Kaufmann, San Francisco, CA, 2004.

23. S. Mahadevan, M. Storgaard, J. Madsen, and K.M. Virk. Arts: A system-level framework for modeling MPSoC components and analysis of their causality. In *13th IEEE International Symposium on Modeling, Analysis, and Simulation of Computer and Telecommunication Systems (MASCOTS)*, Atlanta, GA, 2005. IEEE Computer Society, Septemper 2005.

24. Y. Le Moullec, J.-P. Diguet, N. Ben Amor, T. Gourdeaux, and J.L. Philippe. Algorithmic-level specification and characterization of embedded multimedia applications with design trotter. *VLSI Signal Processing*, 42(2):185–208, 2006.

25. H. Nikolov, M. Thompson, T. Stefanov, A.D. Pimentel, S. Polstra, R. Bose, C. Zissulescu, and E.F. Deprettere. Daedalus: Toward composable multimedia MPSoC design. In L. Fix (editor), *DAC*, pp. 574–579, Anaheim, CA, 2008, ACM, 2008.

26. S. Schliecker, J. Rox, R. Henia, R. Racu, A. Hamann, and R. Ernst. Formal performance analysis for real-time heterogeneous embedded systems. In *Model-Based Design for Embedded Systems*, G. Nicolescu and P.J. Mosterman (editors), Taylor & Francis, Boca Raton, FL, 2009.

27. H. Sun. Timing constraints validation using uppaal: Schedulability analysis. In *DIPES '00: Proceedings of the IFIP WG10.3/WG10.4/WG10.5 International Workshop on Distributed and Parallel Embedded Systems*, pp. 161–172, Deventer, the Netherlands, 2001. Kluwer, B.V. 2001.

28. UPPAAL. http://www.uppaal.com, January 2005.

29. UPPAAL CORA. http://www.cs.aau.dk/~behrmann/cora/, January 2006.

5

Modeling and Analysis Framework for Embedded Systems

Jan Madsen, Michael R. Hansen, and Aske W. Brekling

CONTENTS

5.1 Introduction

Modern hardware systems are moving toward execution platforms made up of multiple programmable and dedicated processing elements implemented on a single chip, known as a multiprocessor system-on-chip (MPSoC). The different parts of an embedded application are executing on these processing elements, but the activities of mapping the parts of an embedded program onto the platform elements are nontrivial. First of all, there may be various and often conflicting resource constraints. The real-time constraint, for example, should be met together with constraints on the uses of memory and energy. There also are huge varieties in the freedom of choices in

121

the mapping of an application to a platform because there are many ways to partition an embedded program into parts, there are many ways these parts can be assigned to processing elements, and there are many ways each processing element can be set up.

As embedded systems become more complex, the interaction between the application and the execution platform becomes more incomprehensible, and problems such as memory overflow, data loss, and missed deadlines become more likely. In the development phase, it is not enough to simply look at the different layers of the system independently, as a minor change at one layer can greatly influence the functionality of other layers. The system-level verification of schedulability, upper limits for memory usage, and power consumption, taking all layers into account, have therefore become central fields of study in the design of embedded systems.

As many important design decisions are made early in the design phase, it is imperative to support the system designer at this level. This chapter presents an abstract embedded system model that is able to capture a set of applications executing on a multicore execution platform. The model of computation for such systems is formalized in [BHM08], which also contains a more refined formalization using timed automata. This refinement into timed automata, which is implemented using UPPAAL [BDL04], gives the ability to model check properties of timing, memory usage, and power consumption.

In order to support designers of industrial applications, the timed-automata model is hidden for the user, allowing the designer to work directly with the abstract system-level model of embedded systems. As outlined in Figure 5.1, the designer provides an application consisting of a set of task graphs, an execution platform consisting of processing elements interconnected by a network, and a mapping of tasks to processing elements. The system model is then translated into a timed-automata model that enables schedulability analysis as well as being able to verify that memory usage and power consumption are within certain limits. In the case where a system is not schedulable, the tool provides useful information about what caused the missed deadline. We do not propose any particular methodology for design space exploration, but provide an analysis framework, MoVES , where embedded systems can be modeled and verified in the early stages of the design process. Thus, the MoVES analysis framework provides tool support for system designers to explore alternatives in an easy and efficient manner.

An important aspect in the design of MoVES is to provide an experimental framework, supporting easy adaptability of the "core model" to capture energy and memory considerations for example, or to experiment with, say, new principles for task scheduling and allocation. Furthermore, the MoVES analysis framework is equipped with different underlying UPPAAL models,

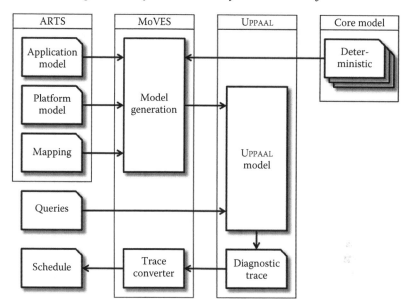

FIGURE 5.1
Overview of the MoVES analysis framework.

aiming at an efficient verification in various situations. For the moment, we are operating with the following underlying models for

- Schedulablity analysis in connection with worst-case execution times only.
- Schedulablity analysis for the full core model (including best- and worst-case execution times).
- Schedulability analysis addressing memory and energy issues as well.
- Schedulability analysis for the full core model on the basis of stopwatch automata. This analysis approach is based on overapproximations, but it has provided exact results in the experiments carried out so far and it appears to be the most efficient UPPAAL implementation.

The chapter is organized as follows. First, we motivate the modeling and analysis of multi-core embedded systems. We then present an embedded system model that consists of an application model, an execution platform model, and a system model, which is a particular mapping of the application onto the execution platform. For an embedded system, we give an informal presentation of the model of computation. We then outline how the model has been captured using timed automata. Finally, we present how the MoVES analysis framework can be used to verify properties of an embedded system through a number of examples, including a smart phone example, showing the ability to handle systems of realistic sizes.

5.2 Motivation

In this work, we aim at models and tools for analysis of properties that must be considered when an application is mapped to an execution platform. Such models are called *system models* [PHL+01] as they comprise a model for the application executing on the platform, and the analysis of such systems is called "cross-layer analysis" as it deals with problems where decisions concerning one layer of abstraction (for instance, concerning the scheduling principle used in a processing element) has an influence on the properties at another level of abstraction (for instance, a task is missing a deadline). One particular challenge of multi-core systems is that of "multiprocessing timing anomaly" [Gra69], where the system is exhibiting a counterintuitive timing behavior.

Example 5.1 *To illustrate this challenge, consider the simple example in Figure 5.2, where the application is specified by five cyclic tasks, $\tau_1, \ldots, \tau_5$, that are mapped onto three processing elements, pe_1, pe_2, and pe_3. The best- and worst-case execution times for each task (bcet and wcet, respectively) are shown in Table 5.1.*

There are causal dependencies between tasks. For example, τ_1 must finish before τ_2 can start. We want to find the shortest period where all tasks meet their deadlines and analyze two different runs corresponding to two possible execution times for τ_1 in Figures 5.3 and 5.4, one where the "best-case execution time", bcet $= 2$, is chosen for τ_1 and another where the "worst-case execution time", wcet $= 4$, is chosen.

In both runs, τ_1 and τ_3 are executing on pe_1 and pe_3, respectively, in the first time step, where no task is executing on pe_2 because of the causal dependencies. The later time steps have similar explanations. Observe that the shortest possible period is $\pi = 8$, corresponding to the case where the best-case execution time, $bcet_{\tau_1} = 2$, is chosen for τ_1. Thus, an analysis based on the worst-case execution time, $wcet_{\tau_1} = 4$, would, in this case, not lead to the worst-case scenario. This is an example of a

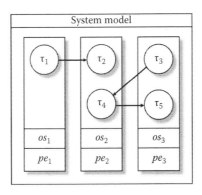

FIGURE 5.2
System model of a simple multicore system.

TABLE 5.1

Characterization of Tasks, for Example, in Figure 5.2

Task	Execution Time (bcet, wcet)	Processor
τ_1	(2,4)	pe_1
τ_2	(2,2)	pe_2
τ_3	(2,2)	pe_3
τ_4	(2,2)	pe_2
τ_5	(2,2)	pe_3

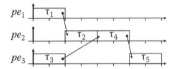

FIGURE 5.3

Execution time for τ_1 is 2.

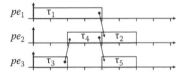

FIGURE 5.4

Execution time for τ_1 is 4.

"multiprocessing timing anomaly" [Gra69] exhibiting a counterintuitive timing behavior. A locally faster execution, either by making the processor faster or by making the algorithm more efficient, may lead to an increase in the execution time of the whole system. The presence of such behavior makes multiprocessor timing analysis particularly difficult [RWT⁺06].

It is easy to check that a period $\pi = 6$ can be achieved for this application, simply by changing the priorities so that τ_4 gets a higher priority than τ_2. But the problems cannot get much larger than the one in Figure 5.2 before the consequences of design decisions cannot be comprehended, and it is necessary to have tool support for the "design space exploration" [HFK⁺07,PEP06]. □

5.3 Embedded Systems Model

In this section, we present a system-level model of an embedded system inspired by ARTS [MVG04,MVM07]. Such a model can be described as a layered structure consisting of three different parts. Figure 5.5 illustrates

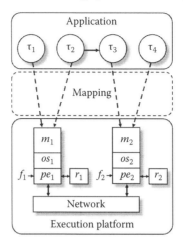

FIGURE 5.5
System-level model of an embedded system.

these layers for a very simple example of an embedded system, which will be used to explain the aspects of the model throughout the chapter.

- The *application* is described by a collection of communicating sequential tasks. Each task is characterized by four timing properties, described later. The dependencies between tasks are captured by an acyclic directed graph (called a "task graph"), which might not be fully connected.
- The *execution platform* consists of several processing elements of possibly different types and clock frequencies. Each processing element will run its own real-time operating system, scheduling tasks in a priority-driven manner (static or dynamic), according to their priorities, dependencies, and resource usage. When a task needs to communicate with a task on another processing element, it uses a network. The setup of the network between processing elements must also be specified, and is part of the platform.
- The "mapping" between the application and the execution platform (shown as dashed arrows in the figure) is done by placing each task on a specific processing element. In our model, this mapping is static, and tasks cannot migrate during run-time.

The top level of the embedded system consists of an application mapped onto an execution platform. This mapping is depicted in Figure 5.5 with dashed arrows. The timing characteristics in Table 5.2 originate from [SL96], while the memory and power figures (in Table 5.4) are created for the purpose of demonstrating parameters of an embedded system. We will elaborate on the various parameters in the following.

TABLE 5.2

Characterization of Tasks

Task	ω	δ	π
τ_1	0	4	4
τ_2	0	6	6
τ_3	0	6	6
τ_4	4	6	6

5.3.1 Application Model

The task graph for the application can be thought of as an abstraction of a set of independent sequential programs that are executed on the execution platform. Each program is modeled as a directed acyclic graph of tasks where edges indicate causal dependencies. Dependencies are shown with solid arrows in Figure 5.5. A task is a piece of a sequential code and is considered to be an atomic unit for scheduling. A task τ_j is periodic and is characterized by a "period" π_j, a "deadline" δ_j, an initial "offset" ω_j, and a fixed priority fp_j (used when an operating system uses fixed priority scheduling). The properties of periodic tasks (except the fixed priority) can be seen in Table 5.2 and are all given in some time unit.

5.3.2 Execution Platform Model

The execution platform is a heterogeneous system, in which a number of processing elements, $pe_1, \ldots, pe_n$, are connected through a network.

5.3.2.1 Processing-Element Model

A processing element pe_i is characterized by a "clock frequency" f_i, a "local memory" m_i with a bounded size, and a "real-time operating system" os_i. The operating system handles synchronization of tasks according to their dependencies using direct synchronization [SL96].

The access to a shared resource r_m (such as a shared memory or a bus) is handled using a resource allocation protocol, which in the current version consists of one of the following protocols: preemptive critical section, nonpreemptive critical section, or priority inheritance. The tasks are in the current version scheduled using either rate monotonic, deadline monotonic, fixed priority, or earliest deadline first scheduling [Liu00]. The properties of a processing element can be seen in Table 5.3. Allocation and scheduling are designed in MoVES for easy extensions, that is, new algorithms can easily be added to the current pool.

The interaction between the operating system and the application model is shown in Figure 5.6. The operating system model consists of a *controller*, a *synchronizer*, an *allocator*, and a *scheduler*. The controller receives `ready` or

TABLE 5.3

Characterization of Processors

	pe_1	pe_2
f_1	1	1
Scheduling	RM	RM
allocation	PRI_INH	PRI_INH

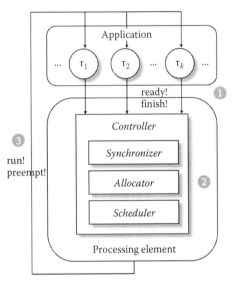

FIGURE 5.6

Interaction of the k tasks, τ_1 to τ_k, with the single processing element to which they are mapped.

`finish` signals from those tasks of the application that are mapped to the processing element (see 1 in Figure 5.6); activates synchronization, allocation, and scheduling to find the task with the highest priority (see 2 in Figure 5.6); and finally sends `run` or `preempt` signals back to the tasks (see 3 in Figure 5.6).

5.3.2.2 Network Model

Inter-processor communication takes place when two tasks with a dependency are mapped to different processing elements. In this case, the data to be transferred is modeled as a message task τ_m. Message tasks have to be transferred across the network between the processing elements. A network is modeled in the same way as a processing element. So far, only busses have been implemented in our model; however, it is shown in [MMV04] how more complicated intercommunication structures, such as meshes or torus networks, can be modeled. As a bus transfer is nonpreemptable, message tasks are modeled as run-to-completion. This is achieved by having all message tasks running on the bus, that is, the processing elements emulating the bus, using the same resource r_m, thereby preventing the preemption of any message task. Intraprocessor communication is assumed to be included in the execution time of the two communicating tasks, and is therefore modeled without the use of message tasks.

5.3.3 Task Mapping

A mapping is a static allocation of tasks to processing elements of the execution platform. This is depicted by the dashed arrows in Figure 5.5. Suppose that the task τ_j is mapped onto the processing element pe_i. The "execution time," e_{ij} measured in cycles, memory footprint ("static memory," sm_{ij} and "dynamic memory," dm_{ij}), and "power consumption," pw_{ij} of a task τ_j, depend on the characteristics of the processing element pe_i executing the task, and can be seen in Table 5.4. In particular, when selecting the operation frequency f_i of the processing element pe_i, the execution time in seconds, ϵ_{ij}, of task τ_j can be calculated as $\epsilon_{ij} = e_{ij} \cdot \frac{1}{f_i}$.

5.3.4 Memory and Power Model

In order to be able to verify that memory and power consumption stay within given bounds, the model keeps track of the memory usage and power costs in each cycle. Additional cost parameters can easily be added to the model as long as the cost can be expressed in terms of the cost of being in a certain state.

The memory model includes both static memory allocation (sm), because of program memory, and dynamic memory allocation (dm), because of data memory of the task. The example in Figure 5.7 illustrates the memory model for a set of tasks executing on a single processor. It shows the scheduling and resulting memory profiles (split into static and dynamic memories). The dynamic part is split into private data memory (pdm) needed while executing the task, and communication data memory (cdm) needed to store data exchanged between tasks. The memory needed for data exchange between

TABLE 5.4

Characterization of Tasks on Processors

Task	e	sm	dm	pw
τ_1	2	1	3	2
τ_2	1	1	7	3
τ_3	2	1	9	3
τ_4	3	1	6	4
τ_m	1			

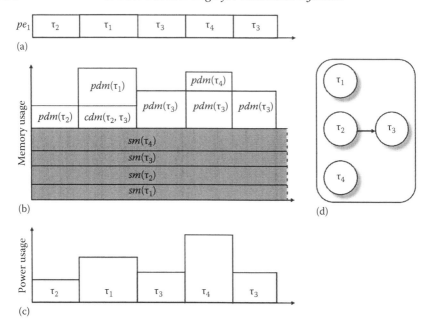

FIGURE 5.7
Memory and power profiles for pe_1 when all four tasks in Figure 5.5 are mapped onto pe_1. (a) Schedule where τ_3 is preempted by τ_4. (b) Memory usage on pe_1: static memory (*sm*), private data memory (*pdm*), and communication data memory (*cdm*). (c) Power usage. (d) Task graph from Figure 5.5.

τ_2 and τ_3 must be allocated until it has been read by τ_3 at the start of τ_3's execution. When τ_3 becomes preempted, the private data memory of the task remains allocated till the task finishes.

Currently, a simple approach for the modeling of power has been taken. When a task is running, it uses power pw. The power usage of a task is zero at all other times. The possible different power usages of tasks can be seen as the heights of the execution boxes in Figure 5.7c. This approach can easily be extended to account for different power contributions depending on the state of the task.

5.4   Model of Computation

In the following, we will give a rather informal presentation of the model of computation. For a formal and more comprehensive description, please refer to [BHM08]. To model the computations of a system, the notion of a "state", which is a snapshot of the state of affairs of the individual processing elements, is introduced. For the sake of argument, we will consider a system consisting of a single processing element pe_i and a set of tasks $\tau_j \in T_{pe_i}$ assigned to pe_i. Furthermore, we shall assume that each τ_j is characterized by "best-case" and "worst-case" execution times, $bcet_j \in \mathbb{N}$ and $wcet_j \in \mathbb{N}$, respectively. At the start of each new period, there is a nondeterministic choice concerning which execution time $e_{ij} \in \{bcet_{\tau_j}, bcet_{\tau_j} + 1, ..., wcet_{\tau_j} - 1, wcet_{\tau_j}\}$ is needed by τ_j to finish its job on pe_i of that period.

For the processing element pe_i, the state component must record which task τ_j (if any) is currently executing, and for every task $\tau_j \in T_{pe_i}$ record the execution time e_{ij} that is needed by τ_j to finish its job in its current period. We denote the state σ, where τ_j is running and where there is a total of n tasks assigned to pe_i, as $\sigma = (\tau_j, (e_{i1}, ..., e_{in}))$. Here, we consider execution time only; other resource aspects, such as memory or power consumption are disregarded.

A trace is a finite sequence of states, $\sigma_1\sigma_2 \cdots \sigma_k$, where $k \geq 0$ is the length of the trace. A trace with length k describes a system behavior in the interval $[0, k]$. For every new period of a task, the task execution time for that period can be any of the possible execution times in the natural number interval $[bcet, wcet]$. If $bcet = wcet$ for all tasks, there is only one trace of length k, for any k. If $bcet \neq wcet$, we may explore all possible extensions of the current trace by creating a new branch for every possible execution time, every time a new period is started for a task. A "computation tree" is an infinite, finitely branching tree, where every finite path starting from the root is a trace, and where the branching of a given node in the tree corresponds to all possible extensions of the trace ending in that node. This is further explained in the following example.

Example 5.2 *Let us consider a simple example consisting of three independent tasks assigned to a single processor. The characteristics of each task are shown in Table 5.5. The computation tree for the first 8 time units is shown in Figure 5.8. Here, we will give a short description of how this initial part of the tree is created.*

> *Time $t = 0$: Only task τ_1 is ready, as τ_2 and τ_3 both have an offset of 2. Hence, τ_1 starts executing, and as $bcet = wcet = 2$, there is only one possible execution time for τ_1. The state then becomes $\sigma_1 = (\tau_1, (2, 0, 0))$.*

TABLE 5.5

Characterization of Tasks

Task	Priority	ω	bcet	wcet	π
τ_1	1	0	2	2	3
τ_2	2	2	1	2	4
τ_3	3	2	1	2	6

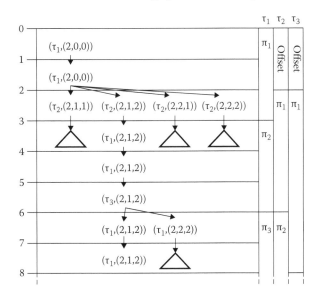

FIGURE 5.8
Possible execution traces. A △ indicates a subtree, the details of which are not further processed in this example.

Time $t = 2$: τ_1 has finished its execution of 2 time units, but a new period for τ_1 has not yet started as $\pi_1 = 3$. Both τ_2 and τ_3 are now ready. Since τ_2 has the highest priority (i.e., the lowest number), it gets to execute. As the execution time interval for both τ_2 and τ_3 is $[1,2]$, there are two different execution times for each, and hence, four different possible states, $(\tau_2, (2,1,1)), (\tau_2, (2,1,2)), (\tau_2, (2,2,1)),$ and $(\tau_2, (2,2,2)),$ which give rise to four branches. In Figure 5.8, we will only continue the elaboration from state $(\tau_2, (2,1,2))$.

Time $t = 3$: τ_2 finishes its execution. τ_3 is still ready and the first period of τ_1 has completed initiating its second iteration, hence, τ_1 is also ready. As τ_1 has the highest priority, it gets to execute. The state becomes $(\tau_1, (2,1,2))$.

Time $t = 5$: τ_1 finishes its execution. τ_3 is the only task ready, as the first period for τ_2 has not yet finished. The state becomes $(\tau_3, (2,1,2))$.

Time $t = 6$: Both τ_1 and τ_2 become ready as a new period starts for each of them. Again, τ_1 has the highest priority and gets executed, preempting τ_3, which then still needs one time unit of execution to complete its job for the current period. Since the execution time of τ_2 can be 1 or 2, and that of τ_1 only 1, we just have two branches, that is, the possible new states are $(\tau_1, (2,1,2))$ and $(\tau_1, (2,2,2))$.

Time $t = 8$: τ_1 has completed its execution allowing τ_2 to take over. However, at this point, the second period of τ_3 starts, while τ_3 has not yet completed its job for the first period. Hence, τ_3 will not meet its deadline and this example is not schedulable. □

This model of computation can easily be extended to a system with multiple processors. The system state then becomes the union of the states for each processor.

A run of a system is an infinite sequence of states. We call a system "schedulable" if for every run, each task finishes its job in all its periods. In [BHM08], we have shown that the schedulability problem is decidable and an upper bound on the depth of the part of the computation tree, which is sufficient to consider when checking for schedulability, is established. An upper bound for that depth is given by

$$\Omega_M + \Pi_H \cdot (1 + \Sigma_{\tau \in \mathcal{T}} wcet_\tau)$$

where $\mathcal{T}$ is the set of all tasks, Ω_M is the maximal offset, Π_H is the hyper-period of the system (i.e., the least common multiple of all periods of tasks in the system), and $\Sigma_{\tau \in \mathcal{T}} wcet_\tau$ an upper bound of the number of hyper-periods after which any traces of the system will reach a previous state.

The reason why it is necessary to "look deeper" than just one hyper-period can be explained as follows: Prior to the time point Ω_M, some tasks may already have started, while others are still waiting for the first period to start. At the time O_M, the currently executing tasks (on various processing elements) may therefore have been granted more execution time in their current periods, than would be the case in periods occurring later than Ω_M—you may say that they have "saved up" some execution time and this saving is bounded by the sum of the worst-case execution times in the system. In [BHM08], we have provided an example where the saving is reduced by one in each hyper-period following Ω_M until a missed deadline is detected. The upper bound above can be tightened:

$$\Omega_M + \Pi_H \cdot (1 + \Sigma_{\tau \in \mathcal{T}_X} wcet_\tau)$$

where $\mathcal{T}_X$ is the set of all tasks that do not have a period starting at Ω_M.

Example 5.3 *Let us illustrate the challenge of analyzing multiprocessor systems by a small example illustrated in Table 5.6.*

We have $\Omega_M = 27, \Pi_H = \mathrm{LCM}\{11, 8, 251\} = 22088$, and $\Sigma_{\tau \in \mathcal{T}_X} wcet_\tau = 3+4 = 7$. The upper bound on the depth of the tree is $\Omega_M + \Pi_H \cdot (1 + \Sigma_{\tau \in \mathcal{T}_X} wcet_\tau) = 176731$. The number of nodes (states) in the computation tree occurring at a depth ≤ 176731 can be calculated to approximately $3.9 \cdot 10^{13}$. For details concerning such calculations we refer to [BHM08]. □

TABLE 5.6

Small Example with a Huge State Space

Task	Execution Time $(bcet_\tau, wcet_\tau)$	Period π_τ	Offset ω_τ
τ_1	$(1, 3)$	11	0
τ_2	$(1, 4)$	8	10
τ_3	$(1, 13)$	251	27

5.5 MoVES Analysis Framework

One aim of our work is to establish a verification framework, called the "MoVES analysis framework" (see Figure 5.1), that can be used to provide guarantees, for example, about the schedulability, of a system-level model of an embedded system. We have chosen to base this verification framework on timed automata [AD94] and, in particular, the UPPAAL [BDL04,LPY97] system for modeling, verification, and simulation. In this section, we will briefly discuss the rationale behind this choice and give a flavor of the framework. We refer to [BHM08] for more details.

First of all, the timed-automata model for an embedded system must be constructed so that the transition system of this model is a refinement of the computation-tree model of Section 5.4, that is, the timed-automata model must be correct with respect to the model of computation.

Another design criteria is that we want the model to be easily extendible in the sense that new scheduling, allocation, and synchronization principles for example, could be added. We therefore structure the timed-automata model in the same way the ARTS [MVG04,MVM07] model of the multi-processor platform is structured (cf. Figure 5.6). This, furthermore, has the advantage that the UPPAAL model of the system can also be used for simulation, because an UPPAAL trace in a direct manner reflects events on the multiprocesser platform.

The timed-automata model is constructed as a parallel composition of communicating timed automata for each of the components of the embedded system. We shall now give a brief overview of the model (details are found in [BHM08]), where an embedded system is modeled as a parallel composition of an application and an execution platform:

$$System = Application \parallel ExecutionPlatform$$
$$Application = \parallel_{\tau \in T} TA(\tau)$$
$$ExecutionPlatform = \parallel_{j=1}^{N} TA(pe_j)$$

where $\parallel$ denotes the parallel composition of timed automata, $TA(\tau)$ the timed automaton for the task τ, and $TA(pe)$ the timed automaton for the processing element, *pe*. Thus, an application consists of a collection of timed automata for tasks combined in parallel, and an execution platform consists of a parallel composition of timed automata for processing elements.

The timed-automata model of a processing element, say pe_j, is structured according to the ARTS model described in Figure 5.6 as a parallel composition of a *controller*, a *synchronizer*, an *allocator*, and a *scheduler*:

$$TA(pe_j) = Controller_j \parallel Synchronizer_j \parallel Allocator_j \parallel Scheduler_j$$

In the UPPAAL model, these timed automata communicate synchronously over channels and over global variables. Furthermore, the procedural language part of UPPAAL proved particularly useful for expressing many algorithms. For example, the implementation of the earliest deadline first scheduling principle is directly expressed as a procedure using appropriate data structures.

Despite that the model of computation in Section 5.4 is a discrete model in nature, the real-time clock of UPPAAL proved useful for modeling the timing in the system in a natural manner, and the performance in verification examples was promising as we shall see in Section 5.6. One could have chosen a model checker for discrete systems, such as SPIN [Hol03], instead of UPPAAL . This would result in a more explicit and less natural modeling of the timing in the system. Later experiments must show whether the verification would be more efficient.

The small example in Table 5.6 shows that verification of "real" systems becomes a major challenge because of the state explosion problem. The MoVES analysis framework is therefore parameterized with respect to the UPPAAL model of the embedded system in order to be able to experiment with different approaches and in order to provide an efficient support for special cases of systems. In the following, we will briefly highlight four of these different models.

1. One model considers the special case where worst-case and best-case execution times are equal. Since scheduling decisions are deterministic, nondeterminism is eliminated, and the computation tree of such a system consists of only one infinite run. Note that for such systems it may still be necessary to analyze a very long initial part of the run before schedulability can be guaranteed. However, it is possible to analyze very large systems. For the implementation of this model, we used a special version of UPPAAL in which no history is saved.

2. Another model extends the previous one by including the notion of resource allocation to be used in the analysis of memory footprint and power consumption.

3. A third model includes nondeterminism of execution times, as described in the model of computation in Section 5.4. In this timed-automata model, the execution time for tasks was made discrete in

order to handle preemptive scheduling strategies. This made the timed-automata model of a task less natural than one could wish.

4. A fourth model used stopwatch automata rather than clocks to model the timing of tasks, which allows preemption to be dealt with in a more natural way. In general, the reachability problem for stopwatch automata is undecidable, and the UPPAAL support for stopwatches is based on overapproximations. But our experiences with using this model were good: In the examples we have tried so far, the results were always exact, the verification was more efficient compared with the previous model (typically 40% faster), and it used less space, and we can thus verify larger systems than with the previous model.

We are currently working toward a model that will reduce the number of clocks used compared to the four models mentioned above. The goal is to have just one clock for each processing element, and achieving this, we expect a major efficiency gain for the verification.

5.6   Using the MoVES Analysis Framework

In order to make the model usable for system designers, details of the timed-automata model are encapsulated in the MoVES analysis framework. The system designer needs to have an understanding of the embedded system model, but not necessarily of the timed-automata model. It is assumed that tasks and their properties are already defined, and, therefore, MoVES is only concerned with helping the system designer configure the execution platform and perform the mapping of tasks on it.

The timed-automata model is created from a textual description that resembles the embedded system model presented in Section 5.3. MoVES uses UPPAAL as back-end to analyze the user's model and to verify properties of the embedded system through model checking, as illustrated in Figure 5.1. UPPAAL can produce a diagnostic trace and MoVES transforms this trace into a task schedule shown as a Gantt chart.

As MoVES is a framework aimed at exploring different modeling approaches, it is possible to change the core model such that the different modeling approaches described in Section 5.5 can be supported. In the following, we will give four examples of using the framework to analyze embedded systems based on the different approaches. The first two examples focus on deterministic models, while the third and the fourth are based on nondeterministic models.

5.6.1   Simple MultiCore Embedded System

To illustrate the design and verification processes using the MoVES analysis framework, consider the simple multi-core embedded system from

```
E<>missedDeadline: true
E<>allFinish(): false                                              5    10
E<>totalCostInSystem(Power)  == 7:true         Task 1: 1100110011
E<>totalCostInSystem(Power)  > 7: false        Task 2: 0010001000
E<>costOnPE[0][Memory]  == 17: true            Task 3: 0000110011
E<>costOnPE[0][Memory]  > 17: false            Task 4: ----001100X
E<>costOnPE[1][Memory]  == 12: true            Task 5: 0001000100
E<>costOnPE[1][Memory]  > 12: false
```

FIGURE 5.9
Queries and the resulting Gantt chart from the analysis of the system in
Figure 5.5 using rate-monotonic scheduling on both processors, and the
memory and power figures from Table 5.4. The notation of the schedule is
0 for idle, 1 for running, - for offset, and X for missed deadline.

Figure 5.5. We will use this example to illustrate cross-layer dependencies
and to show how resource costs can be analyzed. In the first experiment,
we will use rate-monotonic scheduling as the scheduling policy for the real-
time operating system on both processors. Figure 5.9 presents the UPPAAL
queries on schedulability and resource usage, and the resulting schedule of
the system.

The verification results show several properties of the system. First, the
system cannot be scheduled in the given form since it misses a deadline.
Second, at no point does the system use more than 7 units of power, but at
some point before missing the deadline, 7 units of power is used. Finally,
in regard to memory usage, it is verified that pe_1 uses 17 units of memory at some point before missing the deadline but not more, and pe_2 uses
12 units but not more. It is shown that Task 4 misses a deadline after 11
execution cycles. Note that Task 5 is the message task between Task 2 and
Task 3.

In order to explore possible improvements of the system, we attempt verification of the same system where pe_2 uses earliest deadline first scheduling.
The verification results can be seen in Figure 5.10.

First, the system is now schedulable, as can be seen by the
E<>allFinish() query being true. The system still has the same properties for power usage as with rate-monotonic scheduling used on pe_2, but
the verification shows that at no point will the revised system (i.e., where pe_2
uses earliest deadline first) use more than 11 units of memory. Recall that the
system where pe_2 used rate-monotonic scheduling already before missing a
deadline had at some point used 17 units of memory.

5.6.2   Smart Phone, Handling Large Models

As shown in Section 5.4, seemingly simple systems can result in very large
state spaces. In order to analyze a realistic embedded system, we consider
an application that is part of a smart phone. The smart phone includes the

```
E<>missedDeadline: false
E<>allFinish(): true
E<>totalCostInSystem(Power) == 7: true
E<>totalCostInSystem(Power) > 7: false
E<>costOnPE[0][Memory] == 11: true
E<>costOnPE[0][Memory] > 11: false
E<>costOnPE[1][Memory] == 12: true
E<>costOnPE[1][Memory] > 12: false
5    10    15    20    25    30

Task 1:  11001100110011001100110011001100110011
Task 2:  0010001000000001000100000001000
Task 3:  00001100011000001100011000011
Task 4:  ----001110011100001110011100000
Task 5:  00010001000000010001000000000100
```

FIGURE 5.10

Queries and the resulting Gantt chart from the analysis of the system in Figure 5.5 using rate-monotonic scheduling on processor pe_1 and earliest deadline first scheduling on processor pe_2.

following applications: a GSM encoder, a GSM decoder, and an MP3 decoder with a total of 103 tasks, as seen in Figure 5.11. These applications do not together make up the complete functionality of a smart phone, but are used as an example, where the number of tasks, their dependencies, and their timing properties are realistic. The applications and their properties in the smart phone example originate from experiments done by Schmitz [SAHE04]. The timing properties, the period, and the deadline of the tasks are imposed by the application and can be seen in Table 5.7. The smart phone example has been verified using worst-case execution times only. That is, in order to reduce the state space, we have only considered a deterministic version of the application where worst-case execution times equal best-case execution times.

The execution cycles, memory usage, and power consumption of each task depend on the processing element. These properties of the tasks have been measured by simulating the execution of each task on different types of processing elements (the GPP, the FPGA, and the ASIC) as seen in Table 5.7. The execution cycles range from 52 to 266687 and the periods range from 0.02 to 0.025 seconds giving a total number of 504 tasks to be executed in the hyper-period of the system.

The three applications have been mapped onto a platform consisting of four general-purpose processing elements, all of type GPP0 running at 25 MHz, connected by a bus. The parallelism of the MP3-decoder has been exploited to split this application onto two processing elements. The two other applications run on their own processing element.

Having defined the embedded system with the application, the execution platform, and the mapping described above, the MoVES analysis

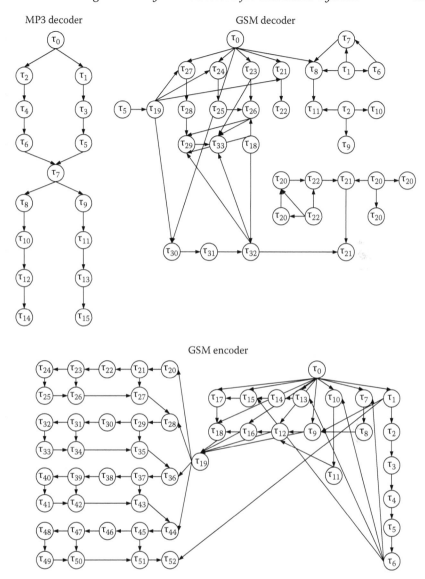

FIGURE 5.11

Task graph for three applications from a smart phone, taken from [SAHE04].

framework is used to verify schedulability, maximum memory usage, and power consumption. In this case, the system is schedulable and the maximum memory usage and power consumption is 1500 bytes and 1000 mW. The verification of this example takes roughly 3 h on a 64 bit Linux server with an AMD dual-core processor with 2 GB of memory.

TABLE 5.7

Application and *pe* Characteristics

Application	Tasks/ Edges	Deadline/ Period (s)	*pe*	Frequency (MHz)
GSM encoder	53/80	0.020	GPP0	25
GSM decoder	34/55	0.020	GPP1	10
MP3 decoder	16/15	0.025	GPP2	6.6
			FPGA	2.5
			ASIC	2.5

It is possible that better designs exist, for instance, where less power is used. A general-purpose processor could, for example, run at a lower frequency, or be replaced by an FPGA or an ASIC. This is, however, not the focus of this case study.

5.6.3 Handling Nondeterministic Execution Times

When we allow a span of execution times between the best-case execution time and the worst-case execution time of a task, the state space grows dramatically, as explained in Section 5.4. We examine the system given in Table 5.6 using an UPPAAL model capturing the nondeterminism in the choices for execution times in each period and using discretization of the running time of tasks. In Section 5.4, it was shown that the maximal depth of the computation tree that is needed when checking for schedulability is $\Omega_M + 8 \cdot \Pi_H$ (i.e., 176731). The number of states in the initial part of the computation tree until that depth is approximately $3.9 \cdot 10^{13}$. The verification used 3.1 GB of memory and took less than 11 min on an AMD CPU of 1.8 MHz and 32 GB of RAM.

If the system is changed slightly by adding an extra choice for the execution time to τ_3 (i.e., $wcet_{\tau_3} = 14$), the number of states in the initial part of the computation tree until depth $\Omega_M + 8 \cdot \Pi_H$ will be approximately $4.2 \cdot 10^{13}$. When attempting verification of this revised system on the same CPU, the verification aborts after 19 min with an "Out of memory" error message after having used 3.4 GB of memory.

5.6.4 Stopwatch Model

Examining the same system (i.e., adding the extra choice for execution time $wcet_{\tau_3} = 14$) using an UPPAAL model with stopwatches, this example can now be analyzed without the "Out of memory" error. Even though the verification with stopwatches is using overapproximations, all the experiments we have conducted so far with this model have provided exact results. Furthermore, the tendency for all these experiments is that memory consumption as well as verification times are reduced by approximately 40% in comparison to the previous model.

5.7 Summary

The classical real-time scheduling theory for single processor systems cannot be applied directly to multiprocessor systems. Already in 1978, Mok and Dertouzos [MD78] showed that the algorithms that are optimal for single processor systems are not optimal for increased numbers of processors, and in this chapter, we have seen that some of the apparently correct methods lead to counterintuitive results, such as timing anomalies.

Hence, one aim of our work has been to establish a verification framework, called the "MoVES analysis framework", that can be used to provide guarantees, for example, about the schedulability, of a system-level model of an embedded system. We have chosen to base this verification framework on timed automata and, in particular, the UPPAAL system for modeling, verification, and simulation.

The framework allows us to model and analyze an embedded system expressed as an application executing on a multiprocessor execution platform, consisting of a set of possible different processors each running its own real-time operating system, and a network connecting the different cores. Furthermore, the framework allows us to experiment with different core-modeling approaches, and hence, allows us to address the challenges of modeling and verification complexities. So far, results are very promising and it is our hope that in the near future we will be able to model and verify realistic systems from our industrial partners.

Acknowledgments

We would like to thank Jens Sten Ellebæk Nielsen and Kristian Stålø Knudsen for their contribution to the simplification of the model and work on the front end of the MoVES analysis framework. Furthermore, we are grateful for comments from Kim G. Larsen on this work. Finally, we are grateful to Jacob Illum and Alexandre David for providing us with versions of UPPAAL under development, in order for us to conduct initial experiments with verifications of larger and realistic systems as well as models with stopwatches.

The work presented in this chapter has been supported by ArtistDesign (FP7 NoE No 214373), MoDES(Danish Research Council 2106-05-0022), and DaNES (Danish National Advanced Technology Foundation).

References

[AD94] R. Alur and D. L. Dill. A theory of timed automata. *Theoretical Computer Science*, 126(2):183–235, 1994.

[BDL04] G. Behrmann, A. David, and K. G. Larsen. A tutorial on UPPAAL. In *Formal Methods for the Design of Real-Time Systems: Fourth International School on Formal Methods for the Design of Computer, Communication, and Software Systems, SFM-RT 2004, LNCS*, Vol. 3185, pp. 200–236, 2004.

[BHM08] A. Brekling, M.R. Hansen, and J. Madsen. Models and formal verification of multipocessor system-on-chips. *The Journal of Logic and Algebraic Programming*, 77(1):1–19, 2008.

[Gra69] R.L. Graham. Bounds on multiprocessor timing anomalies. *SIAM Journal of Applied Mathematics*, 17(2):416–429, March 1969.

[HFK⁺07] C. Haubelt, J. Falk, J. Keinert, T. Schlichter, M. Streubühr, A. Deyhle, A. Hadert, and J. Teich. A systemC-based design methodology for digital signal processing systems. *EURASIP Journal on Embedded Systems*, 2007(1):15–15, 2007.

[Hol03] G. J. Holzmann. *The SPIN Model Checker: Primer and Reference Manual*. Addison-Wesley, Reading, MA, 2003.

[Liu00] J. W.S. Liu. *Real-Time Systems*. Prentice Hall, Upper Saddle River, NJ, 2000.

[LPY97] K.G. Larsen, P. Pettersson, and W. Yi. UPPAAL in a nutshell. *International Journal on Software Tools for Technology Transfer*, 1(1–2):134–152, October 1997.

[MD78] A.K. Mok and M.L. Dertouzos. Multiprocessor scheduling in a hard real-time environment. In *Proceedings of the Seventh IEEE Texas Conference on Computer Systems*, Houston, TX, 1978.

[MMV04] J. Madsen, S. Mahadevan, and K. Virk. Network-centric system-level model for multiprocessor soc simulation. In J. Nurmi, H. Tenhunen, J. Isoaho, and A. Jantsch (editors), *Interconnect-Centric Design for Advanced SoC and NoC*, Chapter 13, pp. 341–365. Kluwer Academic Publishers/Springer Publishers, the Netherlands, July 2004.

[MVG04] J. Madsen, K. Virk, and M.J. Gonzalez. A SystemC-based abstract real-time operating system model for multiprocessor system-on-chip. In A. Jerraya and W. Wolf (editors) *Multi-processor System-on-Chip*, pp. 283–312. Morgan Kaufmann, San Francisco, CA, 2004.

[MVM07] S. Mahadevan, K. Virk, and J. Madsen. ARTS: A systemC-based framework for multiprocessor systems-on-chip modelling. *Design Automation for Embedded Systems*, 11(4):285–311, 2007.

[PEP06] A. D. Pimentel, C. Erbas, and S. Polstra. A systematic approach to exploring embedded system architectures at multiple abstraction levels. *IEEE Transactions on Computers*, 55(2):99–112, 2006.

[PHL⁺01] A. D. Pimentel, L. O. Hertzberger, P. Lieverse, P. van derWolf, and E. F. Deprettere. Exploring embedded-systems architectures with artemis. *IEEE Computer*, 34(11):57–63, 2001.

[RWT⁺06] J. Reineke, B. Wachter, S. Thesing, R. Wilhelm, I. Polian, J. Eisinger, and B. Becker. A definition and classification of timing anomalies. In *Proceedings of Sixth International Workshop on Worst-Case Execution Time (WCET) Analysis*, Dresden, Germany, July 2006, http://drops.dagstuhl.de/portals/WCET06/

[SAHE04] Marcus T. Schmitz, Bashir M. Al-Hashimi, and Petru Eles. *System-Level Design Techniques for Energy-Efficient Embedded Systems*. Kluwer Academic Publishers, Norwell, MA, February 2004.

[SL96] J. Sun and J. W.-S. Liu. Synchronization protocols in distributed real-time systems. In *International Conference on Distributed Computing Systems*, Hong Kong, pp. 38–45, 1996.

6

TrueTime: Simulation Tool for Performance Analysis of Real-Time Embedded Systems

Anton Cervin and Karl-Erik Årzén

CONTENTS

6.1 Introduction

Embedded systems and networked embedded systems play an increasingly important role in today's society. They are often found in consumer products (e.g., in automotive systems and cellular phones), and are therefore subject to hard economic constraints. The pervasive nature of these systems generates further constraints on physical size and power consumption. These product-level constraints give rise to resource constraints on the implementation platform, for example, limitations on the computing speed, memory size, and communication bandwidth. Because of economic considerations, this is true in spite of the rapid hardware development. In many applications, using a processor with a larger capacity than strictly necessary cannot be justified.

Feedback control is a common application type in embedded systems, and many wireless embedded systems are networked control systems, that is, they contain one or several control loops that are closed over a communication network. The latter is particularly common in cars, where several control loops (e.g., engine control, traction control, antilock braking, cruise control, and climate control) are partly or completely closed over a network.

Embedded control systems are also becoming increasingly complex from the control and computer implementation perspectives. Today, even quite simple embedded control systems often contain a multitasking real-time operating system with the controllers implemented as one or several tasks executing on a microcontroller. The operating system typically uses concurrent programming to multiplex the execution of the various tasks. The CPU time and, in the case of networked control loops, the communication bandwidth can, hence, be viewed as shared resources for which the tasks compete.

Sampled control theory normally assumes periodic sampling and negligible or constant input–output latencies. When a controller is implemented as a task in a real-time operating system executing on a computing platform with small resource margins, this can normally not be achieved. Preemptions by higher-priority tasks or interrupt handlers, blockings caused by accesses to mutually exclusive resources, cache misses, etc., cause jitter in sampling intervals and input–output latencies. Likewise, for networked control systems, medium access delays, transmission delays, and network interface delays cause variable communication latencies.

Simulation is a powerful technique that can be used at several stages of system development. For resource-constrained embedded control systems it is important to be able to include the timing effects caused by the implementation platform in the simulation. TrueTime [4,11,17] is a MATLAB®/Simulink®-based (see [23]) simulation tool that has been developed at Lund University since 1999. It provides models of multitasking real-time kernels and networks that can be used in simulation models for networked embedded control systems.

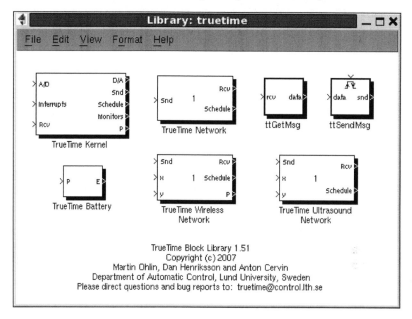

FIGURE 6.1
The TrueTime 1.5 block library. TrueTime is a freeware and can be down-loaded from http://www.control.lth.se/truetime.

In the kernels, controllers and other software components are imple-mented as MATLAB or the C++ code, structured into tasks and interrupt handlers. Support for interprocess communication and synchronization is available similar to a real real-time kernel. In fact, the underlying implemen-tation is very similar to a real kernel, with a ready queue for tasks that are ready to execute, and wait queues for tasks that are waiting for a time inter-val or for access to a shared resource. The network blocks, similarly, provide models of the medium access and transmission delay for a number of differ-ent wired and wireless link-layer protocols. Figure 6.1 shows the Simulink diagram containing the TrueTime library of predefined blocks representing real-time kernels and networks.

TrueTime can be used in a variety of ways in networked embedded con-trol system development. Often it is used to evaluate the influence on the closed-loop control performance of, for example,

- Various task-scheduling policies
- The processor speed
- Various wired or wireless network protocols in networked control
- Different network parameters such as bit rate and maximum packet length
- Disturbance network traffic

TrueTime can also be used as a pure scheduling simulator:

- TrueTime can be used as an experimental testbench for test implementations of new task-scheduling policies and network protocols. Implementing a new policy in TrueTime is often considerably easier than modifying a real kernel.
- TrueTime can be used for gathering various execution statistics (e.g., input–output latency) and various scheduling events (e.g., deadline overruns). Measurements can be logged to a file and then analyzed in MATLAB.

6.1.1 Related Work

There exist today a large number of general network simulators. One of the most well-known is ns-2 [25], which is a discrete-event simulator for both wired and wireless networks with support for, for example, the TCP, the UDP, routing, and multicast protocols. It also supports simple movement models for mobile applications, where the positions and velocities of nodes may be specified in a script. It should be noted that the default radio model in ns-2 is very simplistic (even more simplistic than TrueTime's), although more accurate physical layer models may be implemented by the user [13]. Another discrete-event computer network simulator is OMNeT++ [27]. It contains detailed IP, TCP, and FDDI protocol models and several other simulation models (file system simulator, Ethernet, framework for simulation of mobility, etc.).

Compared to the simulators above, the network simulation part in TrueTime is quite simplistic. However, the strength of TrueTime is the co-simulation facilities that makes it possible to simulate the latency-related aspects of the network communication in combination with the node computations and the dynamics of the physical environment. Rather than basing the co-simulation tool on a general network simulator and then trying to extend this with additional co-simulation facilities, the approach has been to base the co-simulation tool on a powerful simulator for general dynamical systems (i.e., Simulink), and then add support for simulation of real-time kernels and the latency aspects of network communication to this. An additional advantage of this approach is the possibility to make use of the wide range of toolboxes that is available for MATLAB/Simulink, for example, support for virtual reality animation.

There are also some network simulators geared toward the sensor network domain. TOSSIM [20] compiles directly from the TinyOS code and scales very well. The COOJA simulator [28] makes it possible to simulate sensor networks running the Contiki OS. Another example is J-Sim, a general compositional simulation environment that includes a generalized packet-switched network model that may be used to simulate wireless LANs and sensor networks [36]. Again, these types of simulators generally lack the possibility to simulate continuous-time dynamics that is present in TrueTime.

Another type of related tools are complete computer emulators, such as the Simics system [22]. Although systems of this type provide very accurate ways of simulating software, they, generally, have a weak support for networks and continuous-time dynamics. In the real-time scheduling community, a number of task-scheduling simulators have been developed (e.g., STRESS [6], DRTSS [35], RTSIM [10], and Cheddar [34]). Neither of these tools support simulation of what is outside the computer.

A few other tools have been developed that support co-simulation of real-time computing systems and control systems. RTSIM has been extended with a module that allows system dynamics to be simulated in parallel with scheduling algorithms [29]. XILO [18] supports the simulation of system dynamics, CAN networks, and priority-preemptive scheduling. Ptolemy II is a general-purpose multidomain modeling and simulation environment that includes a continuous-time domain and a simple RTOS domain. It has recently been extended in the sensor network direction [8]. In [9], a co-simulation environment based on ns-2 is presented. The ns-2 simulator has been extended with an ODE solver for dynamical simulations of the controller units and the environment. However, this tool lacks support for real-time kernel simulation.

The SimEvents® 2 toolbox [12] is a discrete-event simulator that has been embedded in Simulink in a way that is quite similar to TrueTime. The simulation engine in SimEvents is driven by an event calendar where future events are listed in order of the scheduled times. In addition to the traditional signal-based communication between blocks, SimEvents also adds entities. An entity corresponds to an object that is passed between different blocks, modeling, for instance, a message in a communication network. SimEvents provides blocks for generating entities, queue blocks, server blocks, routing blocks, control-flow control blocks, timer and counter blocks, and blocks for interfacing the SimEvents part of the simulation with the ordinary Simulink model. Using a queue and a server block it is possible to create a simple model of a CPU. It is also possible to model various types of network protocols (e.g., CAN and Ethernet). The major difference between TrueTime and SimEvents is that SimEvents is primarily aimed at discrete queue- and server-system modeling, whereas TrueTime is aimed at models of real-time kernels and real-time networks. SimEvents has no explicit notion of tasks and task codes. On the other hand, TrueTime is not very well suited for modeling of pure queueing systems.

6.1.2 Outline of the Chapter

The rest of this chapter is outlined as follows. In Section 6.2, we introduce the underlying timing and execution models of TrueTime. Sections 6.3 and 6.4 provide introductions to the kernel block and network block functionalities.

We then provide three larger examples in Sections 6.5 through 6.7. Current limitations and possible future extensions of TrueTime are discussed in Section 6.8, and the chapter is concluded with a brief summary in Section 6.9.

6.2 Timing and Execution Models

Below, we first explain how TrueTime interacts with Simulink. Then, the internal structure and logic of the kernel and network blocks are described.

6.2.1 Implementation Overview

The TrueTime Simulink blocks are implemented as variable-step S-functions written in C++. Internally, each block contains a discrete-event simulator. The TrueTime Kernel blocks simulate event-based real-time kernels executing tasks and interrupt handlers, while the TrueTime Network blocks simulate various local-area communication protocols and networks.

There is no global event queue, meaning that each block controls its own timing. A zero-crossing function in each block is used to force the Simulink solver to produce a "major hit" at each internal (scheduled) or external (triggered) event. Events are communicated between the blocks using trigger signals that switch values between 0 and 1. Events that are scheduled or triggered at the same time instant can be processed in any order by the blocks.

At each major time step in the Simulink simulation cycle, the discrete-event simulator is executed and the block outputs are updated. In the minor time steps, the inputs are read and the zero-crossing function is called repeatedly in order for the solver to lock on the next event. The zero-crossing callback function has the following principal structure (the variable nextHit denotes the next scheduled event):

```
void mdlZeroCrossings(SimStruct *S) {
  t = ssGetT(S);
  store all inputs;
  if (any trigger input has changed value) {
    nextHit = t;
  }
  ssGetNonsampledZCs(S)[0] = nextHit - t;
}
```

The timing scheme used introduces a small delay between the block inputs and outputs that depends on the Simulink-solver settings ($1.5 \cdot 10^{-15}$ s by default). At the same time, the scheme allows blocks to be connected in a circular fashion without creating algebraic loops.

6.2.2 Kernel Simulators

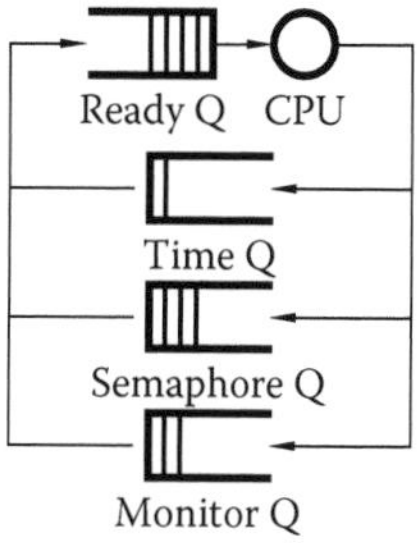

Internally, the kernel simulator is organized in much the same way as a real real-time kernel (see Figure 6.2). Tasks, interrupt handlers, and timers are represented by objects that are moved between various queues. The Ready Q contains all objects that are eligible for execution, while the Time Q contains objects that are scheduled to wake up a later time. As tasks try to get access to semaphores, monitors, or mailboxes, they may be temporarily placed in other wait queues.

Ready Q   CPU

Time Q

Semaphore Q

Monitor Q

FIGURE 6.2
Kernel model.

The tasks and interrupt handlers contain a real code that is executed as the simulation progresses. TrueTime hence implements the "live task model" [35]. The code of a task or an interrupt handler is implemented in a user-defined "code function" that is called repeatedly by the simulator. The code is split into segments, as shown in Figure 6.3. The simulator calls the code function with an increasing input argument, indicating which segment is to be executed. The code function executes nonpreemptively and returns the "simulated" execution time of that segment. The same amount of execution time must then be consumed by the task in the simulated CPU before the task may progress to the next segment.

The task code may contain statements that change the state of the task (e.g., calls to blocking kernel primitives). The task may also communicate with the environment by accessing the A/D–D/A ports of the block or by sending and receiving network messages.

In summary, the kernel simulator performs the following actions every time it executes:

- Check for external interrupts and network interrupts
- Count down the remaining execution time of the running task
- If the current code segment is finished and there are more segments, then execute the next segment
- Check the Time Q for tasks, interrupt handlers, or timers that should be moved to the Ready Q

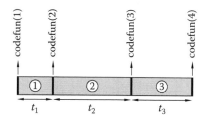

FIGURE 6.3
Model of the task code. The code function returns the simulated execution time of each code segment.

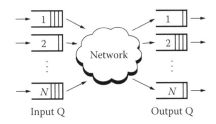

FIGURE 6.4
Network model.

- Dispatch the first object in the Ready Q
- Compute the time of the next expected event

The last item amounts to comparing the remaining execution time of the currently running task to the first object in the Time Q.

6.2.3 Network Simulators

Network simulators have a similar implementation structure as the kernel simulators. The network conceptually consists of a set of the FIFO Input Qs, a shared communication medium, and a set of the FIFO Output Qs (see Figure 6.4). The queues model the send and receive buffers in the nodes connected to the network.

A message that should be transmitted is placed in one of the Input Qs. Messages are moved from the Input Qs, into the network, and into the Output Qs in an order that depends on the simulated network protocol. The transmission time of each message depends on the length of the message. Collisions and retransmissions are simulated in the relevant protocols. A message that arrives in an Output Q potentially triggers an interrupt handler in the receiving node.

In summary, the network simulator performs the following actions every time it executes:

- Count down the remaining transmission time of the current message
- If the current message is finished, then move it to the destination Output Q
- Check for newly arrived messages in the Input Qs
- Check for collisions in the network and take the appropriate action
- Compute the time of the next expected event

6.3 Kernel Block Features

The TrueTime Kernel block simulates a computer node with a generic and flexible real-time kernel, A/D and D/A converters, external interrupt inputs, and network interfaces. The block is configured via an initialization script. The script may be parameterized, enabling the same script to be used for several nodes.

In the initialization script, the programmer may create tasks, timers, interrupt handlers, semaphores, etc., representing the software executing in the computer node. The initialization script and the code functions may be written either in the MATLAB code or in the C++ code. In the C++ case, the initialization script, the code functions, and the kernel are compiled together into a single binary using MATLAB's MEX facility, rendering a much faster simulation.

The TrueTime Kernel block supports various standard preemptive scheduling algorithms including the rate-monotonic (RM) scheduling and the earliest-deadline-first (EDF) scheduling [21]. It is also possible to specify a custom scheduling policy by supplying a sorting function for the Ready Q.

The code below shows how a single node can be configured. In the initialization function, the RM scheduling is selected and the network interface is initialized:

```
function node_init
% Initialize TrueTime kernel
ttInitKernel(0,0,'prioRM'); % nbrOfInputs, nbrOfOutputs,
    RM scheduling
% Initialize network interface
data.u = 0;  % network interrupt handler local variable
ttCreateInterruptHandler('nw_handler',1,'ctrlcode',data);
ttInitNetwork(1,'nw_handler');   % Node #1 in the network
```

The network interrupt handler is connected to a code function that, in this case, implements a simple controller:

```
function [exectime,data] = ctrlcode(segment,data)
switch segment,
  case 1,
    msg = ttGetMsg;          % retrieve msg from network
    y = msg.y;               % extract the measurement data
    data.u = data.u - y;     % compute the control action
    exectime = 0.001;
  case 2,
    newmsg.u = data.u;
    ttSendMsg(3, newmsg, 0); % send result to actuator
    exectime = -1;           % done
end
```

In the code function above, the simulated execution of the first segment is 1 ms. This means that the delay between reading the incoming message and sending the reply will be at least 1 ms (more if there is preemption from higher-priority interrupts).

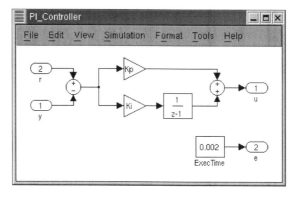

FIGURE 6.5
Controllers represented using ordinary discrete-time Simulink blocks may be called from within the code functions.

In both the C++ and MATLAB cases, it is possible to call Simulink block diagrams from within the code functions. This is sometimes a convenient way to implement controllers. The listing below shows an example where the discrete PI-controller in Figure 6.5 is used in a code function:

```
function [exectime,data] = PIcode(segment,data)
switch segment,
  case 1,
    inp(1) = ttAnalogIn(1);
    inp(2) = ttAnalogIn(2);
    outp = ttCallBlockSystem(2,inp,'PI_Controller');
    data.u = outp(1);
    exectime = outp(2);
  case 2,
    ttAnalogOut(1, data.u);
    exectime = -1;            % done
end
```

TrueTime includes a large library of real-time primitives that may be called from the initialization script and/or the task code. There is support for periodic and aperiodic tasks, periodic and one-shot timers, hardware- or software-triggered interrupts, and task synchronization mechanisms in the forms of semaphores, monitors, and mailboxes. It is possible to read online and modify most task attributes (period, deadline, priority, etc.). Advanced real-time scheduling features include deadline overrun handlers, worst-case execution time (WCET) overrun handlers, and the possibility to abort task jobs. In terms of simulation, the user may log various variables relating to the task schedule. Finally, there are primitives for initializing network interfaces and sending and receiving messages.

6.4   Network Block Features

The TrueTime network block and the TrueTime wireless network block simulate the physical and the medium-access layers of various local-area networks. The types of networks supported are the CSMA/CD (Ethernet), the CSMA/AMP (CAN), the Round Robin (Token Bus), the FDMA, the TDMA (TTP), the Switched Ethernet, the WLAN (802.11b), and the ZigBee (802.15.4). The blocks only simulate the medium access (the scheduling), possible collisions or interferences, and the point-to-point/broadcast transmissions. The higher-layer protocols such as the TCP/IP are not simulated as such but may be implemented as applications in the nodes.

There is also a third network block that simulates the transmission and reception of ultrasound pulses. In this case, no receiver and no data can be specified. This block can be used to simulate ultrasound-based navigation systems for mobile devices.

The network blocks are mainly configured via their block dialogues. Common parameters to all types of networks are the bit rate, the minimum frame size, and the network interface delay. For each type of network, there are a number of further parameters that can be specified. For instance, for wireless networks, it is possible to specify the transmit power, the receiver signal threshold, the pathloss exponent (or a special pathloss function), the ACK timeout, the retry limit, and the error-coding threshold.

A TrueTime model may contain several network blocks, and each kernel block may be connected to more than one network. Each network is identified by a number, and each node connected to a network is addressed by a number that is unique to that network.

The network blocks may be used in two different ways. The first way is to have one kernel block for each node in the network. The tasks inside the kernels can then send and receive arbitrary MATLAB structure arrays over the network using certain kernel primitives. This approach is very flexible but requires some amount of programming to configure the system. The second way is to use the stand-alone network interface blocks. These blocks eliminate the need of kernel blocks, but they restrict the network packets to contain scalar or vector signal values. Finally, it is possible to mix kernel blocks and network interface blocks in the same network.

6.4.1   Wireless Networks

Compared to the wired network block, the wireless network block has additional x and y inputs that represent the actual location of the nodes in the network. These inputs can be connected to further blocks that model the physical movement of the nodes. The current x and y coordinates of the nodes will influence the signal-to-interference ratio at the receiver. The pathloss of radio signals is modeled as $1/d^a$, where d is the distance between the sending and the receiving node, and a is an environment parameter (typically in the range from 2 to 4). If the received energy is below a user-defined threshold, then no reception will take place.

A node that wants to transmit a message will proceed as follows: The node first checks whether the medium is idle. If that has been the case for 50 µs, then the transmission may proceed. If not, the node will wait for a random back-off time before the next attempt. The signal-to-interference ratio in the receiving node is calculated by treating all simultaneous transmissions as an additive noise. This information is used to determine a probabilistic measure of the number of bit errors in the received message. If the number of errors is below a configurable bit-error threshold, then the packet could be successfully received.

6.5 Example: Constant Bandwidth Server

The constant bandwidth server (CBS) [1] is a scheduling server for aperiodic and soft tasks that executes on top of an EDF scheduler. A CBS is characterized by two parameters: a server period T_s and a utilization factor U_s. The server ensures that the task(s) executing within the server can never occupy more than the U_s of the total CPU bandwidth.

Associated with the server are two dynamic attributes: the server budget c_s and the server deadline d_s. Jobs that arrive at the server are placed in a queue and are served on a first-come first-serve basis. The first job in the queue is always eligible for execution, using the current server deadline, d_s. The server is initialized with $c_s := U_s T_s$ and $d_s = T_s$. The rules for updating the server are as follows:

1. During the execution of a job, the budget c_s is decreased at unit rate.
2. Whenever $c_s = 0$, the budget is recharged to $c_s := U_s T_s$, and the deadline is postponed one server period: $d_s := d_s + T_s$.
3. If a job arrives at an empty server at time r and $c_s \geq (d_s - r)U_s$, then the budget is recharged to $c_s := U_s T_s$, and the deadline is set to $d_s := r + T_s$.

The first and second rules limit the bandwidth of the task(s) executing in the server. The third rule is used to "reset" the server after a sufficiently long idle period.

6.5.1 Implementation of CBS in TrueTime

TrueTime provides a basic mechanism for execution-time monitoring and budgets. The initial value of the budget is called the WCET of the task. By default, the WCET is equal to the period (for periodic tasks) or the relative deadline (for aperiodic tasks). The WCET value of a task can be changed by calling `ttSetWCET(value,task)`. The WCET corresponds to the maximum server budget, U_sT_s, in the CBS. The CBS period is specified by setting the relative deadline of the task. This attribute can be changed by calling `ttSetDeadline(value,task)`.

When a task executes, the budget is decreased at unit rate. The remaining budget can be checked at any time using the primitive `ttGetBudget(task)`. By default, nothing happens when the budget reaches zero. In order to simulate that the task executes inside a CBS, an execution overrun handler must be attached to the task. A sample initialization script is given below:

```
function node_init
% Initialize kernel, specifying EDF scheduling
ttInitKernel(0,0,'prioEDF');
% Specify CBS rules for initial deadlines and initial
  budgets
ttSetKernelParameter('cbsrules');
% Specify CBS server period and utilization factor
T_s = 2;
U_s = 0.5;
% Create an aperiodic task
ttCreateTask('aper_task',T_s,1,'codeFcn');
ttSetWCET(T_s*U_s,'aper_task');
% Attach a WCET overrun handler
ttAttachWCETHandler('aper_task','cbs_handler');
```

The execution overrun handler can then be implemented as follows:

```
function [exectime,data] = cbs_handler(seg,data)
% Get the task that caused the overrun
t = ttInvokingTask;
% Recharge the budget
ttSetBudget(ttGetWCET(t),t);
% Postpone the deadline
ttSetAbsDeadline(ttGetAbsDeadline(t)+ttGetDeadline(t),t);
exectime = -1;
```

If many tasks are to execute inside CBS servers, the same code function can be reused for all the execution overrun handlers.

6.5.2 Experiments

The CBS can be used to safely mix hard, periodic tasks with soft, aperiodic tasks in the same kernel. This is illustrated in the following example, where a ball and beam controller should execute in parallel with an aperiodically triggered task. The Simulink model is shown in Figure 6.6.

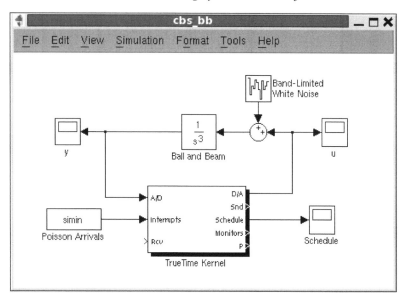

FIGURE 6.6
TrueTime model of a ball and beam being controlled by a multitasking real-time kernel. The Poisson arrivals trigger an aperiodic computation task.

The ball and beam process is modeled as a triple integrator disturbed by white noise and is connected to the TrueTime kernel block via the A/D and D/A ports. A linear-quadratic Gaussian (LQG) controller for the ball and beam has been designed and is implemented as a periodic task with a sampling period of 10 ms. The computation time of the controller is 5 ms (2 ms for calculating the output and 3 ms for updating the controller state). A Poisson source with an intensity of $100/s$ is connected to the interrupt input of the kernel, triggering an aperiodic task for each arrival. The relative deadline of the task is 10 ms, while the execution time of the task is exponentially distributed with a mean of 3 ms.

The average CPU utilization of the system is 80%. However, the aperiodic task has a very uneven processor demand and can easily overload the CPU during some intervals. The control performance in the first experiment, using plain EDF scheduling, is shown in Figure 6.7. A close-up of the corresponding CPU schedule is shown in Figure 6.8. It is seen that the aperiodic task sometimes blocks the controller for several sampling periods. The resulting execution jitter leads to very poor regulation performance.

Next, a CBS is added to the aperiodic task. The server period is set to $T_s = 10$ ms and the utilization to $U_s = 0.49$, implying a maximum budget (WCET) of 4.9 ms. With this configuration, the CPU will never be more than 99% loaded. A new simulation, using the same random number sequences as before, is shown in Figure 6.9. The regulation performance is much

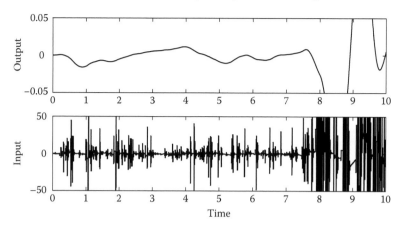

FIGURE 6.7
Control performance under plain EDF scheduling.

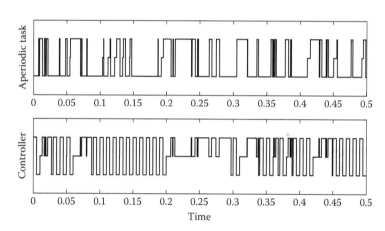

FIGURE 6.8
Close-up of CPU schedule under plain EDF scheduling.

better—this is especially evident in the smaller control input required. The close-up of the schedule in Figure 6.10 shows that the controller is now able to execute its 5 ms within each 10 ms period and the jitter is much smaller.

6.6   Example: Mobile Robots in Sensor Networks

In the EU/IST FP6 integrated project RUNES (reconfigurable ubiquitous networked embedded systems, [32]) a disaster-relief road-tunnel scenario was used as a motivating example [5]. In this scenario, mobile robots were used

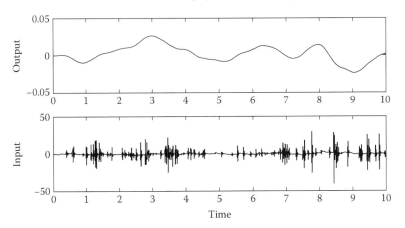

FIGURE 6.9
Control performance under CBS scheduling.

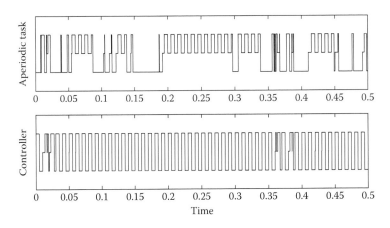

FIGURE 6.10
Close-up of CPU schedule under CBS scheduling.

as mobile radio gateways that ensure the connectivity of a sensor network located in a road tunnel in which an accident has occurred. A number of software components were developed for the scenario. A localization component based on ultrasound was used for localizing the mobile robots and a collision-avoidance component ensured that the robots did not collide (see [2]). A network reconfiguration component [30] and a power control component [37] were responsible for deciding the best position for the mobile robot in order to maximize radio connectivity, and to adjust the radio power transmit level.

In parallel with the physical implementation of this scenario, a TrueTime simulation model was developed. The focus of the simulation was the timing aspects of the scenario. It should be possible to simultaneously simulate the computations that take place within the nodes, the wireless communication between the nodes, the power devices (batteries) in the nodes, the sensor and actuator dynamics, and the dynamics of the mobile robots. In order to model the limited resources correctly, the simulation model must be quite realistic. For example, it should be possible to simulate the timing effects of interrupt handling in the microcontrollers implementing the control logic of the nodes. It should also be possible to simulate the effects of collisions and contention in the wireless communication. Because of simulation time and size constraints, it is at the same time important that the simulation model is not too detailed. For example, simulating the computations on a source-code level, instruction for instruction, would be overly costly. The same applies to simulation of the wireless communication at the radio-interface level or on the bit-transmission level.

6.6.1   Physical Scenario Hardware

The physical scenario consists of a number of hardware and software components. The hardware consists of the stationary wireless communication nodes and the mobile robots. The wireless communication nodes are implemented by Tmote Sky sensor network motes executing the Contiki operating system [14]. In addition to the ordinary sensors for temperature, light, and humidity, an ultrasound receiver has been added to each mote (see Figure 6.11).

The two robots, RBbots, are shown in Figure 6.12. Both robots are equipped with an ultrasound transmitter board (at the top). The robot to the left has the obstacle-detection sensors mounted. This consists of an IR proximity sensor mounted on an RC-servo that sweeps a circle segment in front of the robot and a touch sensor bar.

The RBbots internally consist of one Tmote Sky, one ATMEL AVR Mega128, and three ATMEL AVR Mega16 microprocessors. The nodes communicate internally over an I^2C bus. The Tmote Sky is used for the radio communication as the master. Two of the ATMEL AVR Mega16 processors are used as interfaces to the wheel motors and the wheel encoders measuring the wheel angular velocities. The third ATMEL AVR Mega16 is used as the interface to the ultrasound transmitter and to the obstacle-detection sensors. The AVR Mega128 is used as a compute engine for the software-component code that does not fit the limited memory of the TMote Sky. The structure is shown in Figure 6.13.

6.6.2   Scenario Hardware Models

The basic programming model used for the TI MSP430 processor used in the Tmote Sky systems is event-driven programming with interrupt

FIGURE 6.11
Stationary sensor network nodes with ultrasound receiver circuit. The node is packaged in a plastic box to reduce wear.

FIGURE 6.12
The two Lund RBbots.

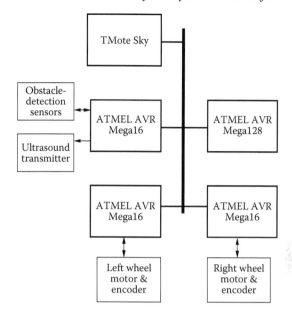

FIGURE 6.13
RBbot hardware architecture.

handlers for handling timer interrupts, bus interrupts, etc. In TrueTime, the same architecture can be used. However, the Contiki OS also supports protothreads [15], lightweight stackless threads designed for severely memory-constrained systems. Protothreads provide linear code execution for event-driven systems implemented in C. Protothreads can be used to provide blocking event-handlers. They provide a sequential flow of control without complex-state machines or full multithreading. In TrueTime, protothreads are modeled as ordinary tasks. The ATMEL AVR processors are modeled as event-driven systems. A single nonterminating task acts as the main program and the event handling is performed in interrupt handlers.

The software executing in the TrueTime processors is written in C++. The names of the files containing the code are input parameters of the network blocks. The localization component consists of two parts. The distance sensor part of the component is implemented as a (proto-)thread in each stationary sensor node. An extended Kalman filter–based data fusion is implemented in the Tmote Sky processor on board each robot. The localization method makes use of the ultrasound network and the radio network. The collision-avoidance component code is implemented in the ATMEL AVR Mega128 processor using events and interrupts. It interacts over the I^2C bus with the localization component and with the robot-position controller, both located in the Tmote Sky processor.

6.6.3 TrueTime Modeling of Bus Communication

The I^2C bus within the RBbots is modeled in TrueTime by a network block. The TrueTime network model assumes the presence of a network interface card or a bus controller implemented either in the hardware or the software (i.e., as drivers). The Contiki interface to the I^2C bus is software-based and corresponds well to the TrueTime model. In the ATMEL AVRs, however, it is normally the responsibility of the application programmer to manage all bus access and synchronization directly in the application code. In the TrueTime model, this low-level bus access is not modeled. Instead, it is assumed that there exists a hardware or a software bus interface that implements this.

Although the I^2C is a multimaster bus that uses arbitration to resolve conflicts, this is not how it is modeled in TrueTime. On the Tmote Sky, the radio chip and the I^2C bus share connection pins. Because of this, it is only possible to have one master on the I^2C bus and this master must be the Tmote Sky. All communication must be initiated by the master. Because of this, bus access conflicts are eliminated. Therefore, the I^2C bus is modeled as a CAN bus with the transmission rate set to match the transmission rate of the I^2C bus.

6.6.4 TrueTime Modeling of Radio Communication

The radio communication used by the Tmote Sky is the IEEE 802.15.4 MAC protocol (the so-called Zigbee MAC protocol) and the corresponding TrueTime wireless network protocol was used. The requirements on the simulation environment from the network reconfiguration and radio power–control components are that it should be possible to change the transmit power of the nodes and that it should be possible to measure the received signal strength, that is, the so-called received signal strength indicator (RSSI). The former is possible through the TrueTime command, `ttSetNetworkParameter('transmitpower',value)`. The RSSI is obtained as an optional return value of the TrueTime function, `ttGetMsg`.

In order to model the ultrasound, a special block was developed. The block is a special version of the wireless network block that models the ultrasound propagation of a transmitted ultrasound pulse. The main difference between the wireless network block and the ultrasound block is that in the ultrasound block it is the propagation delay that is important, whereas in the ordinary wireless block it is the medium access delay and the transmission delay that are modeled. The ultrasound is modeled as a single sound pulse. When it arrives at a stationary sensor node an interrupt is generated. This also differs from the physical scenario, in which the ultrasound signal is connected via an AD converter to the Tmote Sky.

The network routing is implemented using a TrueTime model of the ad hoc on-demand vector (AODV) routing protocol (see [31]) commonly used in sensor network and mobile robot applications. The AODV uses three basic types of control messages in order to build and invalidate routes: route request (RREQ), route reply (RREP), and route error (RERR) messages. These control messages contain source and destination sequence numbers, which are used to ensure fresh and loop-free routes. A node that requires a route to a destination node initiates route discovery by broadcasting an RREQ message to its neighbors. A node receiving an RREQ starts by updating its routing information backward toward the source. If the same RREQ has not been received before, the node then checks its routing table for a route to the destination. If a route exists with a sequence number greater than or equal to that contained in the RREQ, an RREP message is sent back toward the source. Otherwise, the node rebroadcasts the RREQ. When an RREP has propagated back to the original source node, the established route may be used to send data. Periodic hello messages are used to maintain local connectivity information between neighboring nodes. A node that detects a link break will check its routing table to find all routes that use the broken link as the next hop. In order to propagate the information about the broken link, an RERR message is then sent to each node that constitutes a previous hop on any of these routes.

Two TrueTime tasks are created in each node to handle the AODV send and receive actions, respectively. The AODV send task is activated from the application code, as a data message should be sent to another node in the network. The AODV receive task handles the incoming AODV control messages and forwarding of data messages. Communication between the application layer and the AODV layer is handled using TrueTime mailboxes. Each node also contains a periodic task, responsible for broadcasting hello messages and determining local connectivity based on hello messages received from neighboring nodes. Finally, each node has a task to handle the timer expiry of route entries.

The AODV protocol in TrueTime is implemented in such a way that it stores messages to destinations for which no valid route exists, at the source node. This means that when, eventually, the network connectivity has been restored through the use of the mobile radio gateways, the communication traffic will be automatically restored.

6.6.5   Complete Model

In addition to the above, the complete model for the scenario also contains models of the sensors, motors, robot dynamics, and a world model that keeps track of the position of the robots and the fixed obstacles within the tunnel.

The wheel motors are modeled as first-order linear systems plus integrators with the angular velocities and positions as the outputs. From the motor velocities, the corresponding wheel velocities are calculated. The wheel positions are controlled by two PI-controllers residing in the ATMEL AVR processors acting as interfaces to the wheel motors.

The Lund RBbot is a dual-drive unicycle robot. It is modeled as a third-order system

$$\dot{p}_x = \frac{1}{2}(R_1\omega_1 + R_2\omega_2)\cos(\theta)$$

$$\dot{p}_y = \frac{1}{2}(R_1\omega_1 + R_2\omega_2)\sin(\theta) \qquad (6.1)$$

$$\dot{\theta} = \frac{1}{D}(R_2\omega_2 - R_1\omega_1)$$

where the state consists of the x- and y-positions and the heading θ. Inputs to the system are the angular velocities, ω_1 and ω_2, of the two wheels. The parameters R_1 and R_2 are the radii of the two wheels and D is the distance between the wheels.

The top-level TrueTime model diagram is shown in Figure 6.14. The stationary sensor nodes are implemented as Simulink subsystems that internally contain a TrueTime kernel modeling the Tmote Sky mote, and connections to the radio network and the ultrasound communication blocks. In order to reduce the wiring From and To, blocks hidden inside the corresponding subsystems are used for the connections. The block handling the dynamic animation is not shown in Figure 6.14.

The subsystem for the mobile robots is shown in Figure 6.15. The robot dynamics block contains the motor models and the robot dynamics model.

The position of the robots and the status of the stationary sensor nodes (i.e., whether or not they are operational) are shown in a separate animation workspace (see Figure 6.16). The workspace shows one tunnel segment with sensor nodes (out of which some are non-operational) along the walls. Two robots are inside the tunnel together with two obstacles that the robots must avoid.

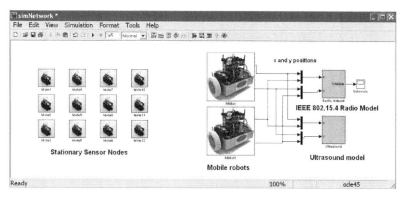

FIGURE 6.14

The TrueTime model diagram. In order to reduce the use of wires From and To, blocks hidden inside the corresponding subsystems are used to connect the stationary sensor nodes to the radio and ultrasound networks.

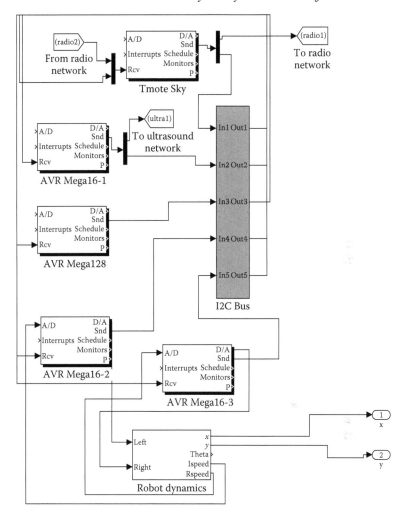

FIGURE 6.15

The Simulink model of the mobile robots. For the sake of clarity, the obstacle-detection sensors have been omitted. These should be connected to AVR Mega16-1.

6.6.6 Evaluation

The implemented TrueTime model contains several simplifications. For example, interrupt latencies are not simulated, only context switch overheads. All execution times are chosen based on experience from the hardware implementation. Also, it is important to stress that the simulated code is only a model of the actual code that executes in the sensor nodes and in the robots. However, since C is the programming language used in both cases the translation is, in most cases, quite straightforward.

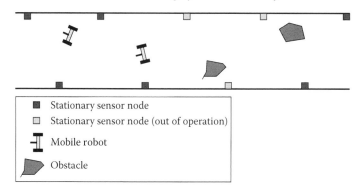

FIGURE 6.16
Animation workspace.

In spite of the above, it is our experience that the TrueTime simulation approach gives results that are close to the real case. The TrueTime approach has also been validated by others. In [7], a TrueTime-based model is compared with a hardware-in-the-loop (HIL) model of a distributed CAN-based control system. The TrueTime simulation result matched the HIL results very well.

An aspect of the model that is extremely difficult, if not impossible, to validate is the wireless communication. Simulation of wireless MANET systems is notoriously difficult (e.g., see [3]). The effects of multipath propagation, fading, and external disturbances are very difficult to model accurately. The approach adopted here is to first start with an idealized exponential decay ratio model and then, when this works properly, gradually add more and more nondeterminism. This can be done either by setting a high probability that a packet is lost, or by providing a user-defined radio model using Rayleigh fading.

The total code size for the model was 3700 lines of C code. Parts of the algorithmic code (e.g., the extended Kalman filter code) were exactly the same as in real robots. The model contained five kernel blocks and one network block per robot, one kernel block per sensor node, with six sensors, one wireless network block for the radio traffic, and one ultrasound block modeling the ultrasound propagation. The simulation rate was slightly faster than real time, executing on an ordinary dual-core MS Windows laptop.

6.7   Example: Network Interface Blocks

The last example illustrates how the stand-alone network interface blocks can be used to simulate time-triggered or event-triggered networked control loops. In this case, because there are no kernel blocks, no initialization scripts or code functions must be written.

The networked control system in this example consists of a plant (an integrator), a network, and two nodes: an I/O device (handling AD and DA conversion) and a controller node. At the I/O node, the process is sampled by a ttSendMsg network interface block, which transmits the value to the controller node. There, the packet is received by a ttGetMsg network interface block. The control signal is computed and the control is transmitted back to the I/O node by another ttSendMsg block. Finally, the signal is received by a ttGetMsg block at the I/O and is actuated to the process.

Two versions of the control loop will be studied. In Figure 6.17, both ttSendMsg blocks are time triggered. The process output is sampled every 0.1 s, and a new control signal is computed with the same interval but with a phase shift of 0.05 s. The resulting control performance and network schedule are shown in Figure 6.18. The process output is kept close to zero

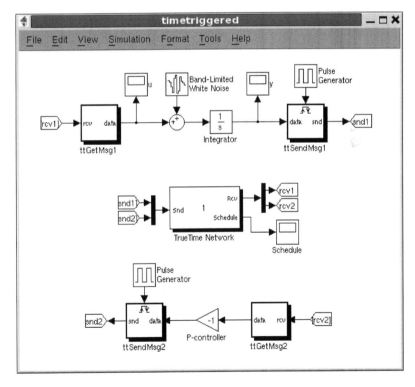

FIGURE 6.17
Time-triggered networked control system using stand-alone network interface blocks. The ttSendMsg blocks are driven by periodic pulse generators.

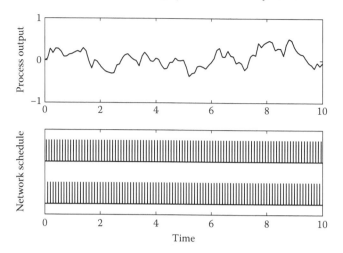

FIGURE 6.18
Plant output and network schedule for the time-triggered control system.

despite the process noise. The schedule shows that the network load is quite high.

In the second version of the control loop, the ttSendMsg blocks are event triggered instead (see Figure 6.19). A sample is generated whenever the magnitude of the process output passes 0.25. The arrival of a measurement sample at the controller node triggers—after a delay—the computation and sending of the control signal back to the I/O node. The resulting control performance and network schedule is shown in Figure 6.20. It can be seen that the process is still stabilized, although much fewer network messages are sent.

6.8 Limitations and Extensions

Although TrueTime is quite powerful, it has some limitations. Some of them could be removed by extending TrueTime in different directions. This will be discussed here.

6.8.1 Single-Core Assumption

Multicore architectures are increasingly common in embedded systems. The TrueTime kernel, however, is single core. Modifying the kernel to instead support a globally scheduled shared-memory multicore platform with a

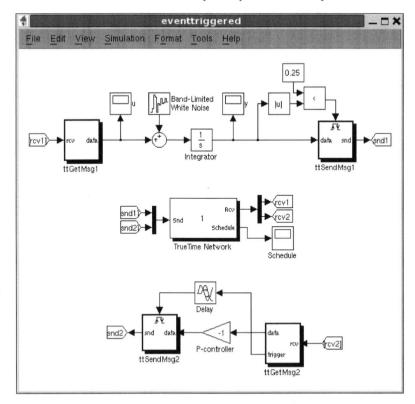

FIGURE 6.19

Event-triggered networked control system using stand-alone network interface blocks. The process output is sampled by the ttSendMsg block when the magnitude exceeds a certain threshold.

single ready queue is probably relatively straightforward. However, to support a partitioned system with separate ready queues, separate caches, and task migration overheads is significantly more complicated.

6.8.2 Execution Times

In TrueTime, it is the user's responsibility to assign the execution times of the different code segments. This should correspond to the amount of time it should take to execute the code on the particular target machine where it should run. For small microcontrollers, it is possible to perform these assessments fairly well. However, for normal-size platforms, it is difficult to get good estimates. The problem can be compared with the problem of performing the WCET analysis.

The idea behind the TrueTime approach is that the execution times should be viewed as design parameters. By increasing or decreasing them,

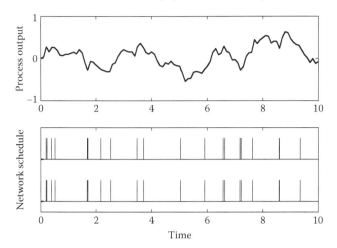

FIGURE 6.20
Plant output and network schedule for the event-triggered control system.

different processor speeds can be simulated. By adding a random element to them, variations in execution times because of code branches and data-dependent execution time statements can be accounted for. However, in a real system, the execution time of a piece of code can be divided into two parts. The first part is the execution of the different instructions in the code. This is fairly straightforward to estimate. The second part is the time caused by the hardware platform. This includes the time caused by cache misses, pipeline breaks, memory access latencies, etc. This time is more difficult to obtain good estimates for. A possible approach is to have this part of the execution time added to the user-provided times automatically by the kernel block based on different parameterized assumptions about the hardware platform.

6.8.3 Single-Thread Execution

Since Simulink simulation is performed by a single-thread execution, the multitasking in the kernel block has to be emulated. One consequence of this is that it is the responsibility of the user that the context of each task is saved and restored in the correct way. This is done by passing the context as an argument to the code functions. Another partly related consequence of this is the segmentation that has to be applied to every task. The latter is the main reason why it is not possible to use the production C code in TrueTime simulations. In addition, a code function may not call other code functions, that is, abstractions on the code function level are not supported.

Preliminary investigations indicate that it should be possible to map the TrueTime tasks onto Posix threads (i.e., to use multiple threads inside each kernel S-function). Using this approach, the problem with the task context and segments would be solved automatically.

6.8.4 Simulation Platform

TrueTime is based on Simulink. This is both an advantage and a disadvantage. It is good since it makes it easy for existing MATLAB/Simulink users to start using it. However, MATLAB/Simulink is still not widely spread in the computer science community. The threshold for a non-Simulink user to start using TrueTime is therefore fairly high. An advantage with building upon MATLAB is the vast availability of other toolboxes that can be combined with TrueTime.

However, it is possible to port TrueTime to other platforms. In [19], a feasibility study is presented where the kernel block of TrueTime is ported to Scilab/Scicos (see [33]). Also, in the new European ITEA 2 project EUROSYS-LIB, the TrueTime network blocks are being ported to the Modelica language (see [24]) and the Dymola simulator (see [16]).

6.8.5 Higher-Layer Protocols

The network blocks only support link-layer protocols. In most cases this suffices, since most real-time networks are local-area networks without any routing or transport layers. However, if higher-layer protocols are needed, these are not directly supported by TrueTime. The examples contain a TCP transport protocol example and an AODV routing protocol example, but these applications are implemented as application codes. It would be interesting to provide built-in support also for some of the most popular higher-order protocols. It would also be useful to have a plug-and-play facility that would make it easy for the user to add new protocols to the network blocks. Currently, this involves modifications of the C++ network block source code.

6.9 Summary

This chapter has presented TrueTime, a freeware extension to Simulink that allows multithreaded real-time kernels and communication networks to be simulated in parallel with the dynamics of the process under control. Having been developed over almost 10 years, TrueTime has several more features than those mentioned in this chapter. For a complete description, please see the latest version of the reference manual (e.g., [26]). In particular, many features related to real-time scheduling are detailed in [26].

References

1. L. Abeni and G. Buttazzo. Integrating multimedia applications in hard real-time systems. In *Proceedings of the 19th IEEE Real-Time Systems Symposium*, Madrid, Spain, 1998.

2. P. Alriksson, J. Nordh, K.-E. Årzén, A. Bicchi, A. Danesi, R. Schiavi, and L. Pallottino. A component-based approach to localization and collision avoidance for mobile multi-agent systems. In *Proceedings of the European Control Conference (ECC)*, Kos, Greece, 2007.

3. T.R. Andel and A. Yasinac. On the credibility of manet simulations. *IEEE Computer*, 39(7), 48–54, July 2006.

4. M. Andersson, D. Henriksson, A. Cervin, and K.-E. Årzén. Simulation of wireless networked control systems. In *Proceedings of the 44th IEEE Conference on Decision and Control and European Control Conference ECC 2005*, Seville, Spain, December 2005.

5. K.-E. Årzén, A. Bicchi, G. Dini, S. Hailes, K.H. Johansson, J. Lygeros, and A. Tzes. A component-based approach to the design of networked control systems. In *Proceedings of the European Control Conference (ECC)*, Kos, Greece, 2007.

6. N. Audsley, A. Burns, M. Richardson, and A. Wellings. STRESS—A simulator for hard real-time systems. *Software—Practice and Experience*, 24(6), 543–564, June 1994.

7. D. Ayavoo, M.J. Pont, and S. Parker. Using simulation to support the design of distributed embedded control systems: A case study. In *Proceedings of First U.K. Embedded Forum*, Brimingham, U.K., 2004.

8. P. Baldwin, S. Kohli, E.A. Lee, X. Liu, and Y. Zhao. Modeling of sensor nets in Ptolemy II. In *IPSN'04: Proceedings of the Third International Symposium on Information Processing in Sensor Networks*, pp. 359–368. ACM Press, 2004.

9. M. Branicky, V. Liberatore, and S.M. Phillips. Networked control systems co-simulation for co-design. In *Proceedings of the American Control Conference*, Denver, CL, 2003.

10. A. Casile, G. Buttazzo, G. Lamastra, and G. Lipari. Simulation and tracing of hybrid task sets on distributed systems. In *Proceedings of the Fifth International Conference on Real-Time Computing Systems and Applications*, Hiroshima, Japan, 1998.

11. A. Cervin, D. Henriksson, B. Lincoln, J. Eker, and K.-E. Årzén. How does control timing affect performance? *IEEE Control Systems Magazine*, 23(3), 16–30, June 2003.

12. M.I. Clune, P.J. Mosterman, and C.G. Cassandras. Discrete event and hybrid system simulation with simEvents. In *Proceedings of the Eighth International Workshop on Discrete Event Systems*, Ann Arbor, MI, 2006.

13. J.-M. Dricot and P. De Doncker. High-accuracy physical layer model for wireless network simulations in NS-2. In *Proceedings of the International Workshop on Wireless Ad-Hoc Networks (IWWAN)*, Oulu, FL, 2004.

14. A. Dunkels, B. Grönvall, and T. Voigt. Contiki — A lightweight and flexible operating system for tiny networked sensors. In *Proceedings of the First IEEE Workshop on Embedded Networked Sensors (Emnets-I)*, Tampa, FL, November 2004.

15. A. Dunkels, O. Schmidt, T. Voigt, and M. Ali. Protothreads: Simplifying event-driven programming of memory-constrained embedded systems. In *Proceedings of the Fourth ACM Conference on Embedded Networked Sensor Systems (SenSys 2006)*, Boulder, CL, November 2006.

16. Dymola. Homepage: http://www.dynasim.se. Visited 2008-09-30.

17. J. Eker and A. Cervin. A Matlab toolbox for real-time and control systems co-design. In *Proceedings of the Sixth International Conference on Real-Time Computing Systems and Applications*, Hong Kong, P.R. China, December 1999. Best student paper award.

18. J. El-Khoury and M. Törngren. Towards a toolset for architectural design of distributed real-time control systems. In *Proceedings of the 22nd IEEE Real-Time Systems Symposium*, London, U.K., December 2001.

19. D. Kusnadi. TrueTime in Scicos. Master's thesis ISRN LUTFD2/TFRT–5799–SE, Department of Automatic Control, Lund University, Sweden, June 2007.

20. P. Levis, N. Lee, M. Welsh, and D. Culler. TOSSIM: Accurate and scalable simulation of entire TinyOS applications. In *Proceedings of the First International Conference on Embedded Networked Sensor Systems*, pp. 126–137, Los Angeles, CA, 2003.

21. C.L. Liu and J.W. Layland. Scheduling algorithms for multiprogramming in a hard-real-time environment. *Journal of the ACM*, 20(1), 40–61, 1973.

22. P.S. Magnusson. Simulation of parallel hardware. In *Proceedings of the International Workshop on Modeling Analysis and Simulation of Computer and Telecommunication Systems (MASCOTS)*, San Diego, CA, 1993.

23. MATLAB. Homepage: http://www.mathworks.com. Visited 2008-09-30.

24. Modelica. Homepage: http://modelica.org. Visited 2008-09-30.

25. ns-2. Homepage: http://www.isi.edu/nsnam/ns. Visited 2008-09-30.

26. Martin Ohlin, Dan Henriksson, and Anton Cervin. *TrueTime 1.5—Reference Manual*, January 2007. Homepage: http://www.control.lth.se/truetime.

27. OMNeT++. Homepage: http://www.omnetpp.org. Visited 2008-09-30.

28. F. Österlind. A sensor network simulator for the Contiki OS. Technical report T2006-05, SICS – Swedish Institute of Computer Science, February 2006.

29. L. Palopoli, L. Abeni, and G. Buttazzo. Real-time control system analysis: An integrated approach. In *Proceedings of the 21st IEEE Real-Time Systems Symposium*, Orlando, FL, December 2000.

30. A. Panousopoulou and A. Tzes. Utilization of mobile agents for Voronoi-based heterogeneous wireless sensor network reconfiguration. In *Proceedings of the European Control Conference (ECC)*, Kos, Greece, 2007.

31. C.E. Perkins and E.M. Royer. Ad-hoc on-demand distance vector (AODV) routing. In *Proceedings of the Second IEEE Workshop on Mobile Computing Systems and Applications*, New Orleans, LA, 1999.

32. RUNES—Reconfigurable Ubiquitous Networked Embedded Systems. Homepage: http://www.ist-runes.org. Visited 2008-09-30.

33. Scilab. Homepage: http://www.scilab.org. Visited 2008-09-30.

34. F. Singhoff, J. Legrand, L. Nana, and L. Marcé. Cheddar: A flexible real time scheduling framework. *ACM SIGAda Ada Letters*, 24(4), 1–8, 2004.

35. M.F. Storch and J.W.-S. Liu. DRTSS: A simulation framework for complex real-time systems. In *Proceedings of the Second IEEE Real-Time Technology and Applications Symposium*, Boston, MA, 1996.

36. H.-Y. Tyan. Design, realization and evaluation of a component-based compositional software architecture for network simulation. PhD thesis, Ohio State University, 2002.

37. B. Zurita Ares, C. Fischione, A. Speranzon, and K.H. Johansson. On power control for wireless sensor networks: Radio model, software implementation and experimental evaluation. In *Proceedings of the European Control Conference (ECC)*, Kos, Greece, 2007.

Part II

Design Tools and Methodology for Multiprocessor System-on-Chip

7

MPSoC Platform Mapping Tools for Data-Dominated Applications

Pierre G. Paulin, Olivier Benny, Michel Langevin, Youcef Bouchebaba,
Chuck Pilkington, Bruno Lavigueur, David Lo, Vincent Gagne, and
Michel Metzger

CONTENTS

7.1 Introduction

The current deep submicron technology era—as it applies to low-cost, high-volume consumer digital convergence products—presents two opposing challenges: rising system-on-chip (SoC) platform development costs and shorter product market windows. Compounding the problem is the rate of change due to evolving specifications and the appearance of multiple standards that need to be incorporated into a single platform.

There are three main causes to the rising SoC platform development costs. The first is the continued rise in gate and memory count. Today's SoCs can have over 100 million transistors—enough to theoretically place the logic of over one thousand 32 bit RISC processors on a single die. Leveraging these capabilities is a major challenge.

The second cause is the increased complexity of dealing with deep submicron effects. These include electro-migration, voltage-drop, and on-chip variations. These effects are having a dampening impact on design productivity. Also, rising mask set costs—currently over one million dollars—compound the problem, and present a nearly insurmountable financial market entry barrier for smaller companies.

The third cause is the rising embedded software development cost in current generation SoCs, driven by an accelerated rate of new feature introduction. This is partly because of the convergence of computing, consumer, and communications domains that implies supporting a broader range of functionalities and standards for a wide set of geographic markets. While the growth of hardware complexity in SoCs has tracked Moore's law, with a resulting growth of 56% in transistor count per year, industry studies [22] show that the complexity of embedded S/W is rising at a staggering 140% per year. This software now represents over 50% of development costs in most SoCs and over 75% in emerging multiprocessor SoC (MP-SoC) platforms.

As a result, the significant investment to develop the platform—typically between 10M$ and 100M$ for *today's* 65 nm platforms—requires to maximize the *time-in-market* for a given platform. On the other hand, the consumer-led product cycles imply increasingly shorter *time-to-market* for the applications supported by the platform.

Finally, customers of a given SoC platform increasingly request to add their own value-added features as a market differentiator. These features are not just superficial additions, such as human-interface and top-level control code. For example, a SoC platform customer may have proprietary multimedia-oriented enhancements that they want to include in the platform (e.g., image noise reduction, face recognition, etc.).

All of these factors lead to the need for a domain-specific flexible platform that can be reused across a wide range of application variants. In addition, time-to-market considerations mean that the platform must come with high-level application-to-platform mapping tools that increase developer productivity. Both of these requirements point in the direction of highly S/W programmable platform solutions. A wide range of general-purpose and domain-specific cores exist and they come with powerful compilation, debug, and analysis tools. This makes them a key component of the flexible SoC of the future.

From the above market trends, it is clear that multiprocessor-based platforms will play a key role. Of course, delivering this flexibility cannot be achieved at any cost or power. In mobile multimedia products, typical power targets for SoCs used in battery-powered products are a few hundred milliwatts [11]. This suggests the use of domain-optimized heterogeneous MP-SoC platforms that will embody a rich mix of general-purpose processor cores, domain- and application-specific processor cores, and H/W processing elements (PEs) to deliver a solution at a competitive cost and power.

A key question is therefore how to effectively exploit this type of platform. We need to tackle this challenge from three main directions:

1. The development of high-level platform programming models
2. The development of effective platform mapping technologies
3. The design of parallel platforms that support the programming models and facilitate the development of the platform mapping tools

This chapter focuses primarily on the first two objectives.

7.1.1 Platform Programming Models

A SoC platform programming model is an abstraction of a heterogeneous system consisting of a range of loosely and tightly coupled processors, local and shared memory, communication channels, various hardware accelerators, and input/output (I/O). A platform programming model must both hide and expose the functionalities offered by the platform. It must hide the heterogeneity of the underlying PEs, the heterogeneity of the tools used to program these PEs, and abstract the low-level communication mechanisms between the PEs, the storage elements, and I/O blocks.

However, the programming model should also expose some top-level characteristics of the underlying platform. It needs to capture the type of high-level parallelism supported by the platform. This is because most platforms are designed to naturally support one main class of high-level programming models. For example, symmetric multiprocessing using shared memory, message-passing, or streaming.

Moreover, in the domain of MP-SoCs, the programming model should not only abstract the programmable processors, it should also allow the exploitation of the abstract functionalities provided by all types of platform components including H/W blocks, communication channels, storage components, and I/O. Figure 7.1 illustrates the programming model as the boundary between the high-level application description and the underlying heterogeneous platform.

We believe that at least three classes of platform programming models are needed:

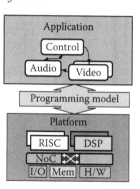

1. A symmetric multiprocessor (SMP) model, in the spirit of Unix POSIX threads [15]. This programming model relies on symmetric processing resources that access a shared memory.
2. A distributed client–server programming model, in the spirit of CORBA [16] or DCOM [17]. In this approach, applications are encapsulated into well-defined components with explicit interfaces. It relies on an abstract message-passing communication scheme where all communication between parallel application components is explicit.

FIGURE 7.1

Application, platform, and programming model.

3. A dataflow-oriented streaming programming model, as illustrated by StreamIt [3] and Brooks [2]. As with the client–server model, this approach encapsulates applications into well-defined S/W components, but implements a dataflow-driven static or dynamic communication semantic. Control is typically fairly simple.

Table 7.1 summarizes the main advantages and drawbacks of these three programming models.

- In the SMP model, the application is organized as a set of processes that share a common operating system (OS) and memory. This model provides the support of current OSs and facilitates the use of legacy code. Moreover, some form of load balancing of resources is usually supported. However, the data coherency has to be maintained. This typically involves expensive cache coherency hardware. In data-dominated applications, this programming model implies high data bandwidth for inter-processor communication unless data movement is controlled carefully. By definition, it is designed for symmetric systems and is hardly applicable for heterogeneous processing resources. In practical implementations of SMP platforms, scalability is limited between two and eight processors.
- In the client–server model, the application is organized as a set of clients and servers; the client makes a service request from the server that fulfills the request. Generally, an object request broker (ORB) acts as an agent between the client request and the completion of this request. This model is appropriate for heterogeneous systems and control-oriented applications and it presents a good potential for scaling and load balancing. However, the client–server model requires data marshaling—the process of gathering data and transforming it into a standard format before it is transmitted over a network—so that the data can transcend network boundaries [8]. This generalization of

TABLE 7.1

Programming Models for MPSoCs

Programming Model	Advantages	Drawbacks
SMP	Natural support of current OS Legacy code support Load balancing	Need to maintain coherence of local, shared data High inter-processor data communication bandwidth Limited scalability No support for heterogeneous systems
Client–server	Supports heterogeneous systems Potential for scaling and load balancing Good support for control-oriented application	Marshalling problem Heavy infrastructure Lack of streamlining
Streaming	Low overhead communications Reduced data bandwidth on communication channels Orthogonal communication and computation Easy to estimate the communication requirements of the application	Timing of control and data Poor support for control-oriented applications

the communication adds to the complexity of the supporting infrastructure and implies some performance overhead.

- In comparison with the client–server and SMP models, the streaming programming model provides poor support for control-oriented computation, and the timing of control and data is difficult. However, this model is more suitable for data-oriented applications. The streaming model enables low overhead communications and the reduction of data bandwidth. Moreover, communication and computation are orthogonal and by analyzing the communication edges in a stream computation, it is possible to obtain precise estimates of the communication requirements for a given application. This greatly simplifies analysis and mapping of application onto parallel architectures [1].

In summary, there is a continuum of characteristics that need to be considered when moving between SMP on one end, client–server in the middle, and streaming on the other end. SMP is the most preferred general-purpose model, it is relatively user-friendly, but this ease of use is at the expense of predictability, performance, and cost. At the opposite end of the continuum, streaming is a more constrained, predictable, and understandable model, but is more specialized toward dataflow and requires more time to express and optimize. The client–server programming model is more general-purpose than streaming, and expresses control applications better. However, automatic load balancing can imply high-communication bandwidth between PEs.

Each of these programming models have their advantages and inconveniences, and we have found that, for the consumer style multimedia and communications SoC platforms we have been working with, we need to use all three—sometimes making use of more than one for a single platform, often in a tightly coupled, interoperable fashion. Due to the tight constraints in the design of MP-SoCs, the designers have to choose the appropriate programming model(s) in order to develop their applications on a particular platform or subsystem.

7.1.1.1 Explicit Capture of Parallelism

A key assumption made here—for all three programming models, as we have defined them—is that the application developer is responsible for identifying and explicitly expressing parallelism. However, in our experience for domain-specific application code in communications, imaging, video, and audio, this is a reasonable assumption. Parallelism is tractable and well understood in many cases. Moreover, designers have been dealing with this type of parallelism in hardware-based platforms for many years. For an application such as an MPEG4 video encoder consisting of 10,000 lines of sequential C reference code, our experience has shown that the parallelization represents less than one or two person-months of work (for a person already familiar with the application and the programming model).

7.1.2 Characteristics of Parallel Multiprocessor SoC Platforms

While our research work is focused primarily on the programming models and platform mapping tools, the characteristics of the target MP-SoC platform have a significant impact on the complexity of the mapping problem, and the efficiency of the end results. From an idealistic mapping tools-only perspective, the MP-SoC platforms would embed a homogeneous set of general-purpose RISC-style processors. This is not realistic for the foreseeable future [20]:

- Domain-specific cores such as DSPs offer 2X–4X performance in their domain of application via instruction specialization and wider instruction words. The combination of SIMD-style word-level parallelism can increase performance by another factor of 2X–8X in certain cases.
- Configurable ASIPs (application-specific instruction-set processors) can offer 10X–100X performance improvements via application-specific instruction sets and tightly coupled H/W coprocessors.
- Hardware coprocessors can offer 100X or more performance advantages and/or significant power and area savings. They will remain essential for highly parallel, regular operations with high data rates. In particular, for data processing operations that are fixed for an

application domain (e.g., direct and inverse discrete cosine transforms—DCT and iDCT—used in video processing).

- Legacy code and general-purpose OS support will often dictate the host processor for the platform. The data representation used in this processor is not likely to be compatible with the parallel processor subsystems, or the hardware coprocessors.
- Some application tasks will not be parallelizable; therefore, fast general-purpose cores will be necessary to support these.

As a result, we believe that a performance and power effective platform for the consumer-dominated convergence platforms will be composed of a heterogeneous composition of the following PE types:

- A medium to high-performance, general-purpose RISC core, typically running a standard general-purpose OS. Increasingly, this host system will consist of a two to four core SMP cluster, as they appear in the marketplace. All the top-level control code will run here. Legacy code that is not performance critical will also run on this processor. Finally, customer-specific developments and controlled access to the domain-specific parallel subsystems will usually occur via this general-purpose processor and OS pair.
- Domain-specific subsystems composed of mostly homogeneous, lightweight multiprocessor clusters. Although homogeneous, the instruction-set of these processors will typically be optimized toward a broad application domain (e.g., video codec, image quality improvement, wireless communications, and 3D graphics).
- Tightly coupled hardware PEs for domain-specific data processing functions.
- Domain-specific I/O blocks, which are becoming increasingly flexible.

7.2   MultiFlex Platform Mapping Technology Overview

This section introduces the MultiFlex technology, which supports the mapping of user-defined parallel applications, expressed in one or more programming models, onto a MP-SoC platform.

The support in MultiFlex of a lightweight SMP programming model was described in [12]. This uses a hardware-assisted concurrency engine to support small grain parallelism dynamically.

In MultiFlex, the client–server programming model is referred to as "DSOC" (Distributed System Object Component), and was also described in [12]. This toolset supports static and dynamic load balancing and supports heterogeneous PEs with potentially different data representations. Dynamic load balancing is achieved using either a lightweight S/W-based kernel to dynamically schedule large-grain tasks, or a hardware-assisted object request broker (HORBA) when the support of small-grain parallelism is needed.

Our most recent developments in MultiFlex are mostly focused on the support of the streaming programming model, as well as its interaction with the client–server model. SMP subsystems are still of interest, and they are becoming increasingly well supported commercially [14,21]. Moreover, our focus is on data-intensive applications in multimedia and communications. For these applications, our focus has been primarily on streaming and client–server programming models for which explicit communication centric approaches seem most appropriate.

This chapter will introduce the MultiFlex framework specialized at supporting the streaming and client–server programming models. However, we will focus primarily on our recent streaming programming model and mapping tools.

7.2.1 Iterative Mapping Flow

MultiFlex supports an iterative process, using initial mapping results to guide the stepwise refinement and optimization of the application-to-platform mapping. Different assignment and scheduling strategies can be employed in this process.

An overview of the MultiFlex toolset, which supports the client–server and streaming programming models, is given in Figure 7.2. The design methodology requires three inputs:

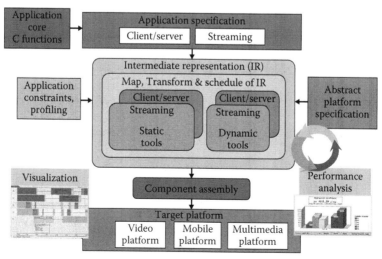

FIGURE 7.2
MultiFlex toolset overview.

- The application specification—the application can be specified as a set of communicating blocks; it can be programmed using the streaming model or client–server programming model semantics.
- Application-specific information (e.g., quality-of-service requirements, measured or estimated execution characteristics of the application, data I/O characteristics, etc.).
- The abstract platform specification—this information includes the main characteristics of the target platform which will execute the application.

An intermediate representation (IR) is used to express the high-level application in a language-neutral form. It is translated automatically from one or more user-level capture environments. The internal structure of the application capture is highly inspired by the Fractal component model [23]. Although we have focused mostly on the IR-to-platform mapping stages, we have experimented with graphical capture from a commercial toolset [7], and a textual capture language similar to StreamIt [3] has also been experimented with.

In the MultiFlex approach, the IR is mapped, transformed, and scheduled; finally the application is transformed into targeted code that can run on the platform. There is a flexibility or performance trade-off between what can be calculated and compiled statically, and what can be evaluated at runtime. As shown on Figure 7.2, our approach is currently implemented using a combination of both, allowing a certain degree of adaptive behaviors, while making use of more powerful offline static tools when possible. Finally, the MultiFlex visualization and performance analysis tools help to validate the final results or to provide information for the improvement of the results through further iterations.

7.2.2 Streaming Programming Model

As introduced above, the streaming programming model [1] has been designed for use with data-dominated applications. In this computing model, an application is organized into streams and computational kernels to expose its inherent locality and concurrency. Streams represent the flow of data, while kernels are computational tasks that manipulate and transform the data. Many data-oriented applications can easily be seen as sequences of transformations applied on a data stream. Examples of languages based on the streaming computing models are: ESTEREL [4], Lucid [5], StreamIt [3], Brooks [2]. Frameworks for stream computing visualization are also available (e.g., Ptolemy [6] and Simulink® [7]).

In essence, our streaming programming model is well suited to a distributed-memory, parallel architecture (although mapping is possible on shared-memory platforms), and favors an implementation using software libraries invoked from the traditional sequential C language, rather than proposing language extensions, or a completely new execution model.

The entry to the mapping tools uses an XML-based IR that describes the application as a topology with semantic tags on tasks. During the mapping process, the semantic information is used to generate the schedulers and all the glue necessary to execute the tasks according to their firing conditions.

In summary, the objectives of the streaming design flow are:

- To refine the application mapping in an iterative process, rather than having a one-way top-down code generation
- To support multiple streaming execution models and firing conditions
- To support both restricted synchronous data-flow and more dynamic data-flow blocks
- To be controlled by the user to achieve the mechanical transformations rather than making decisions for him

We first present the mapping flow in the Section 7.3, and at the end of the section, we will give more details on the streaming programming model.

7.3   MultiFlex Streaming Mapping Flow

The MultiFlex technology includes support for a range of streaming programming model variants. Streaming applications can be used alone or in interoperation with client–server applications. The MultiFlex streaming tool flow is illustrated in Figure 7.3. The different stages of this flow will be described in the next sections.

The application mapping begins with the assignment of the application blocks to the platform resources. The IR transformations consist mainly in splitting and/or clustering the application blocks; they are performed for optimization purposes (e.g., memory optimization); the transformations also imply the insertion of communication mechanisms (e.g., FIFOs, and local buffers).

The scheduling defines the sharing of a processor between several blocks of the application. Most of the IR mapping, transforming, and scheduling is realized statically (at compilation time), rather than dynamically (at run-time).

The methodology targets large-scale multicore platforms including a uniform layered communication network based on STMicroelectronics' network-on-chip (NoC) backbone infrastructure [18] and a small number of H/W-based communication IPs for efficient data transfer (e.g., stream-oriented DMAs or message-passing accelerators [9]). Although we consider our methodology to be compatible with the integration of application-specific hardware accelerators using high-level hardware synthesis, we are not targeting such platforms currently.

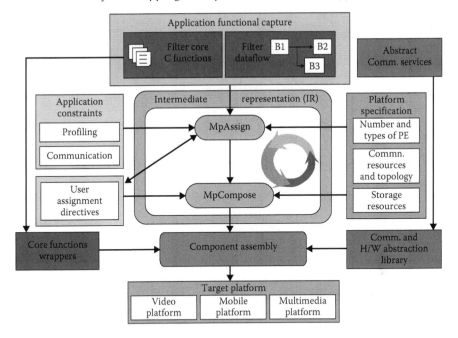

FIGURE 7.3
MultiFlex tool flow for streaming applications.

7.3.1   Abstraction Levels

In the MultiFlex methodology, a data-dominated application is gradually mapped on a multicore platform by passing through several abstractions:

- The application level—at this level, the application is organized as a set of communicating blocks. The targeted architecture is completely abstracted.
- The partitioning level—at this level the application blocks are grouped in partitions; each partition will be executed on a PE of the target architecture. PEs can be instruction-set programmable processors, reconfigurable hardware or standard hardware.
- The communication level—at this level, the scheduling and the communication mechanisms used on each processor between the different blocks forming a partition are detailed.
- The target architecture level—at this level, the final code executed on the targeted platforms is generated.

Table 7.2 summarizes the different abstractions, models, and tools provided by MultiFlex in order to map complex data-oriented applications onto multiprocessor platforms.

TABLE 7.2
Abstraction, Models, and Tools in MultiFlex

Abstraction Level	Model	Refinement Tool
Application level	Set of communicating blocks	Textual or graphical front-end
Partition level	Set of communicating blocks and directives to assign blocks to processors	MpAssign
Communication level	Set of communicating blocks and required communication components	MpCompose
Target architecture level	Final code loaded and executed on the target platform	Component-based compilation back-end

7.3.2 Application Functional Capture

The application is functionally captured as a set of communicating blocks. A *basic (or primitive) block* consists of a behavior that implements a known interface. The implementation part of the block uses streaming application programming interface (API) calls to get input and output data buffers to communicate with other tasks. Blocks are connected through communication channels (in short, channels) via their interfaces. The basic blocks can be grouped in *hierarchical blocks* or *composites*.

The main types of basic blocks supported in MultiFlex approach are

- Simple data-flow block: This type of block consumes and produces tokens on all inputs and outputs, respectively, when executed. It is launched when there is data available at all inputs, and there is sufficient free space in downstream components for all outputs to write the results.
- Synchronous client–server block: This block needs to perform one or many remote procedural calls before being able to push data in the output interface. It must therefore be scheduled differently than the simple data-flow block.
- Server block: This block can be executed once all the arguments of the call are available. Often this type of block can be used to model a H/W coprocessor.
- Delay memory: This type of block can be used to store a given number of data tokens (an explicit state).

Figure 7.4 gives the graphical representation of a streaming application capture which interacts with a client–server application. Here, we focus mostly on streaming applications.

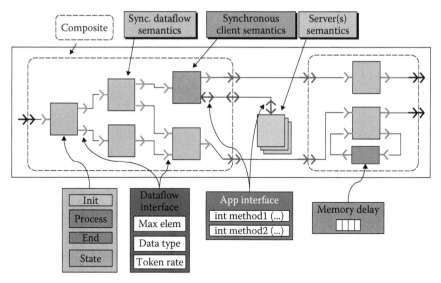

FIGURE 7.4
Application functional capture.

From the point of view of the application programmer, the first step is to split the application into processing blocks with buffer-based I/O ports. User code corresponding to the block behavior is written using the C language. Using component structures, each block has its private state, and implements a constructor (init), a work section (process), and a destructor (end). To obtain access to I/O port data buffers, the blocks have to use a predefined API. A run-to-completion execution model is proposed as a compromise between programming and mapping flexibility. The user can extend the local schedulers to allow the local control of the components, based on application-specific control interfaces. The dataflow graph may contain blocks that use client–server semantics, with application-specific interfaces, to perform remote object calls that can be dispatched to a pool of servers.

7.3.3 Application Constraints

The following application constraints are used by the MultiFlex streaming tools:

1. Block profiling information. For a given block, this represents the average number of clock cycles required for the block execution, on a target processor.
2. Communication volume: the size of data exchanged on this channel.
3. User assignment directives. Three types of directives are supported by the tool:
   a. Assign a block to a specific processor
   b. Assign two blocks to the same processor (can be any processor)
   c. Assign two blocks to any two different processors

7.3.4   The High-Level Platform Specification

The high-level platform specification is an abstraction of the processing, communication, and storage resources of the target platform. In the current implementation, the information stored is as follows:

- Number and type of PEs.
- Program and data memory size constraints (for each programmable PE).
- Information on the NoC topology. Our target platform uses the STNoC, which is based on the "Spidergon" topology [18]. We include the latency measures for single and multihop communication.
- Constraints on communication engines: Number of physical links available for communication with the NoC.

7.3.5   Intermediate Format

MultiFlex relies on intermediate representations (IRs) to capture the application, the constraints, and high-level platform descriptions. The topology of the application—the block declaration and their connectivity—is expressed using an XML-based intermediate format. It is also used to store task annotations, such as the block execution semantics. Other block annotations are used for the application profiling and block assignments. Edges are annotated with the communication volume information.

The IR is designed to support the refinement of the application as it is iteratively mapped to the platform. This implies supporting the multiple abstraction levels involved in the assignment and mapping process described in the next sections.

7.3.6   Model Assumptions and Distinctive Features

In this section, we provide more details about the streaming model. This background information will help in explaining the mapping tools in the next section.

The task specification includes the data type for each I/O port as well as the maximum amount of data consumed or produced on these ports. This information is an important characteristic of the application capture because it is at the foundation of our streaming model: each task has a known computation grain size. This means we know the amount of data required to fire the process function of the task for a single iteration without starving on input data, and we know the maximum amount of output data that can be produced each time. This is a requirement for the nonblocking, or run-to-completion execution of the task, which simplifies the scheduling and communication infrastructure and reduces the system overhead. Finally, we can quantify the computation requirements of each task for a single iteration.

The run-to-completion execution model allows dissociating the scheduling of the tasks from the actual processing function, providing clear scheduling points. Application developers focus on implementing and optimizing the task functions (using the C language), and expressing the functionality in a way that is natural for the application, without trying to balance the task loads in the first place. This means each task can work on a different data packet size and have different computation loads. The assignment and scheduling of the tasks can be done in a separate phase (usually performed later), allowing the exploration of the mapping parameters, such as the task assignment, the FIFO, and buffer sizes, to be conducted without changing the functionality of the tasks: a basic principle to allow correct-by-construction automated refinement.

The run-to-completion execution model is a compromise, requiring more constrained programming but leads to higher flexibility in terms of mapping. However, in certain cases, we have no choice but to support multiple concurrent execution contexts. We use cooperative threading to schedule special tasks that use a mix of streaming and client–server constructs. Such tasks are able to invoke remote services via client–server (DSOC) calls, including synchronous methods (with return values) that cause the caller task to block, waiting for an answer.

In addition, we are evaluating the pros and cons of supporting tasks with unrestricted I/O and very fine-grain communication. To be able to eventually run several tasks of this nature on the same processor, we may need a software kernel or make use of hardware threading if the underlying platform provides it.

To be able to choose the correct scheduler to deploy on each PE, we have introduced *semantic tags*, which describe the high-level behavior type of each task. This information is stored in the IR. We have defined a small set of task types, previously listed in Section 7.3.2. This allows a mix of execution models and firing conditions, thus providing a rich programming environment. Having clear semantic tags is a way to ensure the mapping tools can optimize the scheduling and communications on each processor, rather than systematically supporting all features and be designed for the worst case.

The nonblocking execution is only one characteristic of streaming compared to our DSOC client–server message-passing programming model. As opposed to DSOC, our streaming programming model does not provide data marshaling (although, in principle, this could be integrated in the case of heterogeneous streaming subsystems).

When compared to asynchronous concurrent components, another distinction of the streaming model is the data-driven scheduling. In event-based programming, asynchronous calls (of unknown size) can be generated during the execution of a single reaction, and those must be queued. The quantity of events may result in complex triggering protocols to be defined and implemented by the application programmer. This remains to be a well-known drawback of event-based systems. With the data-flow approach, the clear data-triggered execution semantic, and the specification of I/O data ports resolve the scheduling, memory management, and memory ownership problems inherent to asynchronous remote method invocations.

Finally, another characteristic of our implementation of the streaming programming model, which is also shared with our SMP and DSOC models, is the fact that application code is reused "as is," i.e., no source code transformations are performed. We see two beneficial consequences of this common approach. In terms of debugging, it is an asset, since the programmer can use a standard C source-level debugger, to verify the unmodified code of the task core functions. The other main advantage is related to profiling. Once again, it is relatively easy for an application engineer to understand and optimize the task functions with a profiling report, because his source code is untouched.

7.4   MultiFlex Streaming Mapping Tools

7.4.1   Task Assignment Tool

The main objective of the MpAssign tool (see Figure 7.5) is to assign application blocks to processors while optimizing two objectives:

1. Balance the task load on all processors
2. Minimize the inter-processor communication load

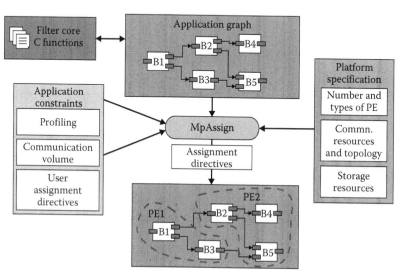

FIGURE 7.5
MpAssign tool.

The inter-processor communication cost is given by the data volume exchanged between two processors, related to each task.

The tool receives as inputs the application capture, the application constraints, and the high-level platform specification. The output of the tool is a set of assignment directives specifying which blocks are mapped on each processor, the average load of each processor, and the cost for each inter-processor communication. The lower portion of Figure 7.5 gives a visual representation of the MpAssign output. The tool provides the visual display of the resulting block assignments to processors.

The implemented algorithm for the MpAssign tool is inspired from Marculescu's research [10] and is based on graph traversal approaches, where ready tasks with maximal 2-minimal cost-variance are assigned iteratively. The two main graph traversal approaches implemented in MpAssign are

- The list-based approach, using mainly the breadth-first principle—a task is ready if all its predecessors are assigned
- The path-based approach, using mainly the depth-first principle—a task is ready if one predecessor is assigned and it is on the critical path

A cost estimator $C(t,p)$ of assigning a task t on processor p is used. This cost estimator is computed using the following equation:

$$C(t,p) = w_1 * C_{proc} + w_2 * C_{comm} + w_3 * C_{succ} \qquad (7.1)$$

where

C_{proc} is the additional average processing cost required when the task t is assigned to processor p

C_{comm} is the communication cost required for the communication of task t with the preceding tasks

C_{succ} represents a look-ahead cost concerning the successor tasks, the minimal cost estimate of mapping a number of successor tasks

This assumes state space exploration for a predefined look-ahead depth. w_i represents the weight associated with each cost factor (C_{proc}, C_{comm}, and C_{succ}) and indicates the significance of the factor in the total cost $C(p,t)$ as compared with the other factors. The factors are weighted by the designer to set their relative importance.

7.4.2   Task Refinement and Communication Generation Tools

The main objective of the MpCompose tool (see Figure 7.6) is to generate one application graph per PE, each graph containing the desired computation blocks from the application, one local scheduler, and the required communication components. To perform this functionality, MpCompose requires the following three inputs:

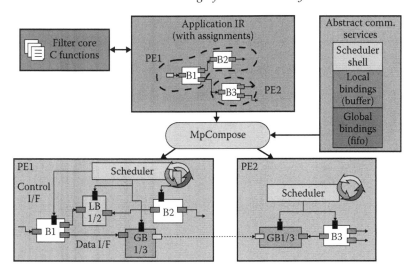

FIGURE 7.6
MpCompose tool.

- The application capture
- The platform description
- The set of directives, optionally generated by the MpAssign tool

The MpCompose tool relies on a library of abstract communication services that provide different communication mechanisms that can be inserted in the application graph. Three types of services are currently supported by MpCompose:

1. Local bindings consisting mainly of a FIFO implemented with memory buffers and enabling the intra-processor communication (e.g., block B1 is connected to block B2 via local buffer LB1/2).
2. Global binding FIFOs, which enable the inter-processor communication (e.g., block B1 on PE1 communicates to block B3 on PE2 via external buffers GB1/3).
3. A scheduler on each PE, which is configurable in terms of number and types of blocks and which enables the sharing of a processor between several application blocks.

A set of libraries are used to abstract part of the platform and provide communication and synchronization mechanism (point-to-point communication, semaphores, access to shared memory, access to I/O, etc.). The various FIFO components have a default depth, but these are configuration values that can be changed during the mapping. Since we support custom data types for I/O port tokens, each element of a FIFO has a certain size that matches the data type and maximum size specified in the intermediate format.

There is no global central controller: a local scheduler is created on each processor. This component is the main controller and has access to the control interface of all the components it is responsible for scheduling. The proper control interface for each filter task is automatically added, based on the type of filter specified in the application IR, and connected to the scheduler. The implementations of the schedulers are partly generated, for example, the list of filter tasks (a static list) and some setup code for the hardware communication accelerators are automatically created. The core scheduling function can be pulled from a library or customized by the application programmer.

The output of MpCompose is a set of component descriptions; one for each processor. From the point of view of the top-level component definitions, these components are not connected together; however, communicating processors use the platform-specific features to actually implement the buffer-based communication at runtime. The set of independent component definitions allow a monoprocessor component–based infrastructure to be used for compilation.

7.4.3 Component Back-End Compilation

Starting from the set of processor graphs, the component back-end generates the targeted code that can run on the platform. MultiFlex tools currently target the Fractal component model, and more specifically its C implementation [19]. Even though this toolset supports features such as a binding controller and a life cycle manager to allow dynamic insertion–removal of components in the graph at runtime, we are not currently using any of the dynamic features of components, such as runtime elaboration, introspection, etc., mainly for code size reasons. Nevertheless, we expect multimedia application requirements to push toward this direction. Until then, we mainly use the component model as a back-end to represent the software architecture to be built on each processor. MpCompose generates one architecture (.fractal) file describing the components and their topology for each CPU. The Fractal tools will generate the required C glue code to bind components, to create block instance structures and will compile all the code into an executable for the specified processor by invoking the target cross-compiler. This build process is invoked for each PE, thus producing a binary for each processor.

7.4.4 Runtime Support Components

The main services provided by the MultiFlex components at runtime are scheduling and communication. The scheduler in fact controls both the communication components and the application tasks.

The scheduler interleaves communication and processing at the block level. For each input port, the scheduler scans if there is available data in the local memory. If not, it checks if the input FIFO is empty. If not, the scheduler orders the input FIFO to perform the transfer into local memory. This is typically done by some coprocessors such as DMA or specialized hardware communication engines. While the transfer occurs, the scheduler can manage other tasks. In the same manner, it can look for previously produced output data ready to be transmitted from local memory to another processor, using an output FIFO. Tasks with more dynamic (data-dependant) behaviors may produce less data than their allowed maximum, including no data at all. If a task is ready to execute, the scheduler simply calls its process function in the same context. The user tasks make use of an API that is based on pointers, thus we avoid data copies between the tasks and the local queues managed by the scheduler.

So in a nutshell, the run-to-completion model allows the scheduler to run ready tasks, manage input and output data consumed or produced by the tasks, while allowing data transfers to take place in parallel, thus overlapping communication and processing without the need for threading. The tasks can have different computation and communication costs: the mapping tools will help to balance the overall tasks load between processors, with the objective to keep the streaming fabric busy and the latency minimized.

7.5   Experimental Results

7.5.1   3G Application Mapping Experiments

In this section, we present mapping results using the MpAssign tool on an application graph having the characteristic of a 3G WCDMA/FDD base-station application from [13].

The block diagram of this application is presented in Figure 7.7 and contains two main chains: transmitter (tx) and receiver (rx). The blocks

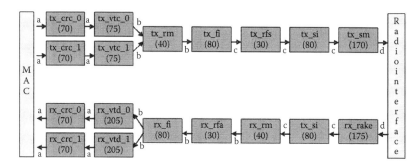

FIGURE 7.7

3G application block diagram. The communication volumes are $a = 260$, $b = 3136$, $c = 1280$, and $d = 768$.

are annotated with numbers that represent the estimated processing load, while each edge has an estimated communication volume given in the figure caption. These numbers are extracted from [13], where the computation cost corresponds to a latency (in microseconds) for a PE to execute one iteration of the corresponding functional block, while the edge cost corresponds to the volume of data (in 16-bit words) transferred at each iteration between the connected functional blocks.

A manual and static mapping of this application is presented in [13], using a 2D-mesh of 46 PEs, where each PE is executing only one of the functional blocks, for which some of them are duplicated to expose more potential parallel processing. We use this example in this chapter mainly for illustrative purposes, to show that MpAssign can be used to explore automatically different mappings where, optionally, multiple functional blocks can be mapped on the same PE to balance the processing load. To expose more potential parallel processing, we create a set of functionally equivalent application graphs of the above reference application in which we duplicate the transmitter and receiver processing chains several times. In our experiments, four versions have been explored:

- v1: 1 transmitter and 1 receiver (original reference application)
- v2: 2 transmitters and 2 receivers
- v3: 3 transmitters and 3 receivers
- v4: 4 transmitters and 4 receivers

The version v1 will be mapped on a 16 processor architecture (v1/16). The version v2 will be mapped on a 16 processor architecture (v2/16) and a 32 processor architecture (v2/32). The version v3 will be mapped on a 32 processor architecture (v3/32) and a 48 processor architecture (v3/48). The version v4 will be mapped on a 48 processor architecture (v4/48). This results in six different mapping configurations (v1/16, v2/16, v2/32, v3/32, v3/48, v4/48) to explore.

For the experiments, we suppose that each PE can execute any of the functional blocks, and that the NoC connecting all the PEs is the STMicroelectronics Spidergon [18].

As described in section "Task Assignment Tool," our mapping heuristic allows exploring different solutions in order to find a good compromise between communication and PE load balancing. These different solutions can be obtained by varying the parameters w1, w2, and w3 (see Equation 7.1): A high value of w1 promotes solutions with good load balancing, while a high value of w2 promotes solutions with minimal communications. The parameter w3, which favors the selection based on an optimistic look-ahead search, will be fixed at 100. For our experiments three combinations of w1 and w2 will be studied:

- c1 (w1 = 1000, w2 = 10): This weight combination tends to maximize the load balancing.
- c2 (w1 = 100, w2 = 100): This weight combination tends to balance load and communications.
- c3 (w1 = 10, w2 = 1000): This weight combination tends to minimize the communications.

Each of the six configurations described above will be tested with these three weight parameter combinations, which results in a total of 18 experiments. For each experiment, we will extract the following statistics:

- Load variance (LV), given by Equation 7.2, where for each mapping solution x, load (PE_i) is the sum of the task costs assigned to the PE_i, avgload is the average load defined by the sum of all task costs divided by the number of PE, and p is the PE number.

$$LV(x) = \sqrt{\frac{\sum_{i=0}^{p-1} \left(load(PE_i) - avgload\right)^2}{p}} \tag{7.2}$$

- Maximal load (ML) defined as max(load(PE_i)), where $0 < i < p - 1$.
- Total communication (TC) given by the sum of each edge cost times the NoC distance of the route related to that edge.
- Maximal communication (MC) the maximum communication cost found between any of two PE.

The LV statistic gives an approximation of the quality of the load balancing. The ML statistic is related to the lower-bound of the application performance after mapping, since the application throughput depends on the slowest PE processing. The MC statistic gives as well a lower-bound on the application performance, but this time with respect to the worst case of communication contention (instead of w.r.t. processing in the case of ML). Finally, the TC indicator gives an approximation of the quality of the communication mapping.

Figure 7.8 shows the resulting LV statistic of the different application configurations and mapping weight combinations. The best results are given by the mapping weight combination c1. This is predictable because c1 promotes solutions with a good load balancing, which means a low LV value.

Figure 7.9 presents the resulting ML statistics of the different application configurations and mapping weight combinations. Following the same logic as with Figure 7.8, the best results here are given by the mapping weight combination c1.

Figure 7.10 presents the resulting TC statistic of the different application configurations and mapping weight combinations. This time, the best results are given by the mapping weight combination c3. This is predictable because c3 promotes solutions with low communication costs.

Figure 7.11 presents the resulting MC statistic for the different application configurations and mapping weight combinations. Contrary to Figure 7.10,

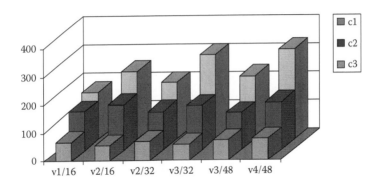

FIGURE 7.8
Load variance.

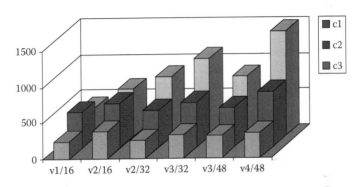

FIGURE 7.9
Maximal load (at any PE).

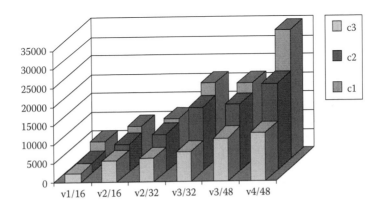

FIGURE 7.10
Total communications.

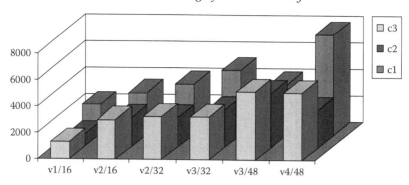

FIGURE 7.11
Maximal communication (through any PE).

the best results are given by the mapping weight combination c2. Since the tool is not trying to optimize this statistic, it appears that when optimizing either for load balancing or TC, the given obtained solution may have a worst case communication contention at a PE. This is one aspect of the mapping heuristics that needs improvement.

These results show that the selection of the final task assignment solution really depends on the target performance, architecture hardware budget, and acceptable bandwidth for communications. Nevertheless, the MpAssign tool, by generating an interesting subset of mapping solutions, allows architects to concentrate on the more detailed and time consuming analysis, rather than trying to find task assignment solutions. At this level, the various costs remain estimates based on platform and application abstractions and assumptions. For a candidate solution, the refinement can continue down to the target architecture level.

7.5.2 Refinement and Simulation

For a given solution, MpAssign provides an output text file that contains the task assignment directives, the resulting average load of each processor and the cost for each inter-processor communication. The mapping results are also available in a graphical representation. For the purpose of a simpler display, we have created a mapping example of only eight processors that is shown in Figure 7.12. We see that the dataflow source and sink blocks have been assigned to dedicated I/O processors (the first and the last respectively), thus keeping the data intensive tasks on the remaining other 6 PEs. The intent was to isolate on different processors the interesting task set that we will want to profile later on, during the simulation.

Starting from the mapping solution presented in Figure 7.12, MpCompose uses those task assignment directives to perform the software synthesis. Final compilation is carried out by the component back-end.

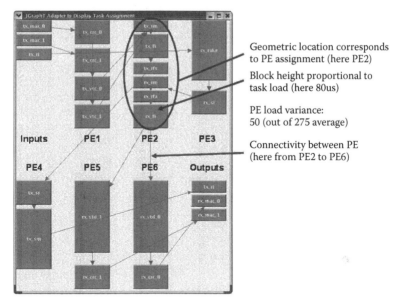

FIGURE 7.12
Sample output of MpAssign for a single solution.

The next step is the execution of the application on an instrumented virtual platform for performance analysis. For each processor, we obtain the function profiling report. By adding up the time spent in the task process functions versus the time spent in the scheduler or waiting for input data to arrive, we obtain the effective processor utilization. Task code optimization can be done orthogonally with the profiling report, in the same way as in a monoprocessor flow. However, in the MultiFlex flow, the user can update the IR with new task profiling information and rerun MpAssign to see how this can influence the suggested task assignments.

The simulations on the virtual platform should additionally provide NoC bandwidth and contention information, given by the instrumented links and routers.

Our actual multicore virtual platform is currently under development, and the accurate STNoC model is in the process of being integrated. Meanwhile, a fast and functional interconnect implementation is used instead.

7.6   Conclusions

The increasing need for flexibility in multimedia SoCs for consumer applications is leading to a new class of programmable, multiprocessor solutions. The high computation and data bandwidth requirements of these applications pose new challenges in the expression of the applications, the platform architectures to support them and the application-to-platform mapping tools.

In this chapter, we elaborated on these challenges and introduced the MP-SoC platform mapping technologies under development at STMicroelectronics called MultiFlex, with emphasis on assignment and scheduling of streaming applications. The use of these tools, integrated in a design flow that proposes a stepwise mapping refinement, was illustrated with a 3G base-station application example.

While keeping the application code unmodified, we have seen how the MultiFlex tools refine an IR of the application, based on information related to the application properties, the high-level platform characteristics, as well as user mapping constraints.

The MpAssign tool provides mapping solutions that minimize (1) the communications on the NoC and (2) the processing Load Variance. By changing the weight factors, the user can direct the heuristics to favor different classes of solutions.

For a chosen solution, the MpCompose tool provides the required intra- and inter-processor communications and local task schedulers, by instantiating generic components with the proper specific configuration.

Finally, for each processor, a self-contained component description generated by MpCompose can be given to the component compilation back-end, which takes care of implementing the low-level component glue code and invoking the compiler for final compilation and linking, ready for simulation and analysis.

This methodology supports a user-driven, iterative approach that automates the mechanical mapping transformations. Performance and profiling information obtained from a given platform mapping iteration can be exploited by the user and the mapping tools to guide the next optimization cycle.

7.6.1 Outlook

We are currently looking at alternative approaches for the implementation of the MpAssign tool, such as evolutionary algorithms, to cope with the scalability problem of list-based heuristics presented in this chapter. In fact, if we want to add several optimization goals to the algorithm, such as minimizing the memory usage, or power optimizations, it will become difficult to implement an efficient list-based algorithm.

Other areas of research include looking at how we can support finer grain streaming with scalar-based I/O (similar to StreamIt [3]), mixed with our buffer-based dataflow approach. We are also evaluating the dynamic elaboration features of the component framework, which could extend our methodology for runtime application deployment.

References

1. G. De Micheli and L. Benini, *Networks on Chip: Technology and Tools*, Morgan Kauffman, San Francisco, CA, 2006.

2. I. Buck, Brook: A Streaming Programming Language, available online at http://graphics.stanford.edu/streamlang/brook_v0.2.pdf

3. W. Thies, M. Karczmarek and S.P Amarasinghe, StreamIt: A language for streaming applications, in *Proceedings of the International Conference on Compiler Construction*, Grenoble, France, April 2002.

4. G. Berry, P. Couronne, and G. Gonthier, *Synchronous Programming of Reactive Systems: An Introduction to ESTEREL*, Elsevier, Amsterdam, the Netherlands, 1988, pp. 35–56.

5. W.W. Wadge and E.A. Ashcroft, *Lucid, the Data-Flow Programming Language*, Academic Press, New York, 1985.

6. J.T. Buck, S. Ha, E.A. Lee, and D.G. Messerschmitt, Ptolemy: A framework for simulating and prototyping heterogeneous systems, *Journal of Computer Simulation*, special issue on "Simulation Software Development," 4, 155–182, April 1994.

7. The MathWorks: Matlab and Simulink for technical computing, available on line at http://www.mathworks.com

8. Data marshalling, http://www.webopedia.com/TERM/D/data_marshalling.html

9. P. Paulin, C. Pilkington, M. Langevin, E. Bensoudane, and D. Lyonnard, A multi-processor SoC platform and tools for communications applications, in *Embedded Systems Handbook*, CRC Press, Boca Raton, FL, 2004.

10. J. Hu and R. Marculescu, Energy-aware communication and task scheduling for network-on-chip architectures under real-time constraints, in *Proceedings of DATE 2004*, Pairs, France.

11. M. Paganini, Nomadik®: A Mobile multimedia application processor platform, in *Proceedings of ASP-DAC (Asia and South Pacific Design Automation Conference)*, Yokohama, Japan, January 2007, pp. 749–750.

12. P.G. Paulin, C. Pilkington, M. Langevin, E. Bensoudane, D. Lyonnard, O. Benny, B. Lavigueur, D. Lo, G. Beltrame, V. Gagné, and G. Nicolescu, Parallel programming models or a multi-processor SoC platform applied to networking and multimedia, *IEEE Transactions on VLSI Journal*, 14(7), July 2006, 667–680.

13. D. Wiklund and D. Liu, Design, mapping, and simulations of a 3G WCD-MA/FDD base station using network on chip, in *Proceedings of the Fifth International Workshop on System-on-Chip for Real-Time Applications*, Banff, Canada, July 2005, pp. 252–256.

14. ARM Cortex-A9 MPCore, available online at http://www.arm.com/products/CPUs/ARMCortex-A9_MPCore.html

15. D.R. Butenhof, *Programming with POSIX Threads*, Addison-Wesley, Reading, MA, 1997.

16. R. Ben-Natan, *CORBA: A Guide to Common Object Request Broker Architecture*, McGraw-Hill, New York, 1995.

17. Thuan L. Thai, *Learning DCOM*, O'Reilly, Sebastopol, CA, 1999.

18. M. Coppola, Spidergon STNoC: The Communication Infrastructure for Multiprocessor Architecture, MPSoC 2008, available on line at http://www.mpsoc-forum.org/slides/2-4%20Coppola.pdf

19. M. Leclercq, O. Lobry, E. Özcan, J. Polakovic, and J.B. Stefani, THINK C implementation of fractal and its ADL tool-chain, *ECOOP 2006, 5th Fractal Workshop*, Nantes, France, July 2006.

20. P. Paulin, Emerging Challenges for MPSoC Design, MPSoC 2006, available online at http://www.mpsoc-forum.org/2006/slides/Paulin.pdf

21. The MIPS32® 34K™ processor overview, available online at http://www.mips.com/products/processors/32-64-bit-cores/mips32-34k/

22. P. Magarshack and P. Paulin, System-on-chip beyond the nanometer wall, in *Proceedings of Design Automation Conference*, Anaheim, CA, 2003, pp. 419–424.

23. E. Bruneton, T. Coupaye, and J.B. Stefani, The Fractal Component Model, 2004, specification available online at http://fractal.objectweb.org/specification/fractal-specification.pdf

8

Retargetable, Embedded Software Design Methodology for Multiprocessor-Embedded Systems

Soonhoi Ha

CONTENTS

8.1 Introduction

As semiconductor and communication technologies improve continuously, we can make very powerful embedded hardware by integrating many processing elements so that a system with multiple processing elements integrated in a single chip, called MPSoC (multiprocessor system on chip), is becoming popular. While extensive research has been performed on the

---

This chapter is an updated version of the following paper: S. Kwon, Y. Kim, W. Jeun, S. Ha, and Y Paek, A retargetable parallel-programming framework for MPSoC, *ACM Transactions on Design Automation of Electronic Systems (TODAES)*, Vol. 13, No. 3, Article 39, July 2008.

design methodology of MPSoC, most efforts have focused on the design of hardware architecture. But the real bottleneck will be software design, as pre-verified hardware platforms tend to be reused in platform-based designs. Unlike application software running on a general purpose computing system, embedded software is not easy to debug at run time. Furthermore, software failure may not be tolerated in safety-critical applications. So the correctness of the embedded software should be guaranteed at compile time.

Embedded software design is very challenging since it amounts to a parallel programming for nontrivial heterogeneous multiprocessors with diverse communication architectures and design constraints such as hardware cost, power, and timeliness. Two major models for parallel programming are the message-passing and shared address-space models. In the message-passing model, each processor has private memory and communicates with other processors via message-passing. To obtain high performance, the programmer should optimize data distribution and data movement carefully, which are very difficult tasks. The message-passing interface (MPI) [1] is a de facto standard interface of this model. In the shared address-space model, all processors share a memory and communicate data through this shared memory. The OpenMP [2] is a de facto standard interface of this model. It is mainly used for a symmetric multiprocessor (SMP) machine. Because OpenMP makes it easy to write a parallel program, there is work such as Sato et al. [3], Liu and Chaudhary [4], Hotta et al. [5], and Jeun and Ha [6] that considers the OpenMP as a parallel programming model on other parallel-processing platforms without shared address space such as system-on-chips and clusters.

While an MPI or OpenMP program is regarded as retargetable with respect to the number of processors and processor type, we consider it as *not* retargetable with respect to task partition and architecture change since the programmer should manually optimize the parallel code considering the specific target architecture and design constraints. If the task partition or communication architecture is changed, significant coding effort is needed to rewrite the optimized code. Another difficulty of programming with MPI and OpenMP is that it is the programmer's responsibility to confirm the satisfaction of the design constraints, such as memory requirements and real-time constraints, in the manually designed code.

The current practice of parallel-embedded software is multithreaded programming with lock-based synchronization, considering all target specific features. The same application should be rewritten if the target is changed. Moreover, it is well known that debugging and testing a multithreaded program is extremely difficult. Another effort of parallel programming is to use a parallelizing compiler that creates a parallel program from a sequential C code. But automatic parallelization of a C code has been successful only for a limited class of applications after a long period of extensive research [7].

In order to increase the design productivity of embedded software, we propose a novel methodology for embedded software design based on a parallel programming model, called a common intermediate code (CIC). In a CIC, the functional and data parallelism of application tasks are specified independent of the target architecture and design constraints. Information on the target architecture and the design constraints is separately described in an xml-style file, called the *architecture information file*. Based on this information, the programmer maps tasks to processing components either manually or automatically. Then, the CIC translator automatically translates the task codes in the CIC model into the final parallel code, following the partitioning decision. If a new partitioning decision is made, the programmer need not modify the task codes, only the partitioning information. The CIC translator automatically generates the newly optimized code from the modified architecture information file.

Thus the proposed CIC programming model is truly retargetable with respect to architecture change and partitioning decisions. Moreover, the CIC translator alleviates the programmer's burden to optimize the code for the target architecture. If we develop the code manually, we have to redesign the hardware-dependent part whenever hardware is changed because of hardware upgrade or platform change. When the lifetime of an embedded system is long, the maintenance of embedded software is very challenging since there will be no old hardware when maintenance is required. In case the lifetime is too short, the hardware platform will change frequently. Automatic code generation will remove such overhead of the software redesign. Thus we increase the design productivity of parallel-embedded software through the proposed methodology.

8.2 Related Work

Martin [8] emphasized the importance of a parallel programming model for MPSoC to overcome the difficulty of concurrent programming. Conventional MPI or OpenMP programming is not adequate since the program should be made target specific for a message-passing or shared address-space architecture. To be suitable for design space exploration, a programming model needs to accommodate both styles of architecture. Recently Paulin et al. [9] proposed the *MultiFlex* multiprocessor SoC programming environment, where two parallel programming models are supported, namely, distributed system object component (DSOC) and SMP models. The DSOC is a message-passing model that supports heterogeneous distributed computing while the SMP supports concurrent threads accessing the shared memory. Nonetheless it is still the burden of the programmer to consider the target architecture when programming the application; thus it is not fully retargetable. On the other hand, we propose here a fully retargetable programming model.

To be retargetable, the interface code between tasks should be automatically generated after a partitioning decision on the target architecture is made. Since the interfacing between the processing units is one of the most important factors that affect system performance, some research has focused on this interfacing (including HW–SW components). Wolf et al. [10] defined a task-transaction-level (TTL) interface for integrating HW–SW components. In the logical model for TTL intertask communication, a task is connected to a channel via a port, and it communicates with other tasks through channels by transferring tokens. In this model, tasks call target-independent TTL interface functions on their ports to communicate with other tasks. If the TTL interface functions are defined optimally for each target architecture, the program becomes retargetable. This approach can be integrated in the proposed framework.

For retargetable interface code generation, Jerraya et al. [11] proposed a parallel programming model to abstract both HW and SW interfaces. They defined three layers of SW architecture: hardware abstraction layer (HAL), hardware-dependent software (HdS), and multithreaded application. To interface between software and hardware, translation to application programming interfaces (APIs) of different abstraction models should be performed. This work is complementary to ours.

Compared with related work, the proposed approach has the following characteristics that make it more suitable for an MPSoC architecture:

1. We specifically concentrate on the retargetability of the software development framework and suggest CIC as a parallel programming model. The main idea of CIC is the separation of the algorithm specification and its implementation. CIC consists of two sections: the tasks codes and the architecture information file. An application programmer writes for all tasks considering the *potential* parallelism of the application itself, independent of the target architecture. Based on the target architecture, we determine how to exploit the parallelism in implementation.

2. We use different ways of specifying functional and data parallelism (or loop parallelism). Data parallelism is usually implemented by an array of homogeneous processors or a hardware accelerator, different from functional parallelism. By considering different implementation practices, we use different specification and optimization methods for functional and data parallelism.

3. Also, we explicitly specify the potential use of a hardware accelerator inside a task code using a pragma definition. If the use of a hardware accelerator is decided after design space exploration, the task code will be modified by a preprocessor, replacing the code segment contained within the pragma section by the appropriate HW interfacing code. Otherwise, the pragma definition will be ignored. Thus the use of hardware accelerators can be determined without code rewriting, which makes design space exploration easier.

8.3   Proposed Workflow of Embedded Software Development

The proposed workflow of MPSoC software development is depicted in Figure 8.1. The first step is to specify the application tasks with the proposed parallel programming model, CIC. As shown in Figure 8.1, there are two ways of generating a CIC program: One is to manually write the CIC program, which is assumed in this chapter. The other is to generate the CIC program from an initial model-based specification such as a dataflow model or UML. Recently, it has become more popular to use a model driven architecture (MDA) for the systematic design of software (Balasubramanian et al. [12]). In an MDA, system behavior is described in a platform-independent model (PIM). The PIM is translated to a platform-specific model (PSM) from which the target software on each processor is generated. The MDA methodology is expected to improve the design productivity of embedded software since it increases the reuse possibilities of platform-independent software modules: The same PIM can be reused for different target architectures.

Unlike other model-driven architectures, the unique feature of the proposed methodology is to allow multiple PIMs in the programming framework. We define an intermediate programming model common to all PIMs including the manual design. Consequently, this programming model is named CIC. The CIC is independent of the target architecture so that we

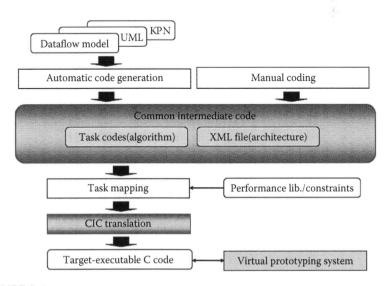

FIGURE 8.1
The proposed framework of software generation from CIC.

may explore the design space at a later stage of design. The CIC program consists of two sections, a task code section and an architecture section.

The next step is to map task codes to processing components, manually or automatically. The optimal mapping problem is beyond the scope of this chapter, so we assume that the mapping is somehow given. We are now developing an optimal mapping technique based on a genetic algorithm, considering three kinds of parallelisms simultaneously: functional parallelism, data (loop) parallelism, and temporal parallelism.

The last step is to translate the CIC program into the target-executable C codes based on the mapping and architecture information. In case more than one task is mapped to the same processor, the CIC translator should generate the run-time kernel that schedules the mapped tasks, or let the OS schedule the mapped tasks to satisfy the real-time constraints of the tasks. The CIC translator also synthesizes the interface codes between processing components optimally for the given communication architecture.

8.4   Common Intermediate Code

The heart of the proposed design methodology is the CIC parallel programming model that separates algorithm specification from architecture information. Figure 8.2a displays the CIC format consisting of the two sections that are explained in this section.

8.4.1   Task Code

A CIC task is a concurrent process that communicates with the other tasks through channels as shown in Figure 8.2b. The "task code" section contains

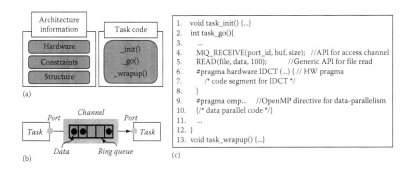

FIGURE 8.2
Common intermediate code: (a) structure, (b) default intertask communication model, and (c) an example of a task code file.

the definitions of CIC tasks that will be mapped to processing components as a unit. An application is partitioned into tasks that represent the potential temporal and functional parallelism. Data, or loop, parallelism is defined inside a task. It is the programmer's decision as to how to define the tasks: As the granularity of a task becomes finer, it will provide more potential for the optimal exploitation of pipelining and functional parallelism at the expense of increasing the burden of the programmer. An intuitive solution is to define a task as reusable for other applications. Such a trade-off should be considered if a CIC is automatically generated from a model-based specification.

Figure 8.2c shows an example of a task–code file (.cic file) that defines a task in C. A task should define three functions: {task name}_init(), {task name}_go(), and {task name}_wrapup(). The {task name}_init() function is called once when the task is invoked to initialize the task. The {task name}_go() function defines the main body of the task and is executed repeatedly in the main scheduling loop. The "{task_name}_wrapup()" function is called before stopping the task to reclaim the allocated resources.

The default channel is a FIFO channel, which is particularly adequate for streaming applications. For target-independent specification, the CIC uses generic APIs: For instance, two generic send–receive APIs are used for inter-task communication as shown in Figure 8.2c, lines 4 and 5. The CIC translator translates the generic API with the appropriate implementations, depending on whether an OS is used or not. By doing so, the same task code will be reused despite architecture variation.

There exist other types of channels among which an array channel is defined to support wave-front parallelism [13]. The producer or the consumer accesses the array channel with an index to the array element. For single-writer-multiple-reader type of communication, a shared memory channel is used.

An example of task specification is shown in Figure 8.3 where an H.263 decoder algorithm is partitioned into six tasks. In this figure, a macroblock decoding task contains three functions: "Dequantize," "Inverse zigzag," and "IDCT." These three functions may be mapped to separate processors only

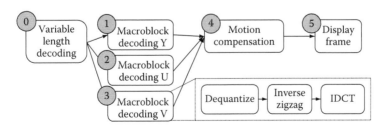

FIGURE 8.3
Task specification example: H.263 decoder. (From Kwon, S. et al., *ACM Trans. Des. Autom. Electron. Syst.*, 13, Article 39, July 2008. With permission.)

if they are specified as separate tasks in the CIC. Note that data parallelism is specified with OpenMP directives within a task code, as shown at line 9 of Figure 8.2c.

If there are HW accelerators in the target platform, we may want to use them to improve the performance. To open this possibility in a task code, we define a special pragma to identify the code section that can be mapped to the HW accelerator, as shown in line 6 of Figure 8.2c. Moreover, information on how to interface with the HW accelerator is specified in an architecture information file. Then, the code segment contained within a pragma section will be replaced with the appropriate HW-interfacing code by the CIC translator.

8.4.2 Architecture Information File

The target architecture and design constraints are separately specified from the task code in the architecture information section. The architecture section is further divided into three sections in an xml-style file, as shown in Figure 8.4. The "hardware" section contains the hardware architecture information necessary to translate target-independent task codes to target-dependent codes. The "constraints" section specifies user-specified constraints such as the real-time constraints, resource limitations, and energy constraints. The "structure" section describes the communication and synchronization requirements between tasks.

The hardware section defines the processor id, the address range and size of each memory segment, use of OS, and the task scheduling policy

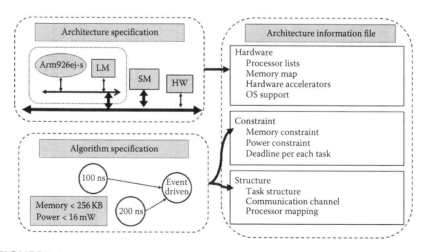

FIGURE 8.4

Architecture information section of a CIC consisting of three subsections that define HW architecture, user-given constraints, and task structure.

for each processor. For shared-memory segments, it indicates which processors share the segment. It also defines information of hardware accelerators, which includes architectural parameters and the translation library of HW-interfacing code.

The constraints section defines the global constraints such as power consumption and memory requirement as well as per task constraints such as period, deadline, and priority. Further, it includes the execution time of tasks. Using this set of information, we will determine the scheduling policies of the target OS or synthesize the run-time system for the processor without OS.

In the structure section, task structure and task dependency are specified. An application task usually consists of multiple tasks that are defined separately in the task–code section of the CIC. The task structure is represented by communication channels between the tasks.

For each task, the structure section defines the file name (with ".cic" suffix) of the task code, and its compile options needed for compilation. Moreover, each task has the index field of the processor to which the task is mapped. This field is updated after the task-mapping decision is made: In other words, task mapping can be changed without modifying the task code, but by changing the processor-mapping id of each task.

8.5 CIC Translator

The CIC translator translates the CIC program into optimized executable C codes for each processor core. As shown in Figure 8.5, the CIC translation consists of four main steps: generic API translation, HW-interface code generation, OpenMP translation if needed, and task-scheduling code generation. From the architecture information file, the CIC translator extracts

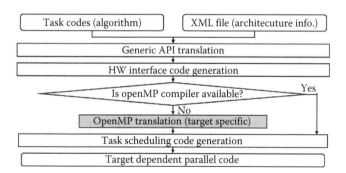

FIGURE 8.5
The workflow of a CIC translator.

the necessary information needed for each translation step. Based on the task-dependency information that tells how to connect the tasks, the translator determines the number of intertask communication channels. Based on the period and deadline information of tasks, the run-time system is synthesized. With the memory map information of each processor, the translator defines the shared variables in the shared region.

To support a new target architecture in the proposed workflow, we have to add translation rules of the generic API to the translator, make a target-specific-OpenMP-translator for data parallel tasks, and apply the generation rule of task scheduling codes tailored for the target OS. Each step of CIC translator will be explained in this section.

8.5.1 Generic API Translation

Since the CIC task code uses generic APIs for target-independent specification, the translation of generic APIs to target-dependent APIs is needed. If the target processor has an OS installed, generic APIs are translated into OS APIs; otherwise, they are translated into communication APIs that are defined by directly accessing the hardware devices. We implement the OS API library and communication API library, both optimized for each target architecture.

For most generic APIs, API translation is achieved by simple redefinition of the API function. Figure 8.6a shows an example where the translator replaces MQ_RECEIVE API with a "read_port" function for a target processor with pthread support. The read_port function is defined using

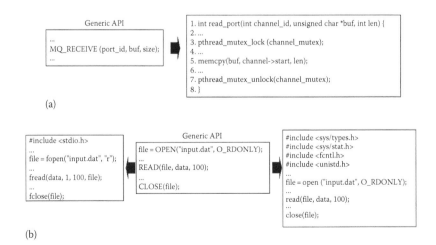

(a)

(b)

FIGURE 8.6
Examples of generic API translation: (a) MQ_RECEIVE operation, (b) READ operation.

pthread APIs and the memcpy C library function. However some APIs need additional treatment: For example, the READ API needs different function prototypes depending on the target architecture as illustrated in Figure 8.6b. Maeng et al. [14] presented a rule-based translation technique that is general enough to translate any API if the translation rule is defined in a pattern-list file.

8.5.2 HW-Interfacing Code Generation

If there is a code segment contained within a HW pragma section and its translation rule exists in an architecture information file, the CIC translator replaces the code segment with the HW-interfacing code, considering the parameters of the HW accelerator and buffer variables that are defined in the architecture section of the CIC. The translation rule of HW-interfacing code for a specific HW is separately specified as a HW-interface library code.

Note that some HW accelerators work together with other HW IPs. For example, a HW accelerator may notify the processor of its completion through an interrupt; in this case an interrupt controller is needed. The CIC translator generates a combination of the HW accelerator and interrupt controller, as shown in the next section.

8.5.3 OpenMP Translator

If an OpenMP compiler is available for the target, then task codes with OpenMP directives can be used easily. Otherwise, we somehow need to translate the task code with OpenMP directives to a parallel code. Note that we do not need a general OpenMP translator since we use OpenMP directives only to specify the data parallel CIC task. But we have to make a separate OpenMP translator for each target architecture in order to achieve optimal performance.

For a distributed memory architecture, we developed an OpenMP translator that translates an OpenMP task code to the MPI codes using a minimal subset of the MPI library for the following reasons: (1) MPI is a standard that is easily ported to various software platforms. (2) Porting the MPI library is much easier than modifying the OpenMP translator itself for the new target architecture. Figure 8.7 shows the structure of the translated MPI program.

As shown in the figure, the translated code has the master–worker structure: The master processor executes the entire core while worker processors execute the parallel region only. When the master processor meets the parallel region, it broadcasts the shared data to worker processors. Then, all processors concurrently execute the parallel region. The master processor synchronizes all the processors at the end of the parallel loop and collects the results from the worker processors. For performance optimization, we have to minimize the amount of interprocessor communication between processors.

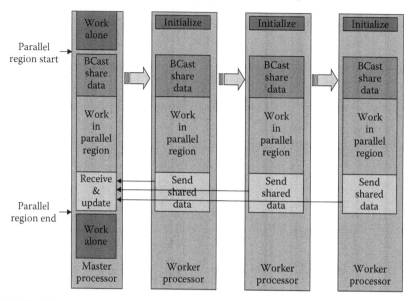

FIGURE 8.7

The workflow of translated MPI codes. (From Kwon, S. et al., *ACM Trans. Des. Autom. Electron. Syst.*, 13, Article 39, July 2008. With permission.)

8.5.4 Scheduling Code Generation

The last step of the proposed CIC translator is to generate the task-scheduling code for each processor core. There will be many tasks mapped to each processor, with different real-time constraints and dependency information. We remind the reader that a task code is defined by three functions: "{task name}_init(), {task name}_go(), and {task name}_wrapup()." The generated scheduling code initializes the mapped tasks by calling "{task name}_init()" and wraps them up after the scheduling loop finishes its execution, by calling "{task name}_wrapup()."

The main body of the scheduling code differs depending on whether there is an OS available for the target processor. If there is an OS that is POSIX-compliant, we generate a thread-based scheduling code, as shown in Figure 8.8a. A POSIX thread is created for each task (lines 17 and 18) with an assigned priority level if available. The thread, as shown in lines 3 to 5, executes the main body of the task, "{task name}_go()," and schedules the thread itself based on its timing constraints by calling the "sleep()" method. If the OS is not POSIX-compliant, the CIC translator should be extended to generate the OS-specific scheduling code.

If there is no available OS for the target processor, the translator should synthesize the run-time scheduler that schedules the mapped tasks. The CIC translator generates a data structure of each task, containing three main functions of tasks ("init(), go(), and wrapup()"). With this data structure, a

```
 1. void * thread_task_0_func(void *argv) {
 2.    ...
 3.    task_0_go();
 4.    get_time(&time);
 5.    sleep(task_0->next_period – time); // sleep for remained time
 6.    ...
 7. }
 8. int main() {
 9.    ...
10.    pthread_t thread_task_0;
11.    sched_param thread_task_0_param;
12.    ...
13.    thread_task_0_param.sched_priority = 0;
14.    pthread_attr_setschedparam(..., &thread_task_0_param);
15.    ...
16.    task_init(); /* {task_name}_init() functions are called */
17.    pthread_create(&thread_task_0,
18.    &thread_task_0_attr, thread_task_0_func, NULL);
19.    ...
20.    task_wrapup(); /* {task_name}_wrapup() functions are called */
21. }
```

(a)

```
 1. typedef struct {
 2.    void (*init)();
 3.    int (*go());
 4.    void (*wrapup)();
 5.    int period, priority, ...;
 6. } task;
 7. task taskInfo[] = { {task 1_init, task 1_go, task 1_wrapup, 100, 0}
 8.    , {task2_init, task2_go, task2_wrapup, 200, 0}};
 9.
10. void scheduler() {
11.    while(all_task_done()==FALSE) {
12.       int taskId = get_next_task();
13.       taskInfo[taskId]->go()
14.    }
15. }
16.
17. int main() {
18.    init();   /* {task_name}_init() functions are called */
19.    scheduler(); /* scheduler code */
20.    wrapup();   /* {task_name}_wrapup() functions are called */
21.    return 0;
22. }
```

(b)

FIGURE 8.8
Pseudocode of generated scheduling code: (a) if OS is available, and (b) if OS is not available. (From Kwon, S. et al., *ACM Trans. Des. Autom. Electron. Syst.*, 13, Article 39, July 2008. With permission.)

real-time scheduler is synthesized by the CIC translator. Figure 8.8b shows the pseudocode of a generated scheduling code. Generated scheduling code may be changed by replacing the function "void scheduler()" or "int get_next_task()" to support another scheduling algorithm.

8.6   Preliminary Experiments

An embedded software development framework based on the proposed methodology, named HOPES, is under development. While it allows the use of any model for initial specification, the current implementation is being done with the PeaCE model. PeaCE model is one that is used in PeaCE hardware–software codesign environment for multimedia embedded systems design [15]. To verify the viability of the proposed programming, we built a virtual prototyping system, based on the Carbon SoC Designer [16], that consists of multiple subsystems of arm926ej-s connected to each other through a shared bus as shown in Figure 8.9. H.263 Decoder as depicted in Figure 8.3 is used for preliminary experiments.

8.6.1   Design Space Exploration

We specified the functional parallelism of the H.263 decoder with six tasks as shown in Figure 8.3, where each task is assigned an index. For data-parallelism, the data parallel region of motion compensation task is specified with an OpenMP directive. In this experiment, we explored the design space of parallelizing the algorithm, considering both functional and data parallelisms simultaneously. As is evident in Figure 8.3, tasks 1 to 3 can be executed in parallel; thus, they are mapped to multiple-processors with three configurations as shown in Table 8.1. For example, task 1 is mapped to processor 1, and the other tasks are mapped to processor 0 for the second configuration.

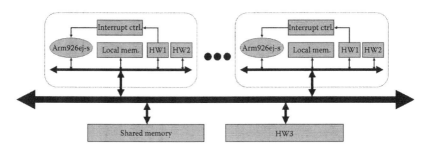

FIGURE 8.9
The target architecture for preliminary experiments. (From Kwon, S. et al., *ACM Trans. Des. Autom. Electron. Syst.*, 13, Article 39, July 2008. With permission.)

TABLE 8.1

Task Mapping to Processors

	The Configuration of Task Mapping		
Processor Id	**1**	**2**	**3**
0	Task 0, Task 1, Task 2, Task 3, Task 4, Task 5	Task 0, Task 2, Task 3, Task 4, Task 5	Task 0, Task 3, Task 4, Task 5
1	N/A	Task 1	Task 1
2	N/A	N/A	Task 2

Source: Kwon, S. et al., *ACM Trans. Des. Autom. Electron. Syst.*, 13, Article 39, July 2008. With permission.

TABLE 8.2

Execution Cycles for Nine Configurations

The Number of Processors for Data-Parallelism	**The Configuration of Task Mapping**		
	1	**2**	**3**
No OpenMP	158,099,172	**146,464,503**	146,557,779
2	167,119,458	152,753,214	153,127,710
4	168,640,527	154,159,995	155,415,942

Source: Kwon, S. et al., *ACM Trans. Des. Autom. Electron. Syst.*, 13, Article 39, July 2008. With permission.

For each configuration of task mapping, we parallelized task 4, using one, two, and four processors. As a result, we have prepared nine configurations in total as illustrated in Table 8.2. In the proposed framework, each configuration is simply specified by changing the task-mapping information in the architecture information file. The CIC translator generates the executable C codes automatically.

Table 8.2 shows the performance result for these nine configurations. For functional parallelism, the best performance can be obtained by using two processors as reported in the first row ("No OpenMP" case). H.263 decoder algorithm uses a 4:1:1 format frame, so computation of Y macroblock decoding is about four times larger than those of U and V macroblocks. Therefore macroblock decoding of U and V can be merged in one processor during macroblock decoding of Y in another processor. There is no performance gain obtained by exploiting data parallelism. This is because the computation workload of motion compensation is not large enough to outweigh the communication overhead incurred by parallel execution.

8.6.2 HW-Interfacing Code Generation

Next, we accelerated the code segment of IDCT in the macroblock decoding tasks (task 1 to task 3) with a HW accelerator, as shown in Figure 8.10a. We use the RealView SoC designer to model the entire system including the

```
#pragma hardware IDCT (output.data, input.data) {
    /* code segments for IDCT */
}
```

(a)

```
1.  <hardware>
2.      <name>IDCT</name>
3.      <protocol>IDCT_slave</protocol>
4.      <param>0x2F000000</param>
5.  </hardware>
```

(b)

```
1.   <hardware>
2.       <name>IDCT</name>
3.       <protocol>IDCT_interrupt</protocol>
4.       <param>0x2F000000</param>
5.   </hardware>
6.   <hardware>
7.       <name>IRQ_CONTROLLER</name>
8.       <protocol>irq_controller</name>
9.       <param>0xA801000</param>
10.  </hardware>
```

(c)

FIGURE 8.10

(a) Code segment wrapped with HW pragma and architecture section information of IDCT, (b) when interrupt is not used, and (c) when interrupt is used. (From Kwon, S. et al., *ACM Trans. Des. Autom. Electron. Syst.*, 13, Article 39, July 2008. With permission.)

HW accelerator. Two kinds of inverse discrete cosine transformation (IDCT) accelerator are used. One uses an interrupt signal for completion notification, and other uses polling to detect the completion. The latter is specified in the architecture section as illustrated in Figure 8.10b, where the library name of the HW-interfacing code is set to *IDCT_slave* and its base address to *0x2F000000*.

Figure 8.11a shows the assigned address map of the IDCT accelerator and Figure 8.11b shows the generated HW-interfacing code. This code is substituted for the code segment contained within a HW pragma section. In Figure 8.11b, bold letters are changeable according to the parameters specified in a task code and in the architecture information file; they specify the base address for the HW interface data structure and the input and output port names of the associated CIC task.

Note that interfacing code uses polling at line 6 of Figure 8.11b. If we use the accelerator with interrupt, an interrupt controller is additionally attached to the target platform, as shown in Figure 8.10c, with information on the code library name, *IRQ_CONTROLLER*, and its base address *0xA801000*. The new IDCT accelerator has the same address map as the previous one, except for

Address (Offset)	I/O Type	Comment
0	Read	Semaphore
4	Write	IDCT start
8	Read	Complete flag
12	Write	IDCT clear
64 ~ 191	Write	Input data
192 ~ 319	Read	Output data

(a)

```
1. int i;
2. volatile unsigned int * idct_base = (volatile unsigned int*) 0x2F000000;
3. while(idct_base[0]==1);   // try to obtain hardware resource
4. for (i=0;i<32;i++)   idct_base[i+16]= ((unsigned int*)(input.data))[i];
5. idct_base[1]= 1;   // send start signal to IDCT accelerator
6. while(idct_base[2]==0);   // wait for completion of IDCT operation
7. for (i=0;i<32;i++) ((unsigned int*)(output.data))[i] = idct_base[i+48];
8. idct_base[3]= 1;   // clear and unlock hardware
```

(b)

FIGURE 8.11
(a) The address map of IDCT, and (b) its generated interfacing code. (From Kwon, S. et al., *ACM Trans. Des. Autom. Electron. Syst.*, 13, Article 39, July 2008. With permission.)

the complete flag. The address of the complete flag (address 8 in Figure 8.11a) is assigned to "interrupt clear."

Figure 8.12a shows the generated interfacing code for the IDCT with interrupt. Note that the interfacing code does not access the HW to check the completion of IDCT, but checks the variable "complete." In the generated code of the interrupt handler, this variable is set to 1 (Figure 8.12b). The initialize code for the interrupt controller ("initDevices()") is also generated and called in the "{task_name}_init()" function.

8.6.3 Scheduling Code Generation

We generated the task-scheduling code of the H.263 decoder while changing the working conditions, OS support, and scheduling policy. At first, we used the eCos real-time OS for arm926ej-s in the RealView SoC designer, and generated the scheduling code, the pseudocode of which is shown in Figure 8.13. In function cyg_user_start() of eCos, each task is created as a thread. The CIC translator generates the parameters needed for thread creation such as stack variable information and stack size (fifth and sixth parameter of cyg_thread_create()). Moreover, we placed "{task_name}_go" in a while loop inside the created thread (lines 10 to 14 of Figure 8.13). Function *{task_name}_init()* is called in init_task().

Note that TE_main() is also created as a thread. TE_main() checks whether execution of all tasks is finished, and calls "{task_name}_wrapup()" in wrapup_task() before finishing the entire program.

```
1. int complete;
2. ...
3. volatile unsigned int * idct_base = (volatile unsigned int*) 0x 2F000000;
4. while(idct_base[0]== 1); // try to obtain hardware resource

5. complete = 0;
6. for (i=0;i<32;i++)   idct_base[i+16]= ((unsigned int*)(input.data))[i];
7. idct_base[1] = 1;   // send start signal to IDCT accelerator
8. while(complete==0);   // wait for completion of IDCT operation
9. for (i = 0; i < 32; i + +) ((unsigned int*)(output.data)[i] = idct_base[i + 48];
10. idct_base[3]= 1;   // clear and unlock hardware
```

(a)

```
1. extern int complete;
2. __irq void IRQ_Handler() {
3.    IRQ_CLEAR(); // interrupt clear of interrupt controller
4.    idct_base[2] =1;   // interrupt clear of IDCT
5.    complete = 1;
6. }
7. void initDevices(){
8.    IRQ_INIT();   // initialize of interrupt controller
9. }
```

(b)

FIGURE 8.12
(a) Interfacing code for the IDCT with interrupt, and (b) the interrupt handler code. (From Kwon, S. et al., _ACM Trans. Des. Autom. Electron. Syst._, 13, Article 39, July 2008. With permission.)

For a processor without OS support, the current CIC translator supports two kinds of scheduling code: default and rate-monotonic scheduling (RMS). The default scheduler just keeps the execution frequency of tasks considering the period ratio of tasks. Figure 8.14a and b show the pseudocode of function get_next_task(), which is called in the function scheduler() of Figure 8.8b, for the default and RMS, respectively.

8.6.4 Productivity Analysis

For the productivity analysis, we recorded the elapsed time to manually modify the software (including debugging time) when we change the target architecture and task mapping. Such manual modification was performed by an expert programmer who is a PhD student.

For a fair comparison of automatic code generation and manual-coding overhead, we made the following assumptions. First, the application task codes are prepared and functionally verified. We chose an H.263 decoder as the application code that consists of six tasks, as illustrated in Figure 8.3. Second, the simulation environment is completely prepared for the initial configuration, as shown in Figure 8.15a. We chose the RealView SoC designer as the target simulator, prepared two different kinds of HW IPs

```
1. void cyg_user_start(void) {
2.    cyg_threaad_create(taskInfo[0]->priority, TE_task_0,
3.       (cyg_addrword_t)0, "TE_task_0", (void*)&TaskStk[0],
4.       TASK_STK_SIZE-1, &handler[0], &thread[0]);
5.    . . .
6.    init_task();
7.    cyg_thread_resume(handle[0]);
8.    . . .
9. }
10. Void TE_task_0(cyg_addrword_t data) {
11.    while(!finished)
12.    if (this task is executable) taskInfo[0]->go();
13.    else cyg_thread_yield();
14. }
15. void TE_main(cyg_addrword_t data) {
16.    while(1)
17.    if (all_task_is_done()) {
18.       wrapup_task();
19.       exit(1);
20.    }
21. }
```

FIGURE 8.13

Pseudocode of an automatically generated scheduler for eCos. (From Kwon, S. et al., *ACM Trans. Des. Autom. Electron. Syst.*, 13, Article 39, July 2008. With permission.)

```
1. int get_next_task() {
2.    a. find executable tasks
3.    b. find the tasks that has the smallest value of time count
4.    c. select the task that is not executed for the longest time
5.    d. add period to the time count of selected task
6.    e. return selected task id
7. }
```

(a)

```
1. int get_next_task() {
2.    a. find executable tasks
3.    b. select the task that has the smallest period
4.    c. update task information
5.    d. return selected task id
6. }
```

(b)

FIGURE 8.14

Pseudocode of "get_next_task()" without OS support: (a) default, and (b) RMS scheduler. (From Kwon, S. et al., *ACM Trans. Des. Autom. Electron. Syst.*, 13, Article 39, July 2008. With permission.)

for the IDCT function block. Third, the software environment for the target system is prepared, which includes the run-time scheduler and target-dependent API library.

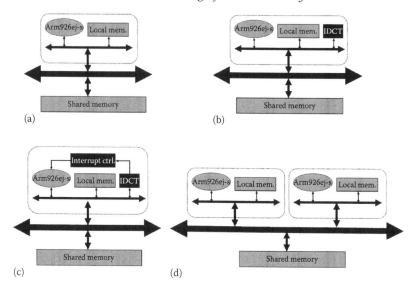

FIGURE 8.15
Four target configurations for productivity analysis: (a) initial architecture, (b) HW IDCT is attached, (c) HW IDCT and interrupt controller are attached, and (d) additional processor and local memory are attached. (From Kwon, S. et al., *ACM Trans. Des. Autom. Electron. Syst.*, 13, Article 39, July 2008. With permission.)

At first, we needed to port the application code to the simulation environment shown in Figure 8.15a. The application code consists of about 2400 lines of C code, in which 167 lines are target dependent. The target-dependent codes should be rewritten using target-dependent APIs defined for the target simulator. It took about 5 h to execute the application on the simulator of our initial configuration (Figure 8.15a). The simulation porting overhead is directly proportional to the target-dependent code size. In addition, the overhead increases as total code size increases, since we need to identify the target-dependent codes throughout the entire application code.

Next, we changed the target architecture to those shown in Figure 8.15b and c by using two kinds of IDCT HW IPs. The interface code between processor and IDCT HW should be inserted. It took about 2–3 h to write and debug the interfacing code with IDCT HW IP, without and with the interrupt controller, respectively. The sizes of the interface without and with the interrupt controller were 14 and 48 lines of code, respectively. Note that the overhead will increase if the HW IP has a more complex interfacing protocol.

Last, we modified the task mapping by adding one more processor, as shown in Figure 8.15d. For this analysis, we needed to make an additional data structure of software tasks to link with the run-time scheduler on each processor. It took about 2 h to make the data structure of all tasks and attach

TABLE 8.3

Time Overhead for Manual Software Modification

	Description	Code Line	Time (h)
Figure 8.15a → Figure 8.15b and c	Initial porting overhead to the target simulator	167 of 2400	5
	Making HW interface code of IDCT (Figure 8.15a → Figure 8.15b)	14	2
	Modifying HW interface code to use interrupt controller (Figure 8.15a → Figure 8.15c)	48	3
Figure 8.15a → Figure 8.15d	Making initial data structure for scheduler	31	2
	Modification of data structure according to the task mapping decision	12	0.5

Source: Kwon, S. et al., *ACM Trans. Des. Autom. Electron. Syst.*, 13, Article 39, July 2008. With permission.

it to the default scheduler. Then, it took about 0.5 h to modify the data structure according to the task-mapping decision. Note that to change the task-mapping configuration, the algorithm part of the software code need not be modified. We summarize the overheads of manual software modification in Table 8.3.

By contrast, in the proposed framework, design space exploration is simply performed by modifying the architecture information file only, not task code. Modifying the architecture information file is much easier than modifying the task code directly, and needs only a few minutes. Then CIC translator generates the target code automatically in a minute. Of course, it requires a significant amount of time to establish the translation environment for a new target. But once the environment is set up for each candidate processing element, we believe that the proposed framework improves design productivity dramatically for design space exploration of various architecture and task-mapping candidates.

8.7 Conclusion

In this chapter, we presented a retargetable parallel programming framework for MPSoC, based on a new parallel programming model called the CIC. The CIC specifies the design constraints and task codes separately. Furthermore, the functional parallelism and data parallelism of application tasks are specified independently of the target architecture and design constraints.

Then, the CIC translator translates the CIC into the final parallel code, considering the target architecture and design constraints, to make the CIC retargetable. Temporal parallelism is exploited by inserting pipeline buffers between CIC tasks and where to put the pipeline buffers is determined at the mapping stage. We have developed a mapping algorithm that considers temporal parallelism as well as functional and data parallelism [17].

Preliminary experiments with a H.263 decoder example prove the viability of the proposed parallel programming framework: It increases the design productivity of MPSoC software significantly. There are many issues to be researched further in the future, which include the optimal mapping of CIC tasks to a given target architecture, exploration of optimal target architecture, and optimizing the CIC translator for specific target architectures. In addition, we have to extend the CIC to improve the expression capability of the model.

References

1. Message Passing Interface Forum, MPI: A message-passing interface standard, *International Journal of Supercomputer Applications and High Performance Computing,* 8(3/4), 1994, 159–416.

2. OpenMP Architecture Review Board, OpenMP C and C++ application program interface, http://www.openmp.org, Version 1.0, 1998.

3. M. Sato, S. Satoh, K. Kusano, and Y. Tanaka, Design of OpenMP compiler for an SMP cluster, in *EWOMP'99,* Lund, Sweden, 1999.

4. F. Liu and V. Chaudhary, A practical OpenMP compiler for system on chips, in *WOMPAT 2003,* Toronto, Canada, June 26–27, 2003, pp. 54–68.

5. Y. Hotta, M. Sato, Y. Nakajima, and Y. Ojima, OpenMP implementation and performance on embedded renesas M32R chip multiprocessor, in *EWOMP,* Stockholm, Sweden, October, 2004.

6. W. Jeun and S. Ha, Effective OpenMP implementation and translation for multiprocessor system-on-chip without using OS, in *12th Asia and South Pacific Design Automation Conference (ASP-DAC'2007),* Yokohama, Japan, 2007, pp. 44–49.

7. R. Eigenmann, J. Hoeflinger, and D. Padua, On the automatic parallelization of the perfect benchmarks(R), *IEEE Transactions on Parallel and Distributed Systems,* 9(1), 1998, 5–23.

8. G. Martin, Overview of the MPSoC design challenge, in *43rd Design Automation Conference,* San Francisco, CA, July, 2006, pp. 274–279.

9. P. G. Paulin, C. Pilkington, M. Langevin, E. Bensoudane, and G. Nicolescu, Parallel programming models for a multi-processor SoC platform applied to high-speed traffic management, in *CODES+ISSS 2004*, Stockholm, Sweden, 2004, pp. 48–53.

10. P. van der Wolf, E. de Kock, T. Henriksson, W. Kruijizer, and G. Essink, Design and programming of embedded multiprocessors: An interface-centric approach, in *Proceedings of CODES+ISSS 2004*, Stockholm, Sweden, 2004, pp. 206–217.

11. A. Jerraya, A. Bouchhima, and F. Petrot, Programming models and HW-SW interfaces abstraction for multi-processor SoC, in *43rd Design Automation Conference*, San Francisco, CA, July 24–28, 2006, pp. 280–285.

12. K. Balasubramanian, A. Gokhale, G. Karsai, J. Sztipanovits, and S. Neema, Developing applications using model-driven design environments, *IEEE Computer*, 39(2), 2006, 33–40.

13. K. Kim, J. Lee, H. Park, and S. Ha, Automatic H.264 Encoder synthesis for the cell processor from a target independent specification, in *6th IEEE Workshop on Embedded Systems for Real-time Multimedia (ESTIMedia'2008)*, Atlanta, GA, 2008.

14. J. Maeng, J. Kim, and M. Ryu, An RTOS API translator for model-driven embedded software development, in *12th IEEE International Conference on Embedded and Real-Time Computing Systems and Applications (RTCSA'06)*, Sydney, Australia, August 16–18, 2006, pp. 363–367.

15. S. Ha, C. Lee, Y. Yi, S. Kwon, and Y. Joo, PeaCE: A hardware-software codesign environment for multimedia embedded systems, *ACM Transactions on Design Automation of Electronic Systems (TODAES)*, 12(3), Article 24, August 2007.

16. Carbon® SoC Designer homepage, http://carbondesignsystems.com/products_socd.shtml

17. H. Yang and S. Ha, Pipelined data parallel task mapping/scheduling technique for MPSoC, in *DATE 2009*, Nice, France, April 2009.

18. S. Kwon, Y. Kim, W. Jeun, S. Ha, and Y Paek, A retargetable parallel-programming framework for MPSoC, *ACM Transactions on Design Automation of Electronic Systems (TODAES)*, 13(3), Article 39, July 2008.

.

9

Programming Models for MPSoC

Katalin Popovici and Ahmed Jerraya

CONTENTS

9.1 Introduction

Multimedia applications impose demanding constraints in terms of time to market and design quality. Efficient hardware platforms do exist for these applications. These feature heterogeneous multiprocessor architectures with specific I/O components in order to achieve computation and communication performance [1]. Heterogeneous MPSoC includes different kinds of processing units (digital signal processor [DSP], microcontroller, application-specific instruction set processor [ASIP], etc.) and different communication schemes (fast links, nonstandard memory organization and access). Typical heterogeneous platforms used in industry are TI OMAP [2], ST Nomadik [3], Philips Nexperia [4], and Atmel Diopis [5]. Next generation MPSoC promises to be a multitile architecture that integrates hundreds of DSP and microcontrollers on a single chip [6]. The software running on these heterogeneous MPSoC architectures is generally organized into several stacks made of different software layers.

Programming heterogeneous MPSoC architectures becomes a key issue because of two competing requirements: (1) Reducing the software development cost and the overall design time requires a higher level programming model. Usually, high level programming models diminish the amount of architecture details that need to be handled by the application software designers, and accelerates the design process. The use of high level programming model also allows concurrent software–hardware design, thus reducing the overall SoC design time. (2) Improving the performance of the overall system requires finding the best matches between the hardware and the software. This is generally obtained through low level programming. Thus, the key challenge is to find a programming environment able to satisfy these two opposing requirements.

Programming MPSoCs means generating software stacks running on the various processors efficiently, while exploiting the available resources of the architecture. Producing efficient code requires that the software takes into account the capabilities of the target platform. For instance, a data exchange between two different processors may use different schemes (global memory accessible by both processing units, local memory of the one of the processors, dedicated hardware FIFO components, etc.). Additionally, different synchronization schemes (polling, interrupts) may be used to coordinate this data exchange. Each of these communication schemes has advantages and disadvantages in terms of performance (e.g., latency, throughput), resource sharing (e.g., multitasking, parallel I/O), and communication overhead (e.g., memory size, execution time).

In an ideal design flow, programming a specific architecture consists of partitioning and mapping, application software code generation, and hardware-dependent software (HdS) code generation (Figure 9.1). The HdS is made of the lower software layers that may incorporate an operating system (OS), communication management, and a hardware abstraction layer (HAL) to allow the OS functions to access the hardware resources of the platform. Unfortunately, we are still missing such an ideal generic flow, which can efficiently map high level programs on heterogeneous MPSoC architectures.

Traditional software development strategies make use of the concept of a software development platform to debug the software before the hardware is ready, thus allowing parallel hardware–software design. As illustrated in Figure 9.2, the software development platform is an abstract model of the architecture in form of a run time library or simulator aimed to execute the software. The combination of this platform with the software code given

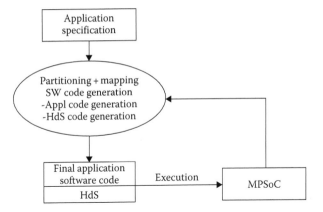

FIGURE 9.1
Software design flow.

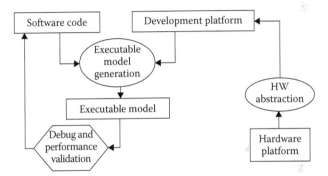

FIGURE 9.2
Software development platform.

as a high level representation produces an executable model that emulates the execution of the final system including hardware and software architecture. Generic software development platforms have been designed to fully abstract the hardware–software interfaces, for example, MPITCH is a run time execution environment designed to execute parallel software code written using MPI [7]. The use of generic platforms does not allow simulating the software execution with detailed hardware–software interaction. Therefore, it does not allow debugging the lower layers of the software stack, for instance, the OS or the implementation of the high level communication primitives. The validation and debug of the HdS is the main bottleneck in MPSoC design [8] because each processor subsystem requires specific HdS implementation to be efficient.

The use of programming models for the software design of heterogeneous MPSoC requires the definition of new design automation methods to enable concurrent design of hardware and software. This also requires new models to deal with nonstandard application specific hardware–software interfaces at several abstraction levels.

In this chapter, we give the definition of the programming models to abstract hardware–software interfaces in the case of heterogeneous MPSoC. Then, we propose a programming environment, which identifies several programming models at different MPSoC abstraction levels. The proposed approach combines the Simulink® environment for high level programming and SystemC design language for low level programming. The proposed methodology is applied to a heterogeneous multiprocessor platform, to explore the communication architecture and to generate efficient executable code of the software stacks for an H.264 video encoder application.

The chapter is composed of seven sections. Section 9.1 gives a short introduction to present the context of MPSoC programming models and environments. Section 9.2 describes the hardware and software organization of the MPSoC, including hardware–software interfaces. Section 9.3 gives the definition of the programming models and MPSoC abstraction levels. Section 9.4 lists several existing programming models. Section 9.5 summarizes the main steps of the proposed programming environment, based on Simulink and SystemC design languages. Section 9.6 addresses the experimental results, followed by conclusion.

9.2   Hardware–Software Architecture for MPSoC

The literature relates mainly two kinds of organizations for multiprocessor architectures. These are called shared memory and message passing [9]. This classification fixes both hardware and software organizations for each class of architectures. The shared memory organization generally assumes multitasking application organized as a single software stack and hardware architecture made of multiple identical processors (CPUs). The communication between the different CPUs is performed through a global shared memory. The message passing organization assumes multiple software stacks running on nonidentical subsystems that may include different CPUs and/or a different I/O systems in addition to specific local memory architectures. The communication between the different subsystems generally proceeds by message passing. Heterogeneous MPSoCs generally combine both models to integrate a massive number of processors on a single chip [10]. Future heterogeneous MPSoC will be made of few heterogeneous subsystems, where each subsystem may include a massive number of the same processor to run a specific software stack.

In the following sections, we describe the hardware organization, software stack composition, and the hardware–software interface for MPSoC architectures.

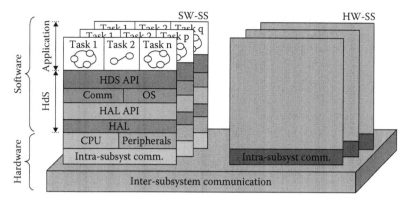

FIGURE 9.3
MPSoC hardware–software architecture.

9.2.1 Hardware Architecture

Generally, MPSoC architectures may be represented as a set of processing subsystems or components that interact via an inter-subsystem communication network (Figure 9.3).

The processing subsystems may be either hardware (HW-SS) or software subsystem (SW-SS). The SW-SS are programmable subsystems that include one or several identical processing units, or CPUs. Different kinds of processing units may be needed for the different subsystems to realize different types of functionality (e.g., DSP for data oriented operations, general purpose processor [GPP], for control oriented operations, and ASIP, for application specific computation). Each SW-SS executes a software stack.

In addition to the CPU, the hardware part of a SW-SS generally includes auxiliary components and peripherals to speed up computation and communication. This may range from simple bus arbitration to sophisticated memory and parallel I/O architectures.

9.2.2 Software Architecture

In classical literature, a subsystem is organized into layers for the purpose of standardization and reuse. Unfortunately, each layer induces additional cost and performances overheads.

In this chapter, we consider that within a subsystem the software stack is structured in only three layers, as depicted in Figure 9.3. The top layer is the software application that may be a multitasking description or a single task function. The application layer consists of a set of tasks that makes use of a programming model or application programming interface (API) to abstract the underlying HdS layer. These APIs correspond the HdS APIs. The separation between the application layer and the underlying HdS layer is required to facilitate concurrent software and hardware development.

The second layer consists in the OS and communication middleware (Comm) layer. This software layer is responsible for providing the necessary services to manage and share resources. The software includes scheduling of the application tasks on top of the available processing elements, inter-task communication, external communication, and all other types of resource management and control services. Conventionally, these services are provided by the OS and additional libraries for the communication middleware. At this level, the hardware dependency is kept functional, i.e., it concerns only high level aspects of the hardware architecture such as the type of available resources. The OS and communication layer make use of HAL APIs to abstract the underlying HAL layer.

Low level details about how to access these resources are abstracted by the third layer, which is the HAL. The separation between OS and HAL makes thereby the architecture exploration for the design of both the CPU subsystem and the OS services easier, enabling easy software portability. The HAL is a thin software layer that not only completely depends on the type of processor that will execute the software stack, but also depends on the hardware resources interacting with the processor. The HAL also includes the device drivers to implement the interface for the communication with the various devices.

9.2.3 Hardware–Software Interface

The hardware–software interface links the software part with the hardware part of the system. As illustrated in Figure 9.4, the hardware–software interface needs to handle two different interfaces: one on the software side using APIs and one on the hardware side using wires [11]. This heterogeneity makes the hardware–software interface design very difficult and time consuming because the design requires both, hardware and software knowledge

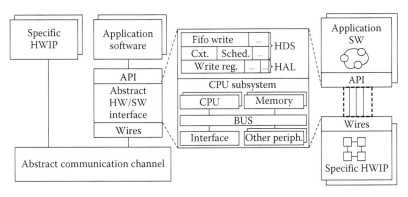

FIGURE 9.4
Hardware–software interface.

as well as their interaction [12]. The hardware–software interface requires handling many software and hardware architecture parameters.

The hardware–software interface has different views depending on the designer. Thus, for an application software designer, the hardware–software interface represents a set of system call used to hide the underlying execution platform, also called programming model. For a hardware designer, the hardware–software interface represents a set of registers, control signals, and more sophisticated adaptors to link the processor to the HW-SS. For a system software designer, the hardware–software interface is defined as the low level software implementation of the programming model for a given hardware architecture. In this case, the processor is the ultimate hardware–software interface. This is a sequential scheme assuming that the hardware architecture is the starting point for the low level software design. Finally, for a SoC designer the hardware–software interface abstracts both hardware and software in addition to the processor.

9.3 Programming Models

Several tools exist for the automatic mapping of sequential programs on homogeneous multiprocessor architectures. Unfortunately, these are not efficient for heterogeneous MPSoC architectures. In order to allow the design of distributed applications, programming models have been introduced and extensively studied by the software communities to allow high level programming of heterogeneous multiprocessor architectures.

9.3.1 Programming Models Used in Software

As long as only the software is concerned, Skillicorn and Talia [13] identifies five key concepts that may be hidden by the programming model, namely concurrency or parallelism of the software, decomposition of the software into parallel threads, mapping of threads to processors, communication among threads, and synchronization among threads. These concepts define six different abstraction levels for the programming models. Table 9.1 summarizes the different levels with typical corresponding programming languages for each of them. All these programming models take into account only the software side. They assume the existence of lower levels of software and a hardware platform able to execute the corresponding model.

9.3.2 Programming Models for SoC Design

In order to allow concurrent hardware–software design, we need to abstract the hardware–software interfaces, including both software and

TABLE 9.1

The Six Programming Levels Defined by Skillicorn

Abstraction Level	Typical Languages	Explicit Concepts
Implicit concurrency	PPP, crystal	None
Parallel level	Concurrent Prolog	Concurrency
Thread level	SDL	Concurrency, decomposition
Agent models	Emerald, CORBA	Concurrency, decomposition, mapping
Process network	Kahn process network	Concurrency, decomposition, mapping, communication
Message passing	MPI, OCCAM	Concurrency, decomposition, mapping, communication, synchronization

TABLE 9.2

Additional Models for SoC Design

Abstraction Level	Typical Programming Languages	Explicit Concepts
System architecture	MPI, Simulink [15]	All functional
Virtual architecture	Untimed SystemC [16]	Abstract communication resources
Transaction accurate architecture	TLM SystemC [16]	Resources sharing and control strategies
Virtual prototype	Cosimulation with ISS	ISA and detailed I/O interrupts

hardware components. Similar to the programming models for software, the hardware–software interfaces may be described at different abstraction levels. The four key concepts that we consider are the following: explicit hardware resources, management and control strategies for the hardware resources, the CPU architecture, and the CPU implementation. These concepts define four abstraction levels, named system architecture level, virtual architecture level, transaction accurate architecture level, and virtual prototype level [14]. The four levels are presented in Table 9.2.

At the system architecture level, all the hardware is implicit similar to the message passing model used for software. The hardware–software partitioning and the resources allocation are made explicit. This level fixes also the allocation of the tasks to the various subsystems. Thus, the model combines both the specification of the application and the architecture and it is also called combined architecture algorithm model (CAAM). At the virtual architecture level, the communication resources, such as global interconnection components and buffer storage components, become explicit. The transaction accurate architecture level implements the resources management and control strategies. This level fixes the OS on the software side. On the hardware side, a functional model of the bus is defined. The software interface is specified at the HAL level, while the hardware communication is defined at the bus transaction level. Finally, the virtual prototype level corresponds to the classical cosimulation with instruction set simulators (ISSs) [17]. At this level the architecture of the CPU is fixed, but not yet its implementation that remains hidden by an ISS.

9.3.3 Defining a Programming Model for SoC

A programming model is made of a set of functions (implicit and/or explicit primitives) that can be used by the software to interact with the hardware. Additionally, the programming model needs to cover the four abstraction levels, previously presented and required for the SoC refinement.

In order to cover different abstraction levels of both software and hardware, the programming model needs to include three kinds of primitives:

- Communication primitives: These are aimed to exchange data between the hardware and the software.
- Task and resources control primitives: These are aimed to handle task creation, management, and sequencing. At the system architecture level, these primitives are generally implicit and built in the language constructs. The typical scheme is the module hierarchy in block structure languages, where each module declares implicit execution threads.
- Hardware access primitives: These are required when the architecture includes specific hardware. The primitives include specific primitives to implement specific protocol or I/O schemes, for example, a specific memory controller allowing multiple accesses. These will always be considered at lower abstraction layers and cannot be abstracted using the standard communication primitives.

The programming models at the different abstraction levels previously described are summarized in Table 9.3. The different abstraction levels may be expressed by a single and unique programming model that uses the same primitives applicable at different abstraction levels or it uses different primitives for each level.

9.4 Existing Programming Models

A number of MP-SoC specific programming models, based on shared memory or message passing, have been defined recently.

The task transaction level interface (TTL) proposed in [18] focuses on stream processing applications in which concurrency and communication

TABLE 9.3

Programming Model API at Different Abstraction Levels

Abstraction Level	Communication Primitives	Task and Resources Control	Hardware Access Primitives
System architecture	Implicit, e.g., Simulink links	Implicit, e.g., Simulink blocks	Implicit, e.g., Simulink links
Virtual architecture	Data exchange, e.g., send–receive(data)	Implicit tasks control, e.g., threads in SystemC	Specific I/O protocols related to architecture
Transaction accurate architecture	Data access with specific addresses e.g., read–write(data, addr)	Explicit tasks control, e.g., create–resume_task(task_id) Hardware management of resources, e.g., test/set(hw_addr)	Physical access to hardware resources
Virtual prototype	Load–store registers	Hardware arbitration and address translation, e.g., memory map	Physical I/Os

are explicit. The interaction between tasks is performed through communication primitives with different semantics, allowing blocking or nonblocking calls, in order or out of order data access, and direct access to channel data. The TTL APIs define three abstraction levels: the *vector_read* and *vector_write* functions are typical system level functions, which combines synchronization with data transfers, the *reAcquireRoom* and *releaseData* functions (*re* stands for relative) grant or release atomic accesses to vectors of data that can be loaded or stored out of order, but relative to the last access (i.e., with no explicit address). This corresponds to virtual architecture level APIs. Finally, the *AcquireRoom* and *releaseData* lock and unlock access to scalars, which requires the definition of explicit addressing schemes. This corresponds to the transaction accurate architecture level APIs.

The Multiflex approach proposed in [10] targets multimedia and networking applications, with the objective of having good performance even for small granularity tasks. Multiflex supports both a symmetric multiprocessing (SMP) approach that is used on shared memory multiprocessors, and a remote procedure call–based programming approach called DSOC (distributed system object component). The SMP functionality is close to the one provided by POSIX, that is, it includes thread creation, mutexes, condition variables, etc. [19] The DSOC uses a broker to spawn the remote methods. These abstractions make no separation between virtual architecture and transaction accurate architecture levels, since they rely on fixed synchronization mechanisms. Hardware support for locks and run queues management is provided by a concurrency engine, and the processors have several hardware contexts to allow context switches in one cycle. DSOC uses a CORBA like approach, but implements hardware accelerators to optimize the performances.

The authors in [11] introduce the concept of service dependency graph to represent HW/SW interface at different abstraction levels and handle application specific API. This model represents the hardware–software interface as a set of interdependent components providing and requiring services. Cheong et al. propose a programming model called TinyGALS, which combines the locally synchronous with the globally asynchronous approach for programming event-driven embedded systems [20].

In the previous section (Table 9.3), we showed that a suitable programming model for MPSoC needs to be defined at several abstraction levels corresponding to different design steps. This hierarchical view of the programming model ensures a seamless implementation of higher level APIs on lower level ones. In order to ensure a better match between the programming model and the underlying hardware architecture, the APIs also have to be made extensible, at each abstraction level, to cope with the broad range of possible hardware components. The existing MPSoC programming models seem either to focus on one aspect or the other. We argue that it is important to consider both aspects, that is, hierarchy and extensibility, when designing an MPSoC oriented programming model.

9.5 Simulink- and SystemC-Based MPSoC Programming Environment

In this section, we apply the concepts previously introduced using a Simulink- and SystemC-based programming environment as a case study. Firstly, we illustrate the adopted MPSoC abstraction levels, modeled using Simulink and SystemC environments, and then we summarize the basic steps required for programming heterogeneous MPSoC.

9.5.1 Programming Models at Different Abstraction Levels Using Simulink and SystemC

The following section gives more details about the programming models used at different MPSoC abstraction levels. Figure 9.5 illustrates the adopted abstraction levels for a simplified application made of three tasks (T1, T2, and T3), that need to be executed on architecture made of 2 processing units and several memory HW-SS. For each level, Figure 9.5 shows the software organization, the hardware–software interface, and the hardware architecture that will be used to execute and validate the software

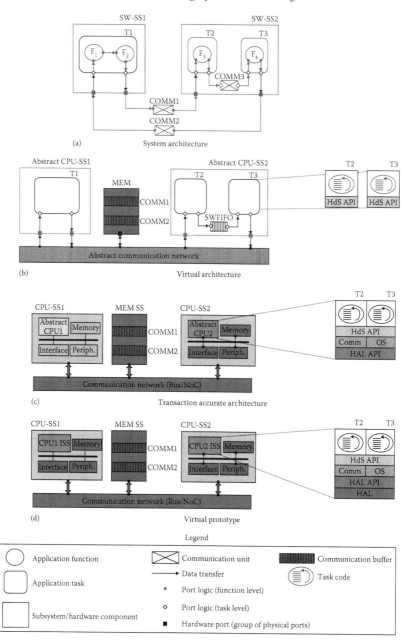

FIGURE 9.5
MPSoC hardware and software at different abstraction levels.

component at the corresponding abstraction level. The key differentiation between these diverse levels is the way of specifying the hardware–software interfaces.

The highest level is the *system architecture level* (Figure 9.5a). In this case, the software is made of a set of functions grouped into tasks. The function is an abstract view of the behavior of an aspect of the application. Several tasks may be mapped on the same SW-SS. The communication between functions, tasks, and subsystems makes use of abstract communication links (e.g., standard Simulink links [15]) or explicit communication units that correspond to specific communication paths of the target platform.

The links and units are annotated with communication mapping information. The corresponding hardware model consists of a set of abstract subsystems. The simulation at this level allows validation of the application's functionality. The programming model relies on implicit primitives for the communication, task control, and hardware accesses based on Simulink semantics.

Figure 9.5a shows the system architecture model with the following symbols: circles for the functions, rounded rectang circles to represent the tasks, rectangles for the subsystems, crossed rectangles for the communication units between the tasks, filled circles for the ports of the functions, diamonds for the logic ports of the tasks and filled rectangles for group of hardware ports. The dataflow is illustrated by unidirectional arrows.

For the considered example, the system architecture is made of 2 abstract software subsystems (SW-SS1, SW-SS2) and 2 inter-subsystem communication units (COMM1, COMM2). The SW-SS1 software subsystem encapsulates task T1, while the subsystem SW-SS2 groups together tasks T2 and T3. The intra-subsystem communication between the tasks T2 and T3 inside SW-SS1 is performed through the communication unit COMM3.

The next abstraction level is called *virtual architecture level* (Figure 9.5b). The hardware–software interfaces are abstracted using an HdS API that hides the OS and the communication layers. The application code is refined into tasks that interact with the environment using explicit primitives of the HdS API. In fact, the HdS APIs forms the programming model at the virtual architecture level, characterized by explicit communication primitives and I/O protocols and implicit tasks control primitives. Each task is refined to sequential C code using static scheduling of the initial application functions. This code is the final application code that will constitute the top layer of the software stacks. The communication primitives of the HdS API access explicit communication components. Each data transfer specifies an end-to-end communication path. For example, the functional primitives *send_mem(ch,src,size)/ recv_mem(ch,dst,size)* may be used to transfer data between the two processors using a global memory connected to the system bus, where *ch* represents the communication channel used for the data transfer, *src/dst* the source or destination buffer, and *size* the number of words to be exchanged. The communication buffers are mapped on explicit hardware resources.

At the virtual architecture level, the software is executed using an abstract model of the hardware architecture that provides an emulation of the HdS

API. The hardware model is composed of these abstract subsystems, explicit interconnection component and storage resources. In this chapter, the virtual architecture platform is considered as a SystemC model where the software tasks are executed as SystemC threads.

In the example illustrated in Figure 9.5b, the virtual architecture model is made of two abstract processor subsystems (CPU1-SS, CPU2-SS) and a global memory (MEM) interconnected through an abstract communication network. The communication units *comm1* and *comm2* are mapped on the global memory, and the communication unit *comm3* becomes a software FIFO (SWFIFO).

The next level is called the *transaction accurate architecture level* (Figure 9.5c). At this level, the hardware–software interfaces are abstracted using a HAL API that hides the processor's architecture. The code of the software task is linked with an explicit OS and specific I/O software implementation to access the communication units. The resulting software makes use of hardware abstraction layer primitives (HAL_API) to access the hardware resources. The programming model at this level uses

- Explicit tasks control primitives provided by the OS, such as *run_scheduler()* to find a new task ready for execution, *context_switch()* to switch the context between the current task and the new task found ready for execution, *create_task()* to set the context and initialize a new task, etc.
- Communication primitives for data transfers with explicit addresses, e.g., *read_mem(addr, dst, size)/write_mem(addr, src, size)*, where *addr* represents the source, the destination address, *src/dst* represents the local address, and *size* the size of the data.
- Explicit primitives for the hardware resources management, such as *enable_interrupt()/disable_interrupt()* to enable or disable specific interrupt vectors, *set_DMA()* to configure a channel for a DMA transfer, etc.

The software is executed using a more detailed development platform to emulate the network component, the explicit peripherals used by the HAL API and an abstract computation model of the processor. The simulation at this level allows validating the integration of the application with the OS and the communication layer. It may also provide precise information about the communication performance. In this work, the transaction accurate architecture is represented by a SystemC model, where the software stacks are executed as external processes communicating with the SystemC simulator through the IPC layer of the Linux OS running on the host machine.

In the example illustrated in Figure 9.5c, the transaction accurate architecture model is made of the two processor subsystems (CPU1-SS, CPU2-SS) and the global memory subsystem (MEM-SS) interconnected through an explicit communication network (bus or NoC). Each processor subsystem includes an abstract execution model of the processor core (CPU1, respectively CPU2), local memory, interface, and other peripherals. Each processor subsystem executes a software stack made of the application tasks code, communication, and OS layers.

Finally, the HAL API and processor are implemented through the use of a HAL software layer and the corresponding processor part for each SW-SS. This represents the *virtual prototype level* (Figure 9.5d). At the virtual prototype level, the communication primitives of the programming model consists of physical I/Os, e.g., *load or store*. The platform includes all the hardware components such as cache memories or scratchpads. The scheduling of the communication and computation activities for the processors becomes explicit. The simulation at this level allows cycle accurate performance validation and it corresponds to classical hardware–software cosimulation models with ISS [21,22] for the processors and RTL components or cycle accurate TLM components for the hardware resources.

In the example illustrated in Figure 9.5d, the two processor subsystems (CPU1-SS, CPU2-SS) include ISS for the execution of the software stack corresponding to CPU1 and CPU2, respectively. Each processor subsystem executes a software stack made of the application tasks code, communication, OS, and HAL layers.

9.5.2 MPSoC Programming Steps

This section describes a programming environment, which employs the programming models at the four MPSoC abstraction levels previously described (system architecture, virtual architecture, transaction accurate architecture, and virtual prototype).

Programming an MPSoC means to generate software running on the MPSoC efficiently by using the available resources of the architecture for communication and synchronization. This involves two aspects: software stack generation and validation for the MPSoC, and communication mapping on the available hardware communication resources and validation for MPSoC.

As shown in Figure 9.6, the software generation flow starts with an application and an abstract architecture specification. The application is made of a set of functions. The architecture specification represents the global view of the architecture, composed of several HW-SS and SW-SS. The main steps in programming the MPSoC architecture are

- Partitioning and mapping the application onto the target architecture subsystems
- Mapping application communication on the available hardware communication resources of the architecture
- Software adaptation to specific hardware communication protocol implementation
- Software adaptation to detailed architecture implementation (specific processors and memory architecture)

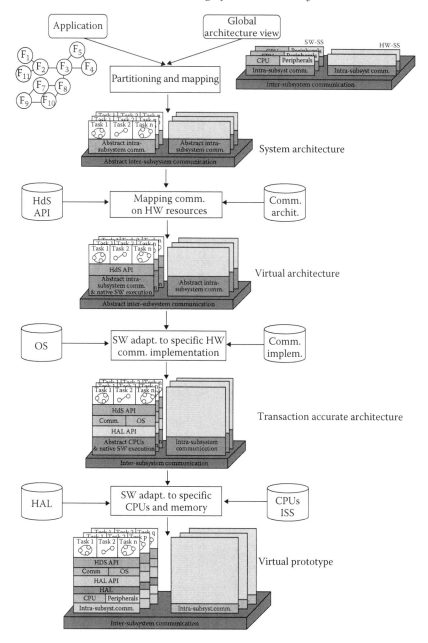

FIGURE 9.6
MPSoC programming steps.

The result of each of these four phases represents a step in the software and communication refinement process. The refinement is an incremental process. At each stage, additional software component and architecture details are integrated with the previously generated and validated components. This results to a gradual transformation of a high level representation with abstract components and high level programming models into a concrete low level executable software code. The transformation has to be validated at each design step. The validation can be performed by formal analysis, simulation, or combining simulation with formal analysis [23]. In the following, we will use simulation-based validation to ensure that the system behavior respects the initial specification.

During the *partitioning* and *mapping* of the application on the target architecture, the relationship between application and architecture is defined. This refers to the number of application tasks that can be executed in parallel, the granularity of these tasks (coarse grain or fine grain), and the association between tasks and the processors that will execute them.

The result of this step is the decomposition of the application into tasks and the association between tasks and processors. The resulting model is the system architecture model. The system architecture model represents a functional description of the application specification, combined with the partitioning and mapping information. Aspects related to the architecture model (e.g., processing units available in the target hardware platform) are combined into the application model (i.e., multiple tasks executed on the processing units). Thus, the system architecture model expresses parallelism in the target application through capturing the mapping of the functions into tasks and the tasks into subsystems. It also makes explicit the communication units to abstract the intra-subsystem communication protocols (the communication between the tasks inside a subsystem) and the inter-subsystem communication protocols (the communication between different subsystems).

The second step implements the *mapping of communication* onto the hardware platform resources. At this phase, the different links used for the communication between the different tasks are mapped on the hardware resources available in the architecture to implement the specified protocol. For example, a FIFO communication unit can be mapped to a hardware queue, a shared memory or some kind of bus-based device. The task code is adapted to the communication mechanism through the use of adequate HdS communication primitives. The resulting model is named virtual architecture model.

The next step of the proposed flow consists of *software adaptation to specific communication protocol implementation*. During this stage, aspects related to the communication protocol are detailed, for example, the synchronization mechanism between the different processors running in parallel becomes explicit. The software code has to be adapted to the synchronization method, such as events or semaphores. This can be done by using the services of OS and communication components of the software stack. The resulting model is the Transaction Accurate Architecture model.

The final step corresponds to *specific adaptation of the software to the target processors and specific memory map*. This includes the integration of the processor dependent software code into the software stack (HAL) to allow low level access to the hardware resources and the final memory mapping. The resulting model is called Virtual Prototype model.

These different steps of the global flow correspond to different software components generation and validation at different abstraction levels.

9.6 Experiments with H.264 Encoder Application

In this section, we apply the proposed programming environment for a complex MPSoC architecture. The target application corresponds to the H.264 encoder, also called AVC (advanced video coding). Firstly, the specification of the target architecture and application are given, and then, the programming steps at the system architecture, virtual architecture, transaction accurate architecture, and virtual prototype levels are described, respectively.

9.6.1 Application and Architecture Specification

The H.264 encoder application is a video processing multimedia application that supports coding and decoding of 4:2:0 YUV video formats [24]. The main functions of the H.264 encoder are illustrated in Figure 9.7. The input image frame (F_n) of a video sequence is processed in units of a macroblock, each consisting of 16 pixels. To encode a macroblock, there are three main steps: (1) prediction, with the main blocks motion estimation-ME, motion compensation-MC, and frame filtering; (2) transformation with quantization (T, Q, and Reorder); and (3) entropy encoding (CABAC in this case). The H.264 standard supports seven sets of capabilities, which are referred to

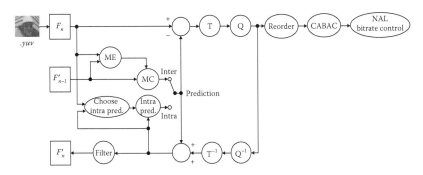

FIGURE 9.7
H.264 encoder.

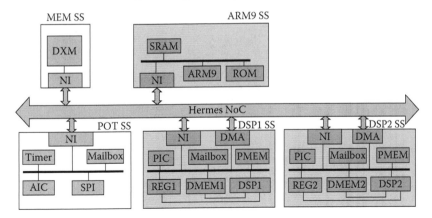

FIGURE 9.8
Diopsis R2DT with Hermes NoC.

as profiles, targeting specific class of applications. In this section, the main profile will be used as an application case study.

The target MPSoC architecture is named Diopsis R2DT (RISC + 2 DSP) tile [25]. As shown in Figure 9.8, it contains three SW-SS: one ARM9 RISC processor subsystem and two ATMEL magicV VLIW DSP processing subsystems.

The hardware nodes represent the global external memory (DXM) and POT (peripherals on tile) subsystem. The POT subsystem contains the peripherals of the ARM9 processor and the I/O peripherals of the tile. All the three processors may access the local memories and registers of the other processors and also the distributed external memory (DXM). The different subsystems are interconnected using the Hermes network on chip (NoC), which supports two types of topologies: Mesh and Torus [26].

9.6.2 Programming at the System Architecture Level

Programming at the system architecture level consists of functional modeling of the application, partitioning the application into the tasks, and mapping them onto the processing subsystems.

Therefore, the H.264 application functions are mapped onto the available SW-SS, as shown in Figure 9.9. Thus, the *DSP1-SS* is responsible for encoding a frame of the video sequence. The *DSP2-SS* compresses the encoded frame. The *ARM9-SS* creates the final bitstream and computes the bit-rate controller. The application executes in pipeline fashion and requires three application data transfers between the processors: *COMM1* between *DSP1* and *DSP2*, *COMM2* between *DSP2* and *ARM9*, and *COMM3* between *ARM9* and *DSP1*.

The resulting system architecture is modeled using the Simulink environment. To validate the H.264 encoder algorithm, the system architecture

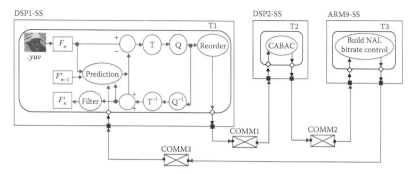

FIGURE 9.9
System architecture model of H.264.

model is simulated using a discrete-time simulation engine. The input test video is a 10 frames video sequence in QCIF YUV 420 format. The simulation requires approximately 30 s on a PC running at 1.73 GHz with 1 GBytes RAM.

The H.264 simulation allowed validating the functionality, but also measuring early execution requirements. Thus, the total number of iterations necessary to decode the 10 frames video sequence was equal with the number of frames. This is because of the fact that all the application functions implemented in Simulink operate at the frame level. The communication between the DSP1 and DSP2 processors uses a communication unit that requires a buffer of 288,585 words to transmit the encoded frame from the DSP1 processor to the DSP2 in order to be compressed. The DSP2 processor and the ARM9 processor communicate through a communication unit that requires a buffer of 19,998 words. The last communication unit between the ARM9 and DSP1 processors requires one word buffer size in order to store the quanta value required for the encoder. The total number of words exchanged between the different subsystems during the encoding process of the 10 frames video sequence, using main profile configuration of the encoder algorithm, was approximately 3085 kWords.

9.6.3 Programming at the Virtual Architecture Level

Programming at the virtual architecture level consists of generating the C code for each task from the system architecture model. The generated tasks code for the H.264 encoder application uses *send_data(. . .)/recv_data(. . .)* APIs for the communication primitives and is optimized in terms of data memory requirements.

Table 9.4 shows the task code and data size of the software at the virtual architecture level. The first two columns represent the code, respectively the data size of the functions that are independent of the design and optimization methods, which are part of an independent library. The third and fourth

TABLE 9.4

Task Code Generation for H.264 Encoder

Library Code Size (Bytes)	Library Data Size (Bytes)	Multitasking Code Size (Bytes)	Multitasking Data Size (Bytes)
270,994	132	366,060	148

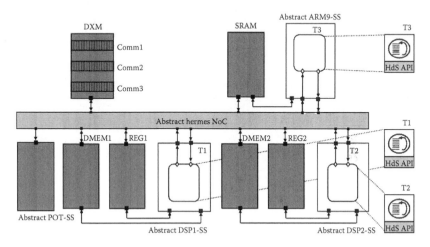

FIGURE 9.10

Global view of Diopsis R2DT running H.264.

columns show the code and data size obtained with memory optimization techniques.

The hardware at the virtual architecture level consists of a SystemC hardware platform, consisting of abstract processor subsystems and interconnect components. Figure 9.10 illustrates a conceptual view of the virtual architecture for the Diopsis R2DT with Hermes NoC.

The virtual architecture can be simulated not only to validate the tasks code, but also to gather important early performance measurements to profile the interconnect charge, for instance, the number of words exchanged between the tasks through the network component or the total packets initiated for the transfer by various subsystems.

Figure 9.11 shows the total words passed through the NoC in case of different communication mapping schemes. Hence, when all the communication buffers are mapped on the DXM memory, as shown in Figure 9.10, the NoC is accessed to transfer 6,171,680 words during the encoding process of the 10 frames. In another case, *comm1* is mapped on DXM, *comm2* on REG2 and *comm3* on DMEM1. This case required 5,971,690 words to be transferred through the NoC. A third case maps *comm1* on DMEM1, *comm2* on DMEM2, and *comm3* on SRAM and it generates 3,085,840 words to be operated by the NoC.

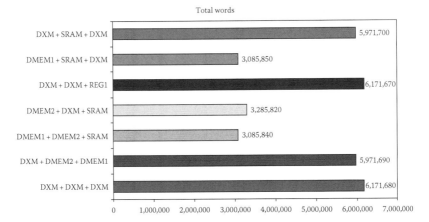

FIGURE 9.11
Words transferred through the Hermes NoC.

TABLE 9.5
Results Captured in Hermes NoC Using DXM as Communication
Scheme

H.264	NoC Address	Read/Write Requests	Total Sent Packets	Sent MBytes
DXM	0×0	0	83,352	17,324
ARM9-SS	1×0	2,426	4,853	68
DSP1-SS	1×1	39,260	78,522	16,167
DSP2-SS	1×2	41,663	83,327	2,090

Table 9.5 summarizes the results captured during the simulation of the
H.264 encoder application in case of the first communication scheme, with all
the buffers mapped on the DXM memory. The first and the second columns
represent the correspondence between the different cores connected to the
NoC and the NoC addresses. The third column represents the total number of reads and writes requested over the NoC. Based on these values, the
designer may define a better mapping of hardware cores over the NoC or
the size of packets. The fourth and the fifth columns (Packets and MBytes
sent) allow evaluating the real amount of communication injected into the
NoC through each network interface. In this example, the DXM was the core
that inserted the biggest amount of data in the NoC. The DXM packets are
originated from read requests and confirmation packets.

In all the communication mapping schemes, the simulation time required
to encode the 10 image frames using QCIF YUV 420 format was approximately 40 s on a PC running Linux OS at 1.73 GHz.

9.6.4 Programming at the Transaction Accurate Architecture Level

Programming at the transaction accurate architecture level means to build each software stack running on the processors. This consists of combining the tasks code with the OS and communication libraries. Thus, the H.264 tasks code previously designed is combined with a tiny OS necessary for the interrupts management and the tasks initialization, and the implementation of the *send_data(...)/recv_data(...)* communication primitives. The processors execute single task on top of the OS.

The transaction accurate architecture of the Diopsis R2DT tile with Hermes NoC is illustrated in Figure 9.12. The hardware platform is composed of the three processor subsystems (ARM9-SS, DSP1-SS, and DSP2-SS), one global MEM-SS, and the peripherals on tile subsystem (POT-SS), all subsystems having the local architecture detailed. The different subsystems are interconnected through an explicit Hermes NoC.

The simulation of the transaction accurate architecture allows validating the integration of the tasks code with the OS and communication libraries, but it also provides better performance estimation, such as communication performances.

At this level, in order to analyze the overall system performance, we experimented with several communication architectures by changing the interconnection component and/or communication mapping scheme. The NoC allows various mapping schemes of the IPs over the NoC with different impact on performance. In this work, two different mappings of the IP cores

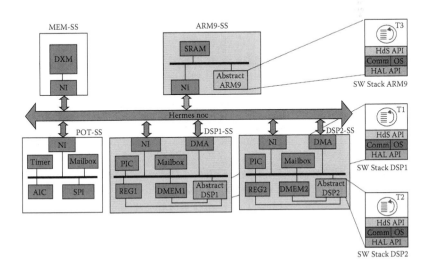

FIGURE 9.12
Global view of the transaction accurate architecture for Diopsis R2DT with Hermes NoC running H.264 encoder application.

Y\ X	0	1	2
0	MEM-SS	POT-SS	-
1	ARM9-SS	DMA1	DMA2
2	-	DSP1-SS	DSP2-SS

Scheme A

Y\X	0	1	2
0	DMA1	-	-
1	ARM9-SS	MEM-SS	DSP2-SS
2	POT-SS	DSP1-SS	DMA2

Scheme B

FIGURE 9.13
IP cores mapping schemes A and B over the NoC.

over the Mesh and Torus NoC are experimented: Scheme A and Scheme B, respectively. Figure 9.13 summarizes these schemes by presenting the correspondence between the Network Interface and the IP core, e.g., the MEM-SS is connected in Scheme B at the network interface with address 1×1 (both x and y coordinates are 1).

Table 9.6 presents the results of the transaction accurate simulations for various interconnection components (AMBA bus, NoC) with different topologies for the NoC (Torus, Mesh), different IP cores mapping over the NoC and diverse communication buffer mapping schemes. The estimated performance indicators are: estimated execution cycles of the H.264 encoder, the simulation time using the different interconnect components on a PC running at 1.73 GHz with 1 GBytes RAM and the total routing requests for the NoC. These results were evaluated for the two considered IP mapping schemes shown in Figure 9.13 (A and B) and for three communication buffer mapping schemes: *DXM+DXM+DXM*, *DMEM1+DMEM2+SRAM* and *DMEM1+SRAM+DXM*. The AMBA had the best performance, as it implied the fewest clock cycles during the execution for all the communication mapping schemes. The Mesh NoC attained the worse performance in case of mapping all the communication buffers onto the *DXM* and similar performance with the Torus in case of using the local memories.

This is explained by the small numbers of subsystems interconnected through the NoC. In fact, NoCs are very efficient in architectures with more than 10 IP cores interconnected, while they can have a comparable performance results with the AMBA bus in less complex architectures. Between the NoCs, the Torus has better path diversity than the Mesh. Thus, Torus reduces network congestion and decreases the routing requests. Also, Scheme A of IP cores mapping provided better results than Scheme B for the *DMEM1+DMEM2+SRAM* buffer mapping. For the other buffer mappings the performance of Scheme A was superior to Scheme B. In fact, the ideal IP cores mapping scheme would have the communicating IPs separated by only one hop (number of intermediate routers) over the network to reduce latency.

9.6.5 Programming at the Virtual Prototype Level

Programming at the virtual prototype level consists of integrating the HAL layer into the software stack for each particular processor subsystem and to

TABLE 9.6

Execution and Simulation Times of the H.264 Encoder for Different Interconnect, Communication, and IP Mappings

Communication Mapping Scheme	Interconnect	IPs Mapping over NoC	Execution Time at 100 MHz (ns)	Simulation Time (min)	Execution Cycles	Simulation Cycles/Second	NoC Routing Requests	Average Interconnect Latency (Cycles/Word)
DXM + DXM + DXM	Mesh	Scheme A	64,028,725	36 min	3,201,436	1482	96,618,508	25
DXM + DXM + DXM	Torus	Scheme A	46,713,986	28 min 29 s	2,335,699	1527	78,217,542	16
DMEM1 + DMEM2 + SRAM	Mesh	Scheme A	28,573,705	12 min 54 s	1,428,685	1846	13,118,044	10
DMEM1 + DMEM2 + SRAM	Torus	Scheme A	26,193,039	12 min	1,309,652	1819	12,674,692	9
DMEM1 + SRAM + DXM	Mesh	Scheme A	26,233,039	14 min 55 s	1,594,237	1466	13,144,538	11
DMEM1 + SRAM + DXM	Torus	Scheme A	26,193,040	14 min 48 s	1,309,652	1475	14,479,723	10
DXM + DXM + DXM	Mesh	Scheme B	35,070,577	18 min 34 s	1,753,529	1574	24,753,610	9
DXM + DXM + DXM	Torus	Scheme B	35,070,587	19 min 8 s	1,753,529	1527	24,753,488	9
DMEM1 + DMEM2 + SRAM	Mesh	Scheme B	31,964,760	17 min 8 s	1,598,238	1555	18,467,386	13
DMEM1 + DMEM2 + SRAM	Torus	Scheme B	31,924,752	16 min 14 s	1,595,238	1639	15,213,557	13
DMEM1 + SRAM + DXM	Mesh	Scheme B	31,964,731	18 min 38 s	1,598,237	1430	18,512,403	15
DMEM1 + SRAM + DXM	Torus	Scheme B	31,924,750	16 min 42 s	1,596,238	1593	18,115,966	14
DXM + DXM + DXM	AMBA	—	17,436,640	8 min 24 s	871,832	1730	—	9
DMEM1 + DMEM2 + SRAM	AMBA	—	17,435,445	7 min 18 s	871,772	1990	—	9
DMEM1 + SRAM + DXM	AMBA	—	17,435,476	7 min 17 s	871,774	1995	—	9

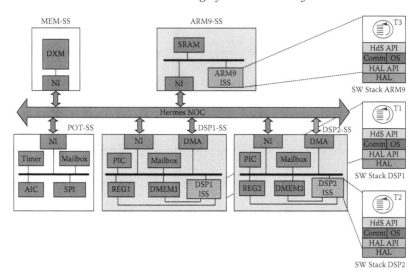

FIGURE 9.14
Global view of the virtual prototype for Diopsis R2DT with Hermes NoC
running H.264 encoder application.

fix the final memory mapping. The H.264 encoder running on the Diopsis
R2DT architecture at the virtual prototype level is illustrated in Figure 9.14.
There are three final software stacks running on the architecture, one per
processor. The hardware platform includes ISS to execute the final software.
The ISS allows determining the execution cycles spent on each task.

9.7   Conclusions

In this chapter, we discussed the use of high level programming for the
abstraction of hardware–software interfaces. A programming model is made
of a set of functions (implicit and/or explicit primitives) that can be used by
the software to interact with the hardware. In case of heterogeneous MPSoC
design, the programming model hides both hardware and software refinements. We proposed a Simulink- and SystemC-based programming environment, which combines high level programming models with the low level
details and involves design at various abstraction levels. The proposed programming environment was applied upon a complex MPSoC architecture to
execute efficiently the H.264 video encoder application and to explore different communication schemes.

References

1. H. Meyr, Application specific processors (ASIP): On design and implementation efficiency, *Proceeding of SASIMI 06*, Nagoya, Japan, 2006.

2. TI OMAP. http://www.omap.com

3. Nomadik. http://www.st.com

4. Nexperia. http://www.nxp.com

5. P.S. Paolucci, A. Jerraya, R. Leupers, L. Thiele, P. Vicini, SHAPES: A tiled scalable software hardware architecture platform for embedded systems, *Proceeding of CODES+ISSS 2006*, Seoul, Korea, 2006, pp. 167–172.

6. J. Turley, Survey says: Software tools more important than chips, *Embedded Systems Design*, April 11, 2005 [Online]. Available: http://www.embedded.com/columns/surveys/160700620?_requestid=177492

7. MPICH—MPI implementation. http://www-unix.mcs.anl.gov/mpi/mpich/index.htm

8. W. Wolf, *High-Performance Embedded Computing: Architectures, Applications, and Methodologies*, Morgan Kaufmann Publishers, Inc., San Francisco, CA, 2006.

9. D. Culler, J.P. Singh, A. Gupta, *Parallel Computer Architecture: A Hardware/Software Approach*, Morgan Kaufmann Publishers, Inc., San Francisco, CA, August 1998, ISBN 1558603433.

10. P. Paulin, C. Pilkington, M. Langevin, E. Bensoudane, D. Lyonnard, O. Benny, B. Lavigueur, D. Lo, G. Beltrame, V. Gagne, G. Nicolescu, Parallel programming models for a multi-processor SoC platform applied to networking and multimedia, *IEEE Transactions on VLSI Journal*, 14(7), 667–680, 2006.

11. A. Bouchhima, X. Chen, F. Pétrot, W. Cesario, A.A. Jerraya, A unified HW/SW interface model to remove discontinuities between HW and SW design, *Proceedings of EMSOFT 2005*, New Jersey City, NJ, September 18–22, 2005.

12. A. Jerraya, W. Wolf, Hardware-software interface codesign for embedded systems, *Computer*, 38(2), 63–69, February 2005.

13. D. Skillicorn, D. Talia, Models and languages for parallel computation, *ACM Computing Surveys*, 30(2), 123–169, 1998.

14. A. Jerraya, A. Bouchhima, F. Petrot, Programming models and HW-SW interfaces abstraction for multi-processor SoC, *Proceeding of DAC 2006*, San Francisco, CA, 2006, pp. 280–285.

15. Simulink, The MathWorks Inc., http://www.mathworks.com

16. F. Ghenassia, *Transaction-Level Modeling with SystemC. TLM Concepts and Applications for Embedded Systems*, Springer, New York, 2005, ISBN 0-387-26232-6.

17. S. Yoo, M.W. Youssef, A. Bouchhima, A.A. Jerraya, M. Diaz-Nava, Multi-processor SoC design methodology using a concept of two-layer hardware-dependent software, *Proceedings of DATE'04*, Paris, France, February 2004.

18. P. van der Wolf, E. de Kock, T. Henriksson, W. Kruijtzer, G. Essink, Design and programming of embedded multiprocessors: An interface-centric approach, Special Session, *Proceeding of CODES+ISSS 2004*, Stockholm, Sweden, September 2004.

19. D.R. Butenhof, *Programming with POSIX Threads*, Addison Wesley, Boston, MA, May, 1997.

20. E. Cheong, J. Liebman, J. Liu, F. Zhao, TinyGALS: A programming model for event-driven embedded systems, *Proceeding of 2003 ACM Symposium on Applied Computing*, Melbourne, FL, March 9–12, 2003, pp. 698–704.

21. J.A. Rowson, Hardware/software cosimulation, *Proceeding of DAC 1994*, San Diego, CA, June 6–10, 1994, pp. 439–440.

22. L. Semeria, A. Ghosh, Methodology for hardware/software co-verification in C/C++, *Proceeding of ASP-DAC 2000*, Yokohama, Japan, 2000, pp. 405–408.

23. S. Kunzli, F. Poletti, L. Benini, L. Thiele, Combining simulation and formal methods for system-level performance analysis, *Proceeding of DATE 2006*, Munich, Germany, March 6–10, 2006, pp. 236–241.

24. J.-W. Chen, C.-Y. Kao, Y.-L. Lin, Introduction to H.264, *Proceeding of ASP-DAC 2006*, Yokohama, Japan, January 24–27, 2006, pp. 736–741.

25. P.S. Paolucci, A.A. Jerraya, R. Leupers, L. Thiele, P. Vicini, SHAPES: A tiled scalable software hardware architecture platform for embedded systems, *Proceeding of CODES+ISSS 2006*, Seoul, Korea, 2006, pp. 167–172.

26. F. Moraes et al., HERMES: An infrastructure for low area overhead packet-switching networks-on-chip integration, *VLSI Journal*, 38(1), 2004, 69–93.

10

Platform-Based Design and Frameworks: Metropolis and Metro II

Felice Balarin, Massimiliano D'Angelo, Abhijit Davare, Douglas Densmore, Trevor Meyerowitz, Roberto Passerone, Alessandro Pinto, Alberto Sangiovanni-Vincentelli, Alena Simalatsar, Yosinori Watanabe, Guang Yang, and Qi Zhu

CONTENTS

10.1 Introduction

System-level design (SLD) means many different things to many different people. In our view, SLD is about the design of a whole that consists of several components where specifications are given in terms of functionality along with

- Constraints on the properties the design has to satisfy
- Constraints on the components that are available for implementation
- Objective functions that express the desirable features of the design when completed

This definition is general since it relates to many application domains from semiconductors to systems such as cars, airplanes, buildings, telecommunications, and biological systems. To deal with system-level problems, our view is that the issue to address is not developing new tools, albeit they are essential to advance the state of the art in design; rather it is the understanding of the principles of system design, the necessary change to design methodologies, and the dynamics of the supply chain. Developing this understanding is necessary to define a sound approach to the needs of the system and component industry as they try to serve their customers better, and to develop their products faster and with higher quality. This chapter is about principles and how a unified methodology together with a supporting software framework, as challenging as it may seem, can be developed to bring the embedded electronics industry to a new level of efficiency.

To demonstrate this view, we will first present the design challenges for future systems and a manifesto espousing the benefits of a unified methodology. We will then summarize a methodology, platform-based design (PBD), that has been developed over the past decade and that we believe can fulfill these needs. Further, we will present METROPOLIS, a software framework supporting this methodology, and METRO II, a second-generation framework tailored to industrial test cases. We conclude the chapter with two test cases in two diverse domains: semiconductor chips (a UMTS multi-chip design) and energy-efficient buildings (an indoor air quality control system).

10.2   Platform-Based Design

10.2.1   Design Challenge

The fundamental design problems are similar enough across the domains exemplifying our work that a common design methodology can be used across these domains (refer to [61] for an extensive review of SLD issues). A design methodology for systems necessarily needs to encompass other business and societal issues: the industrial supply chain participants have to comprehend how to play in a rich environment where there are many opportunities for innovating and optimizing products beyond what has been possible so far. The revolution here is about interfaces and communication patterns between the computing and the physical systems, as well as about the functions that a rich ensemble of heterogeneous entities can collectively implement. Not all the behaviors of these systems are designed on purpose; some are unexpected and emerging from unforeseen interactions among the myriad of components. Recently, malfunctioning of the new automatic baggage-handling system at Heathrow has cost more than $50 million to British Airways in a single day and caused havoc among passengers [35]. A data acquisition system program of the National Census Bureau based on 500,000 wireless handheld devices was canceled because cost overran from $600 million to more than $1.3 billion due to incomplete specifications and faulty design [14]. Design verification becomes a challenge that is difficult to overcome unless new design methods are used where unwanted behaviors are excluded by the design. In addition, safety, security, and privacy concerns will be imposed by regulatory bodies as well as by customers, thus adding new dimensions to the problem of design.

The fundamental issue is understanding the principles of system design, the necessary change to design methodologies, the creation of appropriate abstractions, and the dynamics of the supply chain, i.e., a novel science and engineering for system design. Developing this science and engineering is necessary to define a sound approach to the needs of the system and IC (integrated circuit) companies as they try to serve their customers better, and to develop their products faster and with a higher quality. Major investments are necessary from all constituencies: industry, governments, and research institutions.

10.2.2 Principles of Platform-Based Design

We need to adopt a methodology that can be applied seamlessly at all levels of abstractions and that can capture design constraints and components at each level. In addition, the methodology must favor a system view of the design so that it can deliver an increased productivity and the capability of dealing with multiple design goals, thus always keeping in mind performance, power, reliability, and cost as essential characteristics of the final solution. Productivity can be increased by a number of techniques including

- Design reuse, i.e., the capability of utilizing preexisting designs or components
- Early verification and analysis
- Virtual prototyping
- Automatic traversal of the design hierarchy from specifications to implementation

The overall quality of the final product can also be substantially enhanced by a methodology that includes the possibility of

- Automatic synthesis that move from one level of abstraction to another, thus allowing an effective design-space exploration
- Formally verifying the properties of the design as we progress toward implementation

Our team has developed a design methodology over the past 10 years, PBD [33,54], that in our opinion supports most of the above requirements.

A platform is defined to be a library of components that can be assembled to generate a design at that level of abstraction. This library not only contains computational blocks but also communication components that are used to interconnect the computational components. It is important to keep communication and computation elements well separated as we may want to use different methods for representing and refining these blocks. For example, communication plays a fundamental role in determining the properties of models of computation (MoCs). In addition, designing by aggregation of components requires great care in defining the communication mechanisms as they may help or hurt the reuse of design. In design methodologies based on the IP assembly, communication is the most important aspect. The unexpected behavior of the composition is often due to a negligence in defining the interfaces and the communication among the components.

Each element of the library has a characterization in terms of performance parameters (e.g., timing, power, and dimensions) together with the functionality it can support. The library is a "parameterization" of the space of possible solutions. Not all elements in the library are preexisting components. Some may be placeholders or virtual components to indicate the flexibility of customizing a part of the design into a hardware component (e.g., an ASIC) that is offered to the designer. For this part, we do not have a complete characterization of the element since performance parameters depend upon a lower level of abstraction that, in this case, is not available yet. A platform instance is a set of components that is selected from the library (the platform) and whose parameters are set. In the case of a virtual component, the parameters are set by the requirements rather than by the implementation. In this case, they have to be considered as constraints for the next level of refinement.

10.2.2.1   PBD Flow

The PBD flow concept of platform encapsulates the notion of reuse as a family of solutions that share a set of common features (the elements of the platform). Since we associate the notion of platform to a set of potential solutions to a design problem, we need to define functionality (what the system is supposed to do), the process of mapping a functionality onto the platform elements that will be used to build a platform instance or an "architecture" (how the system does what it is supposed to do), and the constraints and goals that need to be considered when implementing the functionality. This process provides a mechanism to proceed toward implementation in a structured way. We strongly believe that the function and the architecture should be kept separate as these two aspects are often defined independently by different groups (e.g., video encoding and decoding experts versus hardware/software designers for multimedia systems). Too often we have seen designs being difficult to understand and debug because the functionality and the architecture are intermingled during the design capture stage. If the functional aspects are indistinguishable from the implementation aspects, it is very difficult to evolve the design over multiple hardware generations.

Functionality is an implicit or explicit relationship between a set of input signals and a set of output signals defined at a given level of abstraction, i.e., a process [37]. This relationship may be structured as a set of communicating subprocesses. Constraints and goals for the design are expressions that involve variables including physical quantities such as power, latency, and throughput, defined at a multiplicity of levels of abstraction. For example, assuming that the functionality to be implemented by the design is a fast Fourier transform, the designer may have the maximum power consumed by the implementation as a constraint, and area minimization and throughput maximization as goals. These quantities are not defined at the level at which the FFT algorithm is described; they appear when implementation decisions are taken. The goals and constraints guide the selection of the platform instance, since it is a composition of components with a set of characteristics that may involve the variables that enter the constraint and goal expressions. The constraint and goal expressions are called performance indexes of the design and may have to be adjusted (propagated) to reflect the refinements of the platforms as we move towards the final implementation. We then define the specification of a design as the combination of functionality, platforms, and performance indexes. The functionality specifies what the system has to do, the platforms define what can be used to implement the functionality, and the performance indexes guide the mapping process so that the final implementation is feasible and optimized. The PBD design process is not a fully top-down nor a fully bottom-up approach in the traditional sense; rather, it is a meet-in-the-middle process (see Figure 10.1) as it can be seen as the combination of two efforts:

- *Top-down*: Map an instance of the functionality of the design into an instance of the platform and propagate constraints.
- *Bottom-up*: Define a platform by choosing its components and an associated performance abstraction (e.g., timing of the execution of the instruction set for a processor, power consumed in performing an atomic action, number of literals for technology-independent optimization at the logic synthesis level, and area and propagation delay for a cell in a standard cell library).

The "middle" is where functionality meets the platform. Given the original semantic difference between the two, the meeting place must be described with a common domain so that the "mapping" of the functionality to elements of the platform to yield an implementation can be formalized and automated.

10.2.2.2 *"Fractal" Nature of PBD: Successive Refinements*

To better represent the refinement process and to stress that platforms may preexist the functionality of the system, we turn the triangles on their side and represent the "middle" as the mapped functionality. Then, the refinement process takes place on the mapped functionality that becomes the "function" at the lower level of the refinement. Another platform is then considered side by side with the mapped instance and the process is iterated until all the components are implemented in their final form. This process can be applied at all levels of abstraction, thus exposing what we call the fractal nature of design. Note that some of the components may reach their final implementation early in the refinement stage if these elements are fully detailed in the platform. Figure 10.1 exemplifies this aspect of the methodology. It is reminiscent of the Y chart of Gajski, albeit it has a different meaning, since, for us, architecture and functionality are peers and architecture is not necessarily derived from functionality but may exist independently.

To progress in the design, we have to map the new functionality to the new set of architectural components. In case the previous step used an architectural component that was fully instantiated, then that part of the design is considered complete: the mapping process involves only those parts that have not been fully specified as of yet.

In the PBD, the partitioning of the design into hardware and software is not the essence of system design as many think; rather it is a consequence

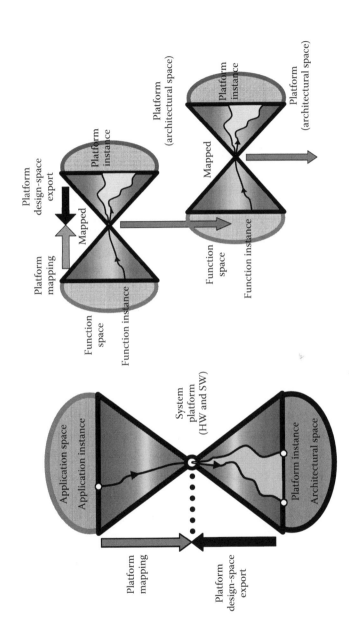

FIGURE 10.1

Hourglass diagram and fractal nature of the PBD. (From Sangiovanni-Vincentelli, A., *Proc. IEEE*, 95, 467, March 2007. With permission.)

of decisions taken at a higher level of abstraction. Critical decisions instead involve the abstraction levels and semantics with which the functionality and the architecture are defined.

10.2.2.3 Design Parameters for PBD

In the PBD refinement-based design process, platforms should be defined to eliminate large loop iterations for affordable designs. They should restrict the design space via new forms of regularity and structure that surrender some design potential for a lower cost and first-pass success. The library of functional and communication components is the design space that we are allowed to explore at the appropriate level of abstraction. Establishing the number, location, and components of intermediate platforms is an essential part of the PBD.

Designs with different requirements and specifications may use different intermediate platforms, hence different layers of regularity and design-space constraints. The trade-offs involved in the selection of the number and characteristics of platforms relate to the size of the design space to be explored and the accuracy of the estimation of the characteristics of the solution adopted. Naturally, the larger the step across platforms, the more difficult it is to predict the performance and provide tight bounds on the performance. In fact, the design space for this approach may actually be smaller than the one obtained with smaller steps because it becomes more difficult to explore meaningful design alternatives and these restrictions on search impede design-space exploration. Ultimately, predictions/abstractions may be so inaccurate that design optimizations are misguided and the bounds are incorrect, leading to multiple design iterations.

The identification of precisely defined layers where the mapping processes take place is an important design decision and should be agreed upon at the top design-management level. Each layer supports a design stage where the performance indexes that characterize the architectural components provide an opaque abstraction of lower layers that allows accurate performance estimations used to guide the mapping process. This approach results in better reuse, because it decouples independent aspects, e.g., a given functional specification to low-level implementation details, or to a specific communication paradigm, or to a scheduling algorithm. It is very important to define only as many aspects as needed at every level of abstraction, in the interest of flexibility and rapid design-space exploration.

Success stories of the applications of these principles abound. For example, in the semiconductor arena there are many examples of platforms and of the PBD approaches. The NXP, ST, TI, and Renesas documented this approach in public presentations and publications [20,23,29,65]. Among system companies, Magneti-Marelli, BMW, GM, Pirelli, United Technologies Corporation, and Telecom Italia have demonstrated applications and research interest in the PBD.

10.3   Metropolis **Design Environment**

The Metropolis design environment was developed to embody the principles of the PBD. Metropolis has its own specification language that is used to capture functionality, architecture, and mapping, as well as a variety of design activities.

10.3.1   Overview

The Metropolis specification language, the Metropolis meta-model (MMM), is based on formal semantics and remains general enough to support existing MoCs [37] and accommodate new ones. The MMM also includes a logic language to capture nonfunctional and declarative constraints. This meta-model can support not only the functionality capture and analysis, but also the architecture description and the mapping of functionality to architectural elements. Because the MMM has a precise semantics, in addition to a simulation Metropolis is able to perform synthesis, and formal analysis for complex electronic-system design.

The Metropolis design environment supports three design activities: specification, analysis, and synthesis.

- *Specification* serves to communicate the design intent and expected results. It focuses on the interactions among people working at different abstraction levels and among people working concurrently at the same abstraction level. The MMM includes constraints that represent, in an abstract form, requirements not yet implemented or assumed to be satisfied by the rest of the system and its environment.
- *Analysis* determines how well an implementation satisfies the requirements through simulation and formal verification. A proper use of abstraction can dramatically accelerate verification. An overuse of detailed representations, on the other hand, can introduce excessive dependencies between developers, reduce understandability, and diminish the efficiency of analysis mechanisms.
- *Synthesis* is supported across the abstraction levels used in a design. The typical problems we address include setting the parameters of architectural elements such as cache sizes, or designing scheduling algorithms and interface blocks. We also deal with synthesis of the final implementations in hardware and software. In Metropolis, a specification may mix declarative and executable constructs of the MMM, which are then automatically translated into the semantically equivalent mathematical models to which the synthesis algorithms are applied.

Metropolis offers syntactic and semantic mechanisms to compactly store and communicate all relevant design information. Designers can plug in algorithms and tools for all possible design activities that operate on the design information. The capability of introducing external algorithms and tools is important, because these needs vary significantly for different application domains. To support this capability, METROPOLIS provides a parser that reads the MMM designs and a standard API that lets developers browse, analyze, modify, and augment additional information within these designs. For each tool integrated into METROPOLIS, a back-end uses the API to generate the required input by the tool from the relevant portion of the design.

10.3.2 METROPOLIS Meta-Model

The MMM is a language that specifies networks of concurrent objects, each taking sequential actions. The behavior of a network is formally defined by the execution semantics of the language [7,59]. A set of networks can be used to represent all the aspects for designs, i.e., function, architecture, mapping, refinement, abstraction, and platforms.

10.3.2.1 Function Modeling

The function of a system is described as a set of objects that concurrently take actions while communicating with each other. Each such object is termed a *process* in the MMM, and associated with a sequential program called a *thread*. A process communicates through *ports* where a port is specified with an *interface*, declaring a set of methods that can be used by the process through the port. In general, one may have a set of implementations of the same interface, and we refer to objects that implement port interfaces as *media*. Any medium can be connected to a port if it implements the interface of the port. This mechanism allows the MMM to separate computation by processes from communication among them. This separation is essential to facilitate the description of the objects to be reused for other designs. Figure 10.2 shows a network of two producer processes and one consumer process that communicate through a medium.

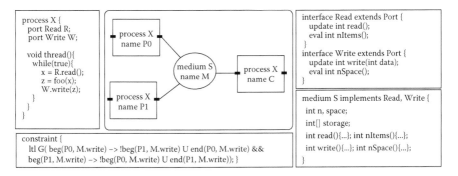

FIGURE 10.2
Functional model of two producers and one consumer. (From Balarin, F. et al., *IEEE Comput.*, 36, 45, April 2003. With permission.)

Once a network of processes is given, the behavior of the network is precisely defined by the MMM semantics as a set of *executions*. First, we define an execution of a process as a sequence of *events*. Events are entries to or exits from a piece of code by a program. For example, for process X in Figure 10.2, the beginning of the call to *R.read()* is an event, as is its termination. The execution of a network is then defined as a sequence of vectors of events, where each vector has at the most one event for each process to define the set of events that happen altogether. The MMM can model nondeterministic behavior, which is useful to abstract a part of the design, and thus there may be more than one possible executions of the network in general.

Constraints, written in logic formulas, further restrict the set of executions defining the set of legal executions [7,64]. For example, the constraint in Figure 10.2 specifies the mutual exclusion of the two producers when one calls the medium's *write* method. Constraints describe the coordination of processes or relate the behavior of networks through mapping or refinement.

10.3.2.2 *Architecture Modeling*

Architectures are distinguished by two aspects: the functionality they can implement and that implementation's efficiency. In the meta-model, the former is modeled as a set of *services* offered by an architecture to the functional model. Services are just methods, bundled into interfaces [56].

To represent the efficiency of an implementation, we need to model the cost of each service. This is done first by decomposing each service into a sequence of events, and then annotating each event with a value representing the cost of the event.

To decompose services into sequences of events, we use networks of media and processes, just as in the functional model. These networks often correspond to physical structures of implementation platforms.

For example, Figure 10.3 shows an architecture consisting of n processes, $T1, \ldots, Tn$, and three media, *CPU, BUS*, and *MEM*. The architecture in Figure 10.3 also contains so-called quantity managers, represented by diamond-shaped symbols. The processes model software tasks executing on a CPU, while the media model the CPU, bus, and memory. The services offered by this architecture are the *execute, read*, and *write* methods implemented in the *Task* processes. The *thread* function of a *Task* process repeatedly and nondeterministically executes one of the three methods. In this way, we model the fact that the *Tasks* are capable of executing these methods in any order. The actual order will become fixed only after the system functionality is mapped to this architecture, when each *Task* implements a particular process of the functional model.

While a *Task* process offers its methods to the functional part of the system, the process itself uses services offered by the *CPU* medium, which, in turn, uses services of the *BUS* medium. In this way, the top-level services offered by the *Tasks* are decomposed into sequences of events throughout the architecture.

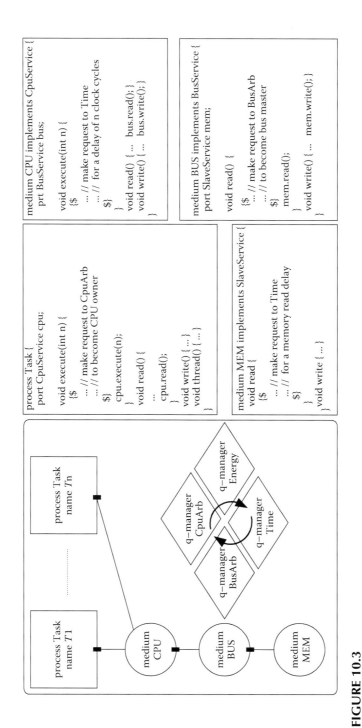

FIGURE 10.3

An architectural model. (From Balarin, F. et al., *IEEE Comput.*, 36, 45, April 2003. With permission.)

The MMM includes the notion of *quantity* used to annotate individual events with values measuring cost. For example, in Figure 10.3 there is a quantity named *energy* used to annotate each event with the energy required to process it. To specify that a given event takes a given amount of energy, we associate with that event a *request* for that amount. These requests are made to the object called *quantity manager*, which collects all requests and fulfills them, if possible.

Quantities can also be used to model shared resources. For example, in Figure 10.3, the quantity *CpuArb* labels every event with the task identifier of the current CPU owner. Assuming that a process can progress only if it is the current CPU owner, the *CpuArb* manager effectively models the CPU scheduling algorithm.

The MMM has no built-in notion of time, but it can be modeled as yet another quantity that puts an annotation, in this case a time stamp, on each event. Managers for common quantities such as time are provided with Metropolis as standard libraries, and are understood directly by some tools (e.g., time-driven simulators) for the sake of efficiency. However, quantity managers can also be written by design flow developers, in order to support quantities that are relevant for a specific application domain. When multiple quantity managers coexist in the same architectural model, they often need to interact with each other. For example, if the *CpuArb* manager decides not to schedule a particular event, other quantity managers, such as *Time* and *Energy*, do not need to annotate the same event. Therefore, the quantity annotation process is iterative, and reaches a fixpoint at the end, where all quantity managers agree on the annotations for all events.

10.3.2.3 *Mapping*

Evaluating a particular implementation's performance requires mapping a functional model to an architectural model. The MMM can do this without modifying the functional and architectural networks. It does so by defining a new network to encapsulate the functional and architectural networks, and relates the two by synchronizing events between them. This new network, called a *mapping network*, can be considered as a top layer that specifies the mapping between the function and the architecture.

The synchronization mechanism roughly corresponds to an intersection of the sets of executions of the functional and architectural models. Functional model executions specify a sequence of events for each process, but usually allow arbitrary interleaving of event sequences of the concurrent processes, as their relative speed is undetermined. On the other hand, architectural model executions typically specify each service as a timed sequence of events, but exhibit nondeterminism with respect to the order in which services are performed, and on what data. Mapping eliminates from the two sets all executions except those in which the events that should be synchronized always appear simultaneously. Thus, the remaining executions represent timed sequences of events of the concurrent processes.

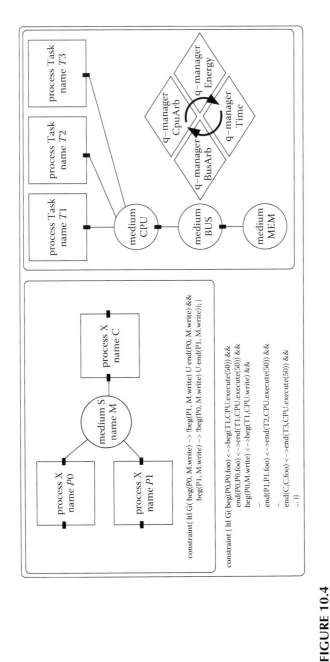

FIGURE 10.4
Mapping of function to architecture. (From Balarin, F. et al., *IEEE Comput.*, 36, 45, April 2003. With permission.)

For example, Figure 10.4 shows a mapping network that combines the functional network from Figure 10.2 with the architectural network from Figure 10.3. The mapping network's constraint clauses synchronize the events of the two networks. For example, executions of *foo*, *read*, and *write* by *P0* have been synchronized with executions of *execute*, *read*, and *write* by *T1*. Since *P0* executes its actions in a fixed order while *T1* chooses its actions nondeterministically, the effect of synchronization is that *T1* is forced to follow the decisions of *P0*, while *P0* "inherits" the quantity annotations of *T1*. In other words, by mapping *P0* to *T1* we make *T1* become a performance model of *P0*. Similarly, *T2* and *T3* are made to be performance models of *P1* and *C*, respectively.

10.3.2.4 *Recursive Paradigm of Platforms*

Suppose that a mapping network such as in Figure 10.4 has been obtained as described above. One can consider such a network itself as an implementation of a certain service. The algorithm for implementing the service is given by its functional network while its performance is defined by the architecture counterpart. In Figure 10.5, the interface on the top defines methods specifying a service, and the medium in the middle implements the interface at the desired abstraction level. Underneath the medium is a mapping network, providing a more detailed description of the implementation of the service. The MMM can relate the medium and each network by using the construct refine and constraints. For example, the constraints may say that the begin event of mpegDecode of the medium is synchronized with the begin event of vldGet of the VLD process in the network, and the end of mpegDecode is synchronized with the end of yuvPut of the OUTPUT process, while the value of the variable YUV at the event agrees with the output value of mpegDecode.

In general, many mapping networks may exist for the same service with different algorithms or architectures. Such a set of networks, together with constraints on event relations for a given interface implementation, constitutes a platform. The elements of the platform provide the same service with different costs, and one is favored over another for given design requirements. This concept of platforms appears recursively in the design process. In general, an implementation of what one designer conceives as the entire system is a refinement of a more abstract model of a service, which is in turn employed as a single component of the larger system. For example, a particular implementation of an MPEG decoder is given by a mapping network, but its service may be modeled by a single medium, where the events generated in the medium are annotated with performance quantities to characterize the decoder. Such a medium may be used as a part of the architecture by a company that designs set-top appliances. This medium serves as a model used by the appliance company to evaluate the set-top box design, while the same is used as design requirements by the provider of the MPEG decoder.

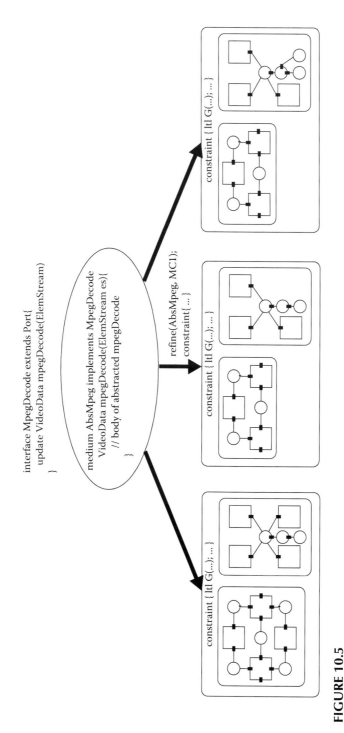

FIGURE 10.5
Multiple mapping netlists refining the same service to form a platform. (From Balarin, F. et al., *IEEE Comput.*, 36, 45, April 2003. With permission.)

Similarly, the MPEG decoder may use in its architecture a model of a bus represented as a medium provided by another company or group. For the latter, the bus design itself is the entire system, which is given by another mapping network that refines the medium. This mapping network may be only one of the candidates in the platform of the communication service, and the designer of the MPEG decoder may explore various options based on his design criteria.

This recursive paradigm of the PBD is described in a uniform way in the MMM. The key is the formal semantics to precisely define the behavior of networks, and the mechanism of relating events of different networks with respect to quantities annotated with the events.

10.3.3 Metropolis Tools

The Metropolis environment includes tools to perform a variety of tasks, some common, such as simulation and formal verification, and some more specialized, such as property verification and synthesis. These tasks are described in this section.

10.3.3.1 Simulation

Nondeterminism, inherent in the MMM, poses unique verification problems. Traditionally, systems are verified by simulating their response to a given set of stimuli. However, with nondeterminism, there may be many valid responses, and it is impossible to know which one will be exhibited by the final implementation. In [6], we have proposed a generic MMM simulation algorithm that allows exhibiting one of the acceptable behaviors for a given input stimulus. Which behavior is selected is to some extent under the user's control. The choice may be driven by different objectives at various design stages. In the beginning, one may want any legal behavior to check for trivial mistakes. At a later stage of the design, one may want to simulate as many legal behaviors as possible in order to evaluate the design more thoroughly.

Another challenge comes from the orthogonalization of concerns principle embodied in the MMM, which requires special treatment to gain efficient simulation. In contrast, the simulation of traditional specifications with monolithic modeling styles can be made efficient with much less effort. In [63], by exploring the intrinsic characteristics of orthogonal concerns, we came up with a series of optimization techniques that boost the simulation performance by two orders of magnitude.

In Metropolis, instead of generating the machine code directly from meta-model descriptions, we translate the meta-model to an executable language, which is combined with the simulation algorithm also implemented in the same language. This feature is important to co-simulate designs captured in the meta-model together with existing designs that have been already specified in other languages. The actual simulation is then carried out in terms of the resulting language. We have tested this simulation approach in SystemC 2.0 [49], Java [3], and C++ with a thread library. In addition, we have devised a service-based formalism [62] that can effectively integrate models specified at different abstraction levels, in different specification languages, and with different MoCs. We also enhanced our simulation tool to support the co-simulation of these heterogeneous models. Further, this service-based formalism became the foundation of the second generation of the METROPOLIS environment, covered in Section 10.4.

10.3.3.2   Formal Property Verification

Both academia and industry have long studied formal property verification, but the state-explosion problem restricts its usefulness to protocols and other high abstraction levels. At the implementation level or other low abstraction levels, hardware and software engineers have used simulation monitors as basic tools to check simulation traces while debugging designs.

Verification languages, such as Promela, which are used by the Spin model checker [28], allow only simple concurrency modeling and are not amenable to the system design specification, which requires complex synchronization and architecture constraints. In contrast, METROPOLIS, with its formal semantics, automatically generates verification models for all the levels of the design [15].

Our translator automatically constructs the Spin verification model from the MMM specification, taking care of all system-level constructs. For example, it can automatically generate a verification model for the example in Figure 10.2 and verify the medium's nonoverwriting properties. Further, as the translator refines the design through structural transformation and architectural mapping, it can prove more properties, including throughput and latency. This kind of property verification typically requires several minutes of computation on a 1.8 GHz Xeon machine with 1 Gbyte of memory. When the state space complexity becomes too high, METROPOLIS uses an approximate verification and provides the user with a confidence factor on the passing result.

10.3.3.3   Simulation Monitor

Simulation monitors offer an attractive alternative to formal property verification. In METROPOLIS, designers can use logic of constraints (LOC) formulas [7] to specify quantitative properties. The system can automatically translate the specification to simulation monitors in C++ [16], thus relieving designers from the tedious and error-prone task of writing monitors in the simulator's language. The monitors analyze the traces and report any LOC formula violations. Like any other simulation-based approach, this one can only disprove an LOC formula if it finds a violation—it can never prove conclusively the formula's correctness because that would require exhaustively analyzing traces. The automatic trace analyzer can be used in concert with model checkers. It can perform property verification on a single trace even when other approaches would fail because of their excessive memory and space requirements.

In our experience with applying the automatic LOC-monitor technique to large designs with complex traces, we have found that in most cases the analysis completes in minutes and consumes only hundreds of bytes of data memory to store the LOC formulas. The analysis time tends to grow linearly with the trace size, while the memory requirement remains constant regardless of the trace size.

10.3.3.4 Quasi-Static Scheduling

We have developed an automatic synthesis technique called quasi-static scheduling (QSS) [19] to schedule a concurrent specification on computational resources that provide limited concurrency. The QSS considers a system to be specified as a set of concurrent processes communicating through the FIFO queues, and generates a set of tasks that are fully and statically scheduled, except for data-dependent controls that can be resolved only at runtime. A task usually results from merging parts of several processes together and shows less concurrency than the initial specification. Moreover, the QSS allows interprocess optimizations that are difficult to achieve if processes remain separated, such as replacing interprocess communication with assignments.

This technique proved particularly effective and allowed us generate a production-quality code with improved performance. Applying the QSS to a significant portion of an MPEG-2 decoder resulted in a 45% increase in the overall performance.

The assumptions that the QSS requires for the input specification form a subset of what the MMM can represent. Therefore, when integrating the QSS into the METROPOLIS framework, we addressed two main problems: how to verify if a design satisfies the required set of rules and how to convey all relevant design information to the QSS tool.

We addressed the first problem by providing a library of interfaces and communication media that implement a FIFO communication model. Those parts of the design optimized with the QSS need to use these communication primitives.

To convey relevant design information to the QSS, we use a back-end tool that translates a design to be scheduled with the QSS into a Petri net specification, which is QSS's underlying model. The QSS then uses the Petri net to produce a new set of processes. These new processes show no interprocess communication because the QSS removes it. The processes communicate with the environment using the same primitives implemented in the library. The new code can thus be directly plugged into the MMM specification as a refinement of the network selected for scheduling.

10.4 METRO II Design Environment

METRO II [21] is the successor to METROPOLIS [8]. METRO II was developed at the University of California, Berkeley, starting in 2006. The following sections introduce the reader to the goals of METRO II, the components of the framework, and the mapping and execution semantics used.

10.4.1 Overview

The second-generation METRO II framework is based on considerations derived from the limitations of METROPOLIS we experienced in a set of designs that were carried out in collaboration with our industrial partners. These considerations are as follows.

1. *Heterogeneous IP import.* IP providers create models using domain-specific languages and tools. Requiring a singular form of design entry in a system-level environment requires complex translation of the original specification into the new language while making sure that semantics is preserved. If different designs or different components within the same design can have different semantics, the heterogeneity has to be supported by the new environment. There are two main challenges that have to be addressed: wrapping and interconnecting the IP.

   First, IPs can be described in different languages and can have different semantics that can be tightly related to a particular simulator. Importing the IP entails providing a way of exposing the IP interface. The user must have the necessary aids to define wrappers that mediate between the IP and the framework such that the behavior can be exposed in an unambiguous way.

   Secondly, wrapped components have to be interconnected. Even if the interfaces are exposed in a unified way, interconnecting them is not usually a straightforward process. The data and the flow of control between IP blocks must be exposed in such a way that the framework has sufficient visibility.

2. *Behavior-performance orthogonalization.* For design frameworks that support multiple abstraction levels, different implementations of the same basic functionality may have the same behavioral representations but different costs. For instance, different processors will be abstracted into the same programmable components. What distinguishes them is the performance vs. cost trade-off. Moreover, not all metrics are considered or optimized simultaneously. It should be possible to introduce performance metrics during the design process, as the design proceeds from specification to implementation.

   The specification of what a component does should be independent of how long it takes or how much power it consumes to carry out a task. This is the reason why we introduce dedicated components, called annotators, to annotate quantities to events.

A distinction has to be made between quantities used just to track the value of a specific metric of interest and quantities whose value is used for synchronization. For instance, time is used to synchronize actions and it is not merely a number that is computed based on the state evolution of the system. For quantities that influence the evolution of the system, special components, called schedulers, are provided to arbitrate shared resources.

The separation of schedulers from annotators allows for a simpler specification and provides a cleaner separation between behavior and performance. As a result, instead of a two-phase execution as in METROPOLIS, the execution semantics become three phase.

3. *Mapping specification.* Mapping relates the functional and architectural models to realize the system model. The specification of this mapping must be carried out such that there is minimal modification to the functional and architectural models themselves.

Following the PBD approach, we want to keep the functionality and the architecture separate. The implementation of the functionality on the architecture is achieved in the mapping step. In order to explore several different implementations with minimal effort, the design environment needs to provide a fast and an efficient way of mapping without modifying either the functional or the architectural model much. In METROPOLIS, this is achieved by event-level synchronization constraints, as shown in [22]. While providing a powerful way to link the models, this approach breaks the encapsulation of the models by allowing constraints between arbitrary pairs of events and allowing access to any local variables in the scope of the events. Also, since there are no special declarative constructs for mapping, this process of finding events and setting up constraints is not easy for designers to manipulate and debug.

In METRO II, we restrict the mapping to be at the service level, i.e., the only accessible events for synchronization constraints are the begin/end events of interface methods in function and architecture models. Also, the only accessible values are parameters and return values of the interface methods. This coarser granularity and a more restrictive mapping approach maintain the IP encapsulation and make mapping more robust for designers.

10.4.2   METRO II Design Elements

An initial implementation of the METRO II framework has been carried out in SystemC 2.2. The framework has been tested under Linux, Solaris, and cygwin.

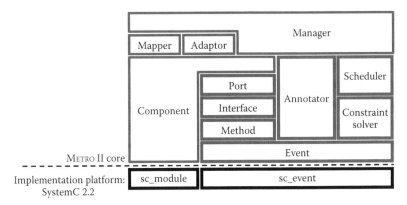

FIGURE 10.6
Implementation of METRO II.

The infrastructure is summarized in Figure 10.6. The sc_event and sc_module classes from SystemC are leveraged directly to derive the corresponding event and component classes in METRO II.

The connection and coordination of components are carried out through events. The event is a key concept in METRO II. It is formally defined as a tuple $< p, T, V >$, where p is a process that generates the event, T is a tag set, and V is a set of associated values. Tags are used to describe the semantics of the system and values are used to represent the states of the system.

Methods, interfaces, and ports are built on the concept of event. A method is characterized by a pair of begin and end events. An interface contains one or more methods. Ports are associated with interfaces, and only ports with compatible interfaces can be connected. A component can have zero or more ports. To handle different aspects of the events, special objects are defined, including annotators, schedulers, and constraint solvers. Annotators annotate events with quantities, schedulers coordinate the execution sequence of events, and constraint solvers resolve the declarative constraints on events. Mappers and adaptors are defined to interconnect components. Mappers bridge the function methods and architecture services. Adaptors interconnect components with heterogeneous MoCs. Finally, the manager coordinates the execution of all the objects using three-phase execution semantics.

Figure 10.7 illustrates the major METRO II elements. We attempt to use the iconography here throughout the work. A snippet of the METRO II code for a reader component, a mapper, a scheduler, and an annotator in a typical producer–consumer design example is shown in the figure. More details of these elements are introduced below.

10.4.2.1 Components

A component is an object that encapsulates an imperative code in a design, either functional or architectural. Components interface with other

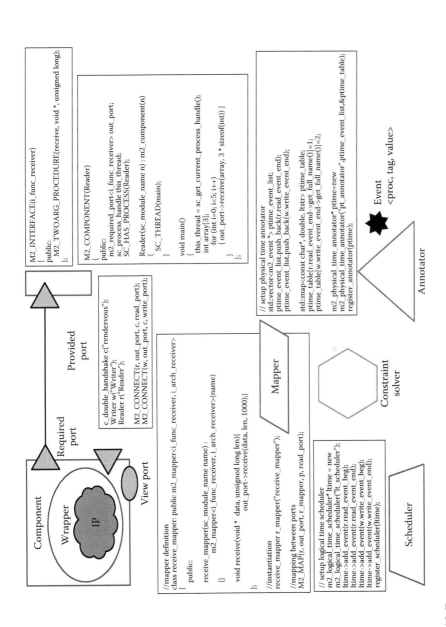

FIGURE 10.7

Overview of METRO II design elements.

components via ports. There are two descriptions of component composition: atomic components and composite components. An atomic component is a block specified in some language and is viewed by the framework as a black box with only its interface information exposed. A composite component is a group of one or more objects as well as any connections between them. When an existing IP is being imported, it will be encapsulated by a wrapper, which translates and exposes the appropriate events and interfaces from the IP. The wrapped IP becomes an atomic component in the framework.

10.4.2.2   Ports

Components can interface with each other via ports. Each port is characterized by an interface that contains a set of methods. A method consists of a sequence of events, with a unique begin/end event pair. Variables in the scope of the begin event are method arguments. Variables in the scope of the end event are return values.

By setting constraints between events associated with the ports of different components, the execution of these components can be coordinated. There are two types of ports: required ports and provided ports. Required ports are used by components to request methods that are implemented in other components. Provided ports are used by components to provide methods to other components. Connections between components are made only between a required port and a provided port with the same interface. The execution semantics that coordinate a pair of required and provided ports will be introduced in Section 10.4.3.

10.4.2.3   Constraint Solvers

Constraints are used to specify the design via declarative means, as opposed to imperative specification which is contained in components. Constraints are described in terms of events: their status (enabled or disabled), their tags, and the values associated with them. The events referenced by constraints must be exposed by ports.

Constraint solvers are objects that resolve these declaration constraints during runtime. Depending on the status, tags, and values of the events, constraint solvers decide whether to enable or disable events, thereby coordinating the execution of components.

Designers can derive various constraint solvers from the base class solver provided by the METRO II infrastructure. The main function to be implemented is the one to resolve the constraints. In METRO II, a synchronization constraint solver is provided. Two events that are specified in a synchronization constraint need to be enabled at the same time—during simulation, they need to be enabled in the same iteration. Further examples will be given in Section 10.4.3. Synchronization constraints are used for mapping between the functionality and the architecture, as is explained later.

10.4.2.4 Annotators and Schedulers

In METROPOLIS, both the performance annotation and the scheduling of events were carried out by a type of special component called a quantity manager. As stated before, to have a more clear separation of design concerns, these two aspects will be handled separately by annotators and schedulers in METRO II.

Annotators annotate events with quantities by writing tags. Each tag that represents some quantity (such as power and physical time) is determined in terms of the parameters supplied to the annotator, the status of the event, and the values of the event. Parameters are given by the designers based on the characterization of the architecture platforms. Only static parameters are permitted for annotators, which may not have their own state. For various quantities or quantities in various systems, designs can derive their own annotators from the annotator base class in METRO II. Currently, a physical time annotator is provided in the METRO II library.

The instantiation of a physical time annotator is shown in Figure 10.7. The *r.read_event_end* and *w.write_event_end* are events associated with a reader and writer component, respectively. These two events are added to a list of events to be considered for annotation. In addition, a table indexed by these events is created along with the assigned time units required for execution (1 and 2 units, respectively). This list and the table are then added to the annotator object itself. If these events are present during the second phase of execution, their tags will be updated accordingly.

Schedulers coordinate the execution of the components by enabling/disabling the events proposed by the processes of the components. Based on the local state of the scheduler, the status of the events, as well as their values and tags, the scheduler determines the scheduling of the events. A base class scheduler is provided in METRO II for designers to derive various schedulers. A logical time scheduler that schedules the events based on the physical time tags, and a round-robin scheduler that schedules the access to shared resources are provided as library schedulers. An example using the logical time scheduler is shown in the code snippet in Figure 10.7.

10.4.2.5 Mappers

As a framework based on the PBD, METRO II supports mapping through mappers, which synchronize the begin and end events of the functional methods and architectural methods. Designers are only allowed to specify mapping at this service level, with access to the parameters and return values of the methods. When the begin/end events in the functional and architectural methods are synchronized, the parameters and return values can be transfered between the two models. For instance, a functional method may have one parameter that the corresponding architectural method is unaware of. During mapping, the value of this parameter can be passed to the architectural method for its usage. METRO II provides an API to specify mappers at the service level. The implementation of mappers is a synchronization constraint solver with value passing of parameters and return values.

An example of a mapper is shown in Figure 10.7. This mapper is called "receive_mapper" and is used to map the consumer in a producer–consumer design example to a processing element, p. During mapping when the receive method is called by the functional model with two arguments, the mapper's *out_port* will call the architectural model's *receive* method that has three arguments. Also shown in Figure 10.7 are the instantiation of the mapper along with how the mapper is connected between the functional model and the architectural model.

10.4.2.6 Adaptors

There are various ways of handling heterogeneous MoCs in a design. One of the most common approaches is the hierarchical composition as in Ptolemy II [38]. With the hierarchical composition, each level of the hierarchy is homogeneous, i.e., a single MoC exists at each level, while different interaction mechanisms are allowed to be specified at different levels in the hierarchy [26]. To allow models in two heterogeneous MoCs to communicate, a third MoC may need to be found within which the two will be embedded.

In our experience, there is a strong need to interconnect heterogeneous models directly at the same level. For instance, the user may want to connect the output of a base-band-processing component (described by a dataflow model) to the input of an RF component (described by a continuous-time model). This way of handling complexity does not require changing the interface of a model in order to behave like another model. This is in line with one of our main concerns: being able to reuse IPs in different contexts.

The complexity of this approach lies in designing the correct interconnections between different MoCs. To bridge the different semantics of heterogeneous components, we use adaptors to modify events as they pass from one component to another. Denotationally, an adaptor is a relation, $A \subseteq (V \times T) \times (V' \times T')$, that maps events from one model to events of another model.

Adaptors are connected with components through specialized adaptor channels. In the PBD methodology, adaptors can be regarded as the bridge between heterogeneous functional components or between heterogeneous architectural components. The METRO II infrastructure provides the base classes of adaptors and adaptor channels. METRO II also includes an example of adaptors between dataflow and finite-state machine (FSM) semantics.

10.4.3 METRO II Semantics

Like METROPOLIS, the semantics of the METRO II framework will be centered around the connection and coordination of components. The execution semantics discussed here are involved in the simulation of a system for design-space exploration.

10.4.3.1   Three-Phase Execution

METRO II has three-phase execution semantics. In order to discuss this semantics, two other concepts must be introduced: process states and event states.

In Figure 10.8, the states that an event can have are shown. Events can be inactive, proposed, and annotated. All events begin as inactive. As the self loop shows, they can remain inactive indefinitely. When a method call on a required port generates an event it becomes proposed. It will then be annotated. If the event is then deemed appropriate to enable (via a variety of scheduling decisions) it will transition to inactive again.

Each process in METRO II has two states: *running* and *suspended*. Processes execute concurrently until an event is proposed on a required port of the component containing the process or until they are blocked on a provided port. At this point they transition to the suspended state. Once the event is enabled or the internal blocking is resolved, the processes return to the running state.

Based on this treatment of events, the design is partitioned into three phases of execution. In the first phase, processes propose possible events; the second phase associates tags with the proposed events; and the third phase allows a subset of the proposed events to execute.

1. **Base model execution.** The base model consists of concurrently executing processes that may suspend only after proposing events or by

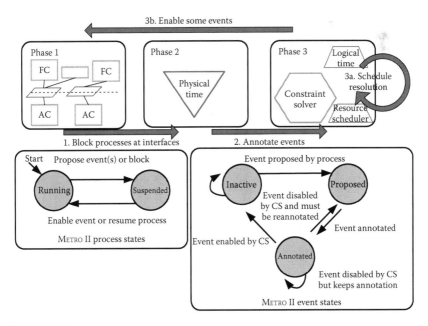

FIGURE 10.8
METRO II three-phase execution semantics.

waiting for (blocking) other processes. A process may atomically propose multiple events—this represents nondeterminism in the system. After all processes in the base model are blocked, the design shifts to the second phase. The execution of processes between blocking points is beyond the control of the framework.

2. **Quantity annotation.** In the second phase, each of the proposed events is annotated with various quantities of interest. For instance, a proposed event may be annotated with local and global time tags. New events may not be proposed during this phase of execution. In this way, events and the methods they correspond to can be associated with cost.

3. **Constraint solving.** In this phase, a subset of the proposed events is enabled and permitted to execute, while the remaining events remain suspended. Events are enabled according to schedulers and constraint solvers. These enabled events then become inactive again while simultaneously allowing their associated processes to resume to the running state. At most one event per process is permitted to execute. Once again, new events may not be proposed during this stage. Constraint solving may be based on the resolution of declarative constraints or on the imperative code.

A collection of three completed phases is referred to as a round. After the constraint solving phase, the states of some processes are switched to running while some others might still be suspended. The execution will then shift to the first phase and start a new round. Those processes that are in the running state will resume their executions. The iterations of these three phases will end when all processes finish their executions. Figure 10.8 illustrates the process states, event states, and the three phases in the execution semantics. Self loops on the inactive and annotated states illustrate that multiple rounds may pass without an update to a particular event's state.

Table 10.1 illustrates the relationships between events and phases. In the first phase (base), events can be proposed and their values can be read or written. In the second phase (annotation), tags can be read and written and values can be read. In the final phase (constraint solving), events can be disabled and their tags and values can be read. The semantics have been carefully designed so that the event manipulation adheres to our separation of

TABLE 10.1

Phase–Event Relationships

Phase	Events		Tags		Values	
	Propose	Disable	Read	Write	Read	Write
Base (1)	Yes				Yes	Yes
Annotation (2)			Yes	Yes	Yes	
Constraint solving (3)		Yes	Yes		Yes	

TABLE 10.2

Metro II Elements and Their Characteristics

Type	Threads	Events	Tags	Values	Hierarchy
Component	0+	Generate	R/W	R/W	Yes
Adaptor	0+	Generate	R/W	R	Yes
Annotator	0	Propagate	R/W	R	No
Scheduler	0	Disable	R	R	Yes

concerns methodology. This is very helpful not only in debugging simulation but also in making sure that the framework functions efficiently.

Table 10.2 indicates the Metro II elements and their characteristics. This details the presence of threads as well as the ability to manipulate events, tags, and values. It also indicates if there is hierarchy. Components and adaptors may have zero or more threads, while annotators and schedulers do not have any threads.

Events, and by extension, services, may be annotated by quantities of interest. Quantities capture the cost of carrying out particular operations and are implemented using quantity managers. Annotators are special components that provide annotation services. Schedulers are similar to quantity managers, but instead of a quantity they provide scheduling and arbitration of shared resources. Adaptors modify tags and provide interfacing between different MoCs. Depending on the MoC used and the needs of the design, different annotators and schedulers can be used.

10.4.3.2 Semantics of Required/Provided Ports

The execution semantics of the required and provided ports are as follows.

For required ports, a component proposes a begin event and associates values with the proposed event that represent the arguments of the method that is requested. When the proposed event is enabled and executed, the control transfers to the component at the other end of the connection, which owns the corresponding provided port. The component waits for the end event to be executed and obtains the return values from the method.

For provided ports, no separate process exists in the component to carry out the provided method. Instead, the component inherits the process from the caller component and executes the events in the provided method using that process. After the method has been executed, the component proposes the end event.

10.4.3.3 Semantics of Mapping

A key feature of Metro II is the ability to separately specify the functional and architectural models. The two are then mapped together to produce a system model with performance metrics. Mapping is realized by adding constraints between events from the functional model and events from the architectural model.

We will present three options for the execution semantics of mapping in METRO II. The "call graph" of the mapping options is shown in Figure 10.9. Option 1 is the first call graph shown, option 2 follows, and option 3 is the last. For options 1 and 2 the structural view (upper right of the figure) is a connection between required ports in the functional model and provided ports in the architectural model. For option 3, the mapping structure is different and is between provided ports in function and provided ports in architecture.

The first option is a sequential option in which the functional model begins execution before the architectural model. Some of the highlights of this option are captured in Table 10.3.

Figure 10.9 shows both a structural and a call-graph view of mapping in the first option. The ports in these and future diagrams are specified with the first letter of the component they belong to. Also, ports are designated as "R" or "P" if they are required or provided. "b" and "e" designate the begin and end events, respectively. These designations can be combined. For example, "FP.e" would indicate the end event of component F's provided port.

Figure 10.9 shows the mapping structure of a system using this option. The functional model contains a method call to G from P. The mapping of this method call occurs by assigning events proposed by FR to events proposed by AP. This is considered a required port to the provided port-mapping structure.

Figure 10.9 also shows the call graph of the system. Boxes with single line borders are events. Boxes that have two line borders are code blocks that may or may not contain events. The arrows indicate program flow (from left to right). If an arrow is dashed it means that two events connected to it are treated as a single event by the framework. The functional component F calls a method from the component G. This is mapped to the architectural component A, which further uses the architectural component B when providing the service.

The execution in this option occurs as follows: component F contains a process. This process is responsible for proposing the event "FR.b." "FR.b" corresponds to "GP.b" (in G). Once these events are enabled, the "G body" (the code body of the function call to G) can now execute. Upon completion, "FP.e" (in G) will be proposed. This event corresponds to "AP.b" in the architecture. The architecture body, "A body," can now execute and culminate with the proposal of "AP.e." As shown, "AP.e" corresponds to "FR.e," which completes the execution.

As shown, the mapping of methods is carried out by invoking the mapped architectural service in the process of the caller after the corresponding functional method has completed the execution.

In option 2, the execution semantics of mapping involve executing mapped architectural services before their functional counterparts. When a

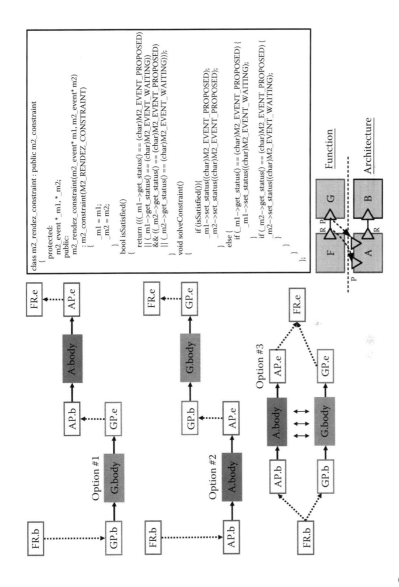

FIGURE 10.9
Metro II mapping semantics options.

TABLE 10.3
Mapping Options Overview

Option	Execution Order in Simulation	Mapping Structure (Func ↔ Arch) Port	Event Correspondence	Requires Blocking
1	Functionality then Architecture	Required ↔ Provided	FR.b, GP.b GP.e, AP.b AP.e, FR.e	Yes
2	Architecture before Functionality	Required ↔ Provided	FR.b, AP.b AP.e, GP.b GP.e, FR.e	Yes
3	Concurrent Functionality and Architecture	Provided ↔ Provided	FR.b, AP.b, FP.b FR.e, AP.e, FP.e	No

mapped method is invoked by a functional process, the begin event of that method is initially proposed, and a phase change is permitted to occur. If this event is enabled, then the architectural service executes first, immediately followed by the invoked functional method. After this, the end event of that method is proposed, with a subsequent phase change. Both the functional method and the architectural service are executed by the functional process; there are no special mapping processes. Additionally, both the functional method and the architectural service may block internally while waiting for other processes.

The functional method is parameterized with arguments and has a return type. The architectural service is also parameterized, but the return value is not used. The correspondence between the architectural service parameters and the functional service parameters is specified at compile time.

This proposal is in some regards the opposite of the first proposal. It is summarized in Table 10.3.

Figure 10.9 for the previous option shows the call graph for execution between the functional and architectural models. Basically, the functional methods need to be completed before the corresponding architecture services start. However, in some cases, this approach may not be able to reflect all the situations in the mapped system.

For instance, let us consider a shared FIFO example. Option #1 cannot assure that the architectural ordering decision impacts the functional execution, since the function methods will finish before the architecture is invoked. Therefore, the shared FIFO example may not work as expected with option #1 if one wants to use the state of the architectural FIFO to block functional processes (i.e., it is full). Essentially, functional nondeterminism cannot be resolved by the architecture. Such operations may be desirable when the architecture is better able to perform given the opportunity to make decisions based on its state (free resources, for example). This also removes some scheduling burden from other areas of the system.

The second option remedies this problem by completing architecture services before the corresponding function method starts. The new call graph is shown in Figure 10.9. This proposal shares the same mapping structure as option # 1.

The third option is summarized in Table 10.3. There is a consensus that the METRO II environment is rooted in the PBD methodology [54], where the functional model and the architectural model meet in the middle with a set of well-defined services as the binding contract. To the architectural model, the middle point represents what services it can provide to implement certain functionalities, or to estimate the implementation cost. To the functional model, the middle point describes its need of services to achieve its entire function. If we look at the design scenarios, the services that are exposed at the middle point include *execute, read_fifo,* and *write_fifo.* Therefore, the architecture model has to provide at least those services. As the three proposals exhibit, there are multiple possibilities in terms of the ports to be mapped. In fact, the syntactic difference does not really matter.

What matters is the role of the mapped architectural component and its relationship to the components on the functional side. Imagine on the functional side, the source component calls *write_fifo* that is provided by FIFO1. No matter which part in the connection (the required port, the provided port, or the connection) is mapped to the architecture, we expect the architectural service at some point to perform *write_fifo.* In that sense, the architectural counterpart corresponds to FIFO1, where both the functional and architectural parts react to the *write_fifo* request and do the job. If we can agree on this correspondence, then any mapping syntax will work. That is, on the functional side, the required port, the provided port, and the connection, each represents a pair of events; on the architectural side, the service is also represented by a pair of events. Then mapping establishes another pair of correspondences between the two pairs of events. However, from the methodology point of view, where we emphasize the meeting point between functional and architectural models, mapping connections or provided ports from the functional side seem to be better choices.

When running the functional and the architectural models together,[*] we would like the mapped services on both sides to finish simultaneously, because this will provide the most information about how an architectural model implements a functional model. However, there are concerns about the fact that suspension of processes on either side would prevent the entire mapped system from progressing. This is primarily caused by the semantics mismatch of the services from both sides. By carefully designing the consistent services, we should be able to make the mapped system work even with blocking behaviors on either or both sides.

---

[*]Note that we can also run the functional model first, recording the service demands, and then drive the execution of the architectural model. But this eliminates the behavior where the feedback from the architectural model would affect the execution of the functional model.

The mapping structure and the call graph for the third proposal are shown in Figure 10.9. Notice that in option #3, the provided ports in the functional model are mapped to the provided ports in the architectural model as well. This is different from the previous two options. Also in the call graph, it is shown that the correspondence points must be created in the form of protocols in order to create a more granular operation at the event level in each model. An example of such a protocol will be shown in more detail in the hand traces for proposal #3.

F and G are two components in the functional model, where F is making a method call on its required port *Req* to G's provided port *Prov*. In this example, the architecture is represented by components A and B, and the provided port of G has the same interface as the provided port of A. In this case, we can say that component G has been mapped into component A. For simplicity, assume that the interface of ports *F.Req*, *G.Prov*, and *A.Prov* contains only one method. The mapping between G and A is realized by placing rendezvous constraints on the begin and end events associated with this method, as shown at the bottom of Figure 10.9. Starting at the bottom left of the figure, one sees the initial event proposal of *F.Req.b*. Moving to the right, the other events are proposed in turn. Left–right arrows indicate causality while vertical arrows indicate the presence of a constraint.

Within the framework, these rendezvous constraints are handled in the same way as any other event constraints during Phase 3 of the execution semantics. Mapping uses the same infrastructure as the rest of the system, and, therefore, the simulation is not burdened with another set of semantics for mapping.

10.5 Related Work

The literature on system-level design and design-space exploration is vast. In the previous work, we presented a broad survey of numerous tools and methodologies in the context of platform-based design [25], both from industry and academia, and categorized them along different axes, such as the ability of supporting functional, architectural, and mapping descriptions, and the depth of the levels of abstraction that are covered. In this section, we focus on the approaches that are directly related to the topics presented.

10.5.1 Origin of Metro II: From Polis to Metropolis

The roots of the METROPOLIS framework can be traced back to the Polis project [5]. The main idea behind Polis is to raise the level of abstraction at which designers work and reason about the system in terms of models that can be then implemented as either hardware or software components.

System optimization is therefore carried out using a hardware/software codesign methodology. The success of the Polis framework was largely due to its well-defined MoC, co-design finite state machines (CFSMs). In CFSMs, each component is described synchronously, using a specification style based on the language Esterel [10] that is well suited for the definition of reactive systems. Using Esterel, the behavior of the system can be described as an instantaneous reaction to the external events, thus abstracting away time and facilitating verification. The synchronous hypothesis is not, however, satisfied by real implementations, especially when the system is deployed on a mixed hardware/software architecture, where delays may be dependent upon the current state of the application and on the particular scheduling policy used by the processors. To explicitly account for this, Polis introduces a globally asynchronous interaction model for the interconnection of the locally synchronous components. Asynchronicity is guaranteed by the presence of buffers that hold the data until the recipient is ready to react. This scheme also facilitates the development of several different ways for the hardware and the software components to interact, through direct communication or shared memory areas.

METROPOLIS was born as an extension of Polis to deal with systems built out of different MoCs, instead of only CFSMs. One of the central ideas of the new framework is to separate between the computation (the process) and the communication (the media). Different media implementations would be used to simulate different interaction semantics. For this reason, the METROPOLIS MoC was named the MMM. Along with the hardware/software codesign technique, METROPOLIS was also intended to implement a full refinement-based design flow. To support this, another innovation in METROPOLIS was introduced, the ability to define the model of the platform or of the architecture in the same formalism used to define the functionality of the system. This made it possible to keep the function and the architecture separate on one side, and to synchronize their execution using a dedicated language of constraints (LoC) on the other. This device was central in the way of mapping development in METROPOLIS. In particular, communication media could be refined to explore different ways of carrying out the communication. In addition, a third element called quantity manager was introduced to account for the evaluation of performance metrics under arbitrary algebras (defined in the methods of the manager) during the very simulation of the system. This technique was also useful to regulate access to shared resources, which introduce an indirect interaction between otherwise unrelated functional blocks.

METRO II builds on the METROPOLIS experience, but focuses less on the model itself, and more on the integration of different existing models. This is achieved with a more lightweight, wrapper-style environment built on top of SystemC. It also greatly simplifies the annotation and scheduling aspects of event management by separating each activity into its own phase. Mapping is also made easier by abstracting the constraints at the service level.

10.5.2 Industrial Approaches

A host of industrial tools have their roots in the model driven architecture (MDA) developed by the OMG [43], which can be cast in the general framework of the PBD. The MDA is an approach to using models in software development. At its basis is the separation between the specification of the operations of a system from the details of the way the system uses the capabilities of a platform. The goal is to achieve portability, interoperability, and reusability of models. In this approach, the development starts with a computation-independent model (CIM) representing the system in the context of the environment in which it will operate, by detailing its requirements with no regard to functionality. This model is later refined into a platform-independent model (PIM), used to specify the functionality of the system without committing to any particular platform. At the same time, platform models are developed as sets of subsystems and technologies that provide a coherent set of functionality through interfaces and specified use patterns. A PIM is transformed into a platform-specific model (PSM) via a mapping that consists of model transformations, i.e., rules or algorithms that take objects in the PIM model language and generate (one or more) objects in the PSM model language. Each mapping is therefore specific to a particular platform. While the basic principles underlying our methodology are similar to the MDA, our objectives are different and geared toward a wider architecture-service exploration. We, therefore, use a mapping that is more generic, and is intended to provide performance metrics rather than generating a detailed implementation. Our notion of mapping makes it easier to adapt to different platforms, which in turn results in a more efficient evaluation of design alternatives.

In addition to the tools inspired by the MDA, industrial approaches can also be classified according to their ability of capturing functionality, describing architecture services, or aiding in the assignment of functionality to services. Four approaches related to our work that follow this classification are presented here.

An industrial tool for creating platform descriptions with mapping capabilities (two of the three Y-chart branches) is the VaST Systems Technology's Comet/Meteor [60]. Comet focuses on creating high performance processor and architecture models at the system level. This tool uses virtual processors, buses, and peripheral devices to create candidate architectures, called virtual system prototypes (VSP), for design-space exploration. The VSP models are provided by VaST in the form of libraries or can be entered by the user in C/C++/SystemC.

Meteor is an embedded software development environment for the VSPs created by Comet. It interacts with VSPs for cycle-accurate simulation and parameter-driven configuration. This process follows much more closely a typical design process for a microprocessor including an optimizing code development, than our approach. A code is developed for a specific

VSP environment as opposed to capturing the pure functionality of an application.

An industrial tool with functional, platform, and mapping capabilities (all three branches of the Y chart) is the MLDesign's MLDesigner [45]. This tool supports discrete event, dynamic dataflow, synchronous dataflow, Boolean dataflow, continuous time, and FSM MoCs. It is intended for a top-down design flow starting from the initial specification to the final implementation. The MLDesigner includes an integrated development environment (IDE) to integrate all aspects in one package. The two major ways in which this work differs from ours are in its inherent top-down nature and in the fact that it supports a finite set (albeit large) of MoCs.

The Mirabilis Design's Visual Sim [44] product family supports the same MoCs natively as MLDesigner and also covers all three branches of the Y chart. The design process in Visual Sim begins by constructing a model of the system using a parameterizable library provided by Mirabilis. This model can be augmented with C, C++, Java, SystemC, Verilog, or VHDL blocks. The library blocks operate semantically using a wide variety of MoCs. The design is then partitioned into software, middleware, and hardware. Finally, the design is optimized by running simulations and adjusting parameters of the library elements. The underlying simulation kernel is the Ptolemy. This tool focuses very much on design-space exploration via the manipulation of the library block parameters. Unlike our approach, it begins with a monolithic design and refines it into its HW and SW components via a manual ad hoc refinement process.

The closest approach to our work is the Cofluent's Systems Studio [17] that provides the transaction-level SystemC models to perform design-space exploration using the Y-chart modeling methodology. The functional description is a set of communicating processes executing concurrently. The platform model is a set of communicating processes and shared memories linked by shared communication nodes. The platform model has performance attributes associated with it as well. This approach is very similar to METROPOLIS but does not support as wide a variety of MoCs or as rich a constraint-verification infrastructure.

10.5.3   Academic Approaches

An approach similar to the MDA is model-integrated computing [32] (MIC). It is based on the use of models for design and representation, and on the use of generators to synthesize and integrate the system. In the MIC, the vehicles for facilitating the design process are the models described in an appropriate modeling language. Unlike the MDA, which uses the UML, the MIC is based on the observation that a single modeling language is not suitable for all embedded systems. Instead, domain-specific modeling languages must be tailored to the needs of each particular domain. Thus, different modeling languages are used to express the functionality, the architecture, and their relation (the mapping). The MIC is supported by a set of tools that can create and manage various modeling languages. For instance, the generic modeling environment (GME) has been designed to facilitate the construction and the manipulation of domain-specific modeling languages, by providing a way of specifying an abstract as well as a concrete syntax (textual or graphical), including well-formed constraints and static semantics. The language design activity is again based on the UML and the Object Constraint Language (OCL) constraints [48], which are used as meta-languages. However, the resulting language need not be related to the UML at all. The manipulation in the GME also includes the possibility to merge and compose languages at the syntactic level, by identifying relationships between elements of different languages.

Languages designed in the GME can be manipulated using GReAT [1] to implement a variety of model transformations based on standard traversal patterns or on graph-rewriting rules (called meta-generators). These tools are used to convert models automatically between languages, or to generate implementation models. Designs can be verified using MILAN [4,36]. MILAN supports the integration of different simulators at various levels of granularity using model interpreters, and integrates the design-space exploration tool, DESERT [47]. DESERT allows the designer to express the flexibility in a platform by specifying structural constraints in the OCL. An efficient symbolic technique is used to explore only the architectures that satisfy the constraints, thus pruning a large part of the design space. Performance evaluation is then carried out using lower-level simulators capable of providing accurate performance measures.

Unlike the GME, our work is not focused on the design of modeling languages; rather we look at ways of using existing modeling languages without the need of introducing new ones. To do so, we use special ports and connections to coordinate components that belong to different domains of computation, possibly expressed with different languages; wrappers bridge between different activation and scheduling protocols. Adaptors represent the interfaces between wrappers. Adaptors are in general written explicitly by users. We developed a theoretical framework where adaptors are interpreted as the result of a mapping of the heterogeneous component wrappers into a common semantic domain where communication can be specified formally and then a mapping back into the original domain of the components. In this sense, our approach is closer to that used in MILAN, where model interpreters can be seen as adaptors. However, MILAN makes no attempt at describing the interaction as a formal relation, and is limited to integrating simulation environments, rather than providing tools to study and define heterogeneous compositions.

Our architecture exploration paradigm differs substantially from that employed in DESERT. In particular, we use functional and scheduling constraints, instead of structural constraints, and we are thus able to relate the execution of a function with its implementation on an architecture without resorting to low-level simulators. However, the combined use of structural and scheduling constraints for fast generation and exploration of architectures is a promising avenue of future research.

Ptolemy II is a design environment for heterogeneous systems that consists of several executable domains of computation that can be mixed in a hierarchy controlled by a global scheduler [12]. Each MoC is described operationally in terms of a common executable interface. For each model, a "director" determines the order of activation of the components (or actors). Similarly, communication is defined in terms of a common interface. A MoC, or domain, in Ptolemy II is a pair composed of a director together with an implementation of the communication interface, called a "receiver." The approach to heterogeneity in Ptolemy II is strictly hierarchical. This implies that each node of the hierarchy contains exactly one domain, and that each component interacts with the rest of the system using the specific communication mechanism selected by the domain for the hierarchy node it belongs to. Domains only interact at the boundary between two different levels of the hierarchy.

SystemC-H is a heterogeneous extension to SystemC that provides additional domains of computation, such as dataflow and hierarchical FSMs [50]. The extension follows the same idea as in Ptolemy II of defining hierarchical directors that follow a predetermined interface and a protocol for scheduling that includes the computation of preconditions for execution, a fixed-point loop, and a wrap-up phase. The authors demonstrate an increase in the simulation efficiency due to the possibility of statically scheduling the dataflow portion.

The hierarchical approach to heterogeneity of Ptolemy II and SystemC-H is nicely structured and is an excellent environment for experimentation. The structure, however, also imposes limitations on the amount of heterogeneity that can be achieved. More importantly, the relationship between different models is implicit in the way the execution protocol schedules the activation of the directors and the transfer of information through the receivers. This makes it difficult to predict the outcome of a hierarchical heterogeneous composition or to study its properties. In addition, the scheduling protocol is hardwired in the framework, and, therefore, it cannot be changed without altering the core of the tools. As a result, the relationship between different models (i.e., the abstraction and the refinement) is fixed, unless boundary components are used to explicitly translate between different modeling conventions. For example, this technique is used in Ptolemy II to translate from the discrete to the continuous domain, and vice versa, through special transducers. This technique, however, appears to relax the requirement for strict hierarchical heterogeneity. Our approach to heterogeneity is instead based on establishing clear abstraction and refinement relationships between models. These are typically constructed through the use of a common semantic domain, although this is not strictly necessary. The idea is then to construct adaptors that coordinate the execution of different models and that are consistent with the chosen abstractions and refinements.

Functional languages have also been used to support SLD. In ForSyDe [53], the system is initially specified as a deterministic network of fully synchronous processes that communicate over sequences of events, a model that facilitates the functional description by abstracting away detailed timing. Compliance with the model is enforced by expressing the basic combinatorial behaviors using functions free of side effects, and by generating processes using higher-order process constructors. Haskell has been chosen as the concrete language for expressing the model, since it natively supports higher-order constructs. The specification is then refined into an implementation by applying a series of network transformations that may or may not preserve the semantics of the network. These transformations can, for example, partition the system into subdomains that run at different speeds (the model is therefore no longer fully synchronous), interfaced through up- and down-converters. When the desired structure has been obtained, processes are converted into hardware or software, depending on the chosen implementation. The basic ForSyDe model was then extended to cover a larger array of MoCs [30], and has been implemented in Standard ML in the SML-Sys project [40], as well as in C++ [41]. There, the initial assumption of a fully synchronous system is dropped in favor of an untimed model similar to Kahn process networks [31]. In addition, synchronous, clocked, and timed models can be used for refinement. However, SML-Sys appears to be more focused on heterogeneous design, rather than on transformational refinement. For this reason, SML-Sys relies on ForSyDe for network transformations, while more complex interfaces have been introduced to bridge the gap between different subdomains. More recently, the same group has developed a front-end to both SML-Sys and ForSyDe, called the EWD [39], which captures their common structure into a GME-based meta-model, and provides some code-generation facilities to target both Standard ML and Haskell. (The behavior code, other than constructors and combinators, must still be manually annotated.) Because the static semantics can be expressed in a GME meta-model, the front-end is able to catch errors at the time of design before compilation is attempted.

Unlike ForSyDe, METROPOLIS supports nondeterministic systems, a choice that renders analysis more complex, but greatly simplifies the description of the environment of the system. As the design is refined, nondeterministic components replace deterministic ones in the system. The transformation-based refinement in ForSyDe has clear advantages in terms of the ability to prove correctness and maintain consistency with the original specification. However, the distinction between the functionality and the architecture is lost, and a change of mapping may require substantial restructuring of the system. Our approach to mapping, instead, makes this task simpler, since only the mapping function must be changed. Heterogeneity is addressed in SML-Sys, using domain interfaces that add or remove events from the event sequences. These interfaces, or adaptors, are somewhat arbitrary. We also employ adaptors, but we justify their use by reasoning on the semantics of the interaction supported by the refinement of communicating models onto a common semantic domain.

Also inspired by functional languages is the heterogeneous SLD language, Rosetta [2,34]. In Rosetta, an MoC, or *domain*, is described declaratively as a set of assertions in a higher-order logic. The definition includes the objects of the discourse, such as variables that represent time and power, or a transition relation, to model behaviors. The assertions then axiomatically determine the interpretation of these quantities as properties that they must satisfy. Different domains can be obtained by extending a definition in a way similar to the subclassing relation of a type system. The extended domain inherits all the assertions (the terms) of the original domain, and adds additional constraints on top of them. Domains that are obtained this way are automatically related by an abstraction/refinement relationship. Domains that are unrelated can still be compared by constructing functions, called interactions that (sometimes partially) express the consequences of the properties and the quantities of one domain onto another. This process is particularly useful for expressing and keeping track of constraints during the refinement of the design.

In contrast to Rosetta, the relationship between the function and the architecture in METROPOLIS is not described explicitly as a function (the interactions), but rather as a mapping and an annotation process at the event level. Annotations at this level are simpler to express, and can be used directly for simulation. Conversely, domain interactions are more difficult to manipulate, since they are defined at the level of the domain, but may, in principle, be used to derive stronger results via formal reasoning. Tools that can take full advantage of the Rosetta representation are, however, still in the development phase.

The separation between computation and coordination is central to the behavior-interaction-priority (BIP) framework [9]. In the BIP, a system specification is divided into three layers. At the bottom layer, the behavior of the system is specified as a collection of independent finite-state transition systems (components), which communicate with the environment through ports. Each transition of a component is activated by an interaction, which is a subset of its ports. At the middle layer, a set of connectors specify the possible interactions of the components. That is, connectors identify the subsets of the ports of the whole system that can participate in interactions, and, therefore, activate a transition. Different connectors, and different ways of linking them, define different kinds of interactions that can be used to model such diverse communication paradigms as asynchronous broadcasts to fully synchronous systems. Connectors often result in nondeterministic systems. Determinism can be recovered by using the third layer, where priorities can be imposed on the interactions to induce a unique choice of transition.

One of the strengths of the BIP framework is the ability to check certain properties, such as deadlock freedom, by composition. This is obtained at the expense of a more complex coordination scheme, involving the connectors, which must determine the global set of possible interactions. This may adversely impact simulation performance, and, for certain communication paradigms, it may require a number of connectors that grow exponentially with the number of components. Recently, Bliudze and Sifakis have proposed symbolic and incremental techniques to get around these problems [11], which are currently being evaluated in the framework. We are also interested in global properties, such as those that have to do with the resource consumption of the architecture. The quantities associated with these properties are, however, determined in centralized components, namely, the annotators and the schedulers. Schedulers can be seen as forms of connectors, since they regulate the execution of the system. Unlike the BIP, we favor an imperative description, as opposed to a declarative one, that is simpler to develop and reuse, and that can be optimized for a high performance. Annotators are not present in the BIP framework, which is not intended for architectural exploration.

FunState [58], an evolution of the SPI system [66], is an internal design representation that supports mixing control and dataflow. This is achieved by a flexible and rich underlying model. In its basic form, the model includes functions that communicate over queues and arrays of registers. Functions produce and consume a specified number of data tokens from the queues and are characterized by defined execution latency. The activation of functions is controlled by an FSM, whose transitions are labeled with conditions on the number of tokens present in the queues. The FSM controls the progress of the execution, so that tokens are consumed and produced at a time consistent with the execution latency of the functions. In addition, the user can set timing constraints on the data path to model deadlines and other performance requirements. Formal verification, based on symbolic model checking and on timed automata techniques, can be applied to check properties of the system, including the satisfaction of the timing constraints and the boundedness of the number of tokens in the queues. This is particularly useful for studying schedulability, for which FunState has been specifically developed.

While the strength of FunState is in formal verification, its design as an internal representation makes it more difficult to use for architectural and design-space explorations. First, FunState lacks the ability to define arbitrary performance metrics (other than time), which is important to model different aspects of an architecture. In addition, the FunState model does not distinguish between the function and the architecture. Instead, it can be seen as a deployment model, where the interaction between the two has already been resolved and annotated by selecting timing and appropriate scheduling policies.

10.6   Case Studies

The following sections will illustrate two case studies each performed in the METRO II design environment. The first discusses a universal mobile telecommunications system (UMTS), which illustrates how easily external models can be imported into METRO II, two distinct architecture-modeling styles, and analyzes the METRO II simulation infrastructure. The other case study is an indoor air quality example where we examine how to integrate METRO II with other design tools for continuous-time systems.

10.6.1   UMTS

We focus on the user equipment domain of the UMTS protocol [52], which is of interest to mobile devices and is subject to stringent implementation constraints. The protocol stack of the UMTS for the user equipment domain has been standardized by the 3rd generation partnership project (3GPP) up to the network layer, including the physical (PHY) and data link layers (DLL). Our model includes the implementation of the DLL layer and the functionality of the PHY layer. We enumerate the 48 different points in the design space that were explored. We detail the estimated execution times and the processing-element utilization, the design effort, the simulation cost breakdown, and an analysis of how events are processed during each simulation phase in METRO II.

10.6.1.1   Functional Modeling

The UMTS DLL layer contains the radio link control (RLC) and the medium access control (MAC) sublayers and performs general packet forming. The RLC communicates with the MAC through different *logical* channels to distinguish between the user, signaling, and the control data. Depending on the required quality of service, the MAC layer maps the logical channels into a set of *transport* channels, which are then passed to the Physical layer. The Physical layer handles lower-level coding and modulation in order to reduce the bit error rate of the transmitted data.

For the purposes of this case study, the UMTS application was largely separated into both receiver and transmitter portions and then further by the RLC and the MAC functionalities. Simulation consists of processing 100 packets, each packet being 70 bytes. The functional model is shown in Figure 10.10. The semantics is dataflow, with blocking read and blocking write semantics for the FIFOs. The functional model is described as a process network or an actor-oriented model, where concurrently executing processes communicate with each other through point-to-point channels.

The communication semantics can be adjusted on a per-channel basis. For instance, the data transfer between a pair of processes may take place with

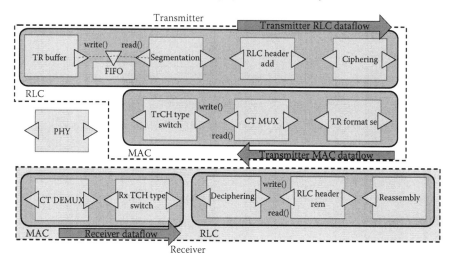

FIGURE 10.10
UMTS METRO II functional model.

rendezvous semantics or blocking read, nonblocking write semantics. In METRO II, there are a number of ways in which this coordination between processes may be specified. Two such mechanisms are detailed in Figure 10.11.

In the explicit synchronization style, an imperative code is written to prevent process P_2 from reading data out of an empty FIFO $F1$, and also preventing it from writing data to a full FIFO $F2$. The code makes use of interface methods provided by the FIFOs as well as events exposed from within the FIFOs through view ports (not shown in the figure). This imperative specification is quite similar to what might be specified in SystemC. These interface methods will themselves propose "begin" and "end" events to the system for scheduling along with any other methods used to query the state of the FIFOs. These additional events have the ability to affect simulation, as will be discussed in Section 10.6.1.4.

The second specification style uses constraints to enforce the same behavior. Now, instead of the imperative code being specified in the Phase 1 components, the corresponding constraints are passed to the Phase 3 constraint solver. This separates computation from coordination, and different coordination models can be used for the same process P_2. This separation is similar to the concept of a "director" within the Ptolemy [38] environment, but within METRO II, the coordination may be specified declaratively.

The first style has more events and may also cause the Phase 1 processes to become suspended due to internal blocking (as opposed to blocking due to event disabling in Phase 3). The benefits of the second style in this regard are described in Section 10.6.1.4.

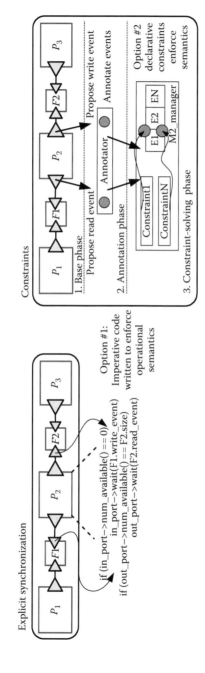

FIGURE 10.11

Two options for expressing coordination.

10.6.1.2 Architectural Modeling

The architecture model assigns one task for each of the 11 UMTS components. (The TR buffer and PHY were not mapped as they represent the environment.) The operating system (OS) employs three different scheduling policies for each processing element. The first is a round-robin scheduler where each processing element is simply selected sequentially. This is a cyclic process beginning with processing element 0 and moving through the number of PEs. If a PE does not have a request pending then the next PE in the list is allowed to proceed. The second algorithm is a priority-based scheduling algorithm where higher priorities are assigned to tasks with higher processing requirements. These requirements are determined during the pre-profiling stage. Preemption is not employed as in the timed functional model. Priority scheduling here examines all the requests for processing in a given round and selects the one with the highest priority. The selected priority is noted and in the next round it cannot be chosen again if there are still events pending from previous rounds. The final algorithm is a first-come-first-serve (FCFS) algorithm, which marks requests with the round that they enter the system and ensures that they are handled in order of appearance. In the event of a tie, this falls back to a round-robin scheme.

The runtime processing elements were supplied with the C code reflecting the kernels of each UMTS component. The runtime processing element available for this case study is a cycle-accurate datapath model of the Leon 3 Sparc processor. The pre-profiled processors use the same code but carry out offline characterization, as detailed in [42] and [24]. The processors profiled were the ARM7, the ARM9, and Xilinx's MicroBlaze. All of the processing elements are common in embedded and SoC applications and are widely documented.

The architecture models to be discussed in this work are composed of the following three portions. Figure 10.12 illustrates these three portions.

1. Tasks—Tasks are lightweight and active components in the architecture model. The thread for each task constantly proposes begin events for its provided services. Mapping creates a rendezvous constraint between the event generated by the task thread (in this case, *request_job*) and the functional event. Therefore, there is a 1:1 mapping between these tasks and functional components. Because of the rendezvous constraint, the task remains blocked until the corresponding event from the functional model is proposed. Step 1 in Figure 10.12 illustrates the task's role in architecture model execution.

2. Operating System—An OS is used to assign tasks to processing elements (in a many-to-one relationship). In addition, it also carries out Phase 1 scheduling—reducing the work to be done in Phase 3. This is done by pruning the events that are proposed in the first phase. An investigation of scheduling policies will be seen in Section 10.6.1.4. An OS is an active component with N threads (where N is the number of

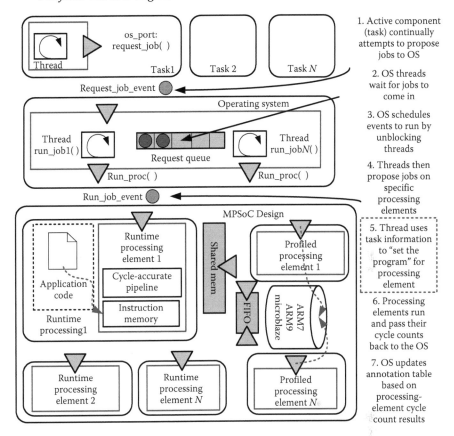

FIGURE 10.12
Topology of architecture service environment.

processing elements it controls). It maintains a queue of requested jobs which processing elements query to decide if they will execute or not. The queue contains events proposed for processing, which processing elements they wish to use, the rounds in which they were proposed, and the statically assigned priorities for the events. Scheduling controls how events are added to and removed from this queue. The access to this queue is coordinated such that there is a limited number of outstanding requests for a given processing element. Steps 2–5 in Figure 10.12 illustrate the OS's role in the architecture model execution.

3. Processing Element—The third piece of the architecture platform consists of the actual processing elements. Once the OS decides to run a task request, it calls the corresponding "run_proc()" function call on one of its N required ports. The interface supported by all processing elements is the same (to provide modularity and flexibility), but there

are different ways in which the cost may be calculated. Steps 6–7 in Figure 10.12 illustrate two different types of processing elements that may be used, and the interface to inform them which processing routine they should compute a cost for. The type of the processing element may be changed easily to provide the necessary balance between the speed of simulation and the required pre-simulation effort.

10.6.1.3 Mapped System

Table 10.4 describes the 48 mappings investigated. These vary from 11 PEs to 1 PE. Partitions are broken down by the Rx, the Tx, the RLC, and the MAC functionalities. Each is categorized into one of nine separate classes based on the number of processing elements and the mix of pre-profiled and runtime processing elements. Mappings are further categorized as purely runtime processing (RTP) elements, purely profiled processing (PP) elements, or a mix (MIX).

10.6.1.4 Results

Results relating to the design effort, the processing time, the framework simulation time, and the event processing are analyzed. Five different models were used: a timed SystemC UMTS model [55], a timed METRO II UMTS model, an untimed METRO II UMTS model, a SystemC runtime processing model, and a METRO II architectural model. In specific configurations, METRO II constraints were used as opposed to explicit synchronization. The selection of constraints, functional model configuration, architectural model parameters, and mapping assignment is all achieved through small changes to the top-level netlist. All results are gathered on a 1.8 GHz Pentium M laptop running Windows XP with 1GB of RAM.

Figure 10.13 shows the UMTS estimated execution times (cycles) along with the average processing-element utilization. Utilization is calculated as the percentage of simulation rounds that an architectural processing element has enabled outstanding functional model event requests for its services. Low utilization indicates that a processing element is idle despite available, outstanding requests. The x-axis (mapping #) is ordered by increasing execution times. The data is collected for each of the three scheduling algorithms.

For round-robin scheduling, the lowest and highest execution times are obtained with mapping #1 (11 Sparcs) and mapping #46 (1 µBlaze), respectively. Mapping #1 is 2167% faster than mapping #46. This shows a large range in potential performances across mappings. It is interesting to note that there are 23 different mappings that offer better performance than the 11 µBlaze or 11 ARM7 cores (mappings #2 and #3). This illustrates that interprocessor communication is a bottleneck for many designs, and despite having more concurrency those designs cannot keep pace with smaller, more heavily-loaded mappings. Among all four processor systems, mapping #14 has the lowest execution time (two ARM9s used for the receiver and two

TABLE 10.4
Mapping Scenarios for the UMTS Case Study

#	Type	Partition	#	Type	Partition	#	Type	Partition
1	1: RTP	11 Sp	17	6: PP	2 μB (2), 2 A9 (3)	33	7: MIX	A7 (4), Sp (5), μB (6), A9 (7)
2	2: PP	11 μB	18	6: PP	2 A9 (2), 2 μB (3)	34	7: MIX	A7 (4), Sp (5), A9 (6), μB (7)
3	2: PP	11 A7	19	6: PP	2 A7 (2), 2 A9 (3)	35	7: MIX	A7 (4), μB (5), Sp (6), A9 (7)
4	2: PP	11 A9	20	6: PP	2 A9 (2), 2 A7 (3)	36	7: MIX	A7 (4), μB (5), A9 (6), Sp (7)
5	3: RTP	4 Sp (1)	21	7: MIX	Sp (4), μB (5), A7 (6), A9 (7)	37	7: MIX	A7 (4), A9 (5), μB (6), Sp (7)
6	4: PP	4 μB (1)	22	7: MIX	Sp (4), μB (5), A9 (6), A7 (7)	38	7: MIX	A7 (4), A9 (5), Sp (6), μB (7)
7	4: PP	4 A7 (1)	23	7: MIX	Sp (4), A7 (5), μB (6), A9 (7)	39	7: MIX	A9 (4), Sp (5), μB (6), A7 (7)
8	4: PP	4 A9 (1)	24	7: MIX	Sp (4), A7 (5), A9 (6), μB(7)	40	7: MIX	A9 (4), Sp (5), A7 (6), μB (7)
9	5: MIX	2 Sp (2), 2 μB (3)	25	7: MIX	Sp (4), A9 (5), A7 (6), μB (7)	41	7: MIX	A9 (4), μB (5), Sp (6), A7 (7)
10	5: MIX	2 μB (2), 2 Sp (3)	26	7: MIX	Sp (4), A9 (5), μB (6), A7 (7)	42	7: MIX	A9 (4), μB (5), A7 (6), Sp (7)
11	5: MIX	2 Sp (2), 2 A7 (3)	27	7: MIX	μB (4), Sp (5), A7 (6), A9 (7)	43	7: MIX	A9 (4), A7 (5), μB (6), Sp (7)
12	5: MIX	2 A7 (2), 2 Sp (3)	28	7: MIX	μB (4), Sp (5), A9 (6), A7 (7)	44	7: MIX	A9 (4), A7 (5), Sp (6), μB (7)
13	5: MIX	2 Sp (2), 2 A9 (3)	29	7: MIX	μB (4), A7 (5), Sp (6), A9 (7)	45	8: RTP	1 Sp
14	5: MIX	2 A9 (2), 2 Sp (3)	30	7: MIX	μB (4), A7 (5), A9 (6), Sp (7)	46	9: PP	1 μB
15	6: PP	2 μB (2), 2 A7 (3)	31	7: MIX	μB (4), A9 (5), A7 (6), Sp (7)	47	9: PP	1 A7
16	6: PP	2 A7 (2), 2 μB (3)	32	7: MIX	μB (4), A9 (5), Sp (6), A7 (7)	48	9: PP	1 A9

(1 = Rx MAC, Tx MAC, Rx RLC, Tx RLC), (2 = Rx MAC, Rx RLC), (3 = Tx MAC, Tx RLC)
(4 = Rx MAC), (5)(Rx RLC), (6)(Tx MAC), (7 = Tx RLC) (Sp = Sparc, μB = Microblaze, A7 = ARM7, A9 = ARM9)

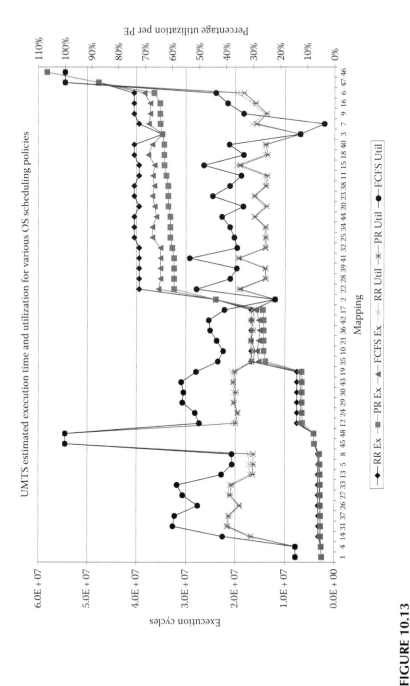

FIGURE 10.13

The UMTS estimated execution time vs. utilization for various OS scheduling policies.

Sparcs used for the transmitter). Mapping #31 has a similar execution time with four different processors (Rx MAC on μBlaze, Rx RLC on ARM9, Tx MAC on ARM7, and Tx RLC on Sparc). Many of the execution times are similar and the graph shows that there are essentially four performance groupings.

The lowest utilization values for round robin occur in the 11 processor setups (an average of 15%). The highest is 100% for all single processor setups. The max utilization before 100% is 39%. This gap points to inefficiency in the round-robin scheduler. It may be a goal of the other scheduling algorithms to close this gap. Also notice that for similar execution times, utilization can vary as much as 28% (mappings #41 and #32, for example).

The priority-based scheduling keeps the same relative ordering amongst the execution times but reduces them on average by 13%. The highest is an 18% reduction (mapping #22, for example) and the smallest reduction is 9% (mapping #8, for example). The utilization numbers are actually reduced as well by an average of 2%. The largest reduction was 7% (in mapping #6, for example) and the smallest was 1% (in mapping #31, for example). As expected there was no change in the utilization or execution times for mappings involving either eleven processing elements (fully concurrent) or those with one element (no scheduling options). The utilization drop results from high-priority, data-dependent jobs running before low-priority, data-independent jobs.

The FCFS scheduling also does not change the relative ordering of execution times but is not as successful at reducing them. The average reduction is only 7%. The maximum reduction is 11% (in mapping #24, for example) and the minimum reduction is 4% (in mapping #5, for example). However, utilization is increased by 27%. The max increase was 45% (in mapping #31, for example) and the minimum improvement was 20% (in mapping #5, for example). The FCFS increases utilization due to the fact that many jobs that would be low priority often request processing in the same round as high-priority jobs. While technically they are both "first," the priority would negate this fact. The FCFS's round-robin tie-breaking scheme helps smaller jobs in this case.

The analysis of execution and utilization for the UMTS shows that high utilization is difficult to obtain due to the data dependencies in the application. Also, some of the partitions explored do not balance computation well amongst the different processing elements in the architecture. Many of the coarser mappings only make this problem worse. A solution is to further refine the functional model to extract more concurrency. From an execution-time standpoint, scheduling can improve the overall execution time but not as much as is needed to make a large majority of these mappings desirable for an actual implementation.

An accuracy comparison was performed with mappings #2, #6, and #46 (pure μBlaze mappings). These designs were created on the Xilinx ML310 development board. For mappings #2 and #46, there was only a 3.1% and a 2% increase, respectively, in execution times in the actual designs. For mapping #6 (when scheduling affects the outcome), the increase was 16.2% (RR), 18% (PR), and 15% (FCFS). Mapping #46 inaccuracy is due to the start-up code and IO operations not captured by the model. Mapping #2 suffers from a slightly oversimplified point-to-point communication scheme in the model as compared to the FSL links used by the MicroBlazes. Finally, mapping #6 requires a more refined OS model to more closely match the scheduling overhead of the actual OS used. This comparison shows that METRO II simulation can closely (within 5%) reflect actual implementations, and in the cases where the differences are greater, a trade-off between the modeling detail, the simulation performance, and the accuracy can be quickly analyzed.

The untimed METRO II UMTS functional model contains 12 processes while the architectural model may contain up to 26 processes. This is a large design, spread across 85 files and 8,300 lines of code. The changing of a mapping is trivial however, which requires only changing a few macros and recompiling two files (2.3% of total; <20 s). All 48 mappings can be done in less than 16 min.

The conversion of the SystemC timed functional model to an untimed METRO II functional model removes 1081 lines of code (related to scheduling and timing—both of which are in the architecture model). METRO II mapping removes much of the overhead associated with the SystemC model synchronization.

METRO II constraints for the read/write semantics of a FIFO only require 60 lines of code, which is 1.4% of the total code cost. The average difference of the entire conversion to METRO II was only 1% per file. More than half of these lines (58%) have to do with registering the constraints with the solvers.

The conversion of a SystemC runtime processing model (the Sparc processing element) to METRO II only requires 92 additional lines. This was a mere 3.4% increase (2773 lines to 2681 lines). This includes adding support for loading a new code at runtime, returning the cost of operation to the netlist, and exposing events for mapping. This result is encouraging for importing code.

Figure 10.14 illustrates the percentage of the actual simulation runtime spent in each of METRO II's simulation phases for the nine classes of mappings. The SystemC entry indicates the time spent in the SystemC simulation infrastructure upon which METRO II is built.

On an average, 61% of the time is spent in Phase 1 (lowest section on the bar graph), 5% in Phase 2 (second section), and 17% in Phase 3 (third section). For models with only runtime processing elements (R), the averages are 93%, 0.9%, and 3%, respectively. This indicates that in runtime processing, the METRO II activities of annotation and scheduling are negligible in the runtime picture. For pure profiled (P) mappings, they are 21%, 7%, and 26%. In this case, one can see that METRO II now accounts for a greater percentage of runtime. (Phase 1 alone is the representative of other

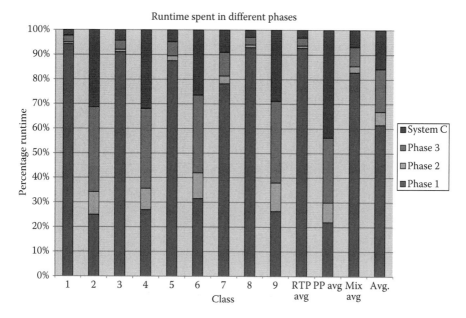

FIGURE 10.14

METRO II phase runtime analysis.

simulation environments.) For mixed classes, the numbers are 82%, 2.6% and 7.6%. Again the runtime processing elements dominate. It should be noted that while Ps have higher averages, the average runtime to process 7000 bytes of data was 54 seconds. The Phase 1 runtime and the SystemC overhead are the main contributors to overall runtime.

If we consider the SystemC timed functional model, the METRO II timed functional model, and the METRO II untimed functional model mapped to an architecture, the METRO II timed functional model had an average increase of 7.4% in runtime for the nine classes while the mapped version had a 54.8% reduction. This reduction is due to the fact that METRO II Phases 2 and 3 have significantly less overheads than the timer- and scheduler-based system required by the SystemC timed functional model.

Table 10.5 shows the average number of event state changes per phase and the average number of phases an event waits.

On an average, only 0.14 events are annotated or scheduled per round. Because of the architectural model integration with the UMTS functional model, there are a limited number of synchronization points (which satisfy a rendezvous constraint, and, hence, an event state change). As shown in Figure 10.14, Phases 2 and 3 do not account for a large portion of the runtime, so, while the event state change activity is low, it does not translate to increased runtime. Runtime is not increased directly by changing an event's state, but rather by the total number of events in Phases 2 and 3.

TABLE 10.5

Metro II Phase Event Analysis

Class	Event/Ph.	Comp. %	Comm. %	Coord. %	Avg Wait
1	0.091	0.083	0.083	0.833	3839.240
2	0.091	0.083	0.083	0.833	3839.240
3	0.169	0.125	0.042	0.833	6276.190
4	0.169	0.125	0.042	0.833	6276.190
5	0.131	0.170	0.114	0.716	5117.003
6	0.169	0.170	0.114	0.716	6276.190
7	0.150	0.101	0.088	0.811	5691.130
8	0.176	0.319	0.043	0.638	6718.550
9	0.176	0.319	0.043	0.638	6718.550
Avg	0.147	0.166	0.072	0.761	5639.143

Events in Classes 1 and 2 on average wait 42% less than the worse case. These classes are precisely those that provide maximum concurrency (11 processing elements). The worst is in Classes 8 and 9 (single processing elements). As one would expect, when the scheduling overhead is lower and more processing elements are available, events wait much less for resource availability.

Finally, it should be noted that runtime processing vs. pre-profiled processing does not impact this aspect of simulation. Comparing Classes 1 with 2 or 3 with 4 confirms this. This contrasts heavily with the runtime of the simulation (in which the PE type is a key factor). The runtime processing in the microarchitectural model is treated as a black box by Metro II such that the internal events are unseen and do not trigger phase changes. This indicates that SystemC components can be imported quite easily into Metro II without affecting the three-phase execution semantics.

The 3rd, 4th, and 5th columns of Table 10.5 categorize the events in Phase 1. Computational events request processing-element services directly. Communication events transfer data between FIFOs, and coordination events maintain correct simulation semantics and operation. The table indicates that events in the system are heavily related to coordination. Classes 8 and 9 have the lowest percentage of coordination events (64%), since these are single-PE systems.

10.6.1.5   Conclusions

We illustrated how an event-based design framework, Metro II, may be used to carry out architectural modeling and design-space exploration. Experimental results show that Metro II is capable of capturing functional modeling, architectural modeling, and mapping for a UMTS case study with limited overhead as compared with a baseline SystemC model. We showed that the design effort involved in carrying out 48 separate mappings with a variety of architectural models is minimal. Within the framework, we detail the runtime spent in the three different METRO II execution phases and provide an idea of how events move throughout the system.

Future work involves identifying and removing events not relevant for annotation or scheduling from METRO II's second and third phases, support for a wider variety of declarative constraints, and the analysis of other applications that may be mapped onto similar architectural platforms.

10.6.2 Intelligent Buildings: Indoor Air Quality

The construction of future energy-efficient commercial buildings will make use of sophisticated control architectures that are able to sense several physical quantities, compute control laws, and apply control actions through actuators. Sensors, actuators, and computation units are physically distributed over the buildings. The control algorithm can be run on either distributed controllers or a central controller. The control performance is critically affected by both computation and communication delays that need to be within precise bounds in order to guarantee energy savings while maintaining the comfort level. Thus, a major challenge in designing such systems is to balance the computation and communication efforts. In particular, a designer needs to decide how to map the control algorithm on a set of controllers and needs to find an optimal communication network, meaning the communication medium and the network topology.

The goal of this case study is to model and simulate the control of the temperature in the rooms of a building at a high level of abstraction. The simulation results will be used to partition the sensor–actuator delay into computation and communication latency requirements. The communication latency requirements are then passed to an optimization tool that finds the best communication network that supports the gathering of data from the sensors and the delivery of commands to actuators.

Our design flow is shown in Figure 10.15. In Step 1, both the functionality of the system and the architecture platform are modeled. The mapping between function and architecture models is carried out where the controllers and the point-to-point communication between sensors, actuators, and controllers are annotated with actual computation delays and virtual communication delays. The performance of the control algorithm is evaluated for different values of the communication delays until the least constraining latency requirements are found. The communication requirements are then passed to an external network synthesis tool—the communication synthesis infrastructure (COSI) [51]. In Step 2, the COSI synthesizes the communication network of the system based on the simulation results. Then, in Step 3, the abstract point-to-point communication channels are mapped to the communication network obtained by COSI.

Both the functionality and the architecture platforms of the control system are modeled in METRO II, while the environment dynamics is modeled in OpenModelica [27], an external simulation tool. OpenModelica interacts

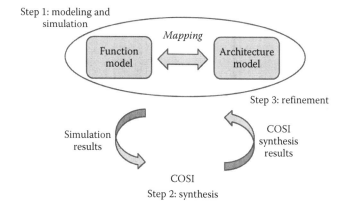

FIGURE 10.15
Design flow of the room temperature control system.

with the function model of the system. The METRO II function model of a two-room example and its interaction with OpenModelica is shown in Figure 10.16. The environment dynamics is described in the Modelica programming language. The Modelica language is designed to allow

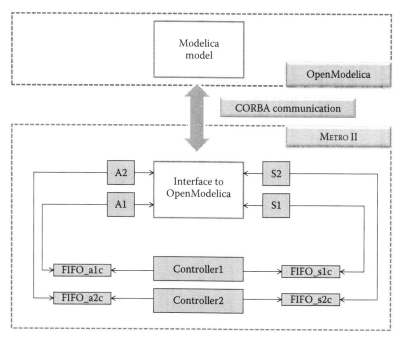

FIGURE 10.16
METRO II function model and OpenModelica.

convenient, component-oriented modeling of complex physical systems, e.g., systems containing mechanical, electrical, electronic, hydraulic, thermal, control, electric power, or process-oriented subcomponents [46]. The Modelica model in the indoor air quality case study deals with pressure and temperature dynamics in an indoor environment. It takes into account the structure of the building, its floorplan, the sizes of the different rooms, and the placement of doors and windows. Moreover, it includes outlet vents that can inject a cold/hot air flow to perform cooling/heating of the environment; they are the actuators of the control system, but expressed in Modelica in terms of their effect on the temperature and pressure dynamics of the system.

The METRO II model and the Modelica model are run together (cosimulation [57]). Sensors and actuators in the functional model interact with the plant to retrieve temperature values in the different rooms and to set the status (closed/open; hot/cold air flow) of the vents. These operations obviously require synchronization and information exchange between the tools. They are managed by the environment functional module, which controls the execution of the Modelica model (start and stop the simulation) and it is able to set and get the value of its parameters. From an implementation point of view, this interaction is performed by the remote calling of a set of services provided by OpenModelica over a CORBA connection [18] established between the tools.

The architecture model includes generic electronic control units (ECUs) communicating with sensors and actuators. During mapping, the controllers in the function model are allocated onto ECUs. If multiple controllers are mapped onto one ECU, a METRO II scheduler is constructed to coordinate their executions. Various scheduling policies can be applied by designing different types of schedulers, while keeping the controller tasks intact. In our example, we use round-robin scheduling. Sensors and actuators in the function model are mapped to architectural sensors and actuators. The communication between ECUs and sensoring/actuating units is modeled at an abstract level in Step 1 of the design flow. The services of sensing, computing control algorithms, and actuating are annotated with time by METRO II annotators. The end-to-end delays from sensing to actuating are computed during simulation. The simulation results are sent to COSI, which synthesizes the communication network in Step 2 of the design flow. Then the synthesis results are utilized to refine the abstract communication network in Step 3 of the flow.

10.7 Conclusions

We discussed the trends and challenges of system design from a broad perspective that covers both semiconductor and industrial segments that use embedded systems. We argued in favor of the need of a unified way of thinking about system design as the basis for a novel system science. One approach was presented, the PBD, that aims at achieving that unifying role. We discussed some of the most promising approaches for chip and embedded system design in the PBD perspective. METROPOLIS and its successor METRO II frameworks were presented. Some examples of METRO II applications to different industrial domains were then described.

While we believe we are making significant inroads, much work remains to be done to transfer the ideas and approaches that are flourishing today in research and in advanced companies to the generality of IC and embedded system designers. To be able to do so,

- We need to further advance the understanding of the relationships among parts of a heterogeneous design and its interaction with the physical environment.
- The efficiency of algorithms and tools must be improved to offer a solid foundation to the users.
- Models and use cases have to be developed.
- The scope of system-level design must be extended to include fault tolerance, security, and resiliency.
- The EDA industry has to embrace the new paradigms and venture into unchartered waters to grow beyond where it is today. It must create the necessary tools to help engineers to apply the new paradigms.
- Academia must develop new curricula (e.g., [13]) that favor a broader approach to engineering while emphasizing the importance of foundational disciplines such as mathematics and physics; embedded system designers require a broad view and the capability of mastering heterogeneous technologies.
- The system and semiconductor industry must recognize the importance of investing in training and tools for their engineers to be able to bring new products and services to market.

Acknowledgments

We wish to acknowledge the support of the Gigascale System Research Center, the support of NSF-sponsored Center for Hybrid and Embedded Software Systems, the support of the EU networks of excellence ARTIST and HYCON, and of the European community project SPEEDS. The past and the present support of General Motors, Infineon, Intel, Pirelli, ST, Telecom Italia (in particular, Marco Sgroi, Fabio Bellifemine, and Fulvio Faraci), UMC, and United Technologies Corporation (in particular, the strong interaction with Clas Jacobson, John F. Cassidy Jr., and Michael McQuade) is also gratefully acknowledged.

References

1. A. Agrawal. Graph rewriting and transformation (GReAT): A solution for the model integrated computing (MIC) bottleneck. In *Proceedings of the 18th IEEE International Conference on Automated Software Engineering (ASE03)*, Montreal, Canada, 2003.

2. P. Alexander. *System Level Design with Rosetta*. Elsevier, San Francisco, CA, 2006.

3. K. Arnold and J. Gosling. *The Java Programming Language*. Addison Wesley, Reading, MA, 1996.

4. A. Bakshi, V. K. Prasanna, A. Ledeczi, V. Mathur, S. Mohanty, C. S. Raghavendra, M. Singh, A. Agrawal, J. Davis, B. Eames, S. Neema, and G. Nordstrom. MILAN: A model based integrated simulation framework for design of embedded systems. In *Proceedings of the Workshop on Languages, Compilers and Tools for Embedded Systems (LCTES 2001)*, Snowbird, UT, June 2001.

5. F. Balarin, M. Chiodo, P. Giusto, H. Hsieh, A. Jurecska, L. Lavagno, C. Passerone, A. Sangiovanni-Vincentelli, E. Sentovich, K. Suzuki, and B. Tabbara. *Hardware-Software Co-Design of Embedded Systems: The Polis Approach*. Kluwer Academic Press, Boston, MA, June 1997.

6. F. Balarin, L. Lavagno, C. Passerone, A. Sangiovanni-Vincentelli, G. Yang, and Y. Watanabe. Concurrent execution semantics and sequential simulation algorithms for the metropolis meta-model. In *Proceedings of the Tenth International Symposium on Hardware/Software Codesign. CODES 2002*, Estes Park, CO, May 6–8, 2002, pp. 13–18. IEEE Computer Society Press, 2002.

7. F. Balarin, L. Lavagno, C. Passerone, A. Sangiovanni-Vincentelli, M. Sgroi, and Y. Watanabe. Modeling and designing heterogenous systems. In J. Cortadella, A. Yakovlev, and G. Rozenberg, editors, *Concurrency and Hardware Design*, pp. 228–273. Springer, Berlin, Heidelberg, 2002. LNCS2549.

8. F. Balarin, H. Hsieh, L. Lavagno, C. Passerone, A. Sangiovanni-Vincentelli, and Y. Watanabe. Metropolis: An integrated environment for electronic system design. *IEEE Computer*, 36(4): 45–52, April 2003.

9. A. Basu, M. Bozga, and J. Sifakis. Modeling heterogeneous real-time components in BIP. In *Proceedings of the Fourth IEEE International Conference on Software Engineering and Formal Methods (SEFM06)*, pp. 3–12, Washington, DC, 2006. IEEE Computer Society.

10. G. Berry and G. Gonthier. The ESTEREL synchronous programming language: Design, semantics, implementation. *Science of Computer Programming*, 19(2):87–152, November 1992.

11. S. Bliudze and J. Sifakis. The algebra of connectors—structuring interactions in BIP. In *Proceedings of the 7th ACM & IEEE International conference on Embedded Software (EMSOFT07)*, Salzburg, Austria, September 30–October 3, 2007.

12. C. Brooks, E. A. Lee, X. Liu, S. Neuendorffer, Y. Zhao, and H. Zheng (eds.). Heterogeneous concurrent modeling and design in Java (Volume 1: Introduction to Ptolemy II). Technical Report UCB/ERL M05/21, University of California, Berkeley, CA, July 2005.

13. A. Burns and A. Sangiovanni-Vincentelli. Editorial. *ACM Transactions on Embedded Computing Systems, Special Issue on Education*, 4(3):472–499, August 2005.

14. San Jose Mercury News (CA). Census counts on pencils, not computers. April 4, 2008.

15. X. Chen, F. Chen, H. Hsieh, F. Balarin, and Y. Watanabe. Formal verification of embedded system designs at multiple levels of abstraction. *International Workshop on High Level Design Validation and Test—HLDVT02*, Cannes, France, September 2002.

16. X. Chen, H. Hsieh, F. Balarin, and Y. Watanabe. Automatic generation of simulation monitors from quantitative constraint formula. *Design Automation and Test in Europe*, Munich, Germany, March 2003.

17. CoFluent Design. *CoFluent Studio*. World Wide Web, http://www.cofluentdesign.com, 2007.

18. Common object request broker architecture. OMG Available Specification 3.1, OMG, January 2008.

19. J. Cortadella, A. Kondratyev, L. Lavagno, C. Passerone, and Y. Watanabe. Quasi-static scheduling of independent tasks for reactive systems. In *Proceedings of the 23rd International Conference on Application and Theory of Petri Nets*, Adelaide, South Australia, June 2002.

20. P. Cumming. The TI OMAP platform approach to SOC. In G. Martin and H. Chang, editors, *Winning the SoC Revolution*, Kluwer Academic, Norwell, MA, 2003.

21. A. Davare, D. Densmore, T. Meyerowitz, A. Pinto, A. Sangiovanni-Vincentelli, G. Yang, and Q. Zhu. A next-generation design framework for platform-based design. In *Design and Verification Conference (DVCON'07)*, San Jose, CA, February 2007.

22. A. Davare, Q. Zhu, J. Moondanos, and A. Sangiovanni-Vincentelli. JPEG encoding on the Intel MXP5800: A platform-based design case Study. In *3rd Workshop on Embedded Systems for Real-time Multimedia*, New York, September 2005.

23. J. A. de Oliveira and H. van Antwerpen. The Philips Nexperia digital video platform. In G. Martin and H. Chang, editors, *Winning the SoC Revolution*, Kluwer Academic, Norwell, MA, 2003.

24. D. Densmore, A. Donlin, and A. L. Sangiovanni-Vincentelli. FPGA architecture characterization for system level performance analysis. In *DATE06*, Munich, Germany, March 6–10, 2006.

25. D. Densmore, R. Passerone, and A. L. Sangiovanni-Vincentelli. A platform-based taxonomy for ESL design. *IEEE Design & Test of Computers*, 23(5):359–374, May 2006.

26. J. Eker, J. W. Janneck, E. A. Lee, J. Liu, X. Liu, J. Ludvig, S. Neuendorffer, S. Sachs, and Y. Xiong. Taming heterogeneity—the Ptolemy approach. *Proceedings of the IEEE*, 91(1):127–144, January 2003.

27. P. Fritzson, P. Aronsson, A. Pop, H. Lundvall, K. Nystrom, L. Saldamli, D. Broman, and A. Sandholm. Openmodelica—a free open-source environment for system modeling, simulation, and teaching. *2006 IEEE International Symposium on Computer-Aided Control Systems Design*, Munich, Germany, pp. 1588–1595, October 2006.

28. G. J. Holzmann. The model checker spin. *IEEE Transactions on Software Engineering*, 23(5):279–258, May 1997.

29. S. Ito. Convergence and divergence in parallel for the ubiquitous era. *Solid-State Circuits Conference, 2007. ASSCC '07. IEEE Asian*, Jeju, Korea, pp. 143–143, November 2007.

30. A. Jantsch. *Modeling Embedded Systems and SOC's: Concurrency and Time in Models of Computation*. Morgan Kaufmann Publishers, San Francisco, CA, 2003.

31. G. Kahn. The semantics of a simple language for parallel programming. In J. L. Rosenfeld, editor, *Proceedings of the IFIP Congress 74, Information Processing 74*, pp. 471–475, North Holland, Amsterdam, the Netherlands, 1974.

32. G. Karsai, J. Sztipanovits, A. Ledeczi, and T. Bapty. Model-integrated development of embedded software. *Proceedings of the IEEE*, 91(1):145–184, January 2003.

33. K. Keutzer, S. Malik, A. R. Newton, J. M. Rabaey, and A. Sangiovanni-Vincentelli. System-level design: Orthogonalization of concerns and

platform-based design. *IEEE Transactions on Computer-Aided Design of Integrated Circuits and Systems*, 19(12):1523–1543, December 2000.

34. C. Kong and P. Alexander. The Rosetta meta-model framework. In *Proceedings of the IEEE Engineering of Computer-Based Systems Symposium and Workshop*, Huntsville, AL, April 7–11, 2003.

35. M. Krigsman. IT failure at Heathrow T5: What really happened. April 7, 2008. http://blogs.zdnet.com/projectfailures/?p=681.

36. A. Ledeczi, J. Davis, S. Neema, and A. Agrawal. Modeling methodology for integrated simulation of embedded systems. *ACM Transactions on Modeling and Compututer Simulation*, 13(1):82–103, 2003.

37. A. Lee and A. Sangiovanni-Vincentelli. A framework for comparing models of computation. *IEEE Transactions on Computer-Aided Design of Integrated Circuits and Systems*, 17(12):1217–1229, December 1998.

38. X. Liu, Y. Xiong, and E. A. Lee. The Ptolemy II framework for visual languages. In *Proceedings of the IEEE 2001 Symposia on Human Centric Computing Languages and Environments (HCC'01)*, Stresa, Italy, p. 50. IEEE Computer Society, 2001.

39. D. Mathaikutty, H. Patel, and S. Shukla. EWD: A metamodeling driven customizable multi-MoC system modeling environment. FERMAT Technical Report 2004-20, Virginia Tech, 2004.

40. D. A. Mathaikutty, H. Patel, and S. Shukla. A functional programming framework of heterogeneous model of computation for system design. In *Forum on Specification and Design Languages (FDL'04)*, Lille, France, September 13–17, 2004.

41. D. A. Mathaikutty, H. D. Patel, S. K. Shukla, and A. Jantsch. UMoC++: A C++-based multi-MoC modeling environment. In A. Vachoux, editor, *Application of Specification and Design Languages for SoCs - Selected paper from FDL 2005*, Chapter 7, pp. 115–130. Springer, Berlin, 2006.

42. T. Meyerowitz, A. Sangiovanni-Vincentelli, M. Sauermann, and D. Langen. Source level timing annotation and simulation for a heterogeneous multiprocessor. In *DATE08*, Munich, Germany, March 10–14, 2008.

43. J. Miller and J. Mukerji, editors. MDA guide version 1.0.1. Technical Report omg/2003-06-01, OMG, 2003.

44. Mirabilis Design. *Visual Sim*. World Wide Web, http://www.mirabilisdesign.com, 2007.

45. MLDesign Technologies. *MLDesigner*. World Wide Web, http://www.mldesigner.com, 2007.

46. http://www.modelica.org/.

47. S. Neema, J. Sztipanovits, and G. Karsai. Constraint-based design-space exploration and model synthesis. In *Proceedings of the Third International Conference on Embedded Software (EMSOFT03)*, Philadelphia, PA, October 13–15 2003.

48. Object constraint language, version 2.0. OMG Available Specification formal/06-05-01, Object Management Group, May 2006.

49. Open SystemC Initiative. *Functional Specification for SystemC 2.0*, September 2001. avaliable at www.systemc.org.

50. H. D. Patel, S. K. Shukla, and R. A. Bergamaschi. Heterogeneous behavioral hierarchy extensions for SystemC. *IEEE Transactions on Computed-Aided Design of Integrated Circuits and Systems*, 26(4):765–780, 2007.

51. A. Pinto, L. Carloni, and A. Sangiovanni-Vincentelli. A communication synthesis infrastructure for heterogeneous networked control systems and its application to building automation and control. In *Proceedings of the Seventh International Conference on Embedded Software (EMSOFT), 2007*, Salzburg, Austria, October 2007.

52. Third Generation Partnership Project. General universal mobile telecommunications system (umts) architecture. Technical Specification TS 23.101, 3GPP, December 2004.

53. I. Sander and A. Jantsch. System modeling and transformational design refinement in ForSyDe. *IEEE Transactions on Computer-Aided Design of Integrated Circuits and Systems*, 23(1):17–32, January 2004.

54. A. Sangiovanni-Vincentelli. Defining platform-based design. *EEDesign*, February 2002.

55. A. Simalatsar, D. Densmore, and R. Passerone. A methodology for architecture exploration and performance analysis using system level design languages and rapid architecture profiling. In *Third International IEEE Symposium on Industrial Embedded Systems (SIES)*, La Grande Motte, France, June 11–13, 2008.

56. S. Solden. Architectural services modeling for performance in HW-SW co-design. In *Proceedings of the Workshop on Synthesis And System Integration of MIxed Technologies SASIMI2001*, Nara, Japan, October 18–19, 2001, pp. 72–77, 2001.

57. Speeds methodology. white paper 1.2, SPEEDS IST European project, April 2008. avaliable at www.speeds.eu.com/downloads/ SPEEDS_WhitePaper.pdf.

58. K. Strehl, L. Thiele, M. Gries, D. Ziegenbein, R. Ernst, and J. Teich. FunState—an internal design representation for codesign. *IEEE Transactions on Very Large Scale Integration (VLSI) Systems*, 9(4): 524–544, August 2001.

59. Metropolis Project Team. The metropolis meta-model - version 0.4. Technical Report UCB/ERL M04/38, EECS Department, University of California, Berkeley, 2004.

60. VaST Systems. *Comet/Meteor*. World Wide Web, http://www.vastsystems.com, 2007.

61. A. Sangiovanni-Vincentelli. Quo Vadis, SLD? Reasoning about trends and challenges of system level design. *Proceedings of the IEEE*, 95(3): 467–506, March 2007.

62. G. Yang, X. Chen, F. Balarin, H. Hsieh, and A. Sangiovanni-Vincentelli. Communication and co-simulation infrastructure for heterogeneous system integration. In *Design Automation and Test in Europe 2006*, Munich, Germany, March 2006.

63. G. Yang, Y. Watanabe, F. Balarin, and A. Sangiovanni-Vincentelli. Separation of concerns: Overhead in modeling and efficient simulation techniques. In *Fourth ACM International Conference on Embedded Software (EMSOFT'04)*, Pisa, Italy, September 2004.

64. G. Yang, H. Hsieh, X. Chen, F. Balarin, and A. Sangiovanni-Vincentelli. Constraints assisted modeling and validation in metropolis framework. In *Asilomar Conference on Signal, Systems and Computers*, Pacific grove, CA, October 2006.

65. J. Yoshida. Philips Semi see payoff in platform-based design. *EE Times*, October 2002.

66. D. Ziegenbein, R. Ernst, K. Richter, J. Teich, and L. Thiele. Combining multiple models of computation for scheduling and allocation. In *Proceedings of the 6th International Workshop on Hardware/Software Codesign (CODES98)*, pp. 9–13, Seattle, WA, March 15–18, 1998. IEEE Computer Society, Los Alamitos, CA.

11

Reconfigurable Multicore Architectures for Streaming Applications

Gerard J. M. Smit, André B. J. Kokkeler, Gerard K. Rauwerda,
and Jan W. M. Jacobs

CONTENTS

323

11.1 Introduction

This chapter addresses reconfigurable heterogenous and homogeneous multicore system-on-chip (SoC) platforms for streaming digital signal processing applications, also called streaming DSP applications. In streaming DSP applications, computations can be specified as a data flow graph with streams of data items (the edges) flowing between computation kernels (the nodes). Most signal processing applications can be naturally expressed in this modeling style [14]. Typical examples of streaming DSP applications are wireless baseband processing, multimedia processing, medical image processing, sensor processing (e.g., for remote surveillance cameras), and phased array radars. In a heterogeneous multicore architecture, a core can either be a bit-level reconfigurable unit (e.g., FPGA), a word-level reconfigurable unit, or a general-purpose programmable unit (digital signal processor (DSP) or general purpose processor (GPP)). We assume the cores of the SoC are interconnected by a reconfigurable network-on-chip (NoC). The programmability of the individual cores enables the system to be targeted at multiple application domains.

We take a holistic approach, which means that all aspects of system design need to be addressed simultaneously in a systematic way (e.g., [24]). We believe that this is key for an efficient overall solution, because an interesting optimization in a small corner of the design might lead to inefficiencies in the overall design. For example, the design of the NoC should be coordinated with the design of the processing cores, and the design of the processing cores should be coordinated with the tile specific compilers. Eventually, there should be a tight fit between the application requirements and the SoC and NoC capabilities.

We first introduce streaming applications and multicore architectures in Sections 11.1.1 and 11.1.2, next we present key design criteria for streaming applications in Section 11.1.3. After that we give a multidimensional classification of architectures for streaming applications in Section 11.2. For each category, one or more sample architectures are presented in Section 11.3. We end this chapter with a conclusion.

11.1.1 Streaming Applications

The focus of this chapter is on multicore SoC architectures for streaming DSP applications where we can assume that the data streams are semi-static and have a periodic behaviour. This means that for a long period of time subsequent data items of a stream follow the same route through the SoC. The common characteristics of typical streaming DSP applications are as follows:

- They are characterized by a relatively simple local processing of a huge amount of data. The trend is that energy costs for data communication dominates energy costs of processing.
- Data arrives at nodes at a rather fixed rate, which causes periodic data transfers between successive processing blocks. The resulting communication bandwidth is application dependent and a large variety of communication bandwidth is required. The size of the data items is application dependent (e.g., 14-bit samples for a sensor system, 64 32-bit words for HiperLAN/2 [15] OFDM symbols, or $8 \times 8 \times 24$-bit macro blocks for a video application). Also the data rate is application dependent (e.g., 100 Msamples/sec after the A/D converter for a sensor system, 200k OFDM symbols per second for HiperLAN/2, and 50 frames/sec for video).
- The data flows through the successive processes in a pipelined fashion. Processes may work in parallel on parallel processors or can be time-multiplexed on one or more processors. Therefore, streaming applications show a predictable temporal and spatial behavior.
- For our application domains, typically throughput guarantees (in data items per sec) are required for communication as well as for processing. Sometimes latency requirements are also given.
- The lifetime of a communication stream is semi-static, which means a stream is fixed for a relatively long time.

11.1.2 Multicore Architectures

Flexible and efficient SoCs can be realized by integrating hardware blocks (called tiles or cores) of different granularities into heterogeneous reconfigurable SoCs. In this chapter the term "core" is used for processor-like hardware blocks and the term "tile" is used for ASICs, fine-grained reconfigurable blocks, and memory blocks. We assume that the interconnected building blocks can be heterogeneous (see Figure 11.1), for instance, bit-level reconfigurable tiles (e.g., embedded FPGAs), word-level reconfigurable cores (e.g., domain-specific reconfigurable cores), general-purpose programmable cores (e.g., DSPs and GPPs), and memory blocks. From a systems point of view these architectures are heterogeneous multiprocessor systems on a single chip. The programmability and reconfigurability of the architecture enables the system to be targeted at multiple application domains. Recently, a number of multicore architectures have been proposed for the streaming DSP application domain. Some examples will be discussed in Section 11.3.

A multicore approach has a number of advantages:

- It is a future-proof architecture as the processing cores do not grow in complexity with technology. Instead, as technology scales, simply the number of cores on the chip grows.

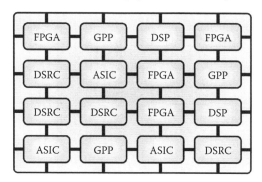

FIGURE 11.1
A heterogenous SoC template.

- A multicore organization can contribute to the energy efficiency of a SoC. The best energy savings can be obtained by simply switching off cores that are not used, which also helps in reducing the static power consumption. Furthermore, the processing of local data in small autonomous cores abides by the locality of reference principle. Moreover, a core processor can be adaptive; it does not have to run at full clock speed to achieve the required QoS at a particular moment in time.

- When one of the cores is discovered to be defect (either because of a manufacturing fault or discovered at operating time by the built-in diagnosis), this defective core can be switched off and isolated from the rest of the design.

- A multicore approach also eases verification of an integrated circuit design, since the design of identical cores has to be verified only once. The design of a single core is relatively simple and therefore a lot of effort can be put in (area/power) optimizations on the physical level of integrated circuit design.

- The computational power of a multicore architecture scales linearly with the number of cores. The more cores there are on a chip, the more computations can be done in parallel (provided that the network capacity scales with the number of cores and there is sufficient parallelism in the application).

- Although cores operate together in a complex system, an individual tile operates quite autonomously. In a reconfigurable multicore architecture, every processing core is configured independently. In fact, a core is a natural unit of partial reconfiguration. Unused cores can be configured for a new task, while at the same time other cores continue performing their tasks. That is to say, a multicore architecture can be reconfigured partly and dynamically.

11.1.2.1 Heterogeneous Multicore SoC

The reason for heterogeneity in a SoC is efficiency, because typically some algorithms run more efficiently on bit-level reconfigurable architectures (e.g., PN-code generation), some on DSP-like architectures, and some perform optimal on word-level reconfigurable platforms (e.g., FIR filters or FFT algorithms). We distinguish four processor types: "GPP," "fine-grained reconfigurable" hardware (e.g., FPGA), "coarse-grained" reconfigurable hardware, and "dedicated" hardware (e.g., ASIC). The different tile processors (TPs) in the SoC are interconnected by a NoC. Both SoC and NoC are dynamically reconfigurable, which means that the programs running on the processing tiles as well as the communication channels are configured at run-time. The idea of heterogeneous processing elements (PEs) is that one can match the granularity of the algorithms with the granularity of the hardware. Application designers or high-level compilers can choose the most efficient processing core for the type of processing needed for a given application task. Such an approach combines performance, flexibility, and energy efficiency. It supports high performance through massive parallelism, it matches the computational model of the algorithm with the granularity and capabilities of the processing entity, and it can operate at minimum supply voltage and clock frequency and hence provides energy efficiency and flexibility at the right granularity only when and where needed and desirable. A thorough understanding of the algorithm domain is crucial for the design of an (energy-)efficient reconfigurable architecture. The architecture should impose little overhead to execute the algorithms in its domain. Streaming applications form a rather good match with multicore architectures: the computation kernels can be mapped on cores and the streams to the NoC links. Interprocessor communication is in essence also overhead, as it does not contribute to the computation of an algorithm. Therefore, there needs to be a sound balance between computation and interprocessor communication. These are again motivations for a holistic approach.

11.1.3 Design Criteria for Streaming Applications

In this section, the key design criteria of multicore architectures for streaming applications are introduced.

11.1.3.1 Predictable and Composable

To manage the complexity of streaming DSP applications, predictable techniques are needed. For example, the NoC as well as the core processors should provide latency and throughput guarantees. One reason for predictability is that the amount of data in streaming DSP applications is so high that even a large buffer would be too small to compensate for unpredictably behaving components and that the latency that these buffers would introduce is not acceptable in typical streaming DSP applications. A second reason for using predictable techniques is composability. This means that in case multiple applications are mapped on the same platform, the behavior of one application should not influence another application. Furthermore, in streaming applications, there are often hard deadlines at the beginning of the chain (e.g., sampling rate of an A/D converter) or at the end of the chain (e.g., fixed rate of the D/A converter, or update rate of the screen). In other applications such as phased array applications, individual paths of signals should be exactly timed before they can be combined. Also in these applications the data rate is so high (e.g., 100 M samples/s) that buffering of data is not useful. Unfortunately, future semiconductor technologies introduce more uncertainty. Design techniques will have to include resiliency at the circuit and microarchitecture level to deal with these uncertainties and the variability at the device technology level. One of the future challenges is to design predictable systems with unpredictable components.

11.1.3.2 Energy Efficiency

Energy efficiency is an important design issue in streaming DSP applications. Because portable devices rely on batteries, the functionality of these devices is strictly limited by the energy consumption. There is an exponential increase in demand for streaming communication and processing for wireless protocol baseband processing and multimedia applications, but the energy content of batteries is increasing at a pace of 10% per year. Also for high-performance computing there is a need for energy-efficient architectures to reduce the costs for cooling and packaging. In addition to that, there are also environmental concerns that urge for more efficient architectures in particular for systems that run 24 h per day such as wireless base stations and search engines (e.g., Google has an estimated server park of 1 million servers that run 24 h per day).

Today, most components are fabricated using CMOS technology. The dominant component of energy consumption (85%–90%) in 130 nm CMOS technology is dynamic power consumption. However, when technology scales to lower dimensions, the static power consumption will become more and more pronounced. A first-order approximation of the dynamic power consumption of CMOS circuitry is given by the formula (see [13]):

$$P_d = \alpha \cdot C_{eff} \cdot f \cdot V^2 \qquad (11.1)$$

where
  P_d is the power in Watts
  C_{eff} is the effective switch capacitance in Farads
  V is the supply voltage in Volts
  α the activity factor
  f is the frequency of operations in Hertz

Equation 11.1 suggests that there are basically four ways to reduce power: reduce the capacitive load C_{eff}, reduce the supply voltage V, reduce the switching frequency f, and/or reduce the activity α. In the context of this chapter, we will mainly address reducing the capacitance.

As shown in Equation 11.1, energy consumption in CMOS circuitry is proportional to capacitance. Therefore energy consumption can be reduced by minimizing the capacitance. This can not only be reached at the technological level, but much profit can be gained by an architecture that exploits locality of reference. Connections to external components typically have much higher capacitance than connections to on-chip resources. Therefore, to save energy, the designer should use few off-chip wires, and have them toggle as infrequently as possible. Consequently, it is beneficial to use on-chip memories such as caches, scratchpads, and registers.

References to memory typically display a high degree of temporal and spatial locality of reference. Temporal locality of reference refers to the observation that referenced data is often referenced again in the near future. Spatial locality of reference refers to the observation that once a particular location is referenced, a nearby location is often referenced in the near future. Accessing a small and local memory is much more energy efficient than accessing a large distant memory. Transporting a signal over a 1 mm wire in a 45 nm technology requires more than 50 times the energy of a 32-bit operation in the same technology (the off-chip interconnect consumes more than a 1000 times the energy of an on-chip 32-bit operation). A multicore architecture intrinsically encourages the usage of small and local on-core memories. Exploiting the locality of the reference principle extensively improves the energy efficiency substantially. Because of the locality of reference principle, the communications within a core are more frequent than between cores.

11.1.3.3 *Programmability*

Design automation tools form the bridge between processing hardware and application software. Design tools are the most important requirement for the viability of multicore platform chips. Such tools reduce the design cycle (i.e., cost and time-to-market) of new applications. The application programmer should be provided with a set of tools that on the one hand hides the architecture details but on the other hand gives an efficient mapping of the applications onto the target architecture. High-level language compilers for (DSP) domain-specific architectures are far more complex than compilers for general-purpose superscalar architectures because of the data dependency analysis, instruction scheduling, and allocation. Besides tooling for application development, tooling for functional verification and debugging is required for programming multicore architectures. In general, such tooling comprises

- General HDL simulation software that provides full insight into the hardware state, but is extremely slow and not suited for software engineers
- Dedicated simulation software that provides reasonable insight into the hardware state, performs better than general hardware simulation software, and can be used by software engineers
- Hardware prototyping boards that achieve great simulation speeds, but provide poor insight into the hardware state and are not suited for software engineers

By employing the tiled SoC approach, as proposed in Figure 11.1, various kinds of parallelism can be exploited. Depending on the core architecture one or more levels of parallellism are supported.

- *Thread-level parallelism* is explicitly addressed by the multicore approach as different tiles can run different threads.
- *Data-level parallelism* is achieved by processing cores that employ parallelism in the data path.
- *Instruction-level parallelism* is addressed by processing cores when multiple data path instructions can be executed concurrently.

11.1.3.4　Dependability

With every new generation of CMOS technology (i.e., 65 nm and beyond) the yield and reliability of manufactured chips deteriorate. To effectively deal with the increased defect density, efficient methods for fault detection, localization, and fault recovery are needed. Besides yield improvement, such techniques also improve the long-term reliability and dependability of silicon-implemented embedded systems. In the ITRS 2003 roadmap (see [1]), it is indicated that "Potential solutions are adaptive and selfcorrecting, self-repairing circuits and the use of on-chip reconfigurability." Modern static and dynamic fault detection and localization techniques and design-for-test (DFT) techniques are needed for advanced multicore designs. Yield and reliability can be improved by (dynamically) circumventing the faulty hardware in deep-submicron chips. The latter requires run-time systems software. This software detects defective cores and network elements and deactivates these resources at run-time. The tests are performed while the chip is already in the field. These self-diagnosis and self-repair hardware and software resources need to be on chip.

11.2　Classification

Different hardware architectures are available in the embedded systems domain to perform DSP functions and algorithms: "GPP, DSP, (re-) configurable hardware, and application-specific hardware." The application-specific hardware is designed for a dedicated function and is usually referred to as ASIC. The ASIC is, as its name suggests, an application-specific processor that has been implemented in an IC.

These hardware architectures have different characteristics in relation to "performance, flexibility" or "programmability," and "energy efficiency." Figure 11.2 depicts the trade-off in flexibility and performance for different hardware architectures. Generally, more flexibility implies a less energy-efficient solution.

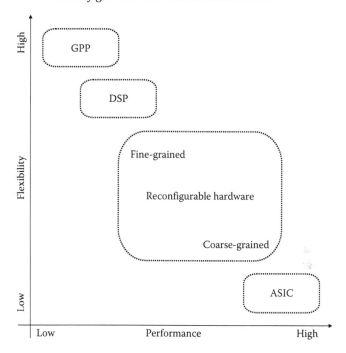

FIGURE 11.2
Flexibility versus performance trade-off for different hardware architectures.

Crucial for the fast and efficient realization of a multiprocessor system-on-chip (MP-SoC) is the use of predesigned modules, the so-called building blocks. In this section, we will first classify these building blocks, and then classify the MP-SoCs that can be designed using these building blocks together with the interconnection structures between these blocks.

A basic classification of MP-SoC building blocks is given in Figure 11.3. The basic PEs of an MP-SoC are run-time reconfigurable cores and fixed cores. The functionality of a run-time reconfigurable core is fixed for a relatively long period in relation to the clock frequency of the cores. Fine-grained reconfigurable cores are reconfigurable at bit level while coarse-grained reconfigurable cores are reconfigurable at word level (8 bit, 16 bit, etc.). Two other essential building blocks are memory and I/O blocks. Designs of MP-SoCs can be reused to build larger MP-SoCs, increasing the designers productivity.

A classification of MP-SoCs is given in Figure 11.4. An MP-SoC basically consists of multiple building blocks connected by means of an interconnect. If an MP-SoC consists of multiple building blocks of a single type, the MP-SoC is referred to as "homogeneous." The homogeneous MP-SoC architectures can be subdivided into single instruction multiple data (SIMD), multiple instruction multiple data (MIMD), and array architectures.

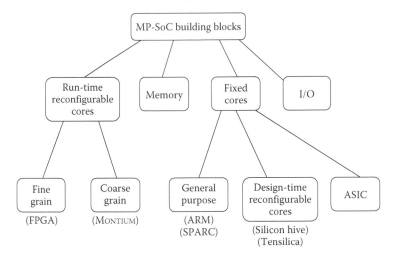

FIGURE 11.3
Classification of MP-SoC building blocks for streaming applications.

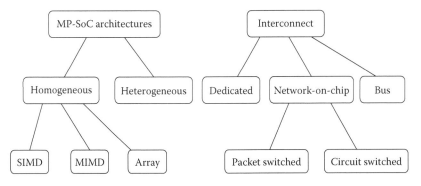

FIGURE 11.4
Classification of MP-SoC architectures and interconnect structures for streaming applications.

Examples of these architectures will be given below. If multiple types of building blocks are used, the MP-SoC is called "heterogeneous."

To interconnect the different building blocks, three basic classes can be identified: bus, NoC, and dedicated interconnects. A bus is shared between different processing cores and is a notorious cause of unpredictability. Unpredictability can be circumvented by an NoC [9]. Two types can be identified: packet-switched and circuit-switched. Besides the use of these more or less standardized communication structures, dedicated interconnects are still widely used. Some examples of different MP-SoC architectures are presented in Table 11.1.

TABLE 11.1

Examples of Different MP-SoC Architectures

Class		Example
	SIMD	Linedancer (see Section 11.3.2)
		Geforce G80 [3]
		Xetal [19]
Homogeneous	MIMD	Tilera (see Section 11.3.4)
		Cell [21]
		Intel Tflop processor [25]
	Array	PACT (see Section 11.3.3)
		ADDRESS [2]
Heterogeneous		ANNABELLE (see Section 11.3.1)
		Silicon Hive [12]

11.3 Sample Architectures

11.3.1 Montium/Annabelle System-on-Chip

11.3.1.1 Montium Reconfigurable Processing Core

The Montium is an example of a coarse-grained reconfigurable processing core and targets the 16-bit DSP algorithm domain. The Montium architecture origins from research at the University of Twente [18,22]. The Montium processing core has been further developed by Recore Systems [23]. A single Montium processing tile is depicted in Figure 11.5. At first glance the Montium architecture bears a resemblance to a very large instruction word (VLIW) processor. However, the control structure of the Montium is very different. The lower part of Figure 11.5 shows the communication and configuration unit (CCU) and the upper part shows the coarse-grained reconfigurable Montium TP.

11.3.1.1.1 Communication and Configuration Unit

The CCU implements the network interface controller between the NoC and the Montium TP. The definition of the network interface depends on the NoC technology that is used in a SoC in which the Montium processing tile is integrated [11]. The CCU enables the Montium TP to run in "streaming" as well as in "block" mode. In "streaming" mode the CCU and the Montium TP run in parallel. Hence, communication and computation overlap in time. In "block" mode, the CCU first reads a block of data, then starts the Montium TP, and finally after completion of the Montium TP, the CCU sends the results to the next processing unit in the SoC (e.g., another Montium

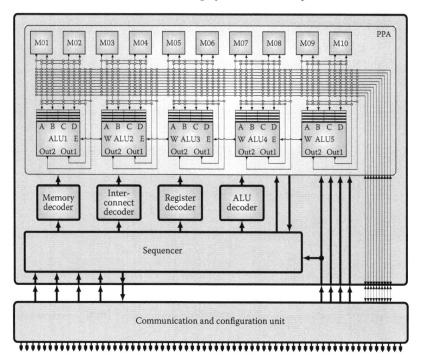

FIGURE 11.5
The Montium coarse-grained reconfigurable processing tile.

processing tile or external memory). Hence, communication and computation are sequenced in time.

11.3.1.1.2 *Montium Tile Processor*

The TP is the computing part of the Montium processing tile. The Montium TP can be configured to implement a particular DSP algorithm. DSP algorithms that have been implemented on the Montium are, for instance, all power of 2 FFTs upto 2048 points, non-power of 2 FFT upto FFT 1920, FIR filters, IIR filters, matrix vector multiplication, DCT decoding, Viterbi decoders, and Turbo (SISO) decoders. Figure 11.5 reveals that the hardware organization of the Montium TP is very regular. The five identical arithmetic logic units (ALU1 through ALU5) in a tile can exploit data level parallellism to enhance performance. This type of parallelism demands a very high memory bandwidth, which is obtained by having 10 local memories (M01 through M10) in parallel. The small local memories are also motivated by the locality of reference principle. The data path has a width of 16 bit and the ALUs support both signed integer and signed fixed-point arithmetic. The ALU input registers provide an even more local level of storage. Locality of reference is one of the guiding principles applied to obtain energy efficiency in the Montium TP.

A vertical segment that contains one ALU together with its associated input register files, a part of the interconnect, and two local memories is called a processing part (PP). The five PPs together are called the processing part array (PPA).

A relatively simple sequencer controls the entire PPA. The sequencer selects configurable PPA instructions that are stored in the decoder blocks of Figure 11.5. For (energy) efficiency it is imperative to minimize the control overhead. The PPA instructions, which comprise ALU, AGU, memory, register file, and interconnect instructions, are determined by a DSP application designer at design time. All MONTIUM TP instructions are scheduled at design time and arranged into a MONTIUM sequencer programme. By statically scheduling the instructions as much as possible at compile time, the MONTIUM sequencer does not require any sophisticated control logic which minimizes the control overhead of the reconfigurable architecture.

The MONTIUM TP has no fixed instruction set, but the instructions are configured at configuration time. During configuration of the MONTIUM TP, the CCU writes the configuration data (i.e., instructions of the ALUs, memories and interconnects, etc., sequencer and decoder instructions) in the configuration memory of the MONTIUM TP. The size of the total configuration memory of the MONTIUM TP is about 2.6 kB. However, configuration sizes of DSP algorithms mapped on the MONTIUM TP are typically in the order of 1 kB. For example, a 64-point fast Fourier transform (FFT) has a configuration size of 946 bytes. By sending a configuration file containing configuration RAM addresses and data values to the CCU, the MONTIUM TP can be configured via the NoC interface. The configuration memory of the MONTIUM TP is implemented as a 16-bit wide SRAM memory that can be written by the CCU. By only updating certain configuration locations of the configuration memory, the MONTIUM TP can be partially reconfigured. In the considered MONTIUM TP implementation, each local SRAM is 16-bit wide and has a depth of 1024 addresses, which results in a storage capacity of 2 kB per local memory. The total data memory inside the MONTIUM TP adds up to a size of 20 kB. A reconfigurable address generation unit (AGU) is integrated into each local memory in the PPA of the MONTIUM TP. It is also possible to use the local memory as a look-up table (LUT) for complicated functions that cannot be calculated using an ALU, such as sine or division (with one constant). The memory can be used in both integer or fixed-point LUT mode.

11.3.1.2 Design Methodology

Development tools are essential for quick implementation of applications in reconfigurable architectures. The MONTIUM development tools start with a high-level description of an application (in C/C++ or MATLAB®) and translate this description to a MONTIUM TP configuration [16]. Applications can be implemented on the MONTIUM TP using an embedded C language, called

MONTIUMC. The MONTIUM design methodology to map DSP applications on the MONTIUM TP is divided into three steps:

1. The high-level description of the DSP application is analyzed and computationally intensive DSP kernels are identified.
2. The identified DSP kernels or parts of the DSP kernels are mapped on one or multiple MONTIUM TPs that are available in a SoC. The DSP operations are programmed on the MONTIUM TP using MONTIUMC.
3. Depending on the layout of the SoC in which the MONTIUM processing tiles are applied, the MONTIUM processing tiles are configured for a particular DSP kernel or part of the DSP kernel. Furthermore, the channels in the NoC between the processing tiles are configured.

11.3.1.3   ANNABELLE *Heterogeneous System-on-Chip*

In this section, the prototype ANNABELLE SoC is described according to the heterogeneous SoC template mentioned before, which is intended to be used for digital radio broadcasting receivers (e.g., digital audio broadcasting, digital radio mondiale). Figure 11.6 shows the overall architecture of the ANNABELLE SoC. The ANNABELLE SoC consists of an ARM926 GPP with a five-layer AMBA AHB, four MONTIUM TPs, an NoC, a Viterbi decoder, two ADCs, two DDCs, a DMA controller, SRAM/SDRAM memory interfaces, and external bus interfaces.

The four MONTIUM TPs and the NoC are arranged in a reconfigurable subsystem, labelled "reconfigurable fabric." The reconfigurable fabric is connected to the AHB bus and serves as a slave to the AMBA system. A configurable clock controller generates the clocks for the individual MONTIUM TPs. Every individual MONTIUM TP has its own adjustable clock and runs at its own speed. A prototype chip of the ANNABELLE SoC has been produced using the Atmel 130 nm CMOS process [8].

The reconfigurable fabric that is integrated in the ANNABELLE SoC is shown in detail in Figure 11.7. The reconfigurable fabric acts as a

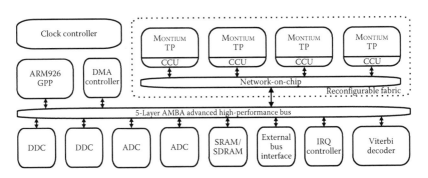

FIGURE 11.6
Block diagram of the ANNABELLE SoC.

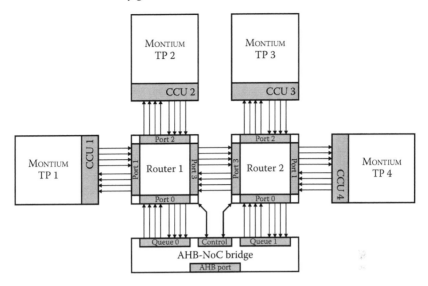

FIGURE 11.7
The ANNABELLE SoC reconfigurable fabric.

reconfigurable coprocessor for the ARM926 processor. Computationally intensive DSP algorithms are typically offloaded from the ARM926 processor and processed on the coarse-grained reconfigurable MONTIUM TPs inside the reconfigurable fabric. The reconfigurable fabric contains four MONTIUM TPs, which are connected via a CCU to a circuit-switched NoC. The reconfigurable fabric is connected to the AMBA system through a AHB–NoC bridge interface. Configurations, generated at design-time can be loaded onto the MONTIUM TPs at run-time. The reconfigurable fabric provides "block mode" and "streaming mode" computation services.

For ASIC synthesis, worst-case military conditions are assumed. In particular, the supply voltage is 1.1 V and the temperature is 125°C. Results obtained with the synthesis are as follows:

- The area of one MONTIUM core is 3.5 mm^2 of which 0.2 mm^2 is for the CCU and 3.3 mm^2 is for the MONTIUM TP (including memory).
- With Synopsys tooling we estimated that the MONTIUM TP, within the ANNABELLE ASIC realization, can implement an FIR filter at about 100 MHz or an FFT at 50 MHz. The worst-case clock frequency of the ANNABELLE chip is 25 MHz.
- With the Synopsys prime power tool, we estimated the energy consumption using placed and routed netlists. The following section provides some of the results.

TABLE 11.2

Dynamic Power Consumption of one
MONTIUM on ANNABELLE

Module	Energy (mW/MHz)		
	FIR-5	**FFT-512**	**FFT-288**
Datapath	0.19	0.24	0.15
Memories	0.0	0.27	0.21
Sequencer	0.02	0.07	0.05
Decoders	0.0	0	0.0
CCU	0.02	0.02	0.02
Total	0.23	0.60	0.43

TABLE 11.3

Energy Comparison of MONTIUM/
ARM926

Algorithm	MONTIUM (μJ)	ARM926 (μJ)	Ratio
FIR-5	0.243	—	—
FFT-112	0.357	9	25
FFT-176	0.616	16	26
FFT-256	0.707	14	20
FFT-288	1.001	23	23
FFT-512	1.563	30	19
FFT-1920	5.054	168	33

11.3.1.4 Average Power Consumption

To determine the average power consumption of the ANNABELLE as accurate as possible, we performed a number of power estimations on the placed and routed netlist using the Synopsys power compiler. Table 11.2 provides the dynamic power consumption in mW/MHz of various MONTIUM blocks for three well-known DSP algorithms. These figures show that the overhead of the sequencer and decoder is low: <16% of the total dynamic power consumption. Finally, Table 11.3 compares the energy consumption of the MONTIUM and the ARM926 on ANNABELLE. For the FIR-5 algorithm the memory is not used.

11.3.1.5 Locality of Reference

As mentioned above, locality of reference is an important design parameter. One of the reasons for the excellent energy figures of the MONTIUM is the use of locality of reference. To illustrate this, Table 11.4 gives the amount of memory references local to the cores compared to the amount of off-core communications. These figures are, as expected, algorithm dependent. Therefore, in this table, we chose three well-known algorithms in the

TABLE 11.4

Internal and External Memory References per Execution of an Algorithm

Algorithm	Number of Memory References		
	Internal	External	Ratio
1024p FFT	51200	4096	12.5
200 tap FIR	405	2	202.5
SISO algorithm (N softbits)	18*N	3*N	6

TABLE 11.5

Reconfiguration of Algorithms on the MONTIUM

Algorithm	Change	Size	# cycles
1024p FFT	Scaling factors	≤150 bit	≤10
to iFFT	Twiddle factors	16384 bit	512
200 tap FIR	Filter coefficients	≤3200 bit	≤80

streaming DSP application domain: a 1024p FFT, a 200 tap FIR filter, and a part of a Turbo decoder (SISO algorithm [17]). The results show that for these algorithms 80%–99% of the memory references are local (within a tile).

11.3.1.6 Partial Dynamic Reconfiguration

One of the advantages of a multicore SoC organization is that each individual core can be reconfigured while the other cores are operational. In the MONTIUM, the configuration memory is organized as a RAM memory. This means that to reconfigure the MONTIUM, the entire configuration memory need not be rewritten, but only the parts that are changed. Furthermore, because the MONTIUM has a coarse-grained reconfigurable architecture, the configuration memory is relatively small. The MONTIUM has a configuration size of only 2.6 kB. Table 11.5 gives some examples of reconfigurations.

To reconfigure a MONTIUM from executing a 1024 point FFT to executing a 1024 point inverse FFT requires updating the scaling and twiddle factors. Updating these factors requires less than 522 clock cycles in total. To change the coefficients of a 200 tap FIR filter requires less than 80 clock cycles.

11.3.2 Aspex Linedancer

The Linedancer [4] is an "associative" processor and it is an example of a homogeneous SoC. Associative processing is the property of instructions to execute only on those PEs where a certain value in their data register matches a value in the instruction. Associative processing is built around an intelligent memory concept: content addressable memory (CAM). Unlike standard computer memory (random access memory or RAM) in which the user supplies a memory address and the RAM returns the data word stored at that address, a CAM is designed such that the user supplies a data word and the CAM searches its entire memory to see if that data word is stored anywhere in it. If the data word is found, the CAM returns a tag list of one or more storage addresses where the word was found. Each CAM line, that contains a word, can be seen as a processor element (PE) and each tag list element as a 1 bit condition register. Dependending on this register, the aggregate associative processor can either instruct the PEs to continue processing on the indicated subset, or to return the involved words subsequently for further processing. There are several implementations possible which vary from bit serial to word parallel, but the latest implementations [4,5] can perform the involved lookups in parallel in a single clock cycle.

In general the Linedancer belongs to the subclass of massively parallel SIMD architectures, with typically more than 512 processors. This SIMD subclass is perfectly suited to support data parallelism, for example, for signal, image, and video processing; text retrieval; and large databases. The associative functions furthermore allow the processor to function like an intelligent memory (CAM), permitting high speed searching and data-dependent image processing operations (such as median filters and object recognition/labeling).

The so called "ASProCore" of the Linedancer, is designed around a very large number—up to 4,096—of simple PEs arranged in a line, see Figure 11.8.

Application areas are diverse but have in common the simple processing of very large amounts of data, from samples in 1D-streams to pixels in 2D or 3D-images. To mention a few: software defined radio (e.g., WiMAX), broadcast (Video compression), medical imaging (3D reconstruction), and in high-end printers—in particular for raster image processing (RIP).

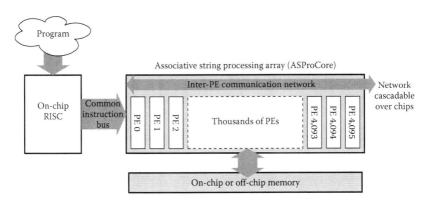

FIGURE 11.8
The scalable architecture of Linedancer.

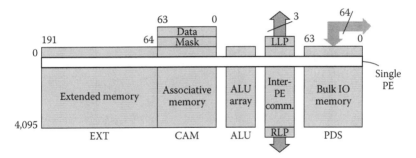

FIGURE 11.9
The architecture of Linedancer's associative string processor (ASProCore).

In the following sections, the associative processor (ASProCore) and the Linedancer family are introduced. At the end, we present the development tool chain and a brief conclusion on the Linedancer application domain.

11.3.2.1 ASProCore Architecture

Each PE has a 1-bit ALU, 32–64 bit full associative memory array, and 128 bit extended memory. See Figure 11.9 for a detailed view on the ASProCore architecture. The processors are connected in a 1D network, actually a 4K bit shift register, in between the indicated "left link port" (LLP) and "right link port" (RLP). The network allows data to be shared between PEs with minimum overhead. The ASProCore also has a separate word serial bit parallel memory, the primary data store (PDS), for high-speed data input. The on-chip DMA engine automatically translates 2D and 3D images into the 1D array (and passed through via the PDS). The 1D architecture allows for linear scaling of performance, memory, and communication, provided the application is expressed in a scalable manner. The Linedancer features also a single or dual bit RISC core (P1, HD, respectively) for sequential processing and controlling the ASProCore.

11.3.2.2 Linedancer Hardware Architecture

The current Linedancers, the P1 and the HD, have been realized in 0.13 µm CMOS process. Both have one or two 32-bit SPARC core(s) with 128 kB internal program memory. System clock frequencies vary from 300, 350, 400 MHz. The Linedancer-P1 integrates an associative processor (ASProCore, with 4K PEs), a single SPARC core with a 4 kB instruction cache, and a DMA controller capable of transferring 64 bit at 66 MHz over a PCI-interface, as shown in Figure 11.10.

It further hosts 128 kB internal data memory. The chip consumes 3.5 W typical at 300 MHz. The Linedancer-HD integrates two associative processors (2 × 2K PEs), two SPARC cores with each 8 kB instruction cache and

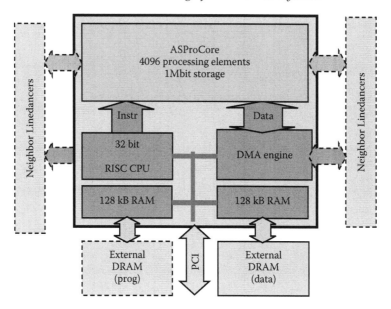

FIGURE 11.10
The Linedancer-P1 layout.

4 kB data cache, four internal DMA engines, and an external data channel capable of transferring 64 bit at 133 MHz over a PCI-X interface, as shown in Figure 11.11. The ASProCore has been extended with a chordal ring inter-PE communication network that allows for faster 2D- and 3D-image processing. It further hosts four external DDR2 DRAM interfaces, eight dedicated streaming data I/O ports (up to 3.2 GB/s), and 1088 kB internal data memory. The chip consumes 4.5 W typical at 300 MHz.

11.3.2.3   Design Methodology

The software development environment for Linedancer consists of a compiler, linker, and debugger. The Linedancer is programmed in C, with some parallel extensions to support the ASProCore processing array. The toolchain is based on the GNU compiler framework, with dedicated pre and postprocessing tools to compile and optimise the parallel extensions to C.

Associative SIMD processing adds an extra dimension to massive parallel processing, enabling new views on problem modeling and the subsequent implementation (for example, in searching/sorting and data-dependent image processing). The Linedancer's 1D-architecture scales better than a 2D array often used in multi-ALU arrays as PACT's XPP [6] or the Tilera's 2D multicore array [7]. Because of the large size of the array, power consumption is relatively high compared to the Montium processor and prevents application into handheld devices.

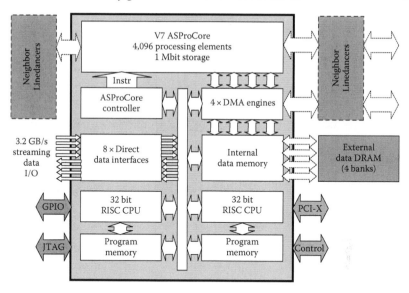

FIGURE 11.11
The Linedancer-HD layout.

11.3.3 PACT-XPP

The eXtreme processing platform (XPP) is an example of a homogeneous array structure. It is a run-time reconfigurable coarse-grained data processing architecture. The XPP provides parallel processing power for high bandwidth data such as video and audio processing. The XPP targets streaming DSP applications in the multimedia and telecommunications domain [10,20].

11.3.3.1 Architecture

The XPP architecture is based on a hierarchical array of coarse-grained, adaptive computing elements, called processing array elements (PAEs). The PAE are clustered in processing array clusters (PACs). All PAEs in the XPP architecture are connected through a packet-oriented communication network. Figure 11.12 shows the hierarchical structure of the XPP array and the PAEs clustered in a PAC.

Different PAEs are identified in the XPP array: "ALU-PAE, RAM-PAE," and "FNC-PAE." The ALU-PAE contains a multiplier and is used for DSP operations. The RAM-PAE contains a RAM to store data. The FNC-PAE is a unique sequential VLIW-like processor core. The FNC-PAEs are dedicated to the control flow and sequential sections of applications. Every PAC contains ALU-PAEs, RAM-PAEs, and FNC-PAEs. The PAEs operate according to a data flow principle; a PAE starts processing data as soon as all required input packets are available. If a packet cannot be processed, the pipeline stalls until the packet is received.

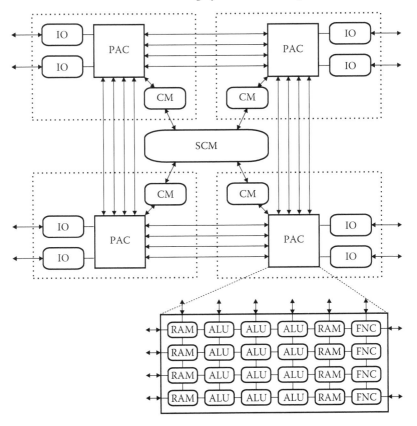

FIGURE 11.12
The structure of an XPP array composed of four PACs. (From Baumgarte, V. et al., *J. Supercomput.*, 26(2), 167, September 2003.)

Each PAC is controlled by a configuration manager (CM). The CM is responsible for writing configuration data into the configurable object of the PAC. Multi-PAC XPP arrays contain additional CMs for concurrent configuration data handling, arranged in a hierarchical tree of CMs. The top CM, called supervising CM (SCM), has an external interface, not shown in Figure 11.12, that connects the supervising CM to an external configuration memory.

11.3.3.2 Design Methodology

DSP algorithms are directly mapped onto the XPP array according to their data flow graphs. The flow graph nodes define the functionality and operations of the PAEs, whereas the edges define the connections between the PAEs. The XPP array is programmed using the native mapping language (NML), see [20]. In NML descriptions, the PAEs are explicitly allocated and the connections between the PAEs are specified. Optionally, the allocated PAEs are placed onto the XPP array. NML also includes statements to support configuration handling. Configuration handling is an explicit part of the application description.

A vectorizing C compiler is available to translate C functions to NML modules. The vectorizing compiler for the XPP array analyzes the code for data dependencies, vectorizes those code sections automatically, and generates highly parallel code for the XPP array. The vectorizing C compiler is typically used to program "regular" DSP operations that are mapped on "ALU-PAEs" and "RAM-PAEs" of the XPP array. Furthermore, a coarse-grained parallelization into several FNC-PAE threads is very useful when "irregular" DSP operations exist in an application. This allows running even irregular, control-dominated code in parallel on several FNC-PAEs. The FNC-PAE C compiler is similar to a conventional RISC compiler extended with VLIW features to take advantage of ILP within the DSP algorithms.

11.3.4 Tilera

The Tile64 [7] is a TP based on the mesh architecture that was originally developed for the RAW machine [26]. The chip consists of a grid of processor tiles arranged in a network (see Figure 11.13), where each tile consists of a GPP, a cache, and a nonblocking router that the tile uses to communicate with the other tiles on the chip.

The Tilera processor architecture incorporates a 2D array of homogenous, general-purpose cores. Next to each processor there is a switch that connects

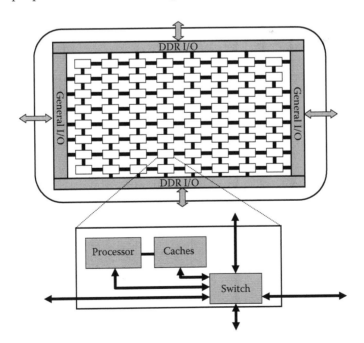

FIGURE 11.13
Tile64 processor.

the core to the iMesh on-chip network. The combination of a core and a switch form the basic building block of the Tilera Processor: the tile. Each core is a fully functional processor capable of running complete operating systems and off-the-shelf "C" code. Each core is optimized to provide a high performance/power ratio, running at speeds between 600 MHz and 1 GHz, with power consumption as low as 170 mW in a typical application. Each core supports standard processor features such as

- Full access to memory and I/O
- Virtual memory mapping and protection (MMU/TLB)
- Hierarchical cache with separate L1-I and L1-D
- Multilevel interrupt support
- Three-way VLIW pipeline to issue three instructions per cycle

The cache subsystem on each tile consists of a high-performance, two-level, non-blocking cache hierarchy. Each processor/tile has a split level 1 cache (L1 instruction and L1 data) and a level 2 cache, keeping the design, fast and power efficient. When there is a miss in the level 2 cache of a specific processor, the level 2 caches of the other processors are searched for the data before external memory is consulted. This way, a large level 3 cache is emulated.

This promotes on-chip access and avoids the bottleneck of off-chip global memory. Multicore coherent caching allows a page of shared memory, cached on a specific tile, to be accessed via load/store references to other tiles. Since one tile effectively prefetches for the others, this technique can yield significant performance improvements.

To fully exploit the available compute power of large numbers of processors, a high-bandwidth, low-latency interconnect is essential. The network (iMesh) provides the high-speed data transfer needed to minimize system bottlenecks and to scale applications. iMesh consists of five distinct mesh networks: Two networks are completely managed by hardware and are used to move data to and from the tiles and memory in the event of cache misses or DMA transfers. The three remaining networks are available for application use, enabling communication between cores and between cores and I/O devices. A number of high-level abstractions are supplied for accessing the hardware (e.g., socket-like streaming channels and message-passing interfaces.) The iMesh network enables communication without interrupting applications running on the tiles. It facilitates data transfer between tiles, contains all of the control and datapath for each of the network connections, and implements buffering and flow control within all the networks.

11.3.4.1 Design Methodology

The TILE64 processor is programmable in ANSI standard C and C++. Tiles can be grouped into clusters to apply the appropriate amount of processing power to each application and parallelism can be explicitly specified.

11.4 Conclusion

In this chapter, we addressed reconfigurable multicore architectures for streaming DSP applications. Streaming DSP applications express computation as a data flow graph with streams of data items (the edges) flowing between computation kernels (the nodes). Typical examples of streaming DSP applications are wireless baseband processing, multimedia processing, medical image processing, and sensor processing. These application domains require flexible and energy-efficient architectures. This can be realized with a multicore architecture. The most important criteria for designing such a multicore architecture are predictability and composability, energy efficiency, programmability, and dependability. Two other important criteria are performance and flexibility. Different types of processing cores have been discussed, from ASICs, reconfigurable hardware, to DSPs and GPPs. ASICs have high performance but suffer from poor flexibility while DSPs and GPPs offer flexibility but modest performance. Reconfigurable hardware combines the best of both worlds. These different processing cores are, together with memory- and I/O blocks assembled into MP-SoCs. MP-SoCs can be classified into two groups: homogeneous and heterogeneous. In homogeneous MP-SoCs, multiple cores of a single type are combined whereas in a heterogeneous MP-SoC, multiple cores of different types are combined.

We also discussed four different architectures: the MONTIUM/ANNABELLE SoC, the Aspex Linedancer, the PACT-XPP, and the Tilera processor. The MONTIUM, a coarse-grain, run-time reconfigurable core has been used as one of the building blocks of the ANNABELLE SoC. The ANNABELLE SoC can be classified as a heterogeneous MP-SoC. The Aspex Linedancer is a homogeneous MP-SoC where a single instruction is executed by multiple processors simultaneously (SIMD). The PACT-XPP is an array processor where multiple ALUs are combined in a 2D structure. The Tilera processor is an example of a homogeneous MIMD MP-SoC.

References

1. The International Technology Roadmap for Semiconductors, ITRS Roadmap 2003. Website, 2003. http://public.itrs.net/Files/2003ITRS/Home2003.htm.

2. A coarse-grained reconfigurable architecture template and its compilation techniques. PhD thesis, Katholieke Universiteit Leuven, Leuven, Belgium, January 2005.

3. Nvidia g80, architecture and gpu analysis, 2007.

4. Aspex Semiconductor: Technology. Website, 2008. http://www.aspex-semi.com/q/technology.shtml.

5. Mimagic 6+ Enables Exciting Multimedia for Feature Phones. Website, 2008. http://www.neomagic.com/product/MiMagig6+_Product_Brief.pdf/.

6. PACT. http://www.pactxpp.com/main/index.php, 2008.

7. Tilera Corporation. http://www.tilera.com/, 2008.

8. Atmel Corporation. ATC13 Summary. http://www.atmel.com, 2007.

9. A. Banerjee, P.T. Wolkotte, R.D. Mullins, S.W. Moore, and Gerard J.M. Smit. An energy and performance exploration of network-on-chip architectures. *IEEE Transactions on Very Large Scale Integration (VLSI) Systems*, 17(3): 319–329, March 2009.

10. V. Baumgarte, G. Ehlers, F. May, A. Nückel, M. Vorbach, and M. Weinhardt. PACT XPP—A self-reconfigurable data processing architecture. *Journal of Supercomputing*, 26(2):167–184, September 2003.

11. M.D. van de Burgwal, G.J.M. Smit, G.K. Rauwerda, and P.M. Heysters. Hydra: An energy-efficient and reconfigurable network interface. In *Proceedings of the International Conference on Engineering of Reconfigurable Systems and Algorithms (ERSA'06)*, Las Vegas, NV, pp. 171–177, June 2006.

12. G. Burns, P. Gruijters, J. Huisken, and A. van Wel. Reconfigurable accelerator enabling efficient sdr for low-cost consumer devices. In *SDR Technical Forum*, Orlando, FL, November 2003.

13. A.P. Chandrakasan, S. Sheng, and R.W. Brodersen. Low-power cmos digital design. *IEEE Journal of Solid-State Circuits*, 27(4):473–484, April 1992.

14. W.J. Dally, U.J. Kapasi, B. Khailany, J.H. Ahn, and A. Das. Stream processors: Progammability and efficiency. *Queue*, 2(1):52–62, 2004.

15. European Telecommunication Standard Institute (ETSI). *Broadband Radio Access Networks (BRAN); HIPERLAN Type 2; Physical (PHY) Layer*, ETSI TS 101 475 v1.2.2 edition, February 2001.

16. Y. Guo. Mapping applications to a coarse-grained reconfigurable architecture. PhD thesis, University of Twente, Enschede, the Netherlands, September 2006.

17. P.M. Heysters, L.T. Smit, G.J.M. Smit, and P.J.M. Havinga. Max-log-map mapping on an fpfa. In *Proceedings of the 2005 International Conference on Engineering of Reconfigurable Systems and Algorithms (ERSA'02)*, Las Vegas, NV, pp. 90–96, June 2002. CSREA Press, Las Vegas, NV.

18. P.M. Heysters. Coarse-grained reconfigurable processors – flexibility meets efficiency. PhD thesis, University of Twente, Enschede, the Netherlands, September 2004.

19. R.P. Kleihorst, A.A. Abbo, A. van der Avoird, M.J.R. Op de Beeck, L. Sevat, P. Wielage, R. van Veen, and H. van Herten. Xetal: A low-power high-performance smart camera processor. *IEEE International Symposium on Circuits and Systems, 2001. ISCAS 2001*, 5:215–218, 2001.

20. PACT XPP Technologies . http://www.pactcorp.com, 2007.

21. D.C. Pham, T. Aipperspach, D. Boerstler, M. Bolliger, R. Chaudhry, D. Cox, P. Harvey et al. Overview of the architecture, circuit design, and physical implementation of a first-generation cell processor. *IEEE Journal of Solid-State Circuits*, 41(1):179–196, January 2006.

22. G.K. Rauwerda, P.M. Heysters, and G.J.M. Smit. Towards software defined radios using coarse-grained reconfigurable hardware. *IEEE Transactions on Very Large Scale Integration (VLSI) Systems*, 16(1):3–13, January 2008.

23. Recore Systems. http://www.recoresystems.com, 2007.

24. G. J. M. Smit, A. B. J. Kokkeler, P. T. Wolkotte, and M. D. van de Burgwal. Multi-core architectures and streaming applications. In I. Mandoiu and A. Kennings (editors), *Proceedings of the Tenth International Workshop on System-Level Interconnect Prediction (SLIP 2008)*, New York, pp. 35–42, April 2008. ACM Press, New York.

25. S.R. Vangal, J. Howard, G. Ruhl, S. Dighe, H. Wilson, J. Tschanz, D. Finan et al. An 80-tile sub-100-w teraflops processor in 65-nm cmos. *IEEE Journal of Solid-State Circuits*, 43(1):29–41, January 2008.

26. E. Waingold, M. Taylor, D. Srikrishna, V. Sarkar, W. Lee, V. Lee, J. Kim et al. Baring it all to software: Raw machines. *Computer*, 30(9):86–93, September 1997.

12

FPGA *Platforms for Embedded Systems*

Stephen Neuendorffer

CONTENTS

12.1 Introduction

Increasingly, programmable logic (such as field programmable gate arrays [FPGAs]) is a critical part of low-power and high-performance signal processing systems. Typically, these systems also include a complex system architecture, along with control processors, digital signal processing (DSP) elements, and perhaps dedicated circuits. In some cases, it is economical to integrate these system components in ASIC technology. As a result, a wide variety of general purpose or application specific standard product (ASSP) system-on-chip (SOC) architectures are available in the market. From the perspective of a system designer, these architectures solve a large portion of the system design problem, typically providing application-specific

I/O interfaces, an operating system for the control processor, processor application programming interfaces (APIs) for accessing dedicated circuits, or communicating with programmable elements such as DSP cores.

As FPGAs have become larger and more capable, it has become possible to integrate a large portion of the system architecture completely within an FPGA, including control processors, communication buses, DSP processing, memory, I/O interfaces, and application-specific circuits. For a system designer, such a System-in-FPGA (SIF) architecture may result in better system characteristics if an appropriate ASSP does not exist. At the same time, designing using FPGAs eliminates the initial mask costs and process technology risks associated with custom ASIC design, while still allowing a system to be highly tuned to a particular application.

Unfortunately, designing a good SIF architecture from scratch and implementing it successfully can still be a risky, time-consuming process. Given that FPGAs only exist in fixed sizes, leveraging all the resources available in a particular device can be challenging. This problem has become even more acute given the heterogeneous nature of current FPGA architectures, making it more important to trade off critical resources in favor of less critical ones. Furthermore, most design is still performed at the register-transfer level (RTL) level, with few mechanisms to capture interface requirements or guarantee protocol compatibility. Constructing radically new architectures typically involves significant code rewriting and under practical design pressures is not an option, given the time required for system verification.

Model-based design is one approach to reducing this risk. By focusing on capturing a designer's intention and providing high-level design constructs that are close to a particular application domain, model-based design can enable a designer to quickly implement algorithms, analyze trade-offs, and explore different alternatives. By raising the level of abstraction, model-based design techniques can enable a designer to focus on key system-level design decisions, rather than low-level implementation details. This process, often called "platform-based design" [10,16], enables higher level abstractions to be expressed in terms of lower level abstractions, which can be more directly implemented.

Unfortunately, in order to provide higher level design abstractions, existing model-based design methodologies must still have access to robust basic abstractions and design libraries. Of particular concern in FPGA systems is the role of the control processor as more complex processor programs, such as an operating system, are used. The low-level interfaces between the processor and the rest of the system can be fragile, since the operating system and hardware must coordinate to provide basic abstractions, such as process scheduling, memory protection, and power management. Architecting, debugging, and verifying this interaction tends to require a wide span of skills and specialized knowledge and can become a critical design problem, even when using traditional design techniques.

One solution to this problem is to separate the control processor subsystem from the bulk of the system and provide it as a fixed part of the FPGA platform. This subsystem can remain simple while being capable of configuring and reconfiguring the FPGA fabric, bootstrapping an operating system, and providing a basis for executing application-specific control code. Historically, several architectures have provided such a platform with the processor system implemented in ASIC technology coupled with programmable FPGA fabric, including the Triscend architecture [20], which was later acquired by Xilinx. Although current FPGAs sometimes integrate hard processor cores (such as in the Xilinx Virtex 2 Pro family), a complete processor subsystem is typically not provided.

This chapter describes the use of the partial reconfiguration (PR) capabilities of some FPGAs to provide a complete processor-based platform using existing general-purpose FPGAs. PR involves the reconfiguration of part of an FPGA (a reconfigurable region) while another part of the FPGA (a static region) remains active and operating. Using PR, the processor subsystem can be implemented as a largely application-independent static region of the FPGA, while the application-specific portion can be implemented in a reconfigurable region. The processor subsystem can be verified and optimized beforehand, combined with an operating system image and distributed as a binary image. From the perspective of a designer or a model-based design tool, the static region of the FPGA becomes part of the FPGA platform, while the reconfigurable region can then be treated as any other FPGA, albeit with some resources reserved.

To understand the requirements for designing such a platform, we will first provide some background of how processors and PR are used to design SIF architectures. Then, we will describe the currently available tools, particularly related to PR, for building a reusable platform. Lastly, we will provide an in-depth design example showing how such a platform can be constructed.

12.2 Background

12.2.1 Processor Systems in FPGAs

Processor-based systems are commonly constructed in FPGAs. An obvious way to build such a system is to take the RTL used for an ASIC implementation and target the RTL toward the FPGA using logic synthesis. In most cases, however, the resulting FPGA design is relatively inefficient (being both relatively large in silicon area and slow). Recent studies suggest that direct FPGA implementation may be around 40 times larger (in silicon area) and one-third of the clock speed of a standard-cell design on small benchmark circuits [9]. Experience with emulating larger processor designs, such as the Sparc V9 core from the OpenSparc T1 [19] and the PowerPC 405 core, in FPGAs suggest a slowdown of at least 20 times compared to ASIC implementations.

The differences arise largely because of the overhead of FPGA programmability, which requires many more transistors than an equivalent ASIC implementation. However, whereas many ASIC processors have complex architectures in order to meet high computation requirements, systems designed for FPGAs tend to make use of FPGA parallelism to meet the bulk of the computation requirements. Hence, only relatively simple control processors are necessary in FPGA systems, when combined with application-specific FPGA design. When a processor architecture can be tuned to match the FPGA architecture, as is typically done with "soft-core" processors, such as the Xilinx Microblaze, reasonable clock rates (100 MHz) can be achieved even in small, relatively slow, cost-optimized Xilinx Spartan 3 FPGAs. Alternatively, somewhat higher clock rates (up to 500 MHz) and performance can be achieved by incorporating the processor core as a "hard-core" in the FPGA, as is done with PowerPC cores in Xilinx Virtex 4 FX FPGAs.

One advantage of a faster control processor is being able to effectively run larger, more complex control programs. Operating systems are often used to mitigate this complexity. An operating system not only provides access to various resources in the system, but also enables multiple pieces of independent code to effectively share those resources by providing locking, memory allocation, file abstractions, and process scheduling. In addition, operating systems are designed to be robust and stable where an application process cannot corrupt the operating system or other processes, making it significantly easier to design and debug large systems.

Such an architecture, which combines a simple control processor hosting an operating system with a high-performance computational engine, is not unique to FPGA-based systems. With the move toward multicore architectures in embedded processing platforms, typically one processor core serves the role of the control processor. This processor typically boots first, and is responsible for configuring and managing the main computational engine(s), which are typically programmable processors tuned for a particular application domain, such as signal processing or networking. Even in platforms where the computational engines are specialized and not programmable processors at the instruction level, such as in low-power cell phone platforms, some initialization and coordination of data transfer must still be performed. The variety in the possible architectures can be seen in Figure 12.1, which summarizes the architecture of several embedded processing platforms.

Platform	Application	Control Proc.	Data Proc.
IBM cell	Media/computing	64-bit PPC	8 128-bit SIMD RISC
Nexperia PNX8526	Digital television	MIPS	1 VLIW and dedicated
Intel IXP2800	Network processing	XScale (ARMv5)	16 multithreaded RISC
TI OMAP2430	Cell phone handset	ARM 1136	dedicated

FIGURE 12.1

Summary of some existing embedded processing platforms with control processors.

Regardless of the processor core architecture, the core must still be integrated into a system in order to access peripherals and external memory. Typically, most system peripherals and device interfaces are implemented in the FPGA fabric, in order to provide the maximum amount of system flexibility. For instance, the Xilinx embedded development kit (EDK) [24] enables FPGA users to assemble existing processor and peripheral IP cores to design a SIF architecture. Application-specific FPGA modules can be imported as additional cores into EDK, or alternatively, the RTL generated by EDK can be encapsulated as a blackbox inside a larger HDL design.

12.2.2 FPGA Configuration and Reconfiguration

FPGAs are designed primarily to implement arbitrary bit-oriented logic circuits. In order to do this, they consist primarily of "lookup tables" (LUTs) for implementing the combinational logic of the circuit, "flip-flops" (FFs) for implementing registers in the circuit, and programmable interconnect for passing signals between other elements. Typically, pairs of LUTs and FFs are grouped together with some additional combinational logic for efficiently forming wide logic functions and arithmetic operations. The Xilinx Virtex 4 slice, which combines two LUTs and two FFs, is shown in Figure 12.2. In the Virtex 4 architecture, four slices are grouped together with routing resources in a single custom design called a configurable logic block (CLB). The layout of FPGAs consists primarily of many tiles of the basic CLB, along with tiles for other other elements necessary for a working system, such as embedded memory (BRAM), external IO pins, clock generation and distribution logic, and even processor cores.

In order to implement a given logic circuit, the logic elements must be configured. Typically, this involves setting the value in a large number of individual SRAM configuration memory cells controlling the logic elements. These configuration cells are often organized in a large shift chain, enabling the configuration bitstream to be shifted in from an external source, such as a nonvolatile PROM. This shift chain is illustrated in Figure 12.3, taken from an early FPGA-related patent [5]. Although this arrangement enables the FPGA configuration to be loaded relatively efficiently, changing any part of the configuration requires loading a completely new bitstream.

In order to increase flexibility, additional logic is often added to the configuration logic of FPGAs that enables portions of the FPGA configuration to be loaded independently. In Xilinx Virtex FPGAs, the configuration shift chain is broken into individually addressed "configuration frames" [26]. The configuration logic contains a register, called the frame address register (FAR), which routes configuration data to the correct configuration frame. The configuration bitstream itself consists of "configuration commands," which can update the FAR and other registers in the configuration logic, load configuration frames, or perform other configuration operations. This architecture enables "partial reconfiguration" of the FPGA,

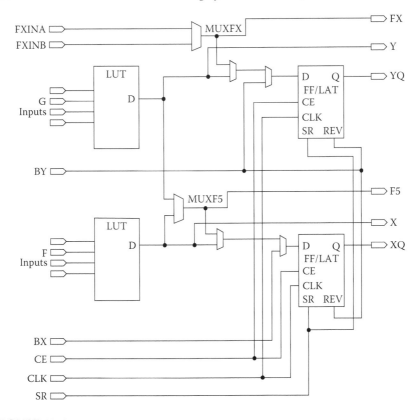

FIGURE 12.2
Simplified architecture of Xilinx Virtex 4 slice [27]. The multiplexers in the middle are primarily used to implement wide multiplexers from several slices. (From Xilinx, *Virtex-4 FPGA User Guide*, ug070 v2.40 edition, April 2008. With permission.)

where some configuration frames are reconfigured while other portions remain active.

In Virtex 4 FPGAs, the configuration frames themselves are organized in columns along the North–South axis of the FPGA. Each configuration frame is the height of 16 CLBs or 4 BRAM memory elements and matches the height of the clock distribution tree. Hence, PR of large portions of the FPGA is best done using rectangular regions that are a multiple of 16 CLBs in that direction. In the East–West direction, the columns are narrow (requiring many configuration frames to configure all of the LUTs in one CLB), which enables the exact size of a reconfigurable region to be more finely controlled. Note that in Virtex 2 and Virtex 2 Pro FPGAs, the configuration frames cross the entire device in the North–South direction, making connectivity between regions more difficult. Although PR is possible in these families, rather complex architectures tend to be used [11].

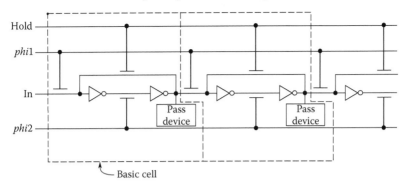

FIGURE 12.3
Early FPGA configuration logic [5]. *phi*1, *phi*2, and"hold" signals control loading of data into the shift chain. Blocks marked "pass device" are controlled by the configuration logic. (From Xilinx, *Virtex-4 FPGA User Guide*, ug070 v2.40 edition, April 2008. With permission.)

Although the logic in a design can often be floorplanned to fit the natural layout of the configuration frames, signal routing is often much more problematic. For instance, the FPGA architecture may require certain external I/O pins to be used for certain purposes, such as clock inputs. It may also be difficult to floorplan a region containing exactly the right number of external pins, while still maintaining a reasonable mix of other elements. These difficulties can be reduced by allowing static signals to be routed through reconfigured regions of the FPGA.

Implementing such "route-overs" require both capabilities in the FPGA architecture and capabilities in the design tools. The FPGA architecture must support the ability to overwrite the configuration of routing resources without causing active signals using those resources to glitch. This capability is supported by Xilinx Virtex 2, Virtex 2 Pro, Virtex 4, and Virtex 5 FPGAs, but not by lower cost Spartan 3 FPGAs. The design tools must have the complementary capability to generate bitstreams for reconfigurable regions where route-overs use exactly the same set of configuration bits to route each signal. This capability is implemented in the Xilinx early access (EA) PR tools using a "merge-based" process [17]. In this process, the static portion of the design is placed and routed first and the routing resources used are stored in a design database. This database, combined with floorplanning constraints, are used to constrain the routing of reconfigurable modules to lie within the boundaries of the reconfigurable region and avoid routing resources used by route-overs. To generate a partial bitstream, the implementation of each reconfigurable module is first merged with the implementation of the static region, ensuring that any route-over uses the same signal routing as the static design. Using this process, each reconfigurable module can be implemented without the knowledge of the implementation of any other reconfigurable module and configured independently, as long as every configuration frame is guaranteed to contain information from the static design and at most one reconfigurable region.

From the perspective of the configuration logic, the process of loading a partial bitstream is handled in exactly the same way. However, from the perspective of building systems, there are several key differences. Primarily, a partial bitstream never contains the configuration commands that are normally present in a bitstream to trigger the initialization and power-on-reset process of the FPGA, since issuing such commands would immediately halt processing in the static region. As a result, a PR design must never rely on the power-on-reset state of flip-flops for proper operation. Secondarily, although routing resources can be reconfigured without glitching in some FPGA architectures, any signal that is sourced by a flip-flop or register that is reconfigured will still glitch during reconfiguration. As a result, extra logic is typically included to ensure that signals driven from the reconfigured region into the static region are forced to a value during reconfiguration.

12.2.3 Partial Reconfiguration with Processors

The PR process itself can be initiated either through an external configuration interface, such as Xilinx SelectMap interface or the joint test action group (JTAG) boundary scan interface, or internally, through the internal configuration access port (ICAP) [26]. The most convenient way to use the ICAP is by using a processor, such as the Xilinx Microblaze processor or PowerPC hard cores found in some FPGAs. A program running on the processor in the static region of the FPGA can make decisions about when reconfiguration should occur and can load an appropriate partial bitstream through the ICAP. When used in this way, the combination of FPGA plus the static design capable of reconfiguration is often called a "self-reconfiguring platform" (SRP) [1,17,22]. The basic architecture of an SRP is shown in Figure 12.4.

One example of how such a system might work is shown in Figure 12.5. This system includes a large number of FPGA computational units in a modular rack-mounted system. Data arrives on the right of the figure and is processed by FPGAs directly connected to A/D converters. Under control of the control workstation, data is routed through a network switch to other FPGA computational units for further processing and data reduction. The processed data is stored or displayed by the control workstation. A similar system, is currently in use at the Allen Telescope Array, using racks of FPGA boards to combine the results from a large number of radio telescopes [13,14].

In order to provide scalability and fault tolerance, each computational unit performs self-checks when it is first powered on. Based on these checks, the unit notifies a centralized server of its availability. When work is available, the centralized server distributes it to any available and unallocated computational units. If any units fail (based on periodic internal checks, or external verification of work results), the centralized server can decide not

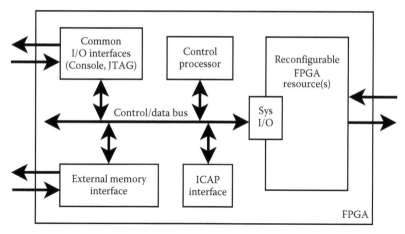

FIGURE 12.4
Basic architecture of a SRP. (From Xilinx, *Virtex-4 FPGA User Guide*, ug070 v2.40 edition, April 2008. With permission.)

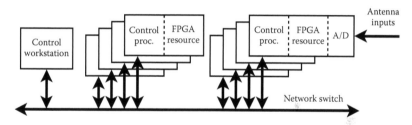

FIGURE 12.5
A radio telescope system architecture based on FPGAs. (From Xilinx, *Virtex-4 FPGA User Guide*, ug070 v2.40 edition, April 2008. With permission.)

to assign additional work to the failed unit and schedule a replacement. This management and coordination task is handled by distributed software executing on the control workstation and the control processors in each FPGA unit.

Another example, based on a software-defined radio is shown in Figure 12.6. In this system, a large number of different communication protocols, called "waveforms," must be implemented in a system although only a small number are active at any one time [21]. In Figure 12.6, waveforms are executed primarily in the reconfigurable FPGA resources on the right. The control processor responds to events initiated by the user of the system through the interfaces on the left, controls reconfiguration of the FPGA resources, and manages the transfer of data between the radio and the other interfaces of the system. When a connection is established, the correct waveform is selected from a library of FPGA implementations and inserted into the system. This type of system also enables a straightforward path toward supporting new

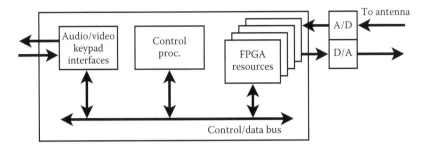

FIGURE 12.6
A software defined radio architecture based on FPGAs.

waveforms through any device that the processor has access to, including a wireless network connection based on an existing waveform supporting data traffic.

12.2.4 Reusable FPGA Platforms for Embedded Systems

Typically, the SRP concept is seen largely as a mechanism for enabling better use of the reconfigurability of the FPGA. Such a system may consume less power, cost less, or be more flexible than an equivalent system without reconfiguration, since only the portion of the system that is active needs to be loaded in the FPGA. In practice, however, these advantages are often difficult to realize, because of the complexity of the resulting system. Compared with a processor, which is typically capable of switching between processes in hundreds of cycles, reconfiguration of a large FPGA may take hundreds of thousands of cycles. In order to leverage FPGA reconfigurability, systems must be capable of accepting this latency. If a task needs to be resumed later, its state must be saved and reloaded, adding not only additional latency but also storage requirements. Even if multiple tasks can be time-shared, realizing a cost savings by fitting a design into a smaller FPGA is difficult since only discrete sizes of FPGAs are available and there is some overhead in using PR techniques.

The SRP concept can also be viewed as a means for enabling faster and more robust system design. The processor is decoupled from the bulk of the FPGA design, enabling it to be designed, verified, and optimized in a working system independent from the FPGA design. Within a given application domain (such as designs requiring DDR2 memory and Gigabit Ethernet as basic infrastructure), the processor system can also be made generic and leveraged across different designs. The processor also becomes centrally involved in how the bulk of the FPGA is configured, enabling flexible programming of the configuration process, rather than relying solely on fixed configuration modes. As a result, new capabilities, such as Built-in Self Test or network-enabled processing resources, can be enabled, which previously required an external processor.

A key benefit of this view is that an SRP can encompass system software, such as an operating system and user libraries, in addition to just the processor subsystem. This can greatly reduce overall risk in the design process, since a system designer does not need to be concerned with verifying the fundamental abstractions of the operating system. By enabling most application code to be managed by the operating system in "user-space," application programming errors can be more easily localized and debugged and the underlying operating system mechanisms can be continually improved. The programmable portion of the FPGA becomes simply another hardware resource managed by the operating system. This management can be implemented once, verified and reused, eliminating the possibility of difficult-to-debug errors that might occur as a result of FPGA reconfiguration. Furthermore, when combined with higher level design techniques, such as C-to-FPGA tools with strong compiler analysis and optimization, application programming in the FPGA can be given the same "user-space" guarantees as application code running on a processor.

The remainder of this chapter will provide a basic introduction into constructing Linux-based FPGA platforms using PR. The first section describes what is required to boot Linux on a PowerPC-based FPGA design. The following section describes the additional constraints of PR in more detail. The final section describes a particular SRP engineered for wireless communication systems.

12.3 EDK Designs with Linux

To a large extent, running Linux on a processor implemented inside an FPGA is very similar to running Linux on any other embedded processor. However, exploring this area can be complex, since FPGAs allow a system designer to change not only the processor code, but also the processor, peripherals, and interconnect architecture in a system. In addition, some interfaces, such as Gigabit ethernet and PCIexpress, are complex pieces of FPGA IP managed by complex subsystems within the Linux kernel. Since this section cannot address all of this complexity, it will focus on some of the general FPGA-specific aspects of working with Linux, focusing on the PowerPC processor embedded in some FPGAs. For a complete introduction to working with embedded Linux, there are many excellent books and online resources.*

12.3.1 Design Constraints

One of the key complexities in using Linux on FPGA-based designs is the great variety of systems that can be constructed using FPGAs. EDK provides

---

*See [3,28], and http://git.xilinx.com.

access to a large number of IP cores, but it is up to the system designer to construct a system correctly to meet a particular design goal. In many FPGA systems, for instance, processors are used directly in a datapath, rather than as a "control processor," and may not require additional IP cores in order to function. However, due to the architecture of the Linux kernel, a number of IP cores must be used in addition to the basic processor core in order to create a Linux-capable processor system. Some of these requirements are fundamental to the way Linux works, such as the need for a root file system. Other requirements are specific to the design of a particular architecture.

In particular, in order to run a Linux on a PowerPC-based EDK design, it is typically necessary to

- Include access to significant amounts of external memory, typically DDR or DDR2 SDRAM).
- Include an interrupt controller, which aggregates the interrupt lines of most IP cores to the processor.
- Include memory at physical address 0, to service the PowerPC trap mechanism. Typically, this will be the external memory.
- Include memory at the reset vector 0xFFFFFFFC. Typically, this will be a small amount of FPGA BRAM.
- Include a console device. In most embedded designs, this is provided by a serial port.
- Include a source for the root filesystem and user applications. A wide variety of options may be used for the root filesystem, including on-board Flash memory, disk drives, and networked booting using network file system (NFS).

Designs generated using EDK base system builder will satisfy these constraints, as long as an external memory IP such as the Xilinx multiported memory controller (MPMC) and some BRAM is included in the design, and "use interrupt" is always selected for other IP.

12.3.2 Device Trees

Historically, 32-bit PowerPC-based architectures and 64-bit PowerPC-based architectures have been supported by two independent code bases in Linux. However, as of Linux kernel version 2.6.27, most 32-bit PowerPC-based architectures in Linux have been merged with the 64-bit architectures, and the older code has been removed. The unifying concept behind this merge is a generic mechanism for exposing hardware device information to the kernel, called the "device tree." In desktop and server applications, the device tree is typically constructed by querying the correct information from Open Firmware. In PowerPC systems without Open Firmware, as often occurs in embedded systems, the device tree is explicitly specified and passed as a binary structure to the kernel at boot time [6,15].

Device trees are a powerful mechanism for FPGA systems, since every FPGA design typically has an application-specific mix of peripherals. For most operating systems, including older versions of Linux, EDK includes a board support package (BSP) generator that generates header files containing compile-time constants (#define statements) describing the IPs and memory map of a particular design. As a result, when the operating system is compiled, it is specific to a particular design. Modifications to the FPGA platform require recompilation of the operating system kernel. Device trees introduce an additional level of indirection in this process, decoupling the Linux kernel from the hardware it is running on. With this level of indirection, it becomes straightforward to run the same kernel binary on different FPGA designs. When combined with the ability to load operating system modules into a running kernel, the indirection also enables a system to react to reconfiguration of the FPGA system.

The simplest way to use a device tree is to link a binary version of it (a "device tree blob") with the Linux kernel, resulting in a kernel that is again specific to particular hardware or FPGA design. More commonly, a boot loader may retrieve the device tree blob from a stored location and pass it to the Linux kernel at boot time. In PowerPC FPGA systems, the device tree blob can be conveniently stored in BRAM. This enables the device tree blob to be strongly associated with the FPGA design it describes and allows the same Linux kernel to be used for different FPGA designs. It is also relatively inexpensive, since a compressed device tree block typically fits in the single BRAM that must exist at the reset vector anyway.

12.4   Introduction to Modular Partial Reconfiguration

In this section, we describe the use of the Xilinx (EA) PR flow, based on Xilinx ISE 9.2.4 and Xilinx EDK 9.2.2 to build a SRP. In this flow, a single partial bitstream is generated for each reconfigurable module and reconfiguration is performed by simply writing this bitstream into the configuration memory of the FPGA. This flow implies some additional floorplanning design constraints in addition to the fundamental architectural constraints implied by the FPGA architecture.

First, any reconfigurable region must be floorplanned within the device and constrained to a particular region of the device. In addition, it is often useful to explicitly floorplan the static region in order to gain more information about resource usage in that portion of the design. Each region is represented by an AREA_GROUP constraint, which must be declared explicitly by the system designer. Typically, each AREA_GROUP consists of all the FPGA resources within a contiguous rectangular region of the FPGA device, including not only LUTs and FFs, but also BRAMs, DSP blocks, and routing resources. However, in some cases, it may be useful to include additional noncontiguous resources in the AREA_GROUP for the static region, such as I/O pins or processor blocks. Remember that routes within a static region can cross over a reconfigurable region. Additionally, no two reconfigurable regions can include the same configuration frame, although a reconfigurable region and a static region can share the same configuration frame.

Furthermore, additional design restrictions are required by the EA PR tools in order to properly route signals across the boundary between the static region and the reconfigured region. For nonclock signals, signals passing between the static region and the reconfigured region must pass through a primitive "bus macro" component. These bus macros must also be constrained to particular locations in the device, ensuring that the portion of signal routed with the static region and the portion routed with the reconfigured region align correctly. Historically, different forms of bus macros, making use of different FPGA resources, such as tristate buffers or slices with varying logic included have been proposed [17]. In the EA PR tools, two types of slice-based bus macros are usually used. An older form consists of two adjacent slices and are required to be placed exactly on the boundary of a reconfigurable region. However, merge-based PR with route-overs enables a simpler bus macro consisting of only a single slice, which is located inside the reconfigured region.

Clock signals are generally distributed by a specialized low-skew clock tree in the FPGA, and are treated independently from other signals. In Virtex 4 devices, these clock signals are sourced at the center of the device and connect to one of eight horizontal wires that distribute the clock in each clock region. Hence only eight clocks can be used in each clock region. Instead of using bus macros, each source of a global clock, represented by a BUFG primitive, is implemented in the static region and constrained to a particular location in the device. In addition, the configuration frames containing the connections to the horizontal clock tree are configured only based on the structure of the static design and are assumed to never be reconfigured. This simplifies the allocation of clocks in a reconfigurable region, although as a result each reconfigurable region is restricted to at most eight global clock signals, in addition to any additional constraints if multiple reconfigurable regions are placed in a clock region.

12.5 EDK Designs with Partial Reconfiguration

As of version 9.2.2, EDK does not directly support the creation and implementation of PR designs. However, by following some simple guidelines, it is possible to construct a design in EDK that can be partially reconfigured. This section describes a procedure for implementing such a design, focusing on the concept of a SRP. This technique relies on two independent EDK designs, one for the static region containing the processor subsystem and a separate design for the reconfigurable module. The interface between the two regions is represented by an IP core, which conceptually encapsulates the reconfigurable region along with bus macros and the interface control logic.

12.5.1 Abstracting the Reconfigurable Socket

Within EDK, all logic must be encapsulated within IP cores. In PR designs, the logic that is instantiated must include bus macros, the module that will be reconfigured, and any control logic for controlling the bus macros and generating other signals. For simplicity, it is easiest to assume that this logic will be encapsulated within a single IP core, which we will call a "reconfigurable socket." Although some other logic is necessary, including generating clocks and interfacing with the ICAP port, it is easiest to simply reuse existing IP cores provided by Xilinx. This enables simple systems with only a single reconfigurable region and more complex systems with multiple regions to be easily constructed by instantiating multiple reconfigurable socket IP cores.

12.5.2 Interface Architecture

The interface of the reconfigurable socket is a critical system-level design decision. Since this interface is fixed with the design of the static system, it must be flexible enough to allow any anticipated applications to be implemented inside the reconfigurable region. For systems where the static design must support a set of reconfigurable modules designed for a particular application, this can be done relatively easily. However, in order to implement an application-independent static design and to enable reuse of the socket IP core, a generic interface must be chosen. Architecting this interface around a standard bus protocol, such as the IBM CoreConnect processor local bus (PLB), provides this flexibility. Most currently available Xilinx IP are based on an FPGA-optimized variant of version 4.6 of this specification [8,25].

However, using this standard directly is somewhat difficult. One difficulty is the large number of signals that is required to implement an arbitrary PLB slave, including up to 128 data signals each way, 64 address signals, plus a large number of additional control signals. In total, this requires over 300 signals to be passed across bus macros, even though most systems are unlikely to implement 128-bit wide slaves. A second difficulty is that some of the widths of the bus control signals depend on the number of masters and slaves. This makes modifying the system difficult, since each signal must be given explicit placement constraints. A final difficulty is that the Xilinx EDK computes the data width of the bus based on the maximum data width of all masters and slaves, and masters and slaves must know the width of the bus in order to include the correct logic to communicate with masters and slaves of different widths. Because of the design flow described above, which uses separate EDK designs to represent the static design and any reconfigurable modules, exposing this width information to EDK is difficult.

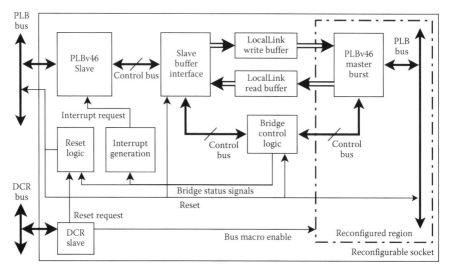

FIGURE 12.7
Reconfigurable socket abstraction based on the "PLBv46 PLBv46 bridge" architecture. The "PLBv46 slave" and "PLBv46 master burst" blocks are standard IP components and all blocks except the *DCR slave* block are part of the bridge. Bus macros are implicitly present on all signals crossing the boundary of the reconfigured region.

An alternative is to architect the interface around a bus bridge, with independent busses in the static region and in the reconfigurable region. The design of the socket is based on partitioning the Xilinx "PLBv46 PLBv46 bridge" IP [23], as shown in the block diagram in Figure 12.7. Internally this core is based around 32-bit fixed-width data FIFOs and a small number of control signals. Most of the bridge is treated as part of the static region, with only a small amount of logic required in the reconfigurable region to complete the bridge. In addition to the bus interface, which is primarily used to interface to the reconfigured region, the socket core also contains a control interface (based on the DCR protocol [7]) which is used to generate an independent reset signal to the reconfigurable region and to force signals driven by the reconfigurable module to stable values during reconfiguration.

12.5.3 Direct Memory Access Interfaces

The bus interface above is a generic and flexible interface, which can be used to communicate with the reconfigured portion of the system in different ways. For instance, it may be used by the processor to both send and receive data from the reconfigured region or as a control interface to set parameter values of IP cores executing in the reconfigured region. However, it does have several disadvantages. Primarily, the bandwidth of data to or from the processor is limited because of the overhead of bus arbitration and the fact that the memory range is treated as uncached I/O transactions. Although performance could be improved somewhat for large transactions by using DMA engines or treating data transfer regions as cached and manually managing cache coherency, this would significantly increase the complexity of the processor software. Secondly, many FPGA algorithms require access to external memory for buffering data until it can be processed. For instance, in a network router, packet data may need to be stored until a routing decision can be made, or in a streaming video system, several frames of video data may need to be stored to analyze object motion between frames.

Because of these limitations, it is best to consider the bus interface above as primarily an interface used for low-bandwidth control and configuration information. In systems that require higher bandwidth communication, or direct access to external memory, the control interface can be augmented with additional interfaces to memory. Although it may seem straightforward to include a complementary bus bridge that can be driven by the reconfigured region to provide this functionality, this tends not to be the highest bandwidth option since performance can be limited by the arbitration logic of the PLB bus. This logic is heavily pipelined in order to maximize the bus throughput under a wide variety of usage, typically incurring three cycles of latency before a slave can respond to a bus access.

One solution is to provide an interface connected directly to the native port interface (NPI) of the Xilinx MPMC IP core, as shown in Figure 12.8.

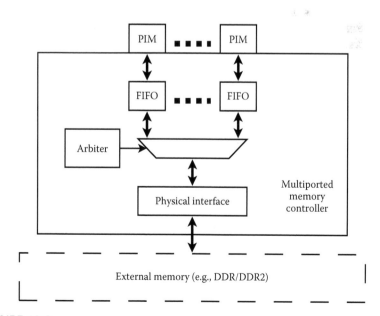

FIGURE 12.8
Architecture of the Xilinx MPMC.

Typically, this interface exhibits both lower latency and higher bandwidth than the PLB bus. Although the MPMC must still arbitrate between different ports attempting to use the memory controller, this arbitration can be performed locally within the memory controller and concurrently with the data being provided. The only disadvantage of connecting directly to the memory controller is that other IP cores in the static region cannot be accessed from the reconfigured region. However, since in the SRP usage model these IP cores are likely being managed by device drivers in the operating system of the processor, it is questionable whether such access should be allowed anyway.

12.5.4 External Interfaces

In addition to communicating with the static region, a reconfigurable module may also communicate with other interfaces external to the FPGA. In order to accomplish this, a reconfigurable region may include external I/O pins and/or high-speed serial transceivers. For the most part, these resources can be treated as any other FPGA primitives and can be placed and routed as usual.

However, there is some complexity with regard to external I/O pins, since in many FPGA designs, the input/output buffer (IOB) primitives representing external I/O pins are not explicitly instantiated in a user design but are inferred in the synthesis process. Normally in a hierarchical design, the netlist can be synthesized using a special option to disable inference of these primitives, since they will be inferred or instantiated during synthesis of the toplevel design. However, when building a generic FPGA platform, relying on this may not be desirable, since the reconfigured region may require more control over the configuration of these primitives. In other cases, exactly which IOB primitives are explicitly instantiated in a reconfigurable module and which ones are not may not be known when the static design is synthesized and implemented. One way to solve this is to not expose any I/O pins of the reconfigurable region as external signals of the static region, implying that synthesis of the static design will never include IOB primitives for these pins. When a reconfigurable module is synthesized, signals interfacing with the static region are individually tagged with the constraint BUFFER_TYPE set to NONE, indicating that no IOB primitives should be inferred for those signals.

High-speed serial transceivers also have additional design complexity, since each transceiver is associated with specialized clock resources in the FPGA. These clock resources typically include phase-locked loops for clock synchronization and dedicated clock distribution paths and may be shared between transceivers. From the perspective of building FPGA platforms, this resource sharing combined with how transceivers are grouped into configuration frames may need to be considered during the floorplanning stage in order to gain maximum usage of the available transceivers.

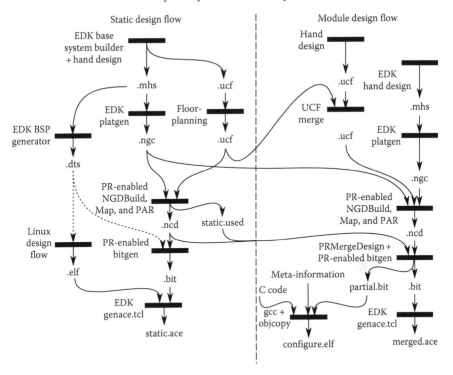

FIGURE 12.9
Design flow for PR systems based on EDK.

12.5.5 Implementation Flow

The implementation flow for the system is shown in Figure 12.9. The static design is implemented first, as shown in the left-hand side of the figure, using the EA PR tools. During this sequence, no netlist for the reconfigurable region is present, and the place and route tools only implement logic for the static region. Design constraints are provided in a .ucf file and must include the required floorplanning constraints for the PR flow. After routing is completed, the routing resources used by the static logic are saved in the file *static.used* for later use. Since by default the interface with the reconfigured region is driven to an idle state, the resulting bitstream can be used in a system without programming the remainder of the FPGA. The device tree for a particular design is generated from the EDK design, and after being converted to a binary device tree blob, can be included in the Linux kernel image, or stored as the initial value of a BRAM in the bitstream. Lastly, EDK is used to package the FPGA bitstream with the Linux kernel binary in a bootable image that can be used with Xilinx SystemAce [24] to boot the kernel.

The right-hand side of Figure 12.9 shows a second pass for the implementation of a reconfigurable module. During this pass, the logic of the reconfigurable module is implemented together with a small portion of the static logic called the "context logic." The context logic is necessary to provide the context of the reconfigurable module, so that hierarchical names in the design and location constraints for clock signals and bus macros can be preserved. The design constraints for implementation are created by merging the design constraints from the static design with any additional design constraints specific to the reconfigurable module, such as pin location constraints. During this pass, the routing resources in the file *static.used* are excluded from use, since these resources are already used in the static design. The final bitstream for the reconfigurable module is generated by first merging the design database (contained in an .ncd file) from both passes, ensuring that the configuration bits used in the static design are programmed correctly. In addition, design rule checks and timing analysis can be applied to the merged design database, to ensure that individual passes were implemented correctly. From the merged design database, it is possible to generate both a partial bitstream that can be used after configuration with the static bitstream and a merged bitstream which can be used as an initial configuration bitstream, with the reconfigurable module already loaded. To enable reconfiguration in a Linux system, the partial bitstream is encapsulated with the Linux code for performing PR and the meta-information about the reconfigurable module, to generate a Linux executable, as described in Section 12.6.

12.6   Managing Partial Reconfiguration in Linux

Two device drivers are used to manage the reconfiguration process. Primarily, the device driver for the ICAP device performs the actual reconfiguration. When a partial bitstream is written to this device (for instance, using the cp command or the write() system call), the bytes are transferred to the ICAP. Since the device driver does not inspect or modify the stream of bytes, the data being written must include the appropriate control words, as expected by the configuration interface [26]. The device driver also includes simple locking of the ICAP resource, in order to prevent different processes from unexpectedly interleaving accesses to the ICAP. Readback is also possible using this device driver by writing the correct readback request bitstream to the ICAP and subsequently reading data (using the read() system call).

The second device driver used to manage reconfiguration is associated with the reconfigurable socket core. This driver exports a character interface to which meta-information about a reconfigurable module can be written. A simple way of representing this meta-information is in the form of an array of struct platform_device, a data structure which is used internally by Linux to represent devices. A more complex, but perhaps more robust

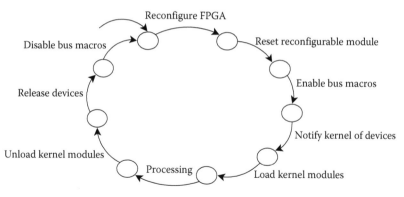

FIGURE 12.10
The reconfiguration process.

representation of meta-information could be an additional device tree blob. This meta-information is parsed and checksummed and, if valid, is used to notify the Linux kernel of the presence of new devices, which can then be bound to other device drivers. An invalid checksum is interpreted as an indication to unbind any previously loaded devices and release ownership of the reconfigured region. Secondarily, this device driver also enables and disables the bus macros between the static region and the reconfigured region, and controls the reset of the reconfigured region. As with the ICAP device driver, the socket device driver includes a simple locking mechanism in order to prevent a process from unexpectedly reconfiguring an active region in use by another process.

The complete process of reconfiguration is shown in Figure 12.10. In the initial state, we assume no module is loaded in the reconfigured region. Next, a reconfigurable module is loaded into the FPGA through the ICAP device driver. Next, meta-information about the reconfigurable module is sent to the socket device driver, which registers the presence of any new devices, resets the newly loaded module, and enables the interface between the static region and the reconfigurable module. At this point, although Linux is aware of the presence of the reconfigured devices, it may not have device drivers appropriate to those devices. Next, device drivers for new devices are provided by loading the appropriate kernel modules and the Linux kernel binds those device drivers to the reconfigured devices. At this point, application code may use the device drivers to communicate with the reconfigured region. A similar sequence of steps in reverse order occurs to unbind the device drivers and release the reconfigured region so that different processing may occur.

Since the ICAP device and the control interface of the socket are exposed through device drivers, it is relatively straightforward to implement reconfiguration through a regular user process. One possibility for implementing this involves linking the bitstream and meta-information into a single executable along with the code for reconfiguration. The process created when this executable is executed can be controlled through any operating system mechanism (such as POSIX signals) to manage the life cycle of the module loaded in the FPGA. The executable can also be linked together with other application code, resulting in a familiar processor-centric usage model for the FPGA fabric. This approach is similar in spirit, but greatly different in implementation from that proposed in [18], which performs essentially the same processes using the Linux kernel's ability to implement new executable formats.

It is important to recognize that although the reconfiguration process is managed by a user process, it must be treated as a privileged operation executed as the root user, since there are many places where both unintended errors and malicious attacks may result in unintended behavior. Some of these places are not specific to the PR process, such as loading kernel modules, whereas others are more subtle vulnerabilities. For instance, as noted before, partial bitstreams have significant constraints on how they are constructed and are specific to a particular implementation of the static system. More directly, it is possible to trigger reconfiguration of the FPGA through the ICAP interface, resulting in the loss of the current state of the system. If the bus macros are enabled during PR, then it is likely that glitching on the interface signals will result in unintended behavior of the static system.

One particularly common usage error is simply attempting to load a partial bitstream that does not correspond to the current implementation of the static design. This may happen during development when a modification is made to the static region, but a designer neglects to reimplement a reconfigured module. One way of avoiding such errors is to prepend each partial bitstream with a hash generated from the static design. This hash can also be stored in the static design, possibly in the device tree blob, and checked before being loaded into the FPGA. If the partial bitstream is not signed properly, then the reconfiguration process can be halted without affecting the operation of the static design. This technique can be simply applied to prevent unintended errors, or adapted using more cryptographically secure techniques to prevent malicious attacks [2,4].

12.7   Putting It All Together

This section illustrates a SRP design targeted at a variant of the WARP Software-defined Radio hardware built by Rice University [12]. Since the original hardware is based on an older Virtex 2 Pro FPGA, we present a design based on an updated Virtex 4 FX 100 device in order to better

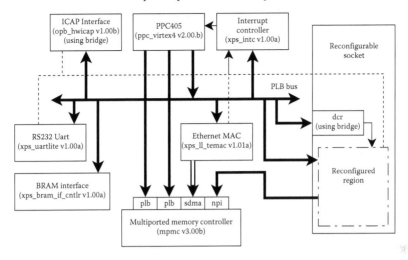

FIGURE 12.11

Architecture of a reconfigurable platform. Some signals and standard cores have not been shown.

represent the PR capabilities of newer FPGA architectures. In particular, we focus on a MIMO OFDM reference design for this board, which implements a bridge from Wired Ethernet to a two-radio MIMO system. The design uses a processor to manage the packet headers and to perform configuration management of the radios, while packet payloads are communicated directly between the wired and wireless network interfaces using direct memory access to a processor-managed memory buffer. In the reference design, the packed payload buffer is implemened in BRAM and communicated through a PLB bus. In the reconfigurable design, we assume that the packet payload buffer is implemented in external DRAM, which must be accessed from the reconfigurable region through a separate port of the memory controller. As a nonreconfigurable system, this design uses approximately 50% of the device (21294 of 42176 slices).

The design of the static subsystem is shown in Figure 12.11. This design is architected around the PowerPC 405 processor core and was largely generated using the Base System Builder capability in Xilinx EDK. Standard serial port and ethernet IP cores provide external connectivity. Access to external 64 bit wide DDR2 SDRAM, including DMA access for the ethernet core, is provided by the Xilinx MPMC IP core. In this system, the processor, memory bus, and memory controller are designed to be "quasi-synchronous," meaning that clocks must be edge-aligned. Based on the speeds of the individual components, a design point was chosen targeting a slow speed grade FPGA (−10) with the memory bus clocked at 83.3 MHz, the memory controller clocked twice as fast (166.6 MHz), and the processor clocked three times as fast (250 MHz).

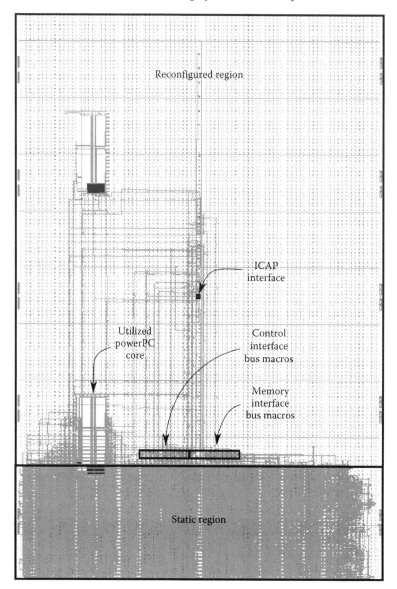

FIGURE 12.12
Placed and routed design of an FPGA processor platform, targeting a Virtex 4 FX 100.

The FPGA layout of the design is shown in Figure 12.12, overlaid with the PR floorplanning constraints. The static region is at the south of the chip, and is exactly two configuration frames tall. This layout provides approximately 8600 slices and 128 external I/O pins, which accommodates both the logic requirements of a simple processor design, and the I/O pins requirements of a 64-bit DDR2 memory interface. A significantly smaller region would fail to provide enough logic cells for the static design, while a larger region would allocate too many pins to the static region, which would be difficult to access from the reconfigurable region.

Note that the majority of the routed signals are contained within the floor-planned area for the static region. The routes entering the top region connect primarily to external I/O pins and FPGA resources, such as clock buffers and the ICAP, located in the center column of the FPGA. Some routes into the top region also connect to the PowerPC cores. Although only one PowerPC is actually used in the static design, current versions of the EA PR tools do not allow PowerPC cores to be part of the reconfigured portion of the design. Hence, this design instantiates both PowerPC cores in the static region, in order to enable use of the JTAG chain, which is assumed to connect through both cores.

The device tree for this design is shown in Figure 12.13. Since the targeted board includes Xilinx SystemACE, this is used to configure the FPGA and initialize external memory with the kernel image. The compressed device tree blob is initialized in the BRAM at address 0xffffff800 and decompressed by the Linux bootwrapper executing out of external memory. The root filesystem is stored on an external file server and loaded over the network interface using the NFS protocol.

12.8   Conclusion

Although high-level algorithmic modeling offers significant promise for increasing design productivity, a common problem with many approaches is representing the environment in which a model exists in a system. A solution to this problem is often to provide platforms that abstract lower level details, provide standardized interfaces, and can be targeted by a high-level design tool. Although this difficulty exists in any embedded system, it is particularly apparent in FPGA systems, which include complex IP blocks, such as processor cores, and where physical interfaces to the rest of the system are highly flexible and incorporate many features that cannot be easily modeled even at the circuit and gate level.

However, using the architectural features of some FPGAs, such as PR, higher level platforms can be constructed that abstract many of these details and are more appropriate for mapping from a high-level design tool. This chapter has particularly shown how this technique can abstract the complexities associated with including a control processor and operating system as part of an FPGA platform.

```
/ {
    #address−cells = <1>; #size−cells = <1>;
    compatible = "xlnx,virtex";
    DDR2_SDRAM: memory@0 {
        device_type = "memory";
        reg = < 0 10000000 >;
    };
    cpus {
        #address−cells = <1>; #size−cells = <0>;
        #cpus = <1>;
        ppc405_0: cpu@0 {
            clock−frequency = <ee6b280>;
            compatible = "PowerPC,405", "ibm,ppc405";
            d−cache−line−size = <20>;
            d−cache−size = <4000>;
            device_type = "cpu";
            i−cache−line−size = <20>;
            i−cache−size = <4000>;
            model = "PowerPC,405";
            reg = <0>;
            timebase−frequency = <ee6b280>;
        };
    };
    plb: plb@0 {
        #address−cells = <1>; #size−cells = <1>;
        compatible = "xlnx,plb−v46−1.00.a";
        ranges ;
        TriMode_MAC_GMII: xps−ll−temac@81c00000 {
            #address−cells = <1>;
            #size−cells = <1>;
            compatible = "xlnx,compound";
            ethernet@81c00000 {
                compatible = "xlnx,xps−ll−temac−1.01.a";
                device_type = "network";
                interrupt−parent = <&xps_intc_0>;
                interrupts = < 2 2 >;
                llink−connected = <&PIM2>;
                local−mac−address = [ 02 00 00 00 00 00 ];
                reg = < 81c00000 40 >;
                xlnx,phy−type = <1>;
                xlnx,temac−type = <1>;
            };
        };
        mpmc@0 {
            #address−cells = <1>; #size−cells = <1>;
            compatible = "xlnx,mpmc−3.00.b";
            PIM2: sdma@84600100 {
                compatible = "xlnx,ll−dma−1.00.a";
                interrupt−parent = <&xps_intc_0>;
                interrupts = < 1 2 0 2 >;
                reg = < 84600100 80 >;
            };
        };
    opb: opb@40000000 {
        #address−cells = <1>; #size−cells = <1>;
        compatible = "xlnx,opb−v20−1.10.c";
        ranges = < 40000000 40000000 10000000 >;
        opb_hwicap_0: opb−hwicap@41300000 {
            compatible = "xlnx,opb−hwicap−1.00.b";
            reg = < 41300000 10000 >;
            xlnx,family = "virtex4";
        };
    };
    plbv46_dcr_bridge_0: plbv46−dcr−bridge@80700000 {
        compatible = "xlnx,plbv46−dcr−bridge−1.00.a";
        dcr−access−method = "mmio";
        dcr−controller ;
        dcr−mmio−range = < 80700000 1000 >;
        dcr−mmio−stride = <4>;
    };
    rs232: serial@84000000 {
        clock−frequency = <4f790d5>;
        compatible = "xlnx,xps−uartlite−1.00.a";
        current−speed = <2580>;
        device_type = "serial";
        interrupt−parent = <&xps_intc_0>;
        interrupts = < 3 0 >;
        port−number = <0>;
        reg = < 84000000 10000 >;
        xlnx,baudrate = <2580>;
        xlnx,data−bits = <8>;
        xlnx,odd−parity = <0>;
        xlnx,use−parity = <0>;
    };
    xps_bram_if_cntlr_1: xps−bram−if−cntlr@fffff000 {
        compatible = "xlnx,xps−bram−if−cntlr−1.00.a";
        reg = < fffff000 1000 >;
    };
    xps_intc_0: interrupt−controller@81800000 {
        #interrupt−cells = <2>;
        compatible = "xlnx,xps−intc−1.00.a";
        interrupt−controller ;
        reg = < 81800000 10000 >;
        xlnx,num−intr−inputs = <5>;
    };
    xps_socket_0: xps−socket@50000000 {
        compatible = "xlnx,xps−socket";
        dcr−parent = <&plbv46_dcr_bridge_0>;
        dcr−reg = < c0 2 >;
        interrupt−parent = <&xps_intc_0>;
        interrupts = < 4 2 >;
        reg = < 50000000 10000000 >;
    };
    };
};
```

FIGURE 12.13
Device tree.

References

1. B. Blodget, P. James-Roxby, E. Keller, S. McMillan, and P. Sundararajaran. A self-reconfiguring platform. In *Proceedings of the International Field Programmable Logic and Applications Conference (FPL)*, Lisbon, Portugal, 2003. *Lecture Notes in Computer Science*, Vol. 2778, Springer-Verlag, September 2003.

2. J. Castillo, P. Huerta, V. Lopez, and J. Martinez. A secure self-reconfiguring architecture based on open-source hardware. In *International Conference on Reconfigurable Computing and FPGAs (ReConFig)*, Puebla City, Mexico, September 2005.

3. J. Corbett, A. Rubini, and G. Kroah-Hartman. *Linux Device Drivers*. O'Reilly, Sebastopol, CA, 3rd edition, 2005.

4. R. Fong, S. Harper, and P. Athanas. A versatile framework for FPGA field updates: An application of partial self-reconfiguration. In *Proceedings of the IEEE International Workshop on Rapid System Prototyping*, San Diego, CA, June 2003.

5. R. Freeman. Configurable electrical circuit having configurable logic elements and configurable interconnects. US Patent 4870302, September 1989.

6. D. Gibson and B. Herrenschmidt. Device trees everywhere. In *Proceedings of linux.conf.au*, Dunedin, New Zealand, January 2006. Available at http://ozlabs.org/people/dgibson/home/papers/dtc-paper.pdf, accessed April 28, 2008.

7. IBM. Device control register bus architecture specifications version 3.5, January 2006.

8. IBM. 128-bit processor local bus architecture specifications version 4.7, May 2007.

9. I. Kuon and J. Rose. Measuring the gap between FPGAs and ASICs. *IEEE Transactions on Computer-Aided Design of Integrated Circuits and Systems*, 26(2):203–215, February 2007.

10. E. A. Lee and S. Neuendorffer. Actor-oriented models for codesign: Balancing re-use and performance. In S. Shukla and J.-P. Talpin (editors), *Formal Methods and Models for System Design: A System Level Perspective*, pp. 33–56, Kluwer, Norwell, MA, 2004.

11. M. Majer, J. Teich, A. Ahmadinia, and C. Bobda. The Erlangen slot machine: A dynamically reconfigurable FPGA-based computer. *Journal of VLSI Signal Processing Systems*, 47(1):15–31, March 2007.

12. P. Murphy, A. Sabharwal, and B. Aazhang. Design of WARP: A flexible wireless open-access research platform. In *Proceedings of the European Signal Processing Conference (EUSIPCO)*, Florence, Italy, 2006.

13. A. Parsons et al. A scalable correlator architecture based on modular FPGA hardware and data packetization. In *Asilomar Conference on Signals, Systems, and Computers*, Pacific Grove, CA, November 2006.

14. A. Parsons et al. A scalable correlator architecture based on modular FPGA hardware and data packetization. Submitted to *IEEE Transactions on Signal Processing*, available at http://casper.berkeley.edu/papers/2008-02_parsons_et_al-correlator.pdf, February 2008.

15. Power.org. Power.org standard for embedded power architecture platform requirements (epapr), July 2008. Version 1.0.

16. A. Sangiovanni-Vincentelli. Defining platform-based design. *EEDesign*, February 2002.

17. P. Sedcole, B. Blodget, T. Becker, J. Anderson, and P. Lysaght. Modular dynamic reconfiguration in Virtex FPGAs. *IEE Proceedings on Computers and Digital Techniques*, 153(3):157–164, May 2006.

18. H. K.-H. So and R. W. Brodersen. Improving usability of FPGA-based reconfigurable computers through operating system support. In *Proceedings of the International Field Programmable Logic and Applications Conference (FPL)*, Madrid, Spain, 2006.

19. Sun. Opensparc web page, available at http://www.opensparc. net, accessed on March 7, 2008.

20. Triscend. Triscend e5 configurable system-on-chip platform datsheet, July 2001, v1.06.

21. M. Uhm and J. Bezile. Meeting software defined radio cost and power targets: Making SDR feasible. in *Military Embedded Systems*, pp. 6–8, May 2005.

22. J. Williams and N. Bergmann. Embedded linux as a platform for dynamically self-reconfiguring systems-on-chip. In *Proceedings of the International Multiconference in Computer Science and Computer Engineering (ERSA)*, Los Vegas, CA, June 2004.

23. Xilinx. *PLBv46 to PLBv6 Bridge Data Sheet*, ds618 edition. Version 1.00.a, available at http:/www.xilinx.com/bvdocs/ipcenter/data_sheet/plbv46_plbv46_bridge.pdf, accessed on March 6, 2008.

24. Xilinx. *Embedded System Tools Reference Manual*, ug111 v9.2 edition, September 2007.

25. Xilinx. *PLBV46 Interface Simplifications,* sp026 edition, October 2007.

26. Xilinx. *Virtex-4 FPGA Confituration User Guide,* ug071 v1.10 edition, April 2008.

27. Xilinx. *Virtex-4 FPGA Guide,* ug070 v2.40 edition, April 2008.

28. K. Yaghmour. *Building Embedded Linux System.* O'Reilly, Sebastopol, CA, 2003.

Part III

Design Tools and Methodology for Multidomain Embedded Systems

13

Modeling, Verification, and Testing Using Timed and Hybrid Automata

Stavros Tripakis and Thao Dang

CONTENTS

13.1 Introduction

Models have been used for a long time to build complex systems, in virtually every engineering field. This is because they provide invaluable help in making important design decisions before the system is implemented. Recently, the term "model-based design" has been introduced to emphasize the use of models and place them in the center of the development process, especially for software-intensive systems. Traditionally, the fact that software is immaterial (contrary, say, to bridges or cars or hardware), has resulted in a software development process that largely blurs the line between design and implementation: a model of the software is the software itself, which is also the implementation. It is "cheap" to write software and test it, or so people used to believe. It is now becoming more and more clear that the costs for software development, testing and maintenance are non-negligible, in fact, they often outweigh the costs of the rest of the system.

As a result of this and other factors, a more rigorous software development process based on formal, high-level models, is becoming widespread, especially in the "embedded software" domain. The term "embedded software" may generally include any type of software that runs on an embedded system. Embedded systems are computer-controlled systems that strongly interact with a physical environment, for example, "x-by-wire" systems for car control, cell phones, multimedia devices, medical devices, defense and aerospace, public transportation, energy and chemical plants, and so on.

In all these systems, timing and other physical characteristics of the environment are essential for system correctness as well as for performance. For instance, in an engine-control system it is critical to ignite the engine at very precise moments in time (in the order of 1 ms). Moreover, the control logic depends on a number of continuously evolving variables that have to do with the combustion in the engine, exhaust, and so on. In order to capture such timing and physical constraints, models of "timed automata" and "hybrid automata" have been developed in the early 1990s [5,9]. Since then, these models have been studied extensively and today there are a number of sophisticated analysis and synthesis methods and tools available for such models. We will discuss some of these methods in this chapter.

Models, in general, play different roles and are used for different tasks, at different phases of the design process. For instance, sometimes a model captures a system that is already built to some extent, while at other times a model serves as a specification that the final system must conform to. In the rest of this chapter, we consider the following tasks in the context of timed or hybrid automata models:

- *Modeling*: We discuss timed and hybrid automata in Section 13.2. Modeling is of course a task by itself, and probably the most crucial one, since it is a creative and to a large extent nonautomatable task.
- *Exhaustive verification*: We use the term exhaustive verification to denote the task of proving that a given model of a system satisfies a given property (also expressed in some modeling language). This task is exhaustive in the sense that it is conclusive: either the proof succeeds or a counter-example is found that demonstrates that the system fails to satisfy the property. We focus on automatic (push-button) exhaustive

verification, also called *model checking*. We discuss exhaustive verification in Section 13.3.

- *Partial verification*: Fundamental issues such as undecidability (in the case of hybrid automata) or state-explosion (in the case of both timed and hybrid automata) place limits on exhaustive verification. In Section 13.4 we discuss an alternative, namely, partial verification, which aims to check a given model as much as possible, given time and resource constraints. This is done using mainly simulation-based methods.

- *Testing*: It is important to test the correctness of a system even after it is built. Testing mainly serves this purpose. Since designing good and correct test cases is itself a time-consuming and error-prone process, one of the main challenges in testing is automatic test generation from a formal specification. We discuss testing in general in Section 13.5, and test generation from timed and hybrid automata models in Sections 13.6 and 13.7, respectively.

Obviously, the topics covered in this chapter are wide and deep, and we can only offer an overview. We attempt an intuitive presentation and omit most of the technical details. Those can be found in the referenced papers. We also restrict ourselves to the topics mentioned earlier and omit many others. For example, we do not discuss discrete-time/state models, theorem proving, controller synthesis, implementability and code generation, as well as other interesting topics. Finally, excellent surveys exist for some of the topics covered in this chapter (e.g., see Alur's survey on timed automata [3] and an overview of hybrid systems in the book [107]), thus we prefer to devote more of our discussion to topics that are more recent and perhaps have been less widely exposed such as testing and partial verification.

13.2   Modeling with Timed and Hybrid Automata

Timed and hybrid automata models are extensions of finite automata with variables that evolve continuously in time. Timed automata are a subclass (i.e., special case) of hybrid automata. In timed automata all variables evolve with rate 1: that is, these variables measure time itself. Hybrid automata are more general, with variables that can in principle obey any type of continuous dynamics, usually expressed by some type of differential equations.

These models were introduced in order to meet the desire to blend the "discrete" world of computers with the "continuous" physical world. Classical models from computer science (e.g., finite-state automata) provide means for reasoning about discrete systems only. Classical models from engineering (e.g., differential equations) provide means for reasoning mostly about continuous systems. Timed and hybrid automata are attempts to bridge the two worlds. Although timed automata are a special class of hybrid automata, their study as a separate model is justified by the fact that many problems that are very difficult or impossible to solve (i.e., undecidable) for hybrid automata are easier or solvable for timed automata.

13.2.1 Timed Automata

Timed automata [9] extend finite automata by adding variables that are able to measure real time: these variables are called "clocks." Standard finite-state automata are able to specify that a certain set of events occur in a specific order; however, they do not typically specify how much time has elapsed between two successive events. Figure 13.1a shows an example of a finite automaton that specifies the order between four events a, b, c, d: event a precedes b, which precedes c, which precedes d. This automaton has five states numbered 0–4, and four transitions labeled by the four events. State 0 is the initial state.

Figure 13.1b shows an example of a timed automaton (TA). It is very similar to the "untimed" version, but its transitions are annotated with additional information, referring to clocks x and y. This TA specifies, in addition to the order a, b, c, d between events, two timing constraints: (1) the time that may elapse between a and d is at most 5 time units; (2) the time that may elapse between b and c is at least 2 time units. In other words, we can view the semantics of this automaton as a set of "all possible" sequences of occurrences of events in time ("timed sequences") that satisfy timing constraints (1) and (2) in addition to the correct order a, b, c, d. This is illustrated in Figure 13.2.

We can also look at the semantics of this automaton in an "operational" way. This is illustrated in Figure 13.3. The automaton starts at state 0 and spends a certain amount of time t_0 there. During this time the value of each clock of the automaton increases by the amount of time that has elapsed,

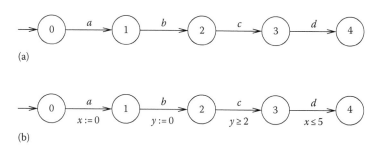

(a)

(b)

FIGURE 13.1
(a) A finite-state automaton. (b) A timed automaton.

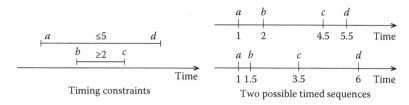

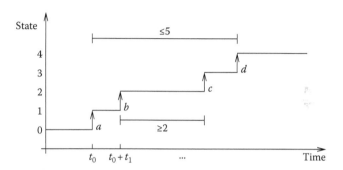

FIGURE 13.2
Behaviors of TA of Figure 13.1.

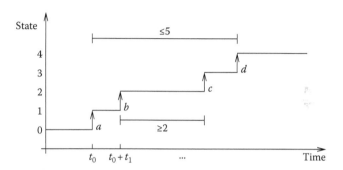

FIGURE 13.3
Operational semantics of TA of Figure 13.1.

that is, t_0 in this case. The automaton then "jumps" to state 1: event a occurs in this jump, which is instantaneous. The automaton proceeds in the same pattern: it spends some time t_1 in state 1, then jumps to state 2, and so on. The automaton alternates between these "timed" and "discrete transitions." During a discrete transition, some clocks may be reset to zero, denoted by $x := 0$, $y := 0$, and so on. Discrete transitions may have "guards," that is, conditions such as $y \geq 2$ or $x \leq 5$, that must be satisfied in order for the transition to be possible.

The operational view reveals that knowing what state the automaton occupies at a given point in time (numbered 0, 1, 2, ..., in the earlier examples) is not enough to predict its future behavior: one must also know the values of its clocks (e.g., in order to check whether the guards are satisfied). This is why we must distinguish between the "discrete" state of the automaton (also called sometimes "control state" or "location") and its "full" state that includes the clock values (sometimes called "configuration" and sometimes simply the "state").

We will use state vs. configuration to distinguish the two.

Notice that the unit of time, although it is assumed to be the same for all the clocks, is not explicit in a TA model. This is often an advantage, especially when we are only interested in the relative magnitude of timing constraints and not their absolute value. In the Alur–Dill model of timed automata, time is "dense": delays can be taken to be positive real or rational numbers. This model is strictly more expressive than a discrete-time model, where delays are the integer multiples of some given quantum of time. For instance, the dense-time model can express that two events a and b can occur arbitrarily close to each other, but not at the same time, using a "strict" constraint of the form $x > 0$. Whether to opt for a dense or discrete TA model depends on the application at hand. Considerations need to take into account not only modeling requirements, but also the complexity of the algorithmic analysis, such as model checking. Dense-time model-checking is more expensive than discrete-time model-checking in theory, and often* in practice as well [24]. The discrete- vs. dense-time debate is a nontrivial topic. In-depth studies can be found in [19,48,81].

The basic model of timed automata described earlier can be extended in various ways (one is by adding more powerful continuous dynamics, which leads to hybrid automata described in Section 13.2.2). Discrete variables can be added to the model, with basic types such as booleans or integers, but also more complex types such as records, queues, and so on. These extensions are very handy when modeling other than very simple examples. As long as the domain of such variables can be restricted to be finite, these extensions do not add to the expressive power of the model, since they can be encoded in the state of the automaton. Note that things become more complicated when attempting to relate such variables to clocks, for instance, resetting a clock x to the value of a variable i, as in $x := i$, or comparing a clock to a variable in a guard, as in $x \leq i$. Some of these extensions can be handled, but others may strictly increase the power of the model, leading even to undecidability. Again this is a nontrivial topic, and the reader is referred to [3,23].

Another interesting extension is modeling "urgency." To motivate this concept, consider the example shown in Figure 13.1b. If the automaton stays in state 2 more than 5 time units, then it can no longer reach state 4. We may want to disallow this behavior, thus, model the assumption that state 4 "will" be reached. We can do this by adding acceptance conditions to the automaton (e.g., making state 4 accepting and the others nonaccepting). But a more convenient way is to state this using clock constraints. For instance, we can impose the constraint $x \leq 5$ at "states" 1, 2, and 3, expressing the fact that the amount of time spent in those states must be such that this constraint is not violated. This is one way of modeling urgency, and these state-associated clock constraints are called "invariants" [49]. Another, more elaborate way is to use "deadlines," associated with transitions [18,93].

Even for relatively simple systems, modeling the entire system as a single automaton can be very tedious. A solution is to build a model by

---

*But not always, as sometimes "symbolic" dense-time model-checking tools can represent timing constraints more effectively than "enumerative" discrete-time methods. For instance, if a guard involves large constants such as $x \leq 10^6$ then a brute-force discrete-time enumeration method with time step 1 may need to represent 10^6 distinct states while a symbolic method can represent an infinite set of states with the symbolic constraint $x \leq 10^6$.

composing other models. In the case of timed automata, the different variants of compositions have been proposed, where the components can communicate through the "rendez-vous" type of action synchronization, FIFO queues, shared variables, and so on. A common assumption in most of these composition frameworks is that the clocks of all automata measure exactly the same time, in other words, that they are perfectly synchronized. This is obviously unrealistic when these clocks model real clocks. Unfortunately, modeling phenomena such as clock drift explicitly (e.g., by defining the rate of a clock x to be $1 \pm \epsilon$ for a fixed $\epsilon > 0$) yields an undecidable model, in general. As an alternative, some researchers studied an asymptotic version of the problem where ϵ can be arbitrarily small [88,109]. This allows to regain decidability while providing a more "robust" semantics. The issue of robustness is especially important when the TA model is to be implemented, for instance, as an embedded controller. However, the problem can also be tackled with standard semantics, using appropriate modeling techniques [2].

Regarding applications, it is fair to say that timed automata have not found as widespread usage as standard, "untimed" models. This is not surprising, given the fact that TA are a more specialized model in the sense that often a discrete-time model is sufficient and this can be captured in a more standard language (e.g., see [24]). Moreover, TA are more expensive to analyze than "untimed" models. Still, timed automata are appealing because of their "declarative" style of specifying timing constraints, that is suitable for capturing high-level models and specifications.* TA have been used to model small- to medium-size systems, such as communication protocols, digital circuits, real-time scheduling systems, robotic controllers, and so on. Up-to-date lists of case studies can be found at the web-sites of timed-automata model-checking tools such as Kronos[†] and Uppaal[‡] as well as in the publications of the authors of these tools.

13.2.2   Hybrid Automata

Hybrid automata [5] can be seen as an extension of timed automata with more general dynamics. A clock c is a continuous variable with time derivative equal to 1, that is $\dot{c}(t) = 1$. In a hybrid automaton, the continuous variables x can evolve according to some more general differential equations, for example $\dot{x} = f(x)$. This allows hybrid automata to capture not only the evolution of time but also the evolution of a wide range of physical entities.

---

*It is perhaps for this reason that some of the concepts in the timed automata model have found their way into the MARTE (modeling and analysis of real-time and embedded systems) profile for UML2. See http://www.omg.org/technology/documents/profile_catalog.htm.

[†]See http://www-verimag.imag.fr/TEMPORISE/kronos and http://www-verimag.imag.fr/tripakis/openkronos.html.

[‡]See http://www.uppaal.com.

The discrete dynamics of hybrid automata can also be more complex and described with more general constraints.

In the following, we present a commonly used version of hybrid automata. The different forms of constraints result in the different variants of this model. A hybrid automaton $\mathcal{A}$ consists of a finite set Q of discrete states and a set of n continuous variables evolving in a continuous state space $\mathcal{X} \subseteq \mathbb{R}^n$. In each discrete state $q \in Q$, the evolution of the continuous variables are governed by a differential equation: $\dot{x}(t) = f_q(x(t), u(t))$ where $u(\cdot) \in \mathcal{U}_q$ is an admissible input function of the form $u : \mathbb{R}^+ \to U_q \subset \mathbb{R}^m$. This input can be used to model some external disturbance or underspecified control. A thermostat is a typical system that can be described by a hybrid automaton. The room temperature x evolves according to $\dot{x}(t) = -x(t) + u(t) + v(t)$ when the thermostat is on, and according to $\dot{x}(t) = -x(t) + v(t)$ when the thermostat is off. The input v is a disturbance input modeling the influence of the outside temperature and u is a control input modeling the heating power.

The invariant of a discrete state q is defined as a subset $\mathcal{I}_q$ of $\mathcal{X}$. The system can stay at q if $x \in \mathcal{I}_q$. The conditions for switching between discrete states are specified by a set of guards such that for each discrete transition $e = (q, q')$, the guard set $\mathcal{G}_e \subseteq \mathcal{I}_q$. Each transition $e = (q, q')$ is additionally associated with a reset map $\mathcal{R}_e : \mathcal{G}_e \to 2^{\mathcal{I}_{q'}}$ that defines how the continuous variables x may change when $\mathcal{A}$ switches from q to q'. For example, a nondeterministic linear reset map can be defined as follows: for each $x \in \mathcal{G}_e$ the new continuous state $x' = Rx + \varepsilon$ where R is a $n \times n$ matrix and $\varepsilon \in P \subseteq \mathcal{X}$. The set P models the uncertainty of this reset map.

The functions f_q are often assumed to be Lipschitz continuous and the admissible input functions $u(\cdot)$ piecewise continuous. This ensures the existence and the uniqueness of solutions of the differential equations of each discrete state. However, because of complicated interactions between the continuous and discrete dynamics, further conditions are needed to guarantee the existence of a global solution of a hybrid automaton [74].

A state (q, x) of $\mathcal{A}$ can change in two ways as follows: (1) by a "continuous evolution," the continuous state x evolves according to the dynamics f_q while the discrete state q remains constant; (2) by a "discrete evolution," x satisfies the guard of an outgoing transition, the system changes discrete state by taking this transition and possibly changing the values of x according to the associated reset map.

Figure 13.4 sketches a hybrid automaton with two discrete states q_1 and q_2 and the continuous state space $\mathcal{X}$ is a 2-dimensional bounding rectangle. The invariant $\mathcal{I}_1$ of q_1 is the upper part of the rectangle limited by the bold line, and the invariant $\mathcal{I}_2$ of q_2 is the rectangle limited by the dashed line. Figure 13.4 also shows a trajectory starting from a hybrid state (q, x^1), which first follows the dynamics f_1 under some input $u(\cdot)$. Under different inputs, the system generates different trajectories* (such as, the dotted curves in this

---

*We use the term "trajectory" instead of "execution" to give a geometric intuition.

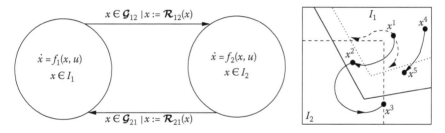

FIGURE 13.4
A hybrid automaton and its trajectories.

Figure 13.4). The infiniteness of the input space results in an infinite number of trajectories starting from the same state, which forms a dense set, often called a "reach tube" [66].

When this trajectory reaches the guard set $\mathcal{G}_{12}$ (which is the band between the bold and the dotted line), the transition from q_1 to q_2 is enabled. At this point, the invariant condition of q_1 is still satisfied and the system can either switch to q_2 or continue with the dynamics of q_1. The former is the case of this example. The system "decides" to switch to q_2 when it reaches point x^2. The reset $\mathcal{R}_{12}$ is the identity function and thus the trajectory starts following the dynamics f_2 from the same point x^2. When the trajectory reaches a point x^3 in the guard $\mathcal{G}_{21}$ (which is the dashed boundary of $\mathcal{I}_2$), it switches back to q_1 and the application of the reset $\mathcal{R}_{21}$ to x^3 results in a new state x^4, from which the system evolves again under the dynamics f_1.

As illustrated with this example, it is important to note that this model allows to capture "non-determinism" in both continuous and discrete dynamics. This nondeterminism is useful for describing disturbances from the environment as well as for taking into account the imprecision in modeling and implementation.

13.3 Exhaustive Verification

We use the term "exhaustive verification" to signify an automated proof that a certain model satisfies a certain property. This problem is also called "model checking." Since what we want is a proof, if we succeed in obtaining it, we can be certain that the model indeed satisfies the property. We contrast this to "partial verification" methods that are discussed in the following section.

In this section, we review exhaustive verification for timed and hybrid automata. The problem is decidable (but expensive) for timed automata and undecidable for hybrid automata in general. In what follows we survey basic methods to tackle the problem for the two models.

13.3.1   Model Checking for Timed Automata

The model checking problem for timed automata can be stated as: given a TA (or a set of communicating timed automata) A, and given a property P, check whether A satisfies P. We will briefly review in this section some methods to answer this question for different types of properties P. This is an extensively studied topic for which tutorials and surveys are already available (for instance, see [3]). For this reason, we will only sketch the basic ideas and refer to the literature for an in-depth study.

The simplest type of property P is "reachability": we want to know whether a given state (or configuration) s of the automaton is "reachable," that is, whether there exists an execution starting at some initial state (or a set of possible initial states) and reaching s. Consider again the TA shown in Figure 13.1b. Is state 4 reachable? It is, and Figure 13.3 presents an example execution that reaches state 4. Suppose, however, that we replaced the condition $y \geq 2$ in this automaton by $y > 5$. In that case, state 4 would become unreachable as can be verified by the reader.

Reachability is not only the simplest, but also the most useful type of property. "Safety" properties (those that state that the system has no "bad" behaviors, informally speaking) can be reduced to reachability with the help of a "monitor." A monitor for a given property is a "passive" component that observes the behavior of the system and checks whether the property is satisfied. If the property is violated then the monitor enters a designated "bad" state. Checking whether the system satisfies the property can then be reduced to checking whether the "bad" state of the monitor is reachable in the composition of the system and the monitor.

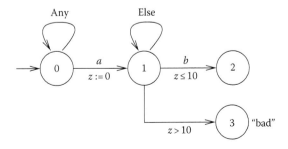

FIGURE 13.5
A TA monitor for checking a bounded-response property.

An example illustrating monitors is shown in Figure 13.5. Figure 13.5 shows a monitor for the property "*a* is always followed by *b* within at most 10 time units." Notice that the monitor tries to capture the violation of the property, that is, its negation. In particular, the monitor synchronizes with the system on common labels. The label "any" in the self-loop transition at state 0 of the monitor is a short-hand for "any label of the system": notice that this includes the label *a*, that is, the monitor is nondeterministic. This is essential, because the monitor should check that the property holds on *any* execution and *every* occurrence of *a*. After picking an *a* at random, the monitor keeps track of the time using its (local) clock *z*. If *b* is observed no later than 10 time units after *a*, the monitor moves to the "pass" state 2. Otherwise, the monitor can move to the "bad" state 3: if the monitor is able to reach this state, then the property is violated. The "else" label stands in this case for "any label except *b*."

How to check reachability for timed automata? In the case of a discrete-time semantics, the problem can be reduced to a problem of checking reachability for a discrete state-transition system: configurations can be seen as vectors of nonnegative integers, where the first element of the vector corresponds to the state and the rest to the values of the clocks.* Moreover, the system can be "abstracted" into a finite-state system by ceasing to increment clocks whose value exceeds a certain constant c_{max}: this is the greatest constant with which a clock is compared in the automaton. For instance, in the example of Figure 13.1b, $c_{max} = 5$. When a clock's value exceeds c_{max} we only need to "remember" this fact, and not the precise value of the clock, since this does not influence the satisfaction of a timing constraint. With this observation, one is left with the task of verifying exhaustively a finite-state system. A vast number of methods exist for fulfilling this task.

The aforementioned reduction does not generally apply to timed automata with dense-time semantics. Since the model is inherently infinite-state, the decidability of the reachability problem is far from obvious. The difficulty has been overcome by Alur and Dill using an elegant technique: the "region graph" abstraction [9]. The idea is to partition the infinite space of clock values (and consequently, the infinite space of configurations) into a finite set of "regions" so that two configurations that belong to the same region are equivalent in terms of their possible future behaviors. The set of regions is carefully constructed so that clock vectors are in the same region iff they satisfy the same constraints and will continue to do so despite time elapsing or some of the clocks being reset.

Regions are illustrated in Figure 13.6. Figure 13.6 shows the partitioning into the regions of the space of two clocks. Roughly speaking, every (x, y) point where x and y assume integer values not greater than 2 is a region: for example, the point $(x = 1, y = 1)$ is a region and so is $(x = 1, y = 0)$.

---

*We can always normalize the constants appearing in the timing constraints of the automaton so that the time quantum is 1. Then clocks assume integer values.

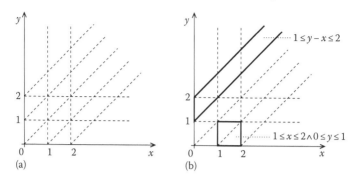

FIGURE 13.6
A partition of the space of two clocks x and y into 78 regions (a); two zones (b).

Moreover, open straight line segments such as $x = 0 \wedge 0 < y < 1$ or $1 < x < 2 \wedge x = y$ are regions. Open triangles such as $1 < x < 2 \wedge 0 < y < 1 \wedge x < y$ are also regions. Finally, unbounded sets such as $1 < y - x < 2 \wedge x > 2$ are also regions. For an exact definition of regions, the reader is referred to [3]. To keep the number of regions bounded, the same idea as the one described earlier in the case of discrete-time is used, namely, abstracting the values of clocks that exceed some maximal constant c_{max}. In Figure 13.6, c_{max} is taken to be 2.

Using the concept of regions, a (dense) TA can again be seen as a finite-state automaton: its states are pairs of discrete (control) states and regions. However, although the region graph is an invaluable tool for proving decidability, it is not very useful in practice. The reason is that the partition into regions is too "fine-grained," resulting in a huge number of regions (exponential in the worst case in both the number of the clocks and the size of the constants used in the timing constraints). Keeping in mind that the size of the discrete state space (excluding the clocks) is often already very large, timed automata suffer from what can be called a "double" "state-explosion" problem. Much of the research in the timed-automata community has attempted to overcome this problem by finding more efficient verification methods. Some of these attempts are described in the following text. It is fair to say that no "silver bullet" has been found to the problem, and TA model checking is still more expensive in practice than "untimed" model checking (which is consistent with the worst-case theoretical complexity of the problem). This is still an active area of research.

One of the ideas was to find partitions that are coarser than the region graph. This led to the ideas of "time-abstract quotient" [4,106,110] and "zone graph" [16,22,33,35,70,105]. The time-abstract quotient of a TA can be seen as a coarse region graph, which still has the same properties. It is obtained by "splitting" sets of configurations depending on their successors, using a classic partition-refinement method [82]. In practice, the refined sets are much coarser than regions (i.e., they are unions of many regions) although in the worst case they can be as fine as individual regions. The zone graph is based on representing sets of configurations as convex polyhedra called "zones." A zone can be seen as a conjunction of simple linear constraints on clocks, for instance, $1 \leq x \leq 2 \wedge 0 \leq y \leq 1$. Examples of zones over two clocks x and y are shown in Figure 13.6 (zones are depicted in thick lines). The zone graph is built by computing all successor zones of a given initial zone in a "forward" manner. Zones may "overlap," so in theory the zone graph can be exponentially larger than the region graph. In practice, however, this does not happen. In fact the zone graph is considerably smaller, and remains to this day the most efficient method of model-checking timed automata.

Both the time-abstract quotient and the zone-graph methods raise a number of interesting problems that have to do with the "symbolic" representation of sets of configurations. Dill [35] proposed an efficient method to represent zones as matrices of size $n \times n$, where n is the number of clocks. These are called "difference bound matrices" or DBMs. DBMs are used in implementations of the time-abstract quotient as well as the zone graph methods. For the former, care must be taken to ensure that partition refinement yields only convex sets of configurations, that is, zones, so that DBMs can be used [106]. In the case of the zone graph, care must be taken to ensure that the graph remains finite, and a set of abstractions have been developed for this purpose [22,34,105].

Reachability covers many of the properties that one usually wishes to check on a model, but not all. In particular, "liveness" properties (those that state, informally speaking, that the system indeed exhibits "good" behaviors) cannot be reduced to reachability. One way to model such properties is by means of "timed Büchi automata" (TBA). TBA are the timed version of Büchi automata: the latter define the sets of infinite behaviors (rather than sets of finite behaviors, defined by standard finite automata). TBA often arise when composing a (network of) plain TA with a monitor of a liveness property: the monitor is often modeled as an "untimed" Büchi automaton; however, the composition of the TA model and the monitor yields an automaton which is both timed and Büchi. Let us provide an example.

A typical example of a liveness property is the "unbounded response property" "a is always followed by b." Notice that this is the "unbounded" version of the property modeled by the monitor of Figure 13.5. It is "unbounded" in the sense that it does not specify how much later b must occur after a. It only requires that b occurs "some" time after a has occurred. In order to check this property, we will again build a monitor that attempts to capture the violation of the property. This monitor is the (untimed) Büchi automaton shown in Figure 13.7. The monitor nondeterministically chooses to monitor an event a, moving from state 0 to state 1. If a b never occurs, the monitor has an infinite execution where it remains at accepting state 1: this is an accepting execution, meaning the property is violated. If b is received,

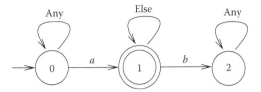

FIGURE 13.7
A Büchi-automaton monitor for checking an unbounded-response property.

on the other hand, the monitor moves to state 2, which is nonaccepting. If no execution is accepting, the property is satisfied.

TBA emptiness can be checked in theory on the region graph, interpreted as a finite (untimed) Büchi automaton. In practice, the time-abstract quotient graph or the zone graph can be used instead. The former can be easily shown to preserve liveness properties [106]. The fact that the zone graph can be used to check TBA emptiness is nontrivial and has been completely proven only recently [101].

13.3.2 Verification of Hybrid Automata

Timed automata are a very special class of hybrid automata. The decidability results for timed automata were generalized to some slightly more complex classes such as multirate automata [6,80] and initialized rectangular automata [89]. In multirate automata, the derivatives of the continuous variables can take constant values other than 1. In rectangular automata, the derivatives are not constant and an allowed to take any value inside some interval. Decidability was also proved for some particular planar systems including 2-dimensional piecewise constant derivatives (PCD) [76], planar multipolynomial systems [26] and nondeterministic planar polygonal systems [14]. Despite these extensions, however, the reachability problem for general hybrid automata is undecidable. In fact, this holds even for classes of systems with constant derivatives, such as linear hybrid automata with three or more continuous variables [47].

Let us illustrate some of the issues that arise in hybrid automata reachability. We will focus only on the problem of computing the reachable sets of hybrid automata where continuous dynamics are defined by nontrivial differential equations. A major difficulty comes with the two-phase evolution of these systems, which requires the ability to compute the successors (or predecessors) of sets of states not only by discrete transitions but also by continuous dynamics. In the continuous phase, this relates to the special problem of characterizing trajectories of continuous systems. For simplicity, we consider a hybrid automaton with only one discrete state and the initial set *Init*. The initial set can be characterized by a formula $\phi_{Init}(x)$ whose truth value is 1 iff $x \in Init$. Suppose further that the differential equation $\dot{x} = f(x)$ of the continuous dynamics admits a closed-form solution $\xi_x(t)$ for every initial condition x; hence the reachable set from *Init* is exactly the set of x for which the formula $r(x) \equiv \exists x' : \phi_{Init}(x') \wedge \exists t \geq 0 : x = \xi_{x'}(t)$ is true. Similarly, proving that the system does not reach a bad state in the set B, represented by a formula $\phi_B(x)$, amounts to proving that the formula

$$\forall x' : \phi_{Init}(x') \Rightarrow \forall t \geq 0 : \neg\phi_B(\xi_{x'}(t)) \tag{13.1}$$

is true, which can be done by eliminating the quantifiers.

When the derivative f is constant, for example $f(x) = c$, we have $\xi_x(t) = x + ct$. For systems with constant derivatives and where invariants and guards are specified by linear inequalities (such as timed automata and linear hybrid automata) the reachable sets can be expressed by linear formulas. Therefore, the quantifiers in Equation 13.1 can be eliminated using linear algebra. A number of tools for systems with PCDs have been developed, such as Kronos [33], Uppaal [70], HyTech [50], and PhaVer [39]. However, the problem becomes more difficult for systems with nontrivial continuous dynamics. On one hand, in many cases we do not know explicit solutions of the differential equations. Even if we know such solutions, such as for a linear system $\dot{x} = Ax$ with a closed-form solution $\xi_x(t) = e^{At}x$, a proof of Equation 13.1 is possible only for a very restricted class of matrices A with special eigenstructure [10,83].

In addition, the successive computations of the states reachable by the continuous dynamics and discrete transitions may not terminate, by alternating indefinitely between two or more discrete states and each time adding more and more successors. Indeed, the reachability problem is undecidable for general hybrid systems, except for the classes with the earlier mentioned special linear continuous dynamics and memoryless switching dynamics [10,83].

Since, in general, there exists no exact reachable set computation method, approximate methods have been developed. In order to be able to compute the reachable sets of a hybrid automaton, we need a finite syntactic representation of these sets. The continuous state space of hybrid automata is in $\mathbb{R}^n$ and hence can only be represented "symbolically," such as by formulas of some logic. Examples of classes of subsets of $\mathbb{R}^n$ which admit a symbolic representation are the polyhedral sets (represented by the Boolean combinations of linear inequalities) and the semialgebraic sets (represented by combinations of polynomial inequalities). Another requirement in choosing a set representation is that it can be efficiently manipulated not only for the computation of continuous successors, but also for the treatment of discrete transitions, such as set intersection for computing the states satisfying the guard conditions and the images by the resets. Polyhedra, ellipsoids, and level sets are the most commonly used representations in hybrid systems reachability algorithms [11,20,27,29,30,40,45,59,67,77,98]. These representations have been used in a variety of tools such as Coho [45], CheckMate [27], d/dt [13], VeriShift [20], HYSDEL [99], MPT [69], HJB toolbox [77], ellipsoidal toolbox (ET) [68]. In the following, we review some techniques for the reachability computation of continuous systems and their extensions to hybrid systems. The field is vast, so we can only provide a brief review and refer the reader to the current literature for a more extensive view.

In order to control the approximation error, as in numerical simulation, most reachability algorithms use a time discretization and operate on a step-by-step basis, for example, $R^{k+1} = \delta(R^k, f, t_k, t_{k+1})$ where δ denotes a function that returns an approximation of all the states reachable from R^k by the dynamics $\dot{x} = f(x)$ during the time interval $[t_k, t_{k+1}]$. The quantity $h_k = t_{k+1} - t_k$ is called the "time step." We use δ_t to denote the set of all states reachable at a "discrete time point" t and $\delta_{[t,t']}$ the set of all states reachable in dense time, that is, for all time points $\tau \in [t, t']$.

13.3.2.1 Autonomous Linear Systems

For a system $f(x) = Ax$, if X^k is a bounded convex polyhedron represented by the convex hull of its vertices $X^k = \text{chull}\{v_1, \ldots, v_m\}$, then the set of all states reachable from X^k at exactly time t_{k+1} can be written as $X^{k+1} = \delta_{t_{k+1}} = \text{chull}\{e^{Ah_k} v_1, \ldots, e^{Ah_k} v_m\}$ where e^{At} denotes the matrix exponentiation (which is a linear operator). Then, the set $\delta_{[t_k, t_{k+1}]}$ of all states reachable from X^k "during the time interval" $[t_k, t_{k+1}]$ can be approximated by "interpolating" the sets X^k and X^{k+1}, for example, $\delta_{[t_k, t_{k+1}]}(X^k) = C^{k+1} = \text{chull}(X^k \cap X^{k+1})$. In order to achieve a conservative approximation, C^{k+1} is then "bloated" by some amount ε that bounds its distance to the exact reachable set. The computation in the next step $(k+1)$ can start from X^{k+1} to compute X^{k+2}, and then from X^{k+1} and X^{k+2} to obtain the bloated convex hull C_0^{k+2} (see Figure 13.8).

This method is indeed the basis of the reachability computation technique for linear systems implemented in the tools CheckMate [27] and d/dt [13]. The tool d/dt additionally overapproximates the convex polyhedra X^k by an orthogonal polyhedron G^k, in order to accumulate all the reachable states in a single orthogonal polyhedron. Orthogonal polyhedra [21] (which can be defined as unions of closed full-dimensional hyper-rectangles) are, unlike convex polyhedra, closed under the union operation. In addition, they admit a canonical representation allowing to perform Boolean operations (in particular the union operation) more efficiently than operations on convex polyhedra. Figure 13.8 illustrates the first two iterations of this method where the initial set X^0 is a 2-dimensional segment and the first time step is r.

13.3.2.2 Linear Systems with Uncertain Input

For a system $\dot{x}(t) = Ax(t) + u(t)$ where $u(\cdot)$ is a piecewise continuous function such that $\|u(\cdot)\| \leq \mu$, and $\|\cdot\|$ is some norm on $\mathbb{R}^m$. The earlier described method for autonomous systems can be extended to these systems, using the "maximum principle" from optimal control [58]. Indeed, given a state x^* on

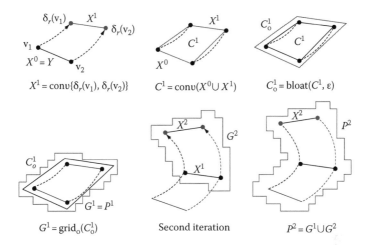

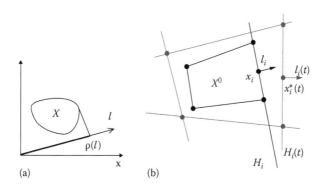

FIGURE 13.8
Illustration of the reachability technique for linear systems using convex polyhedra.

the boundary of the initial set X^0, one can determine an input function $u^*(\cdot)$ such that the trajectory from x^* under this input lies on the boundary of the reachable set.

To exploit this fact, one can represent the reachable set by its support function. The support function of a compact and convex set $X \subset \mathbb{R}^n$ is $\rho_X : \mathbb{R}^n \to \mathbb{R}$ such that for a vector $l \in \mathbb{R}^n$, $\rho(l) = \max_{x \in X} \langle l, x \rangle$ (where $\langle \cdot \rangle$ denotes the inner product). Figure 13.9 illustrates this definition. Therefore, if X is a convex polyhedron, it can be represented as the intersection of its halfspaces $X = \cap_{i=1}^m H_i$ where $H_i = \{x \in \mathbb{R}^n : \langle l_i, x \rangle \le \rho_X(l_i) = \langle l_i, x^* \rangle\}$; l_i is indeed the normal vector of the hyperplane of H_i. The point x^* lies on this

FIGURE 13.9
Illustration of support functions.

hyperplane and is called the support vector of X in the direction l_i. Then, using the maximum principle, we can find for each hyperplane H_i an input function under which the evolutions of $l_i(t)$ and $x^*(t)$ define a new hyperplane $H_i(t) = \{x \in \mathbb{R}^n : \langle l_i, x \rangle \le \rho_X(l_i(t)) = \langle l_i(t), x^*(t) \rangle\}$. Then, from all such hyperplanes $H_i(t)$ ($i = 1, \ldots, m$) we define a polyhedron that overapproximates the reachable set at time t.

This is also the basic principle employed by the reachability technique using ellipsoidal approximations [20,67,68]. An ellipsoid X can be described as $X = \{x \in \mathbb{R}^n : x^T Q^{-1} x \le 1\}$ where Q is positive definite; its support function is $\rho_X(l) = \sqrt{l^T Q l}$ and its support vector in the direction l is $\dfrac{Ql}{\sqrt{l^T Q l}}$.

Then, for a given point on the boundary of the ellipsoid, one can use the maximum principle to track the time evolution of the corresponding support vector l and that of the matrix Q, in order to yield an overapproximation E_o and an underapproximation E_i of the reachable set (see Figure 13.10). This method and its extension to hybrid systems were implemented in the ET [68].

Another way to handle the input uncertainty is to bound its effects in each time step by enlarging the reachable set of the corresponding autonomous system as follows: $X^{k+1} = e^{Ah_k} X^k \oplus B(\eta)$ where $\oplus$ denotes the Minkowski sum, and $B(\eta)$ the ball centered at the origin with the radius $\eta = \dfrac{e^{h_k}\|A\| - 1}{\|A\|}\mu$ (μ is a bound of $\|u\|$). This expansion is similar to the bloating operation used in the method for autonomous systems to cover all the states reachable in dense time.

If the sets X^k are polyhedra, one can use the infinity norm and $B(\eta)$ is thus a box. The computation of the Minkowski sum may generate new sets with high geometric complexity (expressed in terms of the number of vertices and facets). Recently, [40] proposed to use zonotopes, instead of convex polyhedra, as a new set representation. A zonotope X can be described as $X = \{\sum_{i=1}^m \alpha_i g_i \mid \forall i \in \{1, \ldots, m\} : \alpha_i \in [-1, 1]\}$, and the vectors g_i are called

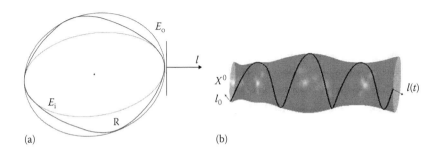

(a)   (b)

FIGURE 13.10
Reachability computation using ellipsoids. (Modified from Tomlin, C. et al., *Proc. IEEE*, 986, 91, 2003.)

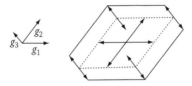

FIGURE 13.11
A zonotope.

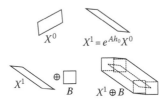

FIGURE 13.12
Reachability computation using zonotopes.

"generators" (see Figure 13.11). An interesting property of zonotopes is that the Minkowski sum of two zonotopes can be obtained by taking the union of their generators. Figure 13.12 illustrate one iteration of the reachability computation using zonotopes. The first zonotope (which is indeed a paral-lelepiped) is the initial set X^0, the second one is $X^1 = e^{Ah_0} X^0$, that is the result of applying the linear transformation e^{Ah_0} to X^0. The Minkowski sum $X^1 \oplus B$ where B is the box representing the effects of uncertainty is the last zonotope shown in Figure 13.11. One can see that the number of generators of the resulting zonotope grows iteration after iteration, but we can overap-proximate it by a zonotope with a smaller number of generators [42]. The main inconvenience of this representation is that Boolean operations over zonotopes are not easy to compute. A method for computing the intersection of a zonotope and a hyperplane was proposed in [41]. Alternatively, oriented boxes can provide a good compromise between the approximation error and computational expenses [94].

13.3.2.3 Nonlinear Systems

While many properties of linear systems can be exploited to develop rel-atively efficient reachability techniques, the situation is more difficult for nonlinear systems. One approach to solving this problem is to use opti-mization to describe the "extremal behaviors." In [30], in each time step, the boundary of an orthogonal polyhedron is lifted outward by some amount that guarantees to cover all the reachable states during that step, as shown in Figure 13.13. For example, a face e is lifted outward by the amount $f_m h$ if $f_m > 0$ where f_m is the maximum of the projection of the derivative on the

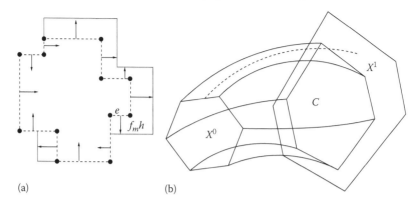

(a)   (b)

FIGURE 13.13
Illustration of reachability computations for nonlinear systems using optimization on orthogonal polyhedra (a) and convex polyhedra (b).

normal of e within some neighborhood of e, and h is the time step. The reachability technique in [27] first computes the convex hull C of the successors at time t from the vertices of the initial polyhedron X^0 (see Figure 13.13), as in the earlier described method for linear systems. However, unlike for linear systems, this convex hull C clearly does not include all the reachable states at t (as illustrated by the dashed trajectory in Figure 13.13). Nevertheless, the directions of its faces can be used to form a "tight"* polyhedral overapproximation X^1 by estimating the distance to the exact set in the directions of the faces. A similar idea has recently been used in [91] where the reachable sets are approximated by template polyhedra (with fixed constraint matrices) using a Taylor expansion of the trajectories.

Another approach is based on a formulation of the evolution of the reachable set, represented by its level set, according to the Hamilton–Jacobi partial differential equation (see for example [98]). This technique was implemented in the HJB toolbox [77].

Polynomial systems have recently received a special interest, partly because of their applications in the modeling and the analysis of biological systems. A method using Bézier techniques for these systems was proposed in [29]. It exploits the fact that by choosing an appropriate basis change, one can exploit the coeficients of the representation of a polymial in order to compute the image of a set by the polynomial.

Most of the earlier described methods are equipped with an error control mechanism, that is, they can produce an approximation as accurate as desired. Nevertheless, it is not always necessary to obtain a very accurate approximation of the reachable set (which is computationally expensive) but only sufficiently accurate to prove the property of interest. Barrier certifi-

---

*If the optimization problem can be exactly solved.

cates [86] and polynomial invariants [97] can be seen as such approxima-tions. A barrier certificate can be intuitively seen as a proof of the existence of an "impermeable" frontier between the reachable set and the bad set. The method in [86] searches for such a frontier in the form of a polynomial, and the "impermeability" can be expressed using the derivatives along the frontier. This results in an optimization problem that can be solved using the sum of squares optimization tool SOSTOOLS [87].

13.3.2.4   *Abstraction*

The main idea of this approach is to start with a rough (conservative and often discrete) approximation of a hybrid system and then iteratively refine it. This refinement is often local in the sense that it uses the pre-vious analysis results to determine where the approximation error is too large to prove the property (see, for example, [8,28,96]). A popular abstrac-tion approach is predicate abstraction where a conservative abstraction can be constructed by mapping the infinite set of states of the hybrid system to a finite set of abstract states using a set of predicates. The property is then verified in the abstract system. If it holds in the abstract system, it also holds in the concrete hybrid system. Otherwise, a counter-example can be generated. If the abstract counter-example corresponds to a con-crete trajectory, then the hybrid system does not satisfy the property; oth-erwise, the abstract counter-example is spurious because the abstraction is too conservative, and the abstraction can then be refined to achieve a better precision.

   In the following, we illustrate this approach by explaining the method using polynomials proposed in [96]. The continuous state space $\mathbb{R}^n$ is par-titioned using the signs of a set of polynomials. As an example, an abstract state s defined by $g_1(x) < 0 \wedge g_2(x) > 0$ corresponds to a (possibly infinite) set $c(s)$ of concrete states. Then, the abstract transition overapproximates the concrete one such that there is a transition from s to s' if there exists a tra-jectory from a concrete state in $c(s)$ to another concrete state in $c(s')$. More precisely, in this method, first the set of polynomials is saturated by adding all the high-order derivatives of the initial polynomials. Then, by looking at the sign of the polynomials, it is possible to decide whether a trajectory can go from one abstract state to another. For example, if there are only two poly-nomials g_1 and g_2 such that $g_2 = \dot{g}_1$. Suppose that the abstract state s satisfies $g_1 = 0$ and $g_2 > 0$, then the new sign of g_1 is positive and from s we add a transition to s' satisfying $g_i > 0$. The abstraction can be refined by adding more polynomials.

   Another abstraction method in [8] uses linear predicates to partition the continuous state space, and thus each abstract $c(s)$ is a convex polyhedron. The abstract transition from s to s' is determined by computing the reachable set from $c(s)$ and check whether it reaches $c(s')$. This is less expensive than the reachability computation on the hybrid system, which requires handling accumulated reachable sets with geometric complexity that grows after successive continuous and discrete evolutions.

Box decompositions are also commonly used to define abstract systems, such as in [59,90]. The abstract system can then be built by exploiting the properties of the system's vector fields over such decompositions. The method proposed in [59] makes use of the following special property of multiaffine systems*: the value of a multiaffine function $f(x)$ with x inside some box can be expressed as a linear combination of the values of f at the vertices of the box. Using this, one can determine whether the derivative vector on the boundary of a box points outward or inward, in order to overapproximate the reachability between adjacent boxes.

While discrete abstractions allow benefiting from the well-developed verification algorithms for discrete systems, they might be too coarse to preserve interesting properties. Timed abstractions can be built by adding bounds on the time for the system to reach from one abstract state to another. A generalization of this idea is called "hybridization" [12] involving approximating a complex system with a simpler system, for which more efficient analysis tools are available. To this end, using a partition of the state space, one can approximate locally the system's dynamics in each region by a simpler dynamics. Globally, the dynamics changes when moving from one region to another, and the resulting approximate system behaves like a hybrid system and this approximation process is therefore called hybridization. Then, the resulting system is used to yield approximate analysis results for the original system. The usefulness of this approach (in terms of accuracy and computational tractability) depends on the choice of the approximate system. For example, the hybridization methods using piecewise affine approximate systems, proposed in [12], allows approximating a nonlinear system with a good convergence rate and, additionally, preserving the attractors of the original system. In addition, the resulting approximate systems can be handled by the existing tools for piecewise affine systems (presented earlier in this section).

13.4 Partial Verification

Exhaustive verification is desirable since, if it succeeds, it guarantees that a model satisfies a property. But exhaustive verification has its limitations as we have seen: state-explosion or even undecidability. In fact, state-explosion is a phenomenon that is also prevalent in the exhaustive verification of much simpler, finite-state models. This phenomenon has so far hindered a wider adoption of exhaustive verification in industrial applications, because the

---

*Multiaffine systems are a particular class of polynomial systems such that if all the variables x_i are constant, the derivatives are linear in x_j with j not equal to i.

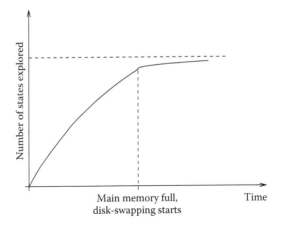

FIGURE 13.14
Hitting the exhaustive verification wall.

size of the problems tackled there is far too big to treat exhaustively. Instead, practitioners use simulation as their main verification tool.* Even though simulation cannot prove that a property is satisfied, it can certainly reveal cases where it is not satisfied, that is, potential bugs of the real system, its model, or its specification.

An advantage of simulation is that it has some time-scalability proper-ties: running 200 simulations is better (i.e., likely to discover more bugs) than running 100 simulations, and running longer simulations is also bet-ter. Moreover, if 100 simulations can be run in one day, say, then in two days we can most likely run 200 simulations. In contrast, most exhaustive verification tools suffer from a "hitting the wall" type of problem. Once they exhaust the main memory of the computer that they run on, they start using disk space, which involves a lot of swapping on the OS side. Disk swapping virtually takes all processing time, leading verification to a halt. This means that the number of new states that are explored per unit of time radically decreases to practically zero, as illustrated in Figure 13.14. Usually this wall is hit after relatively little time, in the order of minutes.† Then, running the tool for many hours will not improve the number of states that are explored when compared to running it for ten minutes. This is not time-scalable.

Time-scalability is obviously critical in an industrial setting, where pre-dictability in terms of allocation of resources vs. expected benefits is highly

---

*The term "verification" usually denotes simulation-based verification in industrial jargon, whereas "formal verification" is used to denote exhaustive verification.

†For example, using a model-checker that can explore 10^5 new states per second, on a model that requires 1000 bytes to represent each state, consumes memory at a rate of approximately 100 MB/s. This means that a main memory of size 8 GB can be filled in about 2 min. Exploration rates in the order of 10^5 states per second are not unusual for an advanced model-checker such as Spin [52].

desirable. Although simulation is more time-scalable, it is still not completely predictable. Running 200 instead of 100 simulations obviously does not guarantee that twice as many bugs will be found. It does not guarantee either that twice as many states will be explored. These are some of the reasons that prompted more systematic methodologies for simulation-based verification (e.g., see [108] for the case of the hardware industry and [44] for work done in a software context). In the hardware case, these methodologies include specialized languages for writing "testbenches" (i.e., simulation environments that allow to specify input-generation policies as well as property monitors), for example, see [53,111].

In this context, the principle of "randomization" is often used as a good aid to uncover corner cases and eventually bugs (e.g., see [64,75,78,92]). We discuss some applications of the randomized state-space exploration principle to embedded system models in the rest of this section. We also introduce the concept of "resource-aware" verification, which goes beyond randomization, and includes all verification methods that are explicitly aware of their memory and time resources. Finally, we examine a particular randomized search algorithm, RRT, and its application to hybrid automata.

13.4.1 Randomized Exploration and Resource-Aware Verification

A simple technique to randomly explore a state space is "random walk": pick randomly an initial state s_0, then pick randomly one of the successors of s_0, say s_1, then pick randomly one of the successors of s_1, and so on. This is obviously a very inexpensive algorithm in terms of space, since it needs only to store a single state at a given time, plus perhaps its set of successor states.*

Basic random walk is limited, especially when bugs lie very "deep" in the state space, that is, the paths to reach an error state are very long. Then, unless the number of such paths is very large, the probability to follow a path that leads to an erroneous state is very small. To alleviate this problem, different variants of "pure" random walk have been proposed. One such variant is the "deep random search" (DRS) algorithm proposed in [46] and applied in the context of timed automata. DRS stores during the random walk a subset of the nodes it visits, called a "fringe," and then randomly backtracks to a node in the fringe when a deadlock (a node with no successors) is reached. DRS can be applied to any model for which forward reachability is available. In the case of timed automata, the "nodes" that DRS visits correspond to symbolic states consisting of discrete state vectors in addition to symbolic representations of sets of clock values, using data structures such as DBMs, as explained earlier.

---

*If there is a number_of_successors() function available, then we do not even need to keep the set of successor states. We can just compute the number of successors, say n, then randomly choose an integer i in the interval $[1 \ldots n]$, and then replace the current state with its ith successor.

As described earlier, DRS maintains a fringe, which is a set of states. For "deep" random walks, this fringe can grow quite large, which means even DRS can suffer from state-explosion and disk-swapping problems, like exhaustive verification methods. In order to alleviate these problems, an idea is to embed the "hard" memory constraints directly into the algorithm itself. This led to the concept of "resource-aware," and in particular "memory-aware," state-space exploration [104]. Memory-aware algorithms are meant to deal with the disk-swapping problem in a rather radical way: by simply using no disk memory, only main memory.

Memory-aware algorithms are by definition "memory-bounded": they use no more than a specified amount of memory. Note, however, that not all memory-bounded algorithms are memory-aware. An example is random walk: it is memory-bounded, since it stores a single state in memory at any given time. But it is not memory-aware, since its behavior does not generally depend on the amount of memory available. Thus, even though main memory could hold more than just one state, the random walk method does not make use of the extra space available.

Many existing verification techniques are memory-aware, including deterministic methods such as "bit-state hashing" [51], as well as randomized ones such as depth-first traversal with replacement [54]. See [104] for a detailed discussion. The idea of memory-aware verification is also exploited in [1], where a class of randomized exploration algorithms are introduced that use a parameter N representing the number of states that the algorithm is allowed to maintain at any given time during its execution. Given a model, N can be computed as follows. If R is the total size of (available) RAM memory (say in bytes) and storing a state of this model costs K bytes, then $N = \frac{R}{K}$.

Having N as an upper bound, many different randomized exploration algorithms can be tried out, depending on how two main policies are defined: how, given the current state of the algorithm, to pick which node to explore next (the "select" function), and how, given a selected node, to pick a successor of this node and update the state (the "update" function). Notice that updating the state does not necessarily mean just adding the state to the current set of visited states. Indeed, if the current set of states already holds N states, then in order to add a new state, at least one of the current states needs to be removed. There are obviously many different policies for choosing the select and update functions, and notice that randomization can be used in both functions.

It would be nice to be able to compare randomized algorithms such as the ones described earlier, so as to pick the "best" one for a given application. What criteria and methods can be used for carrying out such a comparison? In terms of criteria, they can be roughly classified into two classes: "performance" criteria and "coverage" criteria. The former represent the algorithm's performance (both in terms of memory and time) while the latter the algorithm's ability to "cover" the state space. We briefly discuss some criteria of this kind later.

One criterion which is perhaps a hybrid of performance and coverage is the "mean cover time," or average time that it takes for the algorithm to visit all, or a given percentage, of the reachable states. Clearly, the smaller the mean cover time that an algorithm has (for a given percentage), the better it performs. This also means that given more time, the same algorithm is likely to cover more nodes than another algorithm with larger mean cover time. Conversely, one may also be interested in the "mean number of covered states" in a given, fixed, amount of time.

A set of criteria can be defined based on "reachability probabilities" of states. The reachability probability of a given state s can be defined as the probability that a given run of the algorithm (and its associated parameters) visits state s. Then, we could define as comparison criterion, the "minimum" reachability probability over all reachable states.

Note that the aforementioned criteria depend not only on the algorithm, but also on the structure of the state space to be explored. This state space is essentially a directed graph. The characteristics of the graph such as its diameter, the degree of its nodes, whether it is a tree, a DAG (directed acyclic graph), or a graph with cycles, and so on, will generally influence the behavior of an algorithm greatly. Because of this dependence, obtaining analytical formulas for the aforementioned criteria is a very difficult task. Even for simple graphs such as regular trees, it can be nontrivial [1]. On the other hand, experimental results can often be obtained much more easily, e.g., see [1,84]. This is an exciting field of research and we expect it to become more popular in the near future, because of its high relevance in industrial practice.

13.4.2   RRTs for Hybrid Automata

Finding a trajectory of a hybrid automaton violating a safety property can be seen as a path planning problem in robotics, where the goal is to find feasible trajectories in some environment that take a robot from an initial point to a goal point [72]. In the following, we describe a partial verification algorithm based on RRT (rapidly-exploring random trees) [71], a probabilistic path and motion planning technique with a good space-covering property. The RRT algorithm has been used to solve a variety of reachability-related problems such as hybrid systems planning, control, and verification (see, for example, [17,25,37,57,85] and references therein). This approach indeed can be thought of as a simulation-based verification approach. Along this line, one can mention the work on systematic simulation [56] and its extension with sensitivity analysis [36]. For some classes of stable systems, it is possible to use a finite number of simulations and a bisimulation metric to prove a safety property of a hybrid system [43].

The first part of this section will be devoted to the basic RRT algorithm (Figure 13.15). In the second part we extend the RRT algorithm to treat hybrid systems. By "the basic RRT algorithm," we mean the algorithm for a continuous system and without problem-specific optimization. For a thorough

Procedure RRT_Tree_Generation(x_{init}, k_{max})
$\quad$ $\mathcal{T}$.*init*(x_{init}); $k = 1$
$\quad$ **Repeat**
$\qquad$ x_{goal} = RANDOM_STATE($\mathcal{X}$);
$\qquad$ x_{near} = NEIGHBOR($\mathcal{T}$, x_{goal});
$\qquad$ (u, x_{new}) = NEW_STATE(x_{near}, x_{goal}, h);
$\qquad$ $\mathcal{T}$.ADD_VERTEX(x_{new});
$\qquad$ $\mathcal{T}$.ADD_EDGE(x_{near}, x_{new}, u);
$\quad$ **Until**($k \geq k_{max}$ $\vee$ $\mathcal{B} \cap$ Vertices($\mathcal{T}^k$) $\neq \emptyset$)

FIGURE 13.15
The basic RRT algorithm.

description of RRTs and their applications in various domains, the reader is referred to a survey [71] and numerous articles in the RRT literature.

Essentially, the RRT algorithm constructs a tree $\mathcal{T}$, the root of the which corresponds to the initial state x_{init}. Each directed edge of the tree $\mathcal{T}$ is labeled with an input selected from a set of admissible input functions. Hence, an edge labeled with u that connects the vertex x to the vertex x' means that the state x' is reached from x by applying the input u over a duration of h time, called a "time step." When a variable time step is used, each edge is also labeled with the corresponding value of h.

In each iteration, the function RANDOM_STATE samples a "goal state" x_{goal} from the state space $\mathcal{X}$. We call it a goal state because it indicates the direction toward which the tree is expected to evolve. Then, a "neighbor state" x_{near} is determined as a state in the tree closest to x_{goal}, according to some predefined metric. This neighbor state is used as the starting state for the next expansion of the tree. The function NEW_STATE creates a trajectory from x_{near} toward x_{goal} by applying an admissible input function u for some time h. Finally, a new vertex corresponding to x_{new} is added in the tree $\mathcal{T}$ with an edge from x_{near} to x_{new}. In the next iteration, the algorithm samples a new goal state again.

Figure 13.16 illustrates one iteration of the algorithm. One can see that x_{near} is the state closest to the current goal state x_{goal}. From x_{near} the system evolves toward x_{goal} under the input u, which results in a new state x_{new} after h time.

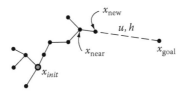

FIGURE 13.16
Illustration of one iteration of the RRT algorithm.

The algorithm terminates after k_{max} iterations or until a bad state in $\mathcal{B}$ is reached. Different implementations of the functions in the basic algorithm and different choices of the metric and of the successor functions in the problem formulation result in different versions of the RRT algorithm. Note that in most versions of the RRT algorithms, the sampling distribution of x_{goal} is "uniform" over $\mathcal{X}$, and the metric ρ is the "Euclidian distance."

Probabilistic completeness is an important property of the RRT algorithm [65,71], which is stated as follows: If "a feasible trajectory from the initial state x_{init} to the goal state x_{goal} exists, then the probability that the RRT algorithm finds it tends to 1 as the number k of iterations tends to infinity." Although the interest of this theorem is mainly theoretical, since it is impossible in practice to perform an infinite number of iterations, this result is a way to explain the good space-covering property of the RRT algorithm.

We now describe an extension of the RRT algorithm to hybrid systems, which we call hRRT. The extension consists of the following points. Since the state space is now hybrid, sampling a state requires not only sampling a continuous state but also a discrete state.

In addition, to determine a nearest neighbor of a state, we need to define a distance between two hybrid states. In a continuous setting where the state space is a subset of $\mathbb{R}^n$, many distance metrics exist and can be used in the RRT algorithms. Nevertheless, in a hybrid setting, defining a meaningful hybrid distance is a difficult problem. Finally, the successor function for a hybrid system should compute not only successors by continuous evolution but also successors by discrete evolution. In the following we briefly describe a hybrid distance, which is not a metric but proved to be appropriate for the purposes of developing guiding strategies discussed in Section 13.5.

13.4.2.1 Hybrid Distance

Given two hybrid states $s = (q, x)$ and $s' = (q', x')$, if they have the same discrete component, that is, $q = q'$, we can use some usual metric in $\mathbb{R}^n$, such as the Euclidian metric. When $q \neq q'$, it is natural to use the average length of the trajectories from one to another, which is explained using an example shown in Figure 13.17. We consider a discrete path γ which is a sequence of two transitions $e_1 e_2$ where $e_1 = (q, q_1)$ and $e_2 = (q_1, q')$.

- The average length of the path γ is some distance between the image of the first guard $\mathcal{G}_{(q,q_1)}$ by the first reset function $\mathcal{R}_{(q,q_1)}$ and the second guard $\mathcal{G}_{(q_1,q')}$. The distance between two sets can be defined as the Euclidian distance between their geometric centroids. This distance is shown in the middle figure.
- The average length of trajectories from $s = (q, x)$ to $s = (q', x')$ following the path γ is the sum of three distances (shown in Figure 13.17 from left to right): the distance between x and the first guard $\mathcal{G}_{(q,q_1)}$, the average length d of the path, and the distance between $\mathcal{R}_{(q_1,q')}(\mathcal{G}_{(q_1,q')})$ and x'.

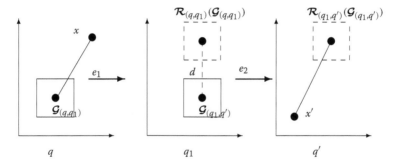

FIGURE 13.17
Illustration of the hybrid distance.

If the set $\Gamma(q, q')$ of all the discrete paths from q to q' is empty, the distance $d_H(s, s')$ from s to s' is equal to infinity. Otherwise, $d_H(s, s') = \min_{\gamma \in \Gamma(q,q')} len_\gamma(s, s')$. It is easy to see that the hybrid distance d_H is only a pseudometric since it does not satisfy the symmetry requirement. Indeed, the underlying discrete structure of a hybrid automaton is a directed graph.

13.5 Testing

Partial verification can be termed "model testing." It is "testing" in the sense that it is generally incomplete. In this section we look at another testing activity, however, not of models, but of physical systems. In particular, we consider the following scenario: we are given a "specification" and a "system under test" (SUT) and we want to check whether the SUT satisfies the specification.

The SUT can be a software system, a hardware system, or a mix of both. Often the SUT is a "black-box" in the sense that we have no knowledge of its "internals" (i.e., how it is built). For example, if the SUT is a SW system, we have no access to the source code. If it is HW, we have no access to the HDL or other model that was used to build the circuit.* Instead, we can "interact" with the SUT by means of inputs and outputs: we can provide the inputs and observe the outputs. A precise, executable description of which inputs to provide and when and how to proceed depending on the observed outputs

---

*Even if the SUT is not entirely black-box, we may still want to treat it as a black-box, because we simply have no effective way of taking advantage of the knowledge we have about its internals. For example, even if we have access to the source code of some piece of SW we want to check, we may still treat this system as black-box because we have no means of analyzing the source code (e.g., with some verification or static analysis method).

is called a "test case." The test case is executed on the SUT* and at the end of the execution it outputs a PASS or FAIL verdict[†]. In the first case the SUT has passed the test, meaning that the test did not discover nonconformance to the specification (however, this clearly does not imply that the SUT meets the specification, as another test may fail). In the case of a FAIL, we know that the SUT does not meet its specification (unless the test case is itself erroneous).

In this context, the problem we are interested in is that of "test generation," namely, synthesizing test cases automatically from a formal description of the specification. The benefits are obvious: test cases do not have to be "manually" written, which is source of errors, as with any other design process. The drawbacks of automatic test generation are similar to many other automatic synthesis techniques: state-explosion problems and competition with human designers who can often "do better." In the case of testing, "better" may mean writing a "minimal" set of test cases that can cover all "important" aspects of the specification. "Minimal" can be made a formal notion (e.g., smallest number of test cases, small size of test cases, etc.). The notion of "importance" is much harder to formalize, however. Coverage has been formalized in many different ways since the beginnings of testing, in terms of statement coverage, condition coverage, and so on (e.g., see [112]). We will briefly return to some of these notions later.

In Sections 13.6 and 13.7, we discuss testing and test generation methods for timed and hybrid automata, respectively.

13.6   Test Generation for Timed Automata

Before we discuss how test cases can be generated automatically from a given formal specification, we must first define what it means for an SUT to "conform" to a specification. The answer to this question depends on the setting, and over the years many different notions of conformance have been proposed by researchers. In this section, we will use a setting based on timed automata. In particular, we will use a model of timed automata with inputs and outputs (TAIO) to formally capture the specification. A TAIO is simply a TA where each one of the events labeling its transitions is distinguished to be either an input or an output (but not both). Some examples are shown in Figure 13.18. Input events are annotated with "?" and outputs with "!." Let us look in particular at TAIO I_1. I_1 models a system that initially awaits input

---

*Often the mere activity of executing a test case is a significant problem by itself. This is the case, for instance, when ensuring right timing on the inputs and observing accurate times of the outputs is crucial. This problem is beyond the scope of this chapter, although it is recognized as an important and practical problem.

[†]generally further verdict types may also appear, e.g., "inconclusive," "error," and "none" (see the standards: UML Testing Profile by the OMG or Testing and Test Control Notation by ETSI).

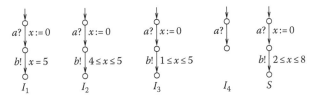

FIGURE 13.18
Timed automata with inputs and outputs.

event a. When (and if) a is received, the system "replies" by producing output event b. The output b is produced exactly 5 time units after a was received. I_2 is similar to I_1, except that its output time is "non-deterministic," although it is guaranteed to be no earlier than 4 and no later than 5 time units from the time a was received. (In these examples we assume that outputs implicitly have an associated notion of urgency: they can be delayed but must be eventually emitted according to the guards specified in the TAIO.) I_3 is a variant of I_2. I_4 receives a but does not "respond."

To capture conformance in a formal way, we use the "timed input–output conformance" relation, or tioco, introduced in [61].* We illustrate tioco in the sequel through an informal description and by providing examples. A formal study can be found in [60,61].

In Figure 13.18, TAIO I_i are given as examples of possible SUTs (they model the behaviors of such SUTs). On the other hand, TAIO S is the formal specification. S states that when/if a is received, b must be produced within 2–8 time units. Which of the four SUTs conform to this specification? It should be clear that I_1 and I_2 conform to S, since all their behaviors satisfy the aforementioned requirement. What about I_3? Some of its behaviors conform to S and others (the ones where b is produced earlier than 2 time units after a) do not. We therefore decide that I_3 does not conform to S: this is because it "may" produce an output too early. It should also be clear that I_4 should not conform to S: it produces no output at all.

tioco captures the earlier informal reasoning in a formal way. It captures the fact that the SUT is allowed to be "more output-deterministic" than the specification. Indeed, the specification generally gives some freedom in terms of what are the "legal" outputs and what are legal times that these outputs may be produced. A given SUT, which can be seen as one of the many possible "implementations" of the specification, may choose to produce any legal output, at any legal time. Different implementations will make different choices, depending on various performances, costs, and other trade-offs.

---

*tioco is inspired by the "untimed" conformance relation ioco introduced by Tretmans and also used in the context of testing [100]. However, tioco differs from ioco in many ways. An important difference is that tioco has no concept of *quiescence*. The latter has been included in ioco as an implicit (and somewhat problematic because it is nonquantified) way of modeling timeouts. In our setting this is unnecessary because time is a "first-class citizen" in our model.

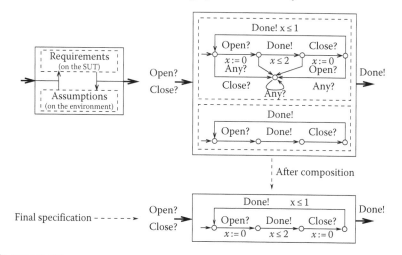

FIGURE 13.19

Specification including assumptions on the environment: generic scheme (left) and example (right).

An input–output system like the SUT is an "open" system, supposed to function in a given environment that generates inputs for the SUT and consumes its outputs. It is often the case that the environment is "constrained" in the sense that it does not behave in an arbitrary way. The SUT is supposed to function correctly in that sort of environment, but not necessarily in another environment, that behaves differently. For instance, a device driver for a given peripheral is supposed to work correctly only for a certain set of devices and may not work for others. An input–output specification must therefore be capable of expressing "assumptions" about the environment. Our modeling framework (and tioco) allows to capture such assumptions in an elegant way, as illustrated in Figure 13.19. The general scheme is shown to the left of Figure 13.9. It consists in modeling the specification as two separate, but communicating, TAIO models. One model captures the requirements on the SUT, that is, the guarantees that the SUT provides on its outputs: this model can be built in such a way that it is "receptive" to any input at any given time, although it may of course ignore inputs that are "illegal" or arrive at "illegal" times. The other model captures the assumptions that the SUT makes on its inputs: these specify formally which inputs are legal and at what times. These assumptions must be satisfied by the environment of the SUT.

We illustrate how this works more precisely through the example shown to the right of Figure 13.19. The specification concerns a system that is supposed to receive a sequence of requests to open or close, say a file. The system executes each request in a given amount of time: it takes at most 2 time units to open and at most 1 time unit to close. During that time, no new request should be received. Moreover, every open request should be followed by a close before a new open request can be issued, and the first request should be an open request.

All these assumptions on the environment are captured formally by the untimed automaton shown in Figure 13.19. This automaton is composed with the input-receptive TAIO shown at the top right, to yield the TAIO shown at the bottom right. The latter represents the final specification. Notice that this final specification is not an input-receptive TAIO: for instance, it does not accept a second open? request until the first one is fulfilled by issuing output done!. Having non-input-receptive specifications is essential in order to model assumptions on the environment, and this is an important feature of the tioco framework.

We could spend many more pages discussing what other properties are desirable from a formal conformance relation such as tioco: transitivity (if A conforms to B and B conforms to C then A conforms to C), compositionality (if A_1 conforms to B_1 and A_2 conforms to B_2, then the composition of A_1 and A_2 conforms to the composition of B_1 and B_2). These properties are satisfied by tioco, under appropriate conditions. We refer the reader to [60] for an in-depth technical study.

Having explained what it means for an SUT to conform to a formal specification, we are almost ready to discuss test generation. However, before doing that, we still need to make more precise what exactly we mean by a test case. A test case is essentially a program that is executed by a "tester." The tester interacts with the SUT through the IO interface of the latter. The tester can be seen as a generic device, capable of running many test cases. So the tester is essentially a computer, with appropriate IO capabilities for the class of SUTs we are interested in.

The execution of a test case by the tester must be as deterministic as possible. This is crucial in order for tests to be reproducible, which in turn is very important for debugging (it is a nightmare to know that some test has failed without being able to reproduce this failure). Nondeterminism can be allowed, for instance, one may allow randomized testing where some choices of the tester can be based on tossing a random coin. In reality, however, this randomness will be generated by a pseudorandom number generator, and the seed of this generator can be saved and reused to achieve reproducibility.

Obviously, the determinism of the execution does not depend only on the tester (and the test case) but also on the SUT: if the SUT itself is nondeterministic (i.e., for the same sequence of inputs, it may produce different sequences of outputs) then determinism cannot be guaranteed. Still, we will require as a minimum the tester/test case to be deterministic (or random but using a pseudorandom generator as explained earlier).

Notice that the behavior of the SUT depends not only on the inputs it receives from the tester, but also on its internal state. In the context of this work, we will assume that it is possible to "reset" the state of the SUT to some given initial state (or set of possible states) after each test case has been executed. A large amount of research is available in the literature for the case where the SUT cannot be reset and "resetting" input sequences must be devised, in addition to the conformance testing sequences (see [73] for an excellent survey). Very little work has been done about this problem in the context of timed automata [63].

Given that a test case is a deterministic program, what does this program do? It essentially interacts with the SUT through inputs and outputs: it generates and issues the inputs to the SUT and consumes its outputs. Since the specification defines not only the legal values of inputs and outputs but also their legal timing, it is very important that the test case be able to capture timing as well. In other words, the test case must specify "not only which input should be generated but also exactly when." Moreover, the test case must specify "how to proceed depending on what output the SUT produces but also on the time in which this output is produced."

For example, consider a specification for a computer mouse that states: "if the mouse receives two consecutive clicks (the input) in less than 0.2 s then it should emit a double-click event to the computer." One can imagine various tests that attempt to check whether a given SUT satisfies the aforementioned specification (and indeed behaves as a proper mouse). One test may consist in issuing two consecutive clicks 0.1 s apart, and waiting to see what happens. If the SUT emits a double-click then it passes the test, otherwise it fails. But there are obviously other tests: issuing two clicks 0.15 s apart, or 0.05 s apart, and so on. Moreover, one may vary the initial waiting time, before issuing the clicks. Moreover, presumably the specification requires that the mouse continues to exhibit the same behavior not only the first time it receives two clicks, but also every time after that. Then a test could try to issue two sets of two clicks and check that the SUT processes both of them correctly. It becomes clear that a finite number of tests cannot ensure that the SUT is correct, at least not in the absence of more assumptions about the SUT. It is also interesting to note some inherent ambiguities in the aforementioned, simple, specification written in English. For instance, does the delay between the two ticks need to be strictly less than 0.2 s or can it be exactly 0.2 s? How much time after the ticks should the mouse respond by emitting an event to the computer? And so on.

In general, a test case in our setting can be cast into the form shown in Figure 13.20. The test case is described in pseudocode. The test case maintains an internal state, which captures the "history" of the execution (e.g., what outputs have been observed, at what times, and so on). The state can also be used to encode whether this history is legal, that is, meets the specification. If it does not, the test stops with the result FAIL. Otherwise, the test can proceed for as long as required.

The test case uses a "timer" to measure time. This timer is an abstract device that can be implemented in different ways in the execution platform of the tester. An important question, however, is what exactly can this timer measure, especially, how precise this timer measures time. For instance, in

```
// test case pseudo-code:
s := initialize state; // this is the state of the tester
while( not some termination condition  ) do
  x := select input in set of legal inputs given s;
  issue x to the SUT;
  set timer to TIMEOUT;
  wait until timer expires or SUT produces an output;
  if ( timer expired ) then
    s := update state s given TIMEOUT;
  end if;
  if ( SUT produced output y, T time units after x ) then
    s := update state s given T and y;
  end if;
  if ( s is not a legal state ) then
    announce that the SUT failed the test and exit;
  end if;
end while;
announce that the SUT passed the test and exit;
```

FIGURE 13.20
Generic description of a test case.

the pseudocode, the timer is set to expire after TIMEOUT time units. One may ask: how critical is it that the timer expires "exactly" after so much time? What if it actually expires a bit late or a bit early? In the pseudocode, the timer is checked to see how much time elapsed from event x until event y: this amount is T time units. But if the timer is implemented as an integer counter, which is typically the case in a digital computer, the value T that the counter reads at any given moment in time is only an approximation of the time that has elapsed since the timer was reset: in reality, the time that has elapsed lies anywhere between T and T+1 time units. To the aforementioned must be added inaccuracies because of processing delays. For example, executing the tester code takes time: this time must be accounted for when updating the state of the tester.

In order to make the issues of time accuracy explicit, we make a distinction between "analog-clock" and "digital-clock" testers (and tests). The former are ideal devices (or programs), assumed to be able to measure time exactly, with an infinite degree of precision. In particular they can be assumed to measure any delay which is a nonnegative rational number. Digital-clock tests have access to a digital clock with finite precision. This clock may suffer from drift, jitter, and so on. Analog-clock tests are not implementable since clocks with infinite precision do not exist in practice. Still, analog-clock tests are worth studying not only because of theoretical interest, but also because they can be used to represent ideal, or "best case" tests, that are independent from a given execution platform. This is obviously useful for test reusability. Analog-clock tests may also be used correctly when real clock inaccuracies or execution delays can be seen as negligible compared to the delays used in the test.

We are now in a position to discuss automatic test generation. The objective is to generate, from a given formal specification, provided in the form of a TAIO, one or more test cases, that can be represented as programs written in some form similar to the pseudocode presented earlier. We briefly describe this quite technical step and illustrate the process through some examples. We refer the reader to [60] for a thorough presentation.

We first describe analog-clock test generation. Suppose the specification is given as a TAIO S. The basic idea is to generate a program that maintains in its memory (the `state` variable in the pseudocode shown in Figure 13.20) the set of all possible legal configurations that S could be in, given the history of inputs and outputs (and their times) so far. Let C be this set of legal configurations. The important thing to note is that C completely captures the set of "all legal future behaviors." Therefore, it is sufficient to determine the future of the test.

The set C is represented symbolically, in much the same way as for reachability analysis used for timed-automata model-checking. C is generally nonconvex and cannot be represented as a single zone, however, it can be represented as a set of zones. C is updated based on the observations received by the test: these observations are events (inputs or outputs) and time delays. Updating C amounts to performing an "on-the-fly subset construction," which can be reduced to reachability. This technique was first proposed in [102] where it was used for monitoring in the context of fault diagnosis. The same technique can be applied to testing with very minor modifications.

Notice that the aforementioned test generation technique is "on-the-fly" (also sometimes called "on-line"). This means that the test state (i.e., C) is generated during the execution of the test, and not a priori. There are good reasons for this in the case of timed automata: since the set of possible configurations of a TA is infinite, the set of all the possible sets of legal configurations is also infinite, thus cannot be completely enumerated.

We illustrate analog-clock on-the-fly test generation and execution on the example specification S shown in Figure 13.18. Suppose the three states of S are numbered 0,1,2, from top to bottom, the initial set of legal configurations can be represented by the predicate $C_0 : s = 0$. Notice that the value of the clock x is unimportant in this case. Next, the test can choose to issue the single input event a to the SUT. The set of legal configurations then becomes $C_1 : s = 1 \wedge x = 0$. Let us suppose that TIMEOUT=2. If the SUT produces output b before the timer expires (i.e., in <2 time units after it received input a), the set of legal configurations becomes empty: this is because there is no configuration in C_1 that can perform a b after <2 time units. An empty C is an illegal state for the test: this means that the SUT fails the test in this case. Indeed, this is correct, since the SUT produces output b too early. On the other hand, if the timer expires before b is received, then C is updated to $C_2 : s = 1 \wedge x = 2$. The timer is reset, and execution continues. Suppose b is not received after four timeouts: the value of C at this point is $C_5 : s = 1 \wedge x = 8$. If a fifth timeout occurs, C becomes empty: this is because there is no state

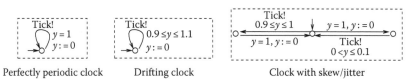

FIGURE 13.21
Models of digital clocks.

in C_5 that can let 10 time units elapse (because of the urgency implied when $x = 8$). Again, the SUT fails in this case, because it does not produce a b by the required deadline.

On-the-fly generation is not the only possibility in the case of analog-clock tests. Another option is to generate analog tests off-line, and represent them as TAIO themselves. However, these TAIO need to be deterministic, and synthesis of deterministic TA that are in some sense equivalent to a nondeterministic TA can be an undecidable problem [103]. Indeed, the test generation of TA testers is generally undecidable. Still, it is possible to restrict the problem so that it becomes decidable. One way of doing this is by limiting the number of clocks that the tester automaton can have. We refer the reader to [60,62] for details.

We now turn to digital-clock test generation. Again, we are given a formal specification in the form of a TAIO S. But in this case, we assume that we are also given a model of the digital clock, also in the form of a TA. The latter is a special TA model, called a "tick" automaton. The reason is that this TA has a single event, named "tick," that represents the discrete tick of a digital clock (e.g., the incrementation of the digital-clock counter). Some possible tick models are shown in Figure 13.21. The left-most automaton models a perfectly periodic clock with period 1 time unit. The automaton in the middle models a clock with drift: its period varies nondeterministically between 0.9 and 1.1 time units. In this model, the kth tick can occur anywhere in the interval $[0.9k, 1.1k]$: as k grows, the uncertainty becomes larger and larger. The right-most automaton models a digital clock where this uncertainty is bounded: the kth tick can occur in the interval $[k − 0.1, k + 0.1]$.

With a Tick model, the user has full control over the assumptions used by the digital-clock test generator. The generator need not make any implicit assumptions about the behavior of the digital clock of the tester: all these assumptions are captured in the tick model. In an application setting, a library of available tick models could be supplied to the user to choose a model from.

Having the specification S and the tick automaton, automatic digital-clock test generation proceeds as follows. First, the product of S and tick is formed, as illustrated in Figure 13.22: this product is again a TAIO, call it $S+$. The tick event is considered an output in $S+$. Next, an "untimed" test is generated from $S+$. An untimed test is one that reacts only to discrete events

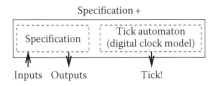

FIGURE 13.22
Specification+ : product of specification and tick model.

and not time. However, time is implicitly captured in $S+$ through the tick
event. Indeed, the tick event represents the time elapse as measured with a
digital clock. This "trick" allows to turn digital-clock test into an "untimed"
test generation problem. The latter can be solved using standard techniques,
such as those developed in [100]. Although these techniques have been originally developed for untimed specifications, they can be applied to TAIO
specifications such as $S+$, because they are based on reachability analysis.
Again, the idea of on-the-fly subset construction is used in this case.

Let us illustrate this process through an example. Suppose we want to
generate digital clock tests from the specification S shown in Figure 13.18
and the left-most, perfectly periodic, tick model shown in Figure 13.21. A
digital-clock test generated by the aforementioned method for this example
is shown in Figure 13.23. Notice that the test is represented as an "untimed"
automaton with inputs and outputs. This is normal, since any reference to
time is replaced by a reference to the tick event of the digital clock. Moreover
notice that inputs and outputs are reversed for the test: a is an output for the
test (an input for the SUT), while b and tick are inputs to the test.

The test of Figure 13.23 starts by issuing a after some nondeterministic
number of ticks (strictly speaking, this is not a valid test since it is nondeterministic: however, it can be seen as representing a set of tests rather than
a single test). It then waits for a b. If a b is received before at least two ticks
are received, the SUT fails the test: indeed, this is because it implies that <2

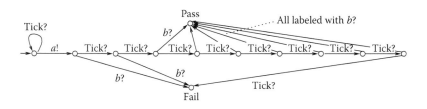

FIGURE 13.23
A digital clock test generated from the specification S shown in Figure 13.18
and the left-most, perfectly periodic, tick model shown in Figure 13.21.

time units have elapsed between the a and the b, which violates the specification S.* If no b is received for 9 ticks, the SUT fails the test: this is because it implies that >8 time units have elapsed between a and b, which again is a violation of S. Otherwise, the SUT passes the test.

We end this section with a few remarks on test selection. The test generation techniques presented earlier can generate a large number of tests: indeed, in the case of multiple inputs, the test must choose which input to issue to the SUT. It must also choose when to issue this input (this can be modeled as a choice between issuing an input or waiting). This represents a huge number of choices, thus a huge number of possible tests that can be generated. This is similar to the state-explosion problem encountered in model checking: in this case it can be called the test explosion problem. It should be noted that this problem arises in any automatic test generation framework, and not just in the timed or hybrid automata case.

The question then becomes, can this large number of tests be somehow limited? Traditionally, there have been different approaches to achieve this. One approach is based on the notion of "coverage": the idea is to generate a "representative" set of tests. One way of defining "representative" is by means of coverage: a set of tests that "covers" all aspects of the specification, in some way. In practice, notions such as state coverage (cover all states of the specification), transition coverage (cover all transitions), and so on can be used. Test generation with coverage objectives is explored in [60]. Another approach to limit test explosion is to "guide" the generation of the test toward some goal: this is done using a so-called "test purpose." A test purpose can specify, for instance, that one of the states of the specification should be reached during the test. The test generator should then produce a test that attempts to reach that state (sometimes reaching a given state cannot be guaranteed since it generally depends on the outputs that the SUT will produce). This approach has been followed by untimed test generation tools such as TGV [38] and can be easily adapted to the timed automata case.

13.7 Test Generation for Hybrid Automata

Concerning hybrid systems, model-based testing is still a new research domain. Previous work [95] proposed a framework for generating test cases by simulating hybrid models specified using the language CHARON [7]. In this work, the test cases are generated by restricting the behaviors of an

---

*We assume here that a, b, and tick cannot occur simultaneously. If this assumption is lifted, then the digital-clock test shown in Figure 13.23 needs to be modified to issue a PASS if b is received exactly one tick after a.

environment automaton to yield a deterministic testing automaton. A test suite can thus be defined as a finite set of executions of the environment automaton. In [57], the testing problem is formulated as one of finding a piecewise constant input that steers the system toward some set, which represents a set of bad states of the systems. The paper [55] addresses the problem of robust testing by quantifying the robustness of some properties under parameter perturbations. This work also considers the problem of how to generate test cases with a number of initial state coverage strategies.

In this section, we present a formal framework for the conformance testing of hybrid automata (including important notions such as conformance relation, test cases, and coverage). We then describe a test generation method, which is a combination of the RRT algorithm (presented in Section 13.4.2) and a coverage-guided strategy.

13.7.1 Conformance Relation

To define the conformance relation for hybrid automata, we need first the notions of inputs. A system input that is controllable by the tester is called a "control input"; otherwise, it is called a "disturbance input."

13.7.1.1 Continuous Inputs

All the continuous inputs of the system are assumed to be controllable. Since we want to implement the tester as a computer program, we are interested in piecewise-constant continuous input functions (a class of functions from reals to reals that can be generated by a computer). Hence, a "continuous control action" $(\bar{u}_q, h)$, where $\bar{u}_q$ is the value of the input and h is the "duration," specifies that the system continues with the continuous dynamics at discrete state q under the input $u(t) = \bar{u}_q$ for exactly h time. We say that $(\bar{u}_q, h)$ is "admissible" at (q, x) if the input function $u(t) = \bar{u}_q$ for all $t \in [0, h]$ is admissible starting at (q, x) for h time.

13.7.1.2 Discrete Inputs

The discrete transitions are partitioned into controllable and uncontrollable discrete transitions. Those that are controllable correspond to discrete control actions, and the others to discrete disturbance actions. The tester emits a discrete control action to specify whether the system should take a controllable transition (among the enabled ones) or continue with the same continuous dynamics. In the latter case, it can also control the values assigned to the continuous variables by the associated reset map. For the simplicity of explanation, we will not consider nondeterminism caused by the reset maps. Hence, we denote a discrete control action by the corresponding transition, such as (q, q').

We then need the notion of "admissible control action sequence," which is formally defined in [32]. Intuitively, this means that an admissible control action sequence, when being applied to the automaton, does not cause it to be blocked.

In the definition of the conformance relation between a SUT $\mathcal{A}_s$ and a specification $\mathcal{A}$, the following assumptions about the inputs and observability are used:

- All the controllable inputs of $\mathcal{A}$ are also the controllable inputs of $\mathcal{A}_s$.
- The set of all admissible control action sequences of $\mathcal{A}$ is a subset of that of $\mathcal{A}_s$. This assumption assures that the SUT can admit all the control action sequences that are admissible by the specification.
- The discrete state and the continuous state of $\mathcal{A}$ and $\mathcal{A}_s$ are observable.

Intuitively, the SUT $\mathcal{A}_s$ is conform to the specification $\mathcal{A}$ iff under every admissible control sequence, the set of all the traces of $\mathcal{A}_s$ is included in that of $\mathcal{A}$. The definition of conformance relation can be easily extended to the case where only a subset of continuous variables are observable by projecting the traces on the observable variables. However, extending this definition to the case where some discrete states are unobservable is more difficult since this requires identifying the current discrete state in order to decide a verdict.

13.7.2   Test Cases and Test Executions

In our framework, a "test case" is represented by a tree where each node is associated with an observation and each path from the root with an observation sequence. Each edge of the tree is associated with a control action. A physical "test execution" can be described as follows:

- The tester applies a test to the SUT.
- An observation (including both the continuous and the discrete state) is measured at the end of "each" continuous control action and after "each" discrete (disturbance or control) action.

This procedure leads to an observation sequence, or a set of observation sequences if multiple runs of the test case are possible (in case nondeterminism is present). In the following, we focus on the case where each test execution involves a single run of a test case. It is clear that the aforementioned test execution process uses a number of implicit assumptions, such as observation measurements take zero time, and in addition, no measurement error is considered. Additionally, the tester is able to realize exactly the continuous input functions (which is often not possible in practice because of actuator imprecision).

After defining the important concepts, it now remains to tackle the problem of generating test cases from a specification model. A hybrid automaton may have an infinite number of infinite traces; however, the tester can only perform a finite number of test cases in finite time. Therefore, we need to select a finite portion of the input space of $\mathcal{A}$ and test the conformance of $\mathcal{A}_s$ with respect to this portion. The selection is done using a coverage criterion that we formally define in the following. Hence, our testing problem is

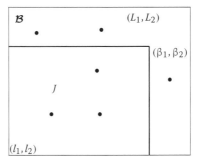

FIGURE 13.24
Illustration of the star discrepancy notion.

formulated so as to automatically generate a set of test cases from the specification hybrid automaton to satisfy this coverage criterion.

13.7.3 Test Coverage

The test coverage we describe here is based on the star discrepancy notion and motivated by the goal of testing reachability properties. It is thus desirable that the test coverage measure can describe how well the states visited by a test suite represent the reachable set. One way to do so is to look at how well the states are equidistributed over the reachable set. However, the reachable set is unknown, we can only consider the distribution of the visited states over the state space (which can be thought of as the potential reachable space).

The star discrepancy is a notion from statistics often used to describe the "irregularities" of a distribution of points with respect to a box. Indeed, the star discrepancy measures how badly a point set estimates the volume of the box. The popularity of this measure is perhaps related to its usage in quasi-Monte Carlo techniques for multivariate integration (see for example [15]).

Let P be a set of k points inside $\mathcal{B} = [l_1, L_1] \times \cdots \times [l_n, L_n]$. Let $\mathcal{J}$ be the set of all sub-boxes J of the form $J = [l_1, \beta_1] \times \cdots \times [l_n, \beta_n]$ with $\beta_i \in [l_i, L_i]$ for all $i \in \{1, \ldots, n\}$ (see Figure 13.24). The local discrepancy of the point set P with respect to the sub-box J is defined as $D(P, J) = |\frac{A(P, J)}{k} - \frac{\text{vol}(J)}{\text{vol}(\mathcal{B})}|$ where $A(P, J)$ is the number of points of P that are inside J, and $\text{vol}(J)$ is the volume of the box J. Then, the star discrepancy of a point set P with respect to the box $\mathcal{B}$ is defined as: $D^*(P, \mathcal{B}) = \sup_{J \in \mathcal{J}} D(P, J)$. The star discrepancy satisfies $0 < D^*(P, \mathcal{B}) \leq 1$. A large value $D^*(P, \mathcal{B})$ means that the points in P are not much equidistributed over $\mathcal{B}$.

Since a hybrid automaton can only evolve within the invariants of its discrete states, one needs to define a coverage with respect to these sets.

For simplicity, all the staying sets are assumed to be boxes. For a set of $\mathcal{P} = \{(q, P_q) \mid q \in Q \wedge P_q \subset \mathcal{I}_q\}$ be the set of states. The coverage of $\mathcal{P}$ is defined as: $\mathrm{Cov}(\mathcal{P}) = \frac{1}{||Q||} \sum_{q \in Q} 1 - D^*(P_q, \mathcal{I}_q)$ where $||Q||$ is the number of discrete states in Q. If an invariant set $\mathcal{I}_q$ is not a box, one can take the smallest oriented box that encloses it and apply the star discrepancy definition to that box after an appropriate coordination change. We can see that a large value of $\mathrm{Cov}(\mathcal{P})$ indicates a good space-covering quality. The star discrepancy is difficult to compute especially for high dimensions; however, it can be approximated (see [79]). Roughly speaking, the approximation considers a finite box decomposition instead of the infinite set of sub-boxes in the definition of the star discrepancy.

13.7.4 Coverage Guided Test Generation

Essentially, the test generation algorithm consists of the following two steps:

- From the specification automaton $\mathcal{A}$, generate an exploration tree using the hRRT algorithm and a guiding tool, which is based on the earlier described coverage measure. The goal of the guiding tool is to bias the evolution of the tree toward the interesting region of the state space, so as to rapidly achieve a good coverage quality.
- Determine the verdicts for the executions in the exploration tree, and extract a set of interesting test cases with respect to the property to verify.

The motivation of the guiding method is as follows. Because of the uniform sampling of goal states, the RRT algorithm is biased by the Voronoi diagram of the vertices of the tree. If the actual reachable set is only a small fraction of the state space, the uniform sampling over the whole state space leads to a strong bias in selection of the points on the boundary of the tree, and the interior of the reachable set can only be explored after a large number of iterations. Indeed, if the reachable was known, sampling within the reachable set would produce better coverage results.

13.7.4.1 Coverage-Guided Sampling

Sampling a goal state $s_{goal} = (q_{goal}, x_{goal})$ in the hybrid state space $\mathcal{S}$ consists of the following two steps: (1) Sample a goal discrete state q_{goal}, according to some probability distribution; (2) Sample a continuous goal state x_{goal} inside the invariant set $\mathcal{I}_{q_{goal}}$.

In each iteration, if a discrete state is not yet well explored, that is, its coverage is low, we give it a higher probability to be selected. Let $\mathcal{P} = \{(q, P_q) \mid q \in Q \wedge P_q \subset \mathcal{I}_q\}$ be the current set of visited states, one can sample the goal discrete state according to the following probability distribution:

$$\Pr[q_{goal} = q] = \frac{1 - \mathrm{Cov}(\mathcal{P}, q)}{||Q|| - \sum_{q' \in Q} \mathrm{Cov}(\mathcal{P}, q')}.$$

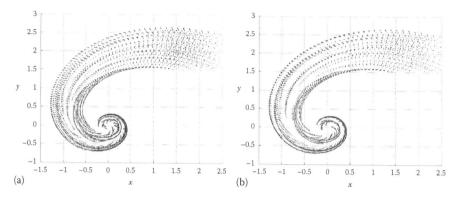

FIGURE 13.25
Results obtained using gRRT (a) and hRRT (b), with the same number of visited states.

Suppose that we have sampled a discrete state $q_{goal} = q$. Since all the staying sets are boxes, the staying set $\mathcal{I}_q$ is denoted by the box $\mathcal{B}$ and called the bounding box.

As mentioned earlier, the coverage estimation is done using a box partition of the state space $\mathcal{B}$, and sampling of a continuous goal state can be done by two steps: first, sample a goal box b_{goal} from the partition, second, "uniformly" sample a point x_{goal} in b_{goal}. Guiding is thus done in the goal box sampling process by defining, at each iteration of the test generation algorithm, a probability distribution over the set of the boxes in the partition. Essentially, we favor the selection of a box if adding a new state in this box allows to improve the coverage of the visited states. This is captured by a potential influence function, which assigns to each elementary box b in the partition a real number that reflects the change in the coverage if a new state is added in b. The current coverage estimation is given in form of a lower and an upper bound. In order to improve the coverage, both the lower and the upper bounds need to be reduced (see more details in [32]).

The hRRT algorithm for hybrid automata in which the goal state sampling is done using this coverage-guided method is now called the gRRT algorithm (which means "guided hRRT"). To illustrate the coverage-efficiency of gRRT, Figure 13.25 shows the results obtained by the hRRT and the gRRT on a linear system after 50,000 iterations. We can see that the gRRT algorithm has a better coverage result. Indeed with the "same number of states," the states visisted by the gRRT are more equi-distributed over the reachable set than those visisted by hRRT.

These algorithms were implemented in the prototype tool HTG, which was successfully applied to treat a number of benchmarks in control applications and in analog and mixed-signal circuits [31,79].

13.8 Conclusions

Embedded systems consist of hardware and software embedded in a physical environment with continuous dynamics. To model such systems, timed and hybrid automata models have been developed and studied extensively in the past two decades. In this chapter we have reviewed the basics of these models and methods of exhaustive or partial verification, as well as testing for these models. We hope that our overview will motivate embedded system designers to use these models in their applications, and that they will find them useful. Timed and hybrid automata are still an active field of research, and we refer the readers to the numerous papers published on these topics, in addition to those referenced in our bibliography section.

Acknowledgments

We would like to thank Eugene Asarin, Olivier Bournez, Saddek Bensalem, Antoine Girard, Moez Krichen, Oded Maler, Tarik Nahhal, Sergio Yovine, and other colleagues for their collaborations and their contributions to the results presented in this chapter.

References

1. N. Abed, S. Tripakis, and J.-M. Vincent. Resource-aware verification using randomized exploration of large state spaces. In *SPIN'08*, Los Angeles, CA, *LNCS*, 5156, 2008.

2. K. Altisen and S. Tripakis. Implementation of timed automata: An issue of semantics or modeling? In P. Pettersson and W. Yi (editors), *3rd International Conference on Formal Modeling and Analysis of Timed Systems (FORMATS'05)*, Uppsala, Sweden, *LNCS*, 3829:273–288, September 2005, Springer, Berlin, Heidelberg.

3. R. Alur. Timed automata. NATO-ASI 1998 Summer School on Verification of Digital and Hybrid Systems, 1998.

4. R. Alur, C. Courcoubetis, N. Halbwachs, D.L. Dill, and H. Wong-Toi. Minimization of timed transition systems. In *Third Conference on Concurrency Theory CONCUR '92*, Stony Brook, NY, *LNCS*, 630:340–354, 1992, Springer-Verlag, New York.

5. R. Alur, C. Courcoubetis, N. Halbwachs, T. Henzinger, P. Ho, X. Nicollin, A. Olivero, J. Sifakis, and S. Yovine. The algorithmic analysis of hybrid systems. *Theoretical Computer Science*, 138:3–34, 1995.

6. R. Alur, C. Courcoubetis, T.A. Henzinger, and P.-H. Ho. Hybrid automata: An algorithmic approach to the specification and verification of hybrid systems. In *Hybrid Systems*, pp. 209–229, 1992.

7. R. Alur, T. Dang, J. Esposito, Y. Hur, F. Ivan, C. Kumar, I. Lee, P. Mishra, G. Pappas, and O. Sokolsky. Hierarchical modeling and analysis of embedded systems. Proceedings of the IEEE, 91(1):11–28, 2003.

8. R. Alur, T. Dang, and F. Ivancic. Counter-example guided predicate abstraction of hybrid systems. *Theoretical Computer Science (TCS)*, 354(2):250–271, 2006.

9. R. Alur and D. Dill. A theory of timed automata. *Theoretical Computer Science*, 126:183–235, 1994.

10. H. Anai and V. Weispfenning. Reach set computations using real quantifier elimination. In M.D. Di Benedetto and A. Sangiovanni-Vincentelli (editors), *Hybrid Systems: Computation and Control*, Rome, Italy, *LNCS*, 2034:63–75, 2001, Springer-Verlag, Berlin, Heidelberg.

11. E. Asarin, O. Bournez, T. Dang, and O. Maler. Approximate reachability analysis of piecewise-linear dynamical systems. In B. Krogh and N. Lynch (editors), *Hybrid Systems: Computation and Control*, Pittsburg, PA, *LNCS*, 1790:20–31, 2000, Springer-Verlag, Berlin, Heidelberg.

12. E. Asarin, T. Dang, and A. Girard. Hybridization methods for the analysis of nonlinear systems. *Acta Informatica*, 43(7):451–476, 2007.

13. E. Asarin, T. Dang, and O. Maler. The d/dt tool for verification of hybrid systems. In *Computer Aided Verification*, Copenhagen, Denmark, *LNCS*, 2404:365–370, 2002, Springer-Verlag, Berlin, Heidelberg.

14. E. Asarin and G. Schneider. Widening the boundary between decidable and unde- cidable hybrid systems. In *CONCUR*, Irno, Czech Republic, 2002.

15. J. Beck and W. W. L. Chen. Irregularities of distribution. In *Acta Arithmetica*, Cambridge, U.K., 1997. Cambridge University Press.

16. B. Berthomieu and M. Menasche. An enumerative approach for analyzing time Petri nets. *IFIP Congress Series*, 9:41–46, 1983.

17. A. Bhatia and E. Frazzoli. Incremental search methods for reachability analysis of continuous and hybrid systems. In *HSCC*, Philadelphia, PA, pp. 142–156, 2004.

18. S. Bornot, J. Sifakis, and S. Tripakis. Modeling urgency in timed systems. In W.P. de Roever, H. Langmaack, and A. Pnueli (editors), *Compositionality: The Significant Difference, International Symposium (COMPOS'97)*, Bad Malente, Germany, *LNCS*, 1536:103–129, September 1998, Springer, Berlin, Heidelberg.

19. D. Bosnacki. Digitization of timed automata. In *Proceedings of the Fourth International Workshop on Formal Methods for Industrial Critical Systems (FMICS '99)*, Berlin, Germany, pp. 283–302, 1999.

20. O. Botchkarev and S. Tripakis. Verification of hybrid systems with linear differential inclusions using ellipsoidal approximations. In B. Krogh and N. Lynch (editors), *Hybrid Systems: Computation and Control*, Pittsburg, PA, *LNCS*, 1790:73–88, 2000, Springer-Verlag, Berlin, Heidelberg.

21. O. Bournez, O. Maler, and A. Pnueli. Orthogonal polyhedra: Representation and computation. In F. Vaandrager and J. van Schuppen (editors), *Hybrid Systems: Computation and Control*, Bergen Dal, the Netherlands, *LNCS*, 1569:46–60, 1999, Springer-Verlag, Berlin, Heidelberg.

22. P. Bouyer. Forward analysis of updatable timed automata. *Formal Methods in System Design*, 24(3):281–320, 2004.

23. P. Bouyer, C. Dufourd, E. Fleury, and A. Petit. Are timed automata updatable? In *CAV'00*, Chicago, IL, *LNCS*, 1855, 2000.

24. M. Bozga, O. Maler, and S. Tripakis. Efficient verification of timed automata using dense and discrete time semantics. In L. Pierre and T. Kropf (editors), *Correct Hardware Design and Verification Methods, 10th IFIP WG 10.5 Advanced Research Working Conference (CHARME '99)*, Bad Herrenalb, Germany, *LNCS*, 1703:125–141, September 1999, Springer, Berlin, Heidelberg.

25. M. Branicky, M. Curtiss, J. Levine, and S. Morgan. Sampling-based reachability algorithms for control and verification of complex systems. In *Thirteenth Yale Workshop on Adaptive and Learning Systems*, New Haven, CI, 2005.

26. K. Cerans and J. Viksna. Deciding reachability for planar multipolynomial systems. In *Hybrid Systems*, pp. 389–400, 1995.

27. A. Chutinan and B.H. Krogh. Verification of polyhedral invariant hybrid automata using polygonal flow pipe approximations. In F. Vaandrager and J. van Schuppen (editors), *Hybrid Systems: Computation and Control*, Bergen Dal, the Netherlands, *LNCS*, 1569:76–90, 1999, Springer-Verlag, Berlin, Heidelberg.

28. E. Clarke, A. Fehnker, Z. Han, B. Krogh, J. Ouaknine, O. Stursberg, and M. Theobald. Abstraction and counterexample-guided refinement in model checking of hybrid systems. *International Journal of Foundations of Computer Science*, 14(4):583–604, 2003.

29. T. Dang. Reachability-based technique for idle speed control synthesis. *International Journal of Software Engineering and Knowledge Engineering IJSEKE*, 15(2):397–404, 2005.

30. T. Dang and O. Maler. Reachability analysis via face lifting. In T.A. Henzinger and S. Sastry (editors), *Hybrid Systems: Computation and Control*, Berkeley, CA, *LNCS*, 1386:96–109, 1998, Springer-Verlag, Berlin, Heidelberg.

31. T. Dang and T. Nahhal. Using disparity to enhance test generation for hybrid systems. In *TESTCOM/FATES*, Tokyo, Japan, *LNCS*, 2008, Springer, Berlin, Heidelberg.

32. T. Dang and T. Nahhal. Model-based testing of hybrid systems. Technical report, Verimag, IMAG, November 2007.

33. C. Daws, A. Olivero, S. Tripakis, and S. Yovine. The tool KRONOS. In R. Alur, T.A. Henzinger, and E.D. Sontag (editors), *Hybrid Systems III: Verification and Control*, *LNCS*, 1066:208–219, 1996, Springer, New York.

34. C. Daws and S. Tripakis. Model checking of real-time reachability properties using abstractions. In B. Steffen (editor), *Fourth International Conference on Tools and Algorithms for the Construction and Analysis of Systems (TACAS'98)*, Lisbon, Portugal, *LNCS*, 1384:313–329, 1998, Springer, Berlin, Heidelberg.

35. D. Dill. Timing assumptions and verification of finite-state concurrent systems. In J. Sifakis (editor), *Automatic Verification Methods for Finite State Systems*, Grenoble, France, *LNCS*, 407:197–212, 1989, Springer.

36. A. Donzé and O. Maler. Systematic simulation using sensitivity analysis. In *HSCC*, Gières, France, 174–189, 2007.

37. J. Esposito, J. W. Kim, and V. Kumar. Adaptive RRTs for validating hybrid robotic control systems. In *Proceedings Workshop on Algorithmic Foundations of Robotics*, Zeist, the Netherlands, July 2004.

38. J.C. Fernandez, C. Jard, T. Jéron, and G. Viho. Using on-the-fly verification techniques for the generation of test suites. In *CAV'96*, New Brunswick, NJ, *LNCS*, 1102, 1996, Springer.

39. G. Frehse, B. Krogh, R. Rutenbar, and O. Maler. Time domain verification of oscillator circuit properties. *Electronics Notes on Theoretical Computer Science*, 153(3):9–22, 2006.

40. A. Girard. Reachability of uncertain linear systems using zonotopes. In *Hybrid Systems: Computation and Control*, Zurich, Switzerland, *LNCS*, 3414:291–305, 2005, Springer, Berlin, Heidelberg.

41. A. Girard and C. Le Guernic. Zonotope/hyperplane intersection for hybrid systems reachability analysis. In *Hybrid Systems: Computation and Control HSCC*, St. Louis, MU, 2008, Springer, Berlin, Heidelberg.

42. A. Girard, C. Le Guernic, and O. Maler. Efficient computation of reachable sets of linear time-invariant systems with inputs. In *Hybrid Systems: Computation and Control HSCC*, Santa Barbara, CA, *LNCS*, 3927:257–271, 2006, Springer, Berlin, Heidelberg.

43. A. Girard and G. Pappas. Verification using simulation. In *HSCC*, Santa Barbara, CA, pp. 272–286, 2006.

44. P. Godefroid, N. Klarlund, and K. Sen. DART: Directed automated random testing. *SIGPLAN Not. (PLDI'05)*, 40(6):213–223, 2005.

45. M.R. Greenstreet and I. Mitchell. Reachability analysis using polygonal projections. In F. Vaandrager and J. van Schuppen (editors), *Hybrid Systems: Computation and Control*, Bergen Dal, the Netherlands, *LNCS*, 1569:76–90, 1999, Springer-Verlag, Berlin, Heidelberg.

46. R. Grosu, X. Huang, S.A. Smolka, W. Tan, and S. Tripakis. Deep random search for efficient model checking of timed automata. In F. Kordon and O. Sokolsky (editors), *Seventh Monterey Workshop on Composition of Embedded Systems*, Paris, France, *LNCS*, 4888, October 2006, Springer.

47. T. Henzinger, P. Kopke, A. Puri, and P. Varaiya. What's decidable about hybrid automata? In *Journal of Computer and System Sciences*, 373–382, 1995, ACM Press.

48. T. Henzinger, Z. Manna, and A. Pnueli. What good are digital clocks? In *ICALP'92*, Vienna, Austria, *LNCS*, 623, 1992.

49. T. Henzinger, X. Nicollin, J. Sifakis, and S. Yovine. Symbolic model checking for real-time systems. *Information and Computation*, 111(2):193–244, 1994.

50. T.A. Henzinger, P.-H. Ho, and H. Wong-Toi. HyTech: A model checker for hybrid systems. *Software Tools for Technology Transfer*, 1:110–122, 1997.

51. G.J. Holzmann. An analysis of bitstate hashing. In *Formal Methods in System Design*, Kluwer, 3(3):287–305, 1998.

52. G.J. Holzmann. *The Spin Model Checker-Primer and Reference Manual*. Addison-Wesley, Reading, MA, 2004.

53. S. Iman and S. Joshi. *The e-Hardware Verification Language*. Springer, New York, 2004.

54. C. Jard and T. Jeron. Bounded-memory algorithms for verification on-the-fly. In *CAV'91*, Aalborg, Denmark, *LNCS*, 575, 1992, Springer, Berlin, Heidelberg.

55. A. A. Julius, G. E. Fainekos, M. Anand, I. Lee, and G. J. Pappas. Robust test generation and coverage for hybrid systems. In *HSCC*, Pisa, Italy, pp. 329–342, 2007.

56. J. Kapinski, B. Krogh, O. Maler, and O. Stursberg. On systematic simulation of open continuous systems. In *HSCC*, Prague, Czech Republic, pp. 283–297, 2003.

57. J. Kim, J. Esposito, and V. Kumar. Sampling-based algorithm for testing and validating robot controllers. *International Journal of Robotics Research*, 25(12):1257–1272, 2006.

58. D. E. Kirk. Optical control theory: An introduction. Dover Publications, May 2004.

59. M. Kloetzer and C. Belta. Reachability analysis of multi-affine systems. In *Hybrid Systems: Computation and Control*, Santa Barbara, CA, pp. 348–362, 2006, Springer, Berlin, Heidelberg.

60. M. Krichen and S. Tripakis. Conformance testing for real-time systems. Formal methods in system design, 34(3):238–304, 2009.

61. M. Krichen and S. Tripakis. Black-box conformance testing for real-time systems. In S. Graf and L. Mounier (editors), *11th International SPIN Workshop on Model Checking Software (SPIN'04)*, Barcelona, Spain, *LNCS*, 2989:109–126, April 2004, Springer, Berlin, Heidelberg.

62. M. Krichen and S. Tripakis. Real-time testing with timed automata testers and coverage criteria. In Y. Lakhnech and S. Yovine (editors), *Joint International Conference on Formal Modelling and Analysis of Timed Systems and Formal Techniques in Real-Time and Fault-Tolerant Systems, FORMATS/FTRTFT 2004*, Grenoble, France, *LNCS*, 3253:134–151, September 2004, Springer.

63. M. Krichen and S. Tripakis. State identification problems for timed automata. In F. Khendek and R. Dssouli (editors), *17th IFIP TC6/WG 6.1 International Conference on Testing of Communicating Systems (Test-Com'05)*, Montreal, QC, *LNCS*, 3502:175–191, May 2005, Springer, Berlin, Germany.

64. A. Kuehlmann, K. McMillan, and R. Brayton. Probabilistic state space search. In *ICCAD'99*, San Jose, CA, 574–579, 1999.

65. J. Kuffner and S. LaValle. RRT-connect: An efficient approach to single-query path planning. In *Proceedings of the IEEE International Conference on Robotics and Automation (ICRA'2000)*, San Francisco, CA, April 2000.

66. A. Kurzhanski and I. Valyi. *Ellipsoidal Calculus for Estimation and Control*. Birkhauser, Boston, MA, 1997.

67. A.B. Kurzhanski and P. Varaiya. Ellipsoidal techniques for reachability analysis. In *Hybrid Systems: Computation and Control*, Pittsburgh, PA, 2000.

68. A. A. Kurzhanskiy and P. Varaiya. Ellipsoidal toolbox (et). In *Proceedings of the 45th IEEE Conference on Decision and Control*, San Diego, CA, 2006.

69. M. Kvasnica, P. Grieder, M. Baoti, and M. Morari. Multi-parametric toolbox (mpt). In *Hybrid Systems: Computation and Control*, Philadelphia, PA, *LNCS*, 2993:448–462, 2004, Springer, Berlin, Heidelberg.

70. K. Larsen, P. Petterson, and W. Yi. Uppaal in a nutshell. *Software Tools for Technology Transfer*, 1(1/2):134–152, October, 1997.

71. S. LaValle and J. Kuffner. Rapidly-exploring random trees: Progress and prospects, 2000. In *Workshop on the Algorithmic Foundations of Robotics*.

72. S. LaValle. *Planning Algorithms*. Cambridge University Press, New York, 2006.

73. D. Lee and M. Yannakakis. Principles and methods of testing finite state machines - A survey. *Proceedings of the IEEE*, 84:1090–1126, 1996.

74. J. Lygeros, K. Johansson, S. Sastry, and M. Egerstedt. the existence of executions of hybrid automata. In *IEEE Conference on Decision and Control*, Phoenix, AZ, 1999.

75. M. Mihail and C. H. Papadimitriou. On the random walk method for protocol testing. In D. L. Dill (editor), *Proceedings of the Sixth International Conference on Computer-Aided Verification CAV*, Stanford, CA, *LNCS*, 818:132–141, 1994, Springer, London, U.K.

76. O. Maler and A. Pnueli. Reachability analysis of planar multilinear systems. In *Proceedings of the 4th Computer-Aided Verification*, Elounda, Greece, volume 697. Springer, 1993.

77. I. M. Mitchell and J. A. Templeton. A toolbox of Hamilton-Jacobi solvers for analysis of nondeterministic continuous and hybrid systems. In *Hybrid Systems: Computation and Control*, Zurich, Switzerland, *LNCS*. Springer-Verlag, 2005, to appear.

78. N. Kitchen and A. Kuehlmann. Stimulus generation for constrained random simulation. In *ICCAD 2007*, San Jose, CA, pp. 258–265, 2007.

79. T. Nahhal and T. Dang. Test coverage for continuous and hybrid systems. In *CAV*, Berlin, Germany, pp. 454–468, 2007.

80. X. Nicollin, A. Olivero, J. Sifakis, and S. Yovine. An approach to the description and analysis of hybrid systems. In *Hybrid Systems*, pp. 149–178, 1992.

81. J. Ouaknine and J. Worrell. Revisiting digitization, robustness, and decidability for timed automata. In *LICS 2003*, Ottawa, ON, 2003, IEEE CS Press, Washington, DC.

82. R. Paige and R. Tarjan. Three partition refinement algorithms. *SIAM Journal on Computing*, 16(6):973–989, 1987.

83. G. Pappas, G. Lafferriere, and S. Yovine. A new class of decidable hybrid systems. In F. Vaandrager and J. van Schuppen (editors), *Hybrid Systems: Computation and Control*, Bergen Dal, the Netherlands, *LNCS*, 1569:29–31, 1999, Springer-Verlag, Berlin, Heidelberg.

84. R. Pelanek and I. Cerna. Enhancing random walk state space exploration. In *Proc. of Formal Methods for Industrial Critical Systems (FMICS'05)*, Lisbon, Portugal, 98–105, 2005, ACM Press, New York.

85. E. Plaku, L. Kavraki, and M. Vardi. Hybrid systems: From verification to falsification. In W. Damm and H. Hermanns (editors), *International Conference on Computer Aided Verification (CAV)*, Berlin, Germany, *LNCS*, 4590:468–481, 2007, Springer-Verlag, Heidelberg, Berlin, Germany.

86. S. Prajna and A. Jadbabaie. Safety verification of hybrid systems using barrier certificates. In R. Alur and G. J. Pappas (editors), *Hybrid Systems: Computation and Control*, Philadelphia, PA, *LNCS*, 2993:477–492, 2004, Springer, Berlin, Heidelberg.

87. S. Prajna, A. Papachristodoulou, P. Seiler, and P. A. Parrilo. *SOSTOOLS: Sum of Squares Optimization Toolbox for MATLAB*, 2004.

88. A. Puri. Dynamical properties of timed automata. *Discrete Event Dynamic Systems*, 10(1–2):87–113, 2000.

89. A. Puri and P. Varaiya. Decidability of hybrid systems with rectangular differential inclusions. In D. L. Dill (editor), *Proceedings of the Sixth International Conference on Computer-Aided Verification CAV*, Stanford, CA, *LNCS*, 818:95–104, 1994. Springer-Verlag, Berlin, Heidelberg.

90. S. Ratschan and Z. She. Safety verification of hybrid systems by constraint propagation-based abstraction refinement. *ACM Transactions on Embedded Computer Systems*, 6(1): 2007.

91. S. Sankaranarayanan, T. Dang, and F. Ivancic. Symbolic model checking of hybrid systems using template polyhedra. In *TACAS'08 — Tools and Algorithms for the Construction and Analysis of Systems*, Budapest, Hungary, 2008, Springer.

92. S. Shyam and V. Bertacco. Distance-guided hybrid verification with GUIDO. In *DATE '06: Proceedings of the Conference on Design, Automation and Test in Europe*, pp. 1211–1216. European Design and Automation Association, Munich, Germany, 2006.

93. J. Sifakis and S. Yovine. Compositional specification of timed systems. In *13th Annual Symposium on Theoretical Aspects of Computer Science, STACS'96*, Grenoble, France, *LNCS*, 1046, 1996, Spinger-Verlag, Berlin, Heidelberg.

94. O. Stursberg and B. Krogh. Efficient representation and computation of reachable sets for hybrid systems. In *Hybrid Systems: Computation and Control HSCC*, Prague, Czech Republic, *LNCS*, 482–497, 2003, Springer, Berlin, Heidelberg.

95. L. Tan, J. Kim, O. Sokolsky, and I. Lee. Model-based testing and monitoring for hybrid embedded systems. In *Proceedings of IEEE Internation Conference on Information Reuse and Integration (IRI'04)*, Los Vegas, NV, 2004.

96. A. Tiwari. Formal semantics and analysis methods for Simulink Stateflow models. Technical report, SRI International, 2002.

97. A. Tiwari and G. Khanna. Nonlinear systems: Approximating reach sets. In *Hybrid Systems: Computation and Control*, Philadelphia, PA, *LNCS*, 2993:600–614, 2004, Springer, Berlin, Heidelberg.

98. C. Tomlin, I. Mitchell, A. Bayen, and M. Oishi. Computational techniques for the verification of hybrid systems. *Proceedings of the IEEE*, 91(7):986–1001, 2003.

99. F. Torrisi and A. Bemporad. HYSDEL—A tool for generating computational hybrid models. *IEEE Transactions on Control Systems Technology*, 12(2):235–249, 2004.

100. J. Tretmans. Testing concurrent systems: A formal approach. In *CONCUR'99*, Eindhoven, the Netherlands, *LNCS*, 1664, 1999, Springer, Berlin, Heidelberg.

101. S. Tripakis. Checking Timed Büchi Automata Emptiness on Simulation Graphs. ACM Transactions on Computational Logic (to appear).

102. S. Tripakis. Fault diagnosis for timed automata. In W. Damm and E.-R. Olderog (editors), *Formal Techniques in Real Time and Fault*

Tolerant Systems, Seventh International Symposium (FTRTFT'02), Oldenburg, Germany, *LNCS*, 2469:205–224, September 2002, Springer, Berlin, Heidelberg.

103. S. Tripakis. Folk theorems on the determinization and minimization of timed automata. *Information Processing Letters*, 99(6):222–226, September 2006.

104. S. Tripakis. What is resource-aware verification? Unpublished document, 2008. Available from the author's web page.

105. S. Tripakis and C. Courcoubetis. Extending promela and spin for real time. In T. Margaria and B. Steffen (editors), *Second International Workshop on Tools and Algorithms for Construction and Analysis of Systems (TACAS'96)*, Passav, Germany, *LNCS*, 1055:329–348, March 1996, Springer, Berlin, Heidelberg.

106. S. Tripakis and S. Yovine. Analysis of timed systems using time-abstracting bisimulations. *Formal Methods in System Design*, 18(1):25–68, January 2001.

107. A. van der Schaft and H. Schumacher. *An Introduction to Hybrid Dynamical Systems. LNCIS*, 251, 2000, Springer, Berlin, Germany.

108. B. Wile, J. Goss, and W. Roesner. *Comprehensive Functional Verification*. Elsevier, San Francisco, CA, 2005.

109. M. De Wulf, L. Doyen, and J.-F. Raskin. Almost ASAP semantics: From timed models to timed implementations. In *Hybrid Systems: Computation and Control (HSCC'04)*, Philadelphia, PA, *LNCS*, 2993, 2004, Springer, Berlin, Heidelberg.

110. M. Yannakakis and D. Lee. An efficient algorithm for minimizing real-time transition systems. In *Fifth International Conference on Computer-Aided Verification*, Elounda, Greece, *LNCS*, 697, June 1993.

111. J. Yuan, C. Pixley, and A. Aziz. *Constraint-Based Verification*. Springer, New York, 2006.

112. H. Zhu, P. Hall, and J. May. Software unit test coverage and adequacy. *ACM Computing Surveys*, 29(4):366–427, 1997.

14

Semantics of Domain-Specific Modeling Languages

Ethan Jackson, Ryan Thibodeaux, Joseph Porter, and Janos Sztipanovits

CONTENTS

14.1   Introduction

Perhaps the most fundamental and persistent difficulty in engineering is misunderstanding between producers and consumers of technology. The computing industry is rife with tales of failed software projects. Bloated projects with obscene cost and schedule overruns mingle with stories of dramatic functional failures due to subtle bugs, incompatibilities, or incompetence. These problems stand in stark contrast to the requirements of embedded system designs, many of which operate in environments that demand total confidence in their proper and timely function. A large number of methodologies claim to address the deficiencies of software design in general [6,26,30,31]. Many have been successful in controlling some of the complexities of development, though notably far fewer have been tailored to address the specific problems of embedded systems design [8,28].

Embedded systems complicate the software development process in a number of important ways:

- Embedded implementations must operate with proven correctness in many environments. The notion of correctness takes on multiple forms. Both hardware and software must be correctly specified, designed, and constructed for the problem at hand. Specifications must correctly characterize the users' intentions, and the relationships between the behaviors of assembled components must not compromise those intentions. Designs and implementations must be verified against the requirements. Safety-critical embedded systems must also conform to additional requirements imposed by government standards and certification processes.

- Embedded systems are heterogeneous. Although we frequently think of embedded systems in terms of small devices, the end result (which is not always small) is the product of large and complex software designs. Even physical interconnections are many and varied between hardware components. Embedded systems require diverse notions of time and data values—a sensor may continuously monitor a process in order to precisely capture the time of occurrence for a desired event; an embedded processor may sample and process discrete streams of data; and analog circuitry may combine with digital logic and embedded software to implement a standard communications protocol. The constraints placed on embedded systems designs are also heterogeneous: power, memory, processor loading, physical dimensions, bandwidths, numbers of I/O lines, and many more. Distribution adds another dimension to design considerations. Engineers create functional designs and validate them through simulation. Implementation of those designs may exhibit unanticipated (even catastrophic) behavior when distributed over a network

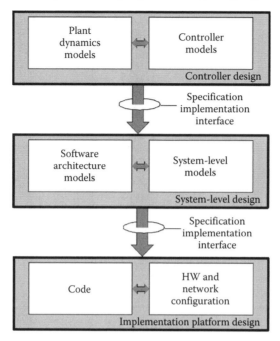

FIGURE 14.1
Simplified design flow for embedded controllers.

of independent processing nodes. In current practice, these issues are resolved by costly and time-consuming testing on a physical prototype.

A simplified design flow for embedded control systems is shown in Figure 14.1. Heterogeneity of the design objectives (e.g., dynamics, safety, and power consumption) and the need for mitigating design complexity dictates that design progresses along abstraction layers, or "design platforms" [8]. The objective of controller design is the construction and verification of Controller Models that meet performance and safety requirements. This step requires modeling plant dynamics, controller dynamics, and verifying the performance and safety criteria using simulation and verification tools. System-level design takes the next step toward implementation. The objective is to select (or design) a software component model and a system architecture that are consistent with the implementation requirements in the controller design. This step requires careful considerations on the effects of the selected interaction model of the software component platform and the execution model of the system platform on the required controller dynamics. The last stage of the design flow is implementation platform design, which includes code generation for the software components from controller models, design of the assignment of the software components and their interactions to the computation, and communication resources in the form of a Deployment Model and verification of the implemented system.

In each of the stages of the design flow, the actual state of the design is expressed using domain-specific modeling languages (DSML). These languages comprise the required heterogeneous abstractions for expressing controller dynamics, software and system architecture, component behavior, and deployment effects. The models expressed in these DSMLs need to be precisely related to each other via the specification/implementation interfaces. They need to be analyzable and their fidelity must be sufficiently precise to accurately predict the behavior of the implemented embedded controller. In addition, the design flow is supported by heterogeneous tools including modeling tools, formal verification tools, simulators, test generators, language design tools, code generators, debuggers, and performance analysis tools that must all cooperate to assist developers and engineers struggling to construct the required systems. If the DSMLs are only informally specified then mismatched tool semantics may introduce mismatched interpretations of requirements, models, and analysis results. This is particularly problematic in the safety critical real-time and embedded systems domain, where semantic ambiguities may produce conflicting results across different tools.

The goal of this chapter is to discuss the fundamental problems, methods, and techniques for specifying the semantics of DSMLs.

14.2   Domain-Specific Modeling Languages

Formal specification of DSMLs promises to extend the reach of DSML-based development techniques to ensure consistent analysis of designs, reuse of models between tools, and to increase the extent to which models can be constructed correctly during design. Numerous studies have shown the benefits of dealing with design flaws early in the development process [30]. As a first step, we discuss current techniques used for DSML specification, show examples for the different specification styles, and discuss the key concepts required for the formal specification of DSMLs.

14.2.1   DSML Specification: Informal and Formal

Current practice of specifying DSMLs covers a wide range of methods from formal to informal. Starting with the conceptualization of Harel and Rumpe [19], a DSML specification can be expressed as a 5-tuple $L =< A, C, S, M_S, M_C >$ consisting of abstract syntax (A), concrete syntax (C), syntactic mapping (M_C), semantic domain (S), and semantic mapping (M_S). The abstract syntax A defines the language concepts, their relationships, and well-formedness rules available in the language. The concrete syntax C defines the specific notations used to express models, which may be graphical, textual, or mixed. The syntactic mapping, $M_C : C \rightarrow A$, assigns syntactic constructs to elements in the abstract syntax. The DSML semantics are defined in two parts: a semantic domain S and a semantic mapping $M_S : A \rightarrow S$. The semantic domain S is usually defined in some formal, mathematical framework, in terms of which the meaning of the models is explained. The semantic mapping relates syntactic concepts to those of the semantic domain. First, we consider informal and widely used approaches.

- Least formally, the specification is implicit: natural language constructs and notations are chosen to represent concepts familiar to users. This approach is the simplest, but also most dangerous. Individual interpretations of language details are guaranteed to differ. Few would argue the adequacy of the fully informal approach for any sizable project or development team.
- A common technique used quite extensively in practice is the specification of the abstract syntax A and the $M_C : C \rightarrow A$ syntactic mapping using a model, called a metamodel. The modeling language used for metamodeling is frequently UML Class Diagrams and OCL [32,33] or some other variations of the metamodeling languages (for an overview, see [15]). While this approach represented a major step ahead and had opened up the possibility for developing metaprogrammable tool suites for model-based design [15,27], specification of precise semantics both for the metamodeling languages and for the DSMLs has remained an open problem.
- Specification of semantics for various DSMLs have been most frequently done by means of natural language (possibly interspersed with mathematical notations). At a minimum, writing down the semantic ideas and expressing their mathematical meaning reduces misunderstandings between individual developers, though characterization of completeness or consistency of the specification is impossible as languages grow in size and complexity.
- Informal semantics for modeling languages can also be defined implicitly at a lower level by creating code generators for models defined in the language. We loosely define a code generator as any model interpreter that provides useful machine-executable output. A code generator explicitly defines an executable representation of behavior (or other system aspects, such as data and configuration) for any model. At least four problems mar the practicality of this approach:

  1. Observing generated behavior for embedded systems may be difficult—especially when deployed in highly integrated or distributed configurations.

2. Changes made to the language often require significant and detailed changes to the code generators, which often contain the detailed description of the DSML semantics.

3. Code generators may be developed separately by different developers or teams. Problems arising from inconsistency or incompleteness of language interpretations have not been eliminated, only deferred to a later stage of development. This is a step backward from the goal to resolve inconsistencies early in the design process.

4. Source code templates are poor documentation, frequently tending toward incomprehensibility. Further, the unrestricted expressiveness of general-purpose programming languages often leads to undisciplined (but executable) specifications.

Clearly, there is a need for the explicit, formal specification of DSML semantics. A formal specification is not only precise, but also manipulable. As such, formal specifications remove ambiguity and enable automated analysis for many design issues at a cost of higher detail. Formal specification of semantics is particularly important in translating models between languages recognized by different tools. In the rest of the chapter we will focus on the formal and explicit specification of semantics of DSMLs.

14.2.2   Framework for Formal Semantics of DSMLs

The first step in developing a framework for the formal specification of DSMLs is a more precise definition of modeling languages. Following the discussion in [24], we define a "domain" as the set of all "structurally well-formed models." A *model* is a "point" in the domain. A "well-formed model" is a model that satisfies all the constraints imposed on its construction. The set of all well-formed models contains all the meaningful structures of a domain, that is, to say the set of well-formed models defines the structural semantics of a domain. Formally, a domain is

1. A set Υ of concepts, components, or primitives from which models are built
2. A set R_Υ of all possible model realizations
3. A set of constraints C over R_Υ

The model realizations are all the ways that models can be built from the available primitives. The set of well-formed models in a domain is the set of all models that satisfy the constraints. This set construction is written as

$$D(\Upsilon, C) = \{r \in R_\Upsilon | r \models C\} \tag{14.1}$$

where the notation $r \models C$ can be read as "r satisfies the constraints C." Domains frequently carry meaning beyond that of the structure. However, to express meaning in terms of behavior, there must exist a mapping from models in one domain to models in another domain with existing behavioral semantics. This mapping is called an *interpretation* $[\![]\!]$.

$$[\![]\!] : R_\Upsilon \mapsto R_{\Upsilon'} \tag{14.2}$$

A single domain may have many different interpretations, and these form a family of mappings $([\![]\!]_i)_{i \in I}$. For some model $r \in R_\Upsilon$, we denote the ith interpretation of r as $[\![r]\!]_i$. The interpretations together with the behavioral semantics of the target domain define the behavioral semantics of the domain. Notice this approach ties model transformation closely to the specification of semantics; in fact any framework that supports model transformations also supports specification of semantics. Based on these notions of domains and interpretations, a DSML L is defined as a 4-tuple comprised of its domain and a possibly empty set of interpretations.

$$L = \left(\Upsilon, R_\Upsilon, C, ([\![]\!]_i)_{i \in I} \right) \tag{14.3}$$

Every domain has at least one interpretation, which is the *structural interpretation*, which defines the *structural semantics*. DSMLs may have behavioral interpretations and their specification provides the *behavioral semantics*. In the next two sections, we discuss methods for defining structural and behavioral semantics of modeling languages.

14.3 Specification of Structural Semantics of DSMLs

The interpretation $[\![]\!]_{struc}$ maps model realizations of a source domain to the Boolean domain where model r of domain R_Υ is mapped to {*true*} upon the satisfaction of Equation 14.1 and {*false*} otherwise. The structures and interpretations of domains can be extended to describe the process of defining DSMLs via *metamodeling*. A *metamodel* is expressed in a DSML called the *metamodeling language*. A metamodeling language L_{meta} is a DSML with a special interpretation $[\![]\!]_{meta}$, the *metamodeling semantics*, which maps models to *domains*:

$$L_{meta} = \left(\Upsilon_{meta}, R\Upsilon_{meta}, C_{meta}, ([\![]\!]_{struc}, [\![]\!]_{meta}) \right) \tag{14.4}$$

The interpretation $[\![]\!]_{struc}$ is the same structural interpretation that indicates whether a metamodel r is a well-formed model. If r is well formed, then $[\![r]\!]_{meta}$ maps r to a new domain. Note that all other interpretations map models in a domain to models of another domain. To make a mapping from models to domain, as $[\![r]\!]_{meta}$ does, a *domain of domains* must be created. This is beyond the scope of this chapter. See [23] for a more detailed explanation of metamodeling semantics.

14.3.1   Structural Semantics in DSMLs

DSMLs require a notation (syntax) for expressing models, as is the case for traditional programming languages. However, the use of syntax in DSMLs goes beyond notational convenience. Syntax is used to characterize models with unwanted behavioral properties. It also serves as the basis for model transformations between DSMLs. With these applications in mind, DSML syntax is better understood as a *constraint system* that identifies behaviorally meaningful models. We call this constraint system the *structural semantics* of a DSML. Unlike traditional programming languages, the structural semantics of DSMLs may be more general than simple, regular, or context-free languages.

The various uses of DSMLs demand a number of key operations on structural semantics:

1. **Conformance testing**: Test if a model satisfies the structural semantics (constraints) of a DSML. This is analogous to testing if some program text conforms to the syntax of a programming language.
2. **Non-emptiness checking**: Check if there exists any model that satisfies the structural semantics. This amounts to finding contradictions in the definitions of structural semantics.
3. **DSML composing**: Given two DSMLs with structural semantics X and Y, a composition operator $\oplus$, and a property P. Create a composite DSML with structural semantics $(X \oplus Y)$ such that the composite constraint system satisfies property P. A formal modeling framework may provide many different composition operators as building blocks for creating composite DSMLs.
4. **Model finding**: Given a DSML X and property P, automatically construct a set of models S, each of which satisfies the property P and conforms to the structural semantics of X. This basic operation can be used for design space exploration and to calculate platform mappings.
5. **Transforming**: Given DSMLs X and Y and a model transformation T from X to Y, find the constraint system $T(X)$ such that a model m' satisfies $T(X, Y)$ if and only if there exists a model m that satisfies the structural semantics X and $m' = T(m)$ satisfies Y. In other words, m' is the result of transforming m with T; both m and m' are well formed.

These requirements necessitate formalisms that uniformly address structural semantics and model transformations. For example, we must be able to reason about the effects of model transformations on structural semantics. However, model transformations are not constraint systems, but (semi-)operational rewriting procedures. Thus, a purely algebraic formalism is not an ideal candidate. Additionally, the non-emptiness checking and model finding operations suggest formalisms for which *finite model finding* algorithms are known.

In this chapter, we describe one formalization of structural semantics based on structured *logic programming* (LP). LP allows formal constraint systems to be specified in a programmatic style [1,13]. Logic programs have the advantage of being mathematically precise and executable. Therefore model transformations can be captured in the same framework. Our specification language, called FORMULA (**For**mal **M**odeling **U**sing **L**ogic **A**nalysis), provides structuring and composition operators for formalizing and composing structural semantics and model transformations. At its core is a well-studied class of LP called *non-recursive and stratified*.

14.3.2   Formal Foundations

14.3.2.1   Signatures and Terms

A function symbol, for example, $f(\cdot)$, is a symbol denoting a unary function over a universe U. We say that f stands for an *uninterpreted function* if f satisfies no additional equalities other than $\forall x \in U,\ f(x) = f(x)$. Let Σ be an infinite alphabet of constants, then a *term* is either a constant or an application of some uninterpreted function to a term. For example, $\{1, f(2), f(f(3))\}$ are all terms assuming $\{1, 2, 3\} \subseteq \Sigma$. Henceforth, our function symbols will be n-ary to capture relations and other constructs. Constructing terms generalizes for arbitrary arity.

Uninterpreted functions form a flexible mechanism for capturing sets, relations, and relations over relations without assigning any deeper interpretations (behavioral semantics) to the syntax. A *finite signature* Υ is a finite set of n-ary function symbols. The *term algebra* $\mathcal{T}_\Upsilon(\Sigma)$ is an algebra where all symbols of Υ stand for uninterpreted functions. The universe of the term algebra is inductively defined as the set of all terms that can be constructed from Σ and Υ. It is standard to let $\mathcal{T}_\Upsilon(\Sigma)$ either denote the term algebra or its universe.

A structural semantics provides a term algebra whose function symbols characterize the key sets and relations through uninterpreted functions. A *syntactic instance* of some structural semantics is a finite set of terms over its term algebra $\mathcal{T}_\Upsilon(\Sigma)$. The set of all syntactic instances is then the power set of its term algebra: $\mathcal{P}(\mathcal{T}_\Upsilon(\Sigma))$.

14.3.2.2   Terms with Types

The power set of terms contains many unintended instances of the DSML syntax. For example, we can encode directed graphs with the two function symbols $\{vertex(\cdot), edge(\cdot, \cdot)\}$. However, the syntactic instance

```
edge(edge(1,2),edge(3,4))
```

is not a meaningful directed graph, because the edge function was not intended to be applied to edges. Erroneous terms can be eliminated by *typing* the arguments of uninterpreted functions. In FORMULA we write

```
vertex : (Integer).     edge : (vertex, vertex).
```

These statements declare a unary function `vertex` and a binary function `edge`. The argument to `vertex` must be an integer, while the arguments to `edge` must be `vertex` terms. These typed functions are undefined when applied to badly typed values, otherwise they behave exactly like uninterpreted functions.

This enrichment of the term algebra semantics with types leads naturally to an *ordersorted*-type system. We formalize this type system now. An *order-sorted alphabet* $\Sigma_\subseteq$ is a structure:

$$\Sigma_\subseteq = \langle I, \preceq, (\Sigma_i)_{i \in I} \rangle \tag{14.5}$$

The set I, called the *index set*, is a set of sort names (alphabet names). Associated with each sort name $i \in I$ is a set of constants Σ_i called the *carrier* of i. An order-sorted alphabet has the following properties:

$$\Sigma = \bigcup_{i \in I} \Sigma_i, \quad (i \preceq j) \Leftrightarrow (\Sigma_i \subseteq \Sigma_j) \tag{14.6}$$

In other words, Σ is the union of smaller alphabets and alphabets are ordered by set inclusion; the sub-typing relation $\preceq$ is set inclusion. A *type* τ is a term constructed from function symbols and elements of I or the special top-type $\top$. Each type τ identifies a subset $[\![\tau]\!] \subseteq T_\Upsilon(\Sigma)$ according to

1. The top type is the entire term algebra:

$$[\![\top]\!] = T_\Upsilon(\Sigma) \tag{14.7}$$

2. A sort name $\tau \in I$ is just the carrier set Σ_τ:

$$\forall \tau \in I, \quad [\![\tau]\!] = \Sigma_\tau \tag{14.8}$$

3. Otherwise $\tau = f(\tau_1, \tau_2, \ldots, \tau_n)$ where f is an n-ary function symbol:

$$[\![\tau]\!] = \left\{ v \in T_\Upsilon(\Sigma) \;\middle|\; \begin{array}{c} v = f(v_1, v_2, \ldots, v_n) \wedge \\ \bigwedge_{1 \leq j \leq n} v_j \in [\![\tau_j]\!] \end{array} \right\} \tag{14.9}$$

The sub-typing relation $\preceq$ is extended to arbitrary types:

$$\forall \tau_p, \tau_q \quad (\tau_p \preceq \tau_q) \Leftrightarrow ([\![\tau_p]\!] \subseteq [\![\tau_q]\!]) \tag{14.10}$$

14.3.2.3 *Expressive Constraints with Logic Programming*

Structural semantics often contain complex conformance rules; these rules cannot be captured by simple-type systems. One common solution to this problem is to provide an additional constraint language for expressing syntactic rules such as the Object Constraint Language (OCL) [33]. Unlike other approaches, we choose LP to represent syntactic constraints because

1. LP extends our term algebra semantics while supporting declarative rules.
2. The fragment of LP supported by FORMULA is equivalent to full first-order logic over term algebras thereby providing expressiveness [13].
3. Unlike purely algebraic specifications, there is a clear execution semantics for logic programs making it possible to specify model transformations in the same framework.
4. Many analysis techniques are known for logic programs; we have adapted these to analyze FORMULA specifications [25].

FORMULA supports a class of logic programs with the following properties: (1) expressions may contain uninterpreted function symbols, (2) the semantics for negation is *negation as finite failure*, and (3) all logic programs must be *non-recursive* and *stratified*. We summarize this class now.

Definitions. Let V be a countably infinite alphabet of *variables* disjoint from basic constants: $V \cap \Sigma = \emptyset$. Let the term algebra $\mathcal{T}_v$ be an extension of a term algebra with these variables: $\mathcal{T}_\Upsilon(\Sigma \cup V)$. For simplification, we write $\mathcal{T}_g$ for $\mathcal{T}_\Upsilon(\Sigma)$. A term t is called a *ground term* if it does not contain any variables; $\mathcal{T}_g$ is the set of all ground terms. A *substitution* ϕ replaces variables with ground terms. Formally, ϕ is a homomorphism from terms (with variables) to ground terms that fixes constants. We write $\phi(t)$ for the ground term formed by replacing every variable x in t with $\phi(x)$. Two terms s and t *unify* if there exists a substitution ϕ such that $\phi(s) = \phi(t)$.

Expressions. Logic programs are built from expressions of the form:

$$h \leftarrow t_1, t_2, \ldots, t_n, \quad \neg s_1, \neg s_2, \ldots, \neg s_m.$$

where $h \in \mathcal{T}_v$ is a term called the *head*. The sets $\{t_1, \ldots, t_n\} \subseteq \mathcal{T}_v$ and $\{s_1, \ldots, s_m\} \subseteq \mathcal{T}_v$ are sets of terms collectively called the *body*. Each t_i is called a *positive term* and each s_j is called a *negative term*. In the kth expression an implicit relation symbol R_k surrounds each body term; an implicit relation symbol R'_k surrounds the head term h.

$$R'_k(h) \leftarrow R_k(t_1), R_k(t_2), \ldots, R_k(t_n), \quad \neg R_k(s_1), \neg R_k(s_2), \ldots, \neg R_k(s_m).$$

Intuitively, this expression means the following: If there exists a substitution ϕ so that each $\phi(t_i)$ is in relation R_k and for each s_j there is no substitution ϕ' that places $\phi'(s_j)$ in R_k, then add $\phi(h)$ to the relation R'_k [1].

The semantics of LP languages vary on how this intuition is formalized. The formalization must take into account the generality of allowed expressions and the mechanism by which the implicit relations are calculated. The fragment we utilize is a well-behaved fragment called non-recursive and stratified LP with negation as finite failure. Logic programs of this fragment always terminate and have expressive power equivalent to first-order logic.

Semantics. Let $\prec$ be a relation on expressions. Expressions e_i and e_j are related $(e_i \prec e_j)$ if the head h of e_i unifies with some term in the body of expression e_j, regardless of whether the body term is positive or negative. A finite collection of expressions $(e_i)_{i \in E}$ is non-recursive and stratified if $\prec$ is an acyclic relation.

Let $o : E \to \mathbb{Z}^+ \cup \{0\}$ be an ordering of non-recursive and stratified expressions that respects $\prec$:

$$\forall i, j, \in E, \ (e_i \prec e_j) \Rightarrow o(i) < o(j). \tag{14.11}$$

Using this ordering, the kth expression tests for the presence and absence of body terms in a relation $R_{o(k)}$. Whenever these tests succeed for some substitution ϕ, then the substituted head term $\phi(h)$ is added to the relation $R_{o(k)+1}$.

$$R_{o(k)+1}(h) \leftarrow R_{o(k)}(t_1), \ldots, \neg R_{o(k)}(s_1), \ldots \tag{14.12}$$

This rule is used in conjunction with the general rule: Whatever can be found in relation R_i can also be found in relation R_{i+1}.

$$\forall i \geq 0, (t \in R_i) \Rightarrow (t \in R_{i+1}). \tag{14.13}$$

Additionally, LP uses the *closed world assumption*: $t \in R_i$ if and only if it is in R_0 or it is placed in R_i by the application of rules (14.12) or (14.13). The *input* to a logic program is the initial relation R_0. The program executes by working from smallest-to-largest expressions, as ordered by o, building the contents of the relations along the way. Note that for non-recursive and stratified programs the choice of o does not affect the results of the logic program.

Negation as Failure. *Negation as failure* (NAF) allows an expression e_k to test for the absence of some terms in the relation $R_{o(k)}$. Developing a semantics for NAF in arbitrary logic programs has been one of the biggest challenges for the LP community. Fortunately NAF is well behaved in our fragment. The only question is how to interpret variables appearing in negative terms. For example:

$$f(x) \leftarrow g(x); \neg h(x, y). \tag{14.14}$$

We use the interpretation consistent with the SLDNF-resolution (Selective Linear Definite clause resolution with NAF) procedure found in standard PROLOG: The variable y is effectively a wild card, so the body succeeds if there exists a substitution ϕ such that $\phi(g(x)) \in R_{o(k)}$ and for all other substitutions ϕ', $\phi'(\phi(h(x, y))) \notin R_{o(k)}$. Substitutions fix constants, so this is equivalent to the condition that $\exists \phi, \forall \phi', f(\phi(x)) \in R_{o(k)} \wedge h(\phi(x), \phi'(y)) \notin R_{o(k)}$. More generally, let P be the set of all positive terms and N be the set of all negative terms in an expression. We write $\phi(T)$ for the application of a substitution to each term $t \in T$. The body of an expression succeeds if

$$\exists \phi, \forall \phi', \ \phi(P) \subseteq R_{o(k)} \wedge \phi'(\phi(N)) \cap R_{o(k)} = \emptyset. \tag{14.15}$$

We are interested in finite relations R_k, so we require that the variables in the head term h appear in some positive term $t \in P$.

Queries. A query is just an expression that does not add new terms to any relation. In FORMULA each query q has a name, which is a boolean variable that is true whenever the body of the query is satisfied, otherwise it is false:

$$qname \ :? \ t_1, t_2, \ldots, t_n, \quad \neg s_1, \neg s_2, \ldots, \neg s_n. \tag{14.16}$$

Queries are identified by the special operator ":?". Since queries do not modify any relations, it is never the case that $q_i \prec e_j$. Otherwise, queries are ordered like clauses for the purpose of execution. For the sake of convenience, queries can be composed with standard boolean operators into new queries, but this is only syntactic sugar.

Domain Constraints. We use LP to capture the complex rules of DSML structural semantics. This is done by writing a special query called `conforms`, which evaluates to true exactly when a syntactic instance satisfies all the rules. The process for testing whether a syntactic instance X conforms to the syntax is

1. Set $R_0 = X$
2. Run the logic program
3. Check if `conforms` evaluates to true

14.3.3 An Introduction to Domains and Models

LP provides the basis for capturing structural semantics. However, in its raw form it lacks the concepts of encapsulation and composition required by a formal modeling framework. FORMULA also provides these key encapsulation and composition concepts. In the following sections, we present these key concepts through examples.

FORMULA specifications are separated into logical units called *domains*. A domain encapsulates a set of function symbols, LP expressions, and queries all used to describe a structural semantics. Declarations within a domain are scoped to that domain and are not visible to other parts of the specification. Structuring the specification into domains is not optional. A *model* is a a finite set of ground terms built from the uninterpreted functions of some domain. A model is well-formed if its set of terms satisfies the constraints of the domain used to construct it. Intuitively, a domain represents a family of models by providing functions and constraints.

In this introduction we will develop a FORMULA specification for the structural semantics of a finite state machine (FSM) DSML. The FSM abstraction is a well-studied formalism that abstracts away from the idiosyncrasies of programming languages. Using FSMs we can view a computation as a progression through a finite sequence of *states*. This progression is triggered by external stimuli called *events*. The FSM abstraction is still important to many engineering communities because of the verification and optimization techniques known for FSMs.

Figure 14.2 shows a sample FSM with two states S_1 and S_2. The small black circle indicates that S_1 is the initial (starting) state for the FSM. The arrow from S_1 to S_2 is a *transition* that is taken if the FSM is in state S_1 and receives event e. The transition from S_2 to S_1 is triggered by the event f.

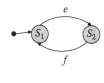

FIGURE 14.2
Example of a FSM.

We begin by (partially) specifying the non-deterministic version of the FSM abstraction, called the *non-deterministic finite automaton* (NFA) abstraction. NFAs may have states with many transitions triggered by the same event. In this case, the NFA non-deterministically chooses which transition to take. Figure 14.3 shows the partial FORMULA specification. Line 2 declares a domain called NFA, which acts as a container for the NFA abstraction. The building blocks of an NFA are the functions declared in Lines 4–7. Line 4 defines the Event function, which takes one argument providing the name of the event. The form of data used to represent this name is irrelevant, so we give it the type Any. The argument for the State function is also the name of the state. Lines 6 and 7 give functions for transitions and initial states. Finally, Line 11 closes the NFA domain; the code appearing after Line 11 does not belong to the NFA specification.

```
     /// Non-deterministic Finite Automaton Domain
2.   domain NFA {
       /// Constructors for NFAs
4.     Event      : ( name : Any ).
5.     State      : ( name : Any ).
6.     Transition : ( current : State, trigger : Event, next : State ).
7.     Initial    : ( state : State ).
       /// More to come later
       ...
11.  }

13.  model ExampleNFA1 : NFA {
14.    /// ***** Events *****
15.    Event("e"),
16.    Event("f"),
17.    ///***** States *****
18.    State("S1"),
19.    State("S2"),
20.    ///***** Initial States *****
21.    Initial(State("S1")),
22.    /// ***** Initial States *****
23.    Transition(State("S1"), Event("e"), State("S2")),
24.    Transition(State("S2"), Event("f"), State("S1"))
     }
```

FIGURE 14.3
Specifying the non-deterministic finite automaton (NFA) abstraction.

A model is simply a set of ground terms built using the functions of a domain. The syntax for declaring a model is

$$\textbf{model } ModelName \; : \; DomainName \; \{ \dots \} \qquad (14.17)$$

Line 13 declares a model equivalent to the FSM in Figure 14.2. The contents of the model is a comma-delimited list of NFA functions applied to some basic values. It is illegal to use any other functions than those of the NFA domain. Also, it is illegal to use basic values that do not appear inside of a function.

14.3.3.1 The Type of a Domain

A domain D also has an associated type τ_D, which is the set of all finite models (term sets) that belong to the domain D. There is a minimum requirement for a model to belong to a domain: Every term in the model must be built from functions of the domain. Typically, a domain has more complex constraints limiting which term sets are legal models of the domain. We call these rules *domain constraints*, and discuss them shortly. Right now our NFA domain only imposes the minimum domain constraint.

The notation:

$$\textbf{model } \texttt{ModelName} \; : \; \texttt{DomainName} \, \{ \dots \}$$

is making the assertion that the data inside of the model belongs to the domain. FORMULA will type-check the model against the definition of the domain to ensure the accuracy of this assertion. From this perspective, the order-sorted type system also extends to domains. Two domains are subtype related if the sets of conforming models are subset related.

14.3.4 Examining the Contents of Models

Models contain important information that must be extracted. This is accomplished by adding rules to a domain that locate key information within a model. For example, we might want to know which states are the initial states of a FSM. Such questions are called *queries*; this query finds the initial states of an FSM:

$$hasInit \; :? \; Initial(x). \qquad (14.18)$$

A query is evaluated over a model; it is satisfied if there exists some assignment of variables to values such that each item in the body exists in R_∞. For example, if this query is evaluated against the model ExampleNFA1 in Figure 14.3 and variable x is assigned as follows:

$$x = State(\texttt{"S1"}) \qquad (14.19)$$

then the query is satisfied because the term $Initial(State(\texttt{"S1"}))$ exists in the model. FORMULA will find all assignments of variables satisfying the query, or report that no such assignment exists.

On the other hand, the query

$$hasInit2 :? \ Initial(State(\text{``S2''})). \tag{14.20}$$

fails, because $S2$ is not an initial state. Queries can contain more than one pattern in the body; each pattern must be satisfied in order for the query to be satisfied:

$$hasTrans :? \ Initial(x), \ Transition(x, e, y). \tag{14.21}$$

This query is satisfied whenever there exists a transition from an initial state x to another state y triggered by event e. Evaluating this query on ExampleNFA1 yields success for the following assignment of variables:

$$x = State(\text{``S1''}) \tag{14.22}$$
$$y = State(\text{``S2''}) \tag{14.23}$$
$$e = Event(\text{``e''}) \tag{14.24}$$

Note that the order in which items occur in the body does not affect the results of a query.

FORMULA also allows the arguments of functions to be labeled. Terms built using functions with labeled arguments can be accessed like a record structure. For example, query 14.21 can be rewritten to:

$$hasInit2 :? \ x \ \textbf{is} \ Initial, \ t \ \textbf{is} \ Transition, \ t.current = x.state. \tag{14.25}$$

In order to access data through the argument labels of function S, a variable x must first be "cast" to type S using the `is` operator. Note that this syntax is not an extension of the language.

14.3.4.1 Examples of Negation as Failure

Queries with only positive body terms cannot express all interesting queries. For example, we might want to know if there are two states x and y such that there are *no transitions* from x to y. Expressing this query requires the use of NAF. Intuitively, a negative term forces the absence of certain terms in the input model. Here is an example:

$$noTrans :? \ x \ \textbf{is} \ State, \ y \ \textbf{is} \ State, \ e \ \textbf{is} \ Event, \ \textbf{fail} \ Transition(x, e, y). \tag{14.26}$$

The first three terms find two states and an event. The last terms uses the keyword **fail** to indicate NAF:

$$\textbf{fail} \ Transition(x, e, y)$$

This part of the query succeeds if the logic program does not include a Transition from x to y triggered by event e in R_∞. It is important to understand that variables occurring in positive terms are bound first. The values that satisfy the positive terms constrain the negative terms. For example, when evaluating the query on ExampleNFA1, the positive part of the query is satisfied by

$$x = State(\text{``S1''})$$
$$y = State(\text{``S2''})$$
$$e = Event(\text{``e''})$$

These values constrain the negative pattern, resulting in a test for

$$Transition(State(\text{``S1''}), Event(\text{``e''}), State(\text{``S2''}))$$

This test determines that there is a transition from S_1 to S_2 so the negative pattern fails and the overall query fails for this assignment of variables. However, there are more assignments of variables to try before deciding that the query completely fails. For example, here is another assignment:

$$x = State(\text{``S1''})$$
$$e = Event(\text{``f''}),$$
$$y = State(\text{``S2''})$$

Now, the event is f. In this case, there is a test for

$$Transition(State(\text{``S1''}), Event(\text{``f''}), State(\text{``S2''}))$$

Indeed, the model does not contain this transition, so the query succeeds for this assignment of variables.

Figures 14.4 through 14.6 show some examples of NAF. The dashed lines in the figures represent tests for the absence of terms. Note that every occurrence of an underscore (_) stands for a uniquely named variable that does not appear anywhere else in the query. Each example shows several equivalent ways of writing the query.

```
/// Queries testing if there is the absence of
/// a transition between states x and y triggered by e.
4. ?: State(x1,x2), State(y1,y2), Event(e),
5.    fail Transition(State(x1,x2), Event(e), State(y1,y2)).

7. ?: x is State, y is State, e is Event, fail Transition(x,e,y).

9. ?: x is State, y is State, e is Event, fail t is Transition,
10.   t.current = x, t.event = e, t.next = y.
```

FIGURE 14.4
Find states x and y without a transition triggered by event e.

```
     /// Queries testing if there are states x and y
     /// such that there are no transitions from x to y.

4. ?: State(x1,x2), State(y1,y2),
5.     fail Transition(State(x1,x2), e, State(y1,y2)).

7. ?: x is State, y is State, fail Transition(x,_,y).

9. ?: x is State, y is State, fail t is Transition,
10.    t.current = x, t.next = y.
```

FIGURE 14.5

Find states x and y without any transitions from x to y.

```
     /// Queries testing if there are blocking states, i.e.
     /// states with no outgoing transitions.

4. ?: State(x1,x2), fail Transition(State(x1,x2), e, y).

6. ?: x is State, fail Transition(x,_,_).

8. ?: x is State, fail t is Transition, t.current = x.
```

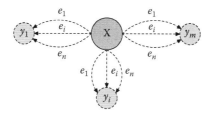

FIGURE 14.6

Find blocking states x, that is, states without any outgoing transitions.

14.3.4.2 Boolean Composition of Queries

A query is either satisfied (i.e., `true`) or not satisfied (i.e., `false`) when evaluated against a model. Therefore, queries return boolean values, so new queries can be built from old ones using standard boolean operators: & (and), | (or), and ! (not).

$$queryC \; :? \; queryA \mid !queryB. \tag{14.27}$$

There are some strict rules that apply when composing queries with boolean operators:

1. A query must be given a name before it can be used in a composition.
2. A query cannot be defined in terms of itself, as in: *queryC :? queryC & queryB.*

3. The body of a query is either a list of terms or a boolean composition of queries, but never both. This is illegal: *queryC* :? *State(x)*, *queryB*.

Note that a boolean composition of queries does not allow information to be passed between subqueries. Each subquery is evaluated independently from the others, and then the results are aggregated.

14.3.5   Adding Domain Constraints

Domains contains rules for distinguishing the data sets that are legal models from those that are not. Earlier, we explained that there is a minimum requirement for a model M to belong to a domain D: Each term in M must be built from functions of D. Typically that requirement does not capture all the rules of a particular abstraction level. For example, the rule

Every NFA must have at least one initial state

is not captured by this minimum requirement.

FORMULA supports additional domain constraints through a special query called conforms. If this query is defined, then a data set must satisfy the query to be a valid model of the domain. For example:

$$\textbf{conforms} \ :? \ Initial(x). \tag{14.28}$$

captures the rule that every NFA must have at least one start state. Figure 14.7 shows the domain constraints needed for the NFA abstraction. Lines 10 and 11 defines queries that are satisfied whenever there is a dangling transition to a state not in the model. The query at Line 12 finds transition that are triggered by undefined events and the query at Line 14 identifies undefined initial states. Line 16 declares the hasInitial query that guarantees the presence of an initial state. Finally, Line 18 defines the **conforms** query as a boolean combination of these queries.

```
...
     /// Domain constraints for NFA models
10. missingSrc  :? t is Transition, fail s is State, s = t.current.
11. missingDst  :? t is Transition, fail s is State, s = t.next.
12. missingTrig :? t is Transition, fail e is Event, e = t.event.

14. missingInit :? x is Initial, fail s is State, s = x.state.

16. hasInitial  :? x is Initial.

18. conforms :? hasInitial & !(missingSrc | missingDst |
                          missingTrig | missingInit).
...
```

FIGURE 14.7
Domain constraints for NFAs; continued from Figure 14.3.

14.3.5.1 Derived Functions and Logic Programs

A query reduces some property of a model to a boolean value: Either the property holds (the query is satisfied) or it does not hold. Sometimes this reduction loses information that should persist. For example, suppose we want to know which states can be reached after one, two, and three transitions of a FSM, as shown in Figure 14.8. This can be done with the queries shown in Figure 14.9. Notice that the query findTwo, which finds all states after two steps, must first discover all the states one step away. Similarly, the findThree query must find the states reachable in one and two steps, before finding the states reachable in three steps. The Nth find query does all the work of the $N-1$

FIGURE 14.8
Calculating reachable states from initial state 0.

previous queries, which is wasteful and makes the queries unnecessarily verbose.

This problem can be solved by temporarily storing the results of queries into data structures. These temporary results are created using *derived functions*. **A derived function is a function beginning with a lowercase letter**. Derived functions are only used to temporarily remember some intermediate results and can never appear in the data elements of an input model.* For example:

$$reach1 : (\, state : State\,). \qquad\qquad (14.29)$$

declares a derived function for remembering those states reachable from an initial state after one step. Temporary results are created by writing the following *expression*:

$$reach1(y) :\!-x\ \textbf{is}\ Initial,\ y\ \textbf{is}\ State,\ t\ \textbf{is}\ Transition,$$
$$t.current = x.state,\ t.next = y.$$

```
    /// Queries searching for reachable states
3. findOne ?: x is Initial, y is State,
4.              t is Transition,
5.              t.current = x.state, t.next = y.

7. findtwo ?: x is Initial, y is State,
8.              t is Transition, s is Transition,
9.              t.current = x.state, t.next = s.current, s.next = y.

11. findThree ?: x is Initial, y is State,
12.             t is Transition, s is Transition, u is Transition,
13.             t.current = x.state, t.next = s.current,
14.             s.next = u.current, u.next = y.
```

FIGURE 14.9
Queries for locating reachable states after one, two, and three steps.

---

*The upper/lowercase syntax for functions and derived functions is similar to the notation used in BNF grammars to distinguish terminal from non-terminal elements.

Note that this expression has the semantics explained in Section 14.3.2. For example, this rule adds the *reach1* term to all relations $R_{k>0}$ for some k, when evaluated against the example FSM:

$$reach1(State(\texttt{"S2"}, \texttt{""}))$$

The information from *reach1* can be used to calculate the states reachable in two steps. First, we introduce another derived function:

$$reach2 : (state : State). \tag{14.30}$$

and then the expression:

$$reach2(y) :\!-x \textbf{ is } reach1, y \textbf{ is } State, t \textbf{ is } Transition,$$
$$t.current = x.state, t.next = y.$$

Now *reach1* is used as a term in the body of this expression. Note that the ordering of expressions ensures that all the *reach1* terms were calculated before evaluating this expression.

The following set of expressions are recursive and because our formalism requires non-recursiveness, they cannot be expressed in it.

$$reachN(y) :\!- x \textbf{ is } Initial, y \textbf{ is } State, t \textbf{ is } Transition,$$
$$t.current = x.state, t.next = y. \tag{14.31}$$

$$reachN(y) :\!- x \textbf{ is } reachN, y \textbf{ is } State, t \textbf{ is } Transition,$$
$$t.current = x.state, t.next = y. \tag{14.32}$$

The rules above calculate all of the reachable states of a FSM. Rule 14.31 creates a *reachN* instance for each state immediately reachable from an initial state. Rule 14.32 creates a *reachN* instance for each state that is reachable in one step from a *reachN* instance. Rule 14.32 depends on Rule 14.31 and on itself, thereby creating a dependency cycle.

This example shows that recursive logic programs are useful, but they may also result in an infinite loop of data creation. Furthermore, in general it is impossible to determine whether an infinite loop truly exists or not. In order to support analysis of the specifications, FORMULA disallows all recursive logic programs. Note that it is possible to write logic programs that appear recursive, but can be unrolled into non-recursive logic programs. FORMULA supports this form of recursion, but it is beyond the scope of this chapter. The non-recursive restriction also means there is always a good evaluation order for expressions:

1. Evaluate the expression that has not yet been evaluated and depends on no other expressions.
2. For an expression e_i that has just been evaluated, remove all dependencies on e_i from other expressions e_j for $j > i$.
3. Repeat (1) until no more clauses are left to evaluate.

We have presented logic expressions as a convenient way to reuse work, which might give the (wrong) impression that they only serve this purpose. Some queries *cannot* be expressed without using logic expressions. Here is an example:

> *Is there a state x for which no transition goes to a blocking in state?*

This query must locate a state x and test that every outgoing transition does not end on a state with no outgoing transitions. Neither a single query nor a boolean composition of queries can keep track of all this information. Instead, an expression is needed that calculates those states that reach a blocking state in one step:

$$blocksOne(x, y) \text{ :---} x \textbf{ is } State, y \textbf{ is } State, Transition(x, \_\_, y),$$
$$\textbf{fail } Transition(y, \_\_, \_\_).$$

The query uses this intermediate data to find states that do not go directly to blocking states:

$$\text{:? } x \textbf{ is } State, \textbf{fail } blocksOne(x, \_\_).$$

14.3.6 Domains and Compositions of Domains

Bringing these concepts together, a domain D is simply a triple:

$$D = \langle \Upsilon_P, \Upsilon_R, E \rangle \tag{14.33}$$

where
  Υ_P is a signature of primitive function symbols
  Υ_R is a signature of derived function symbols
  E is a nonrecursive stratified logic program defining the `conforms` query

The set *models(D)* is the set of all models that satisfy the `conforms` query:

$$models(D) = \left\{ X \in \mathcal{P}(\mathcal{T}_{\Upsilon_P}(\Sigma)) \,\middle|\, X \text{ satisfies conforms} \right\}, X \text{ is finite.} \tag{14.34}$$

This simple semantics admits powerful composition operators for building new domains with known properties. Table 14.1 lists these composition operators.

Includes.   The `includes` operator is used to import the declarations of one domain into another domain:

```
domain D' includes D { ... }.
```

The resulting domain D' has

$$\Upsilon'_P \supseteq \Upsilon_P, \ \Upsilon'_R \supseteq \Upsilon_R, \ E' \supseteq E[conforms/D.conforms] \tag{14.35}$$

TABLE 14.1

Basic Set of Composition Operators

Operator	Usage	Description
includes, restricts, extends operators	`D' includes D,` `D' restricts D,` `D' extends D,`	Imports the declarations of D into D' while renaming the `conforms` query of D to D.`conforms`.
renaming operator "as"	`D as X`	Produces a new domain from D by replacing every occurrence of a function symbol $f(\ldots)$ with $X.f(\ldots)$ and every query name q with $X.q$.
pseudo-product operator "∗"	$D_1 * D_2$	Produces a new domain D' by combining the specifications of D_1 and D_2, and then adding the query (`conforms :? ` D_1`.conforms &.` D_2`.conforms`).
pseudo-coproduct operator "+"	$D_1 + D_2$	Produces a new domain D' by combining the specifications of D_1 and D_2, and then adding the query (`conforms :? ` D_1`.conforms XOR .` D_2`.conforms`).

The notation $E[x_1/x_1', \ldots, x_n/x_n']$ denotes the expressions formed by replacing every occurrence of x_i in E with x_i'. Thus, domain D' has direct access to the declarations in D, but does not necessarily utilize the conformance rules of D because it is renamed to D.*conforms*. The `includes` operation is defined if the signatures of D' do not contain contradictory function symbol definitions, and the expressions E are non-recursive and stratified.

There are several variants of includes that make stronger statements about D'. The `restricts` keyword requires that no new primitives are introduced in D'. Additionally, D.*conforms* is implicitly conjuncted onto the *conforms* of D'. The `restricts` operator enforces that

$$models(D') \subseteq models(D). \qquad (14.36)$$

The `extends` variant implicitly disjuncts D.*conforms* onto the *conforms* of D', therefore

$$models(D') \supseteq models(D). \qquad (14.37)$$

Renaming.   The *renaming operator* "as" gives new names to the function symbols and queries in a domain. The expression

$$(D \text{ as } X)$$

produces a domain D' with the same signatures and expressions as D, except that every occurrence of a function symbol and query name is prepended by "X."

Pseudo-product.   The *pseudo-product operator* "∗" is a precursor for building the *categorical product* of domains. The expression

$$(D_1 \star D_2)$$

defines a domain D' where

$$
\begin{aligned}
\Upsilon'_P &= \Upsilon^1_P \cup \Upsilon^2_P,\ \Upsilon'_R = \Upsilon^1_R \cup \Upsilon^2_R, \\
E' &= E^1[conforms/D_1.conforms] \cup \\
&\quad E^2[conforms/D_2.conforms] \cup \\
&\quad \{conforms \ :? \ D_1.conforms \ \& \ D_2.conforms.\}
\end{aligned}
\tag{14.38}
$$

The pseudo-product has the property that if D_1 and D_2 have disjoint signatures and query names, then

$$models(D') \cong models(D_1) \times models(D_2). \tag{14.39}$$

This is called the *categorical product*; it means that every model $X \in models(D')$ can be uniquely partitioned into two subsets X_1 and X_2 so that $X_i \in models(D_i)$. This construct is important, because it combines two domains into a larger one while guaranteeing no nontrivial interactions.

Pseudo-coproduct. The *pseudo-coproduct operator* "+" is a precursor for building the *categorical coproduct* of two domains. The expression

$$(D_1 + D_2)$$

defines a domain D' where

$$
\begin{aligned}
\Upsilon'_P &= \Upsilon^1_P \cup \Upsilon^2_P,\ \Upsilon'_R = \Upsilon^1_R \cup \Upsilon^2_R, \\
E' &= E^1[conforms/D_1.conforms] \cup \\
&\quad E^2[conforms/D_2.conforms] \cup \\
&\quad \{conforms \ :? \ D_1.conforms \ \text{XOR} \ D_2.conforms.\}
\end{aligned}
\tag{14.40}
$$

Let the $models(D_i)$ be the set of all finite syntactic instances that satisfy the `conforms` query of domain D_i. The pseudo-product has the property that if D_1 and D_2 have disjoint signatures and query names, then

$$models(D') \cong models(D_1) \uplus models(D_2). \tag{14.41}$$

This is called the *categorical coproduct*; it means that every model $X \in models(D')$ is either in $models(D_1)$ or $models(D_2)$, but never both. Again, this construct is important, because it combines two domains into a larger one while guaranteeing no nontrivial interactions.

14.3.6.1 Properties of Compositions

Regardless of the approach, complex modeling processes are often plagued by the nonlocal effects of composition. In our framework, compositions may yield unexpected results because of interactions between declarations and logic programs. A minimum requirement to ensure that composition does not introduce inconsistencies is to check non-emptiness of *models(D)*. The model finding procedure of FORMULA is suited for this task: Perform model finding on the `conforms` query to check non-emptiness.

However, checking non-emptiness of models is only one of the tools available in FORMULA. Many of the composition operators guarantee relationships between domains. For example, the composition operators can also be combined to guarantee relationships by construction. For example, given a family of domains $(D_i)_{i \in I}$ and an one-to-one renaming function $r : I \mapsto \Sigma$, the categorical product can always be built by the construction:

$$\left(D_1 \text{ as } r(1) * D_2 \text{ as } r(2) * \ldots * D_n \text{ as } r(n) \right) \tag{14.42}$$

where renaming is used to ensure disjointness of declarations. The categorical coproduct can be formed by a similar construction.

Figure 14.10 shows the specification of the deterministic finite automaton (DFA) abstraction by restricting the NFA domain. Line 2 declares DFA as a restriction of NFA. Lines 3–6 define the `isNonDeter` query that is satisfied if there exists a state with two distinct outgoing transitions triggered by the same event. The `conforms` query is satisfied if `isNonDeter` is not satisfied. Note that the `restricts` keyword implicitly conjuncts `NFA.conforms` onto the `conforms` query of DFA.

14.3.7 Summary

The structural semantics of DSMLs serve as interfaces to the users of modeling languages and to underlying tool flows. Beyond this, structural semantics facilitate reuse and composition of DSMLs, as we describe in the next section. Therefore, it is important to formally specify and to provide tool support for analysis of structural semantics. In this section, we have provided a general framework for understanding the relationship between structural semantics (domains), DSMLs, and metamodels. We have also described a concretization of this framework using structured LP with carefully chosen composition operators. This approach allows formal analysis and correct-by-construction of structural semantics. Please see [25] for a complete example, including formal analysis.

```
    /// Deterministic Finite Automaton Domain
2.  domain DFA restricts  (NFA){
3.     isNonDeter :?
4.        s is State, t1 is Transition, t2 is Transition,
5.        t1.current = s, t2.current = s, t1.trigger = t2.trigger,
6.        t1.next != t2.next.
7.     conforms :? !isNonDeter.
8. }
```

FIGURE 14.10
DFA as a restriction of NFA.

14.4   Specification of Behavioral Semantics of DSMLs

As defined by their structural semantics, models are well-formed structures that can represent domains (in metamodeling) or specific points in domains. While structural semantics of DSMLs are important, they are not sufficient for expressing all essential meanings associated with models. The most important semantic category that needs to dealt with is behavior. For example, in the simplified design flow of embedded controllers (see Figure 14.1), the model of a proportional/integral/derivative (PID) controller [34] would be represented in Simulink® [21] as a structure of simple differentiator, integrator, adder, and multiplier nodes that form a point in the domain of all well-formed Simulink(R) Models. At the same time the PID controller has a well-defined dynamics that can be described using different mathematical formalisms such as differential equations, an impulse response, or a function in the Laplace domain [34]. In general, behavioral semantics need to be represented as an interpretation of the model in a mathematical domain that is sufficiently rich for capturing essential aspects of the behavior (such as dynamics).

Explicit representation of structural and behavioral semantics in DSMLs conforms well to all engineering disciplines where the relationship between structure and behavior is extensively studied. Formalization of these two closely related aspects of modeling enables the exploration of fundamental issues in modeling:

1. **DSML composition:** Given two DSMLs L_1 and L_2 with behavioral semantics X and Y and a composition operator $\oplus$, create a composite DSML with structural semantics $(L_1 \oplus L_2)$ such that the constraint system of the composed domain is consistent (the domain of the composed DSML is not empty) and the composed DSML exhibits both X and Y behavioral semantics (there are two behavioral interpretations of the models in the composed DSML).

2. **Orthogonality:** Given a DSML $L = (\Upsilon, R_\Upsilon, C, (\llbracket \rrbracket_{\text{struc}}, \llbracket \rrbracket_X, \llbracket \rrbracket_Y))$ with behavioral semantics X and Y and two sets of structural operators $op_X \in OP_X$ and $op_Y \in OP_Y$ that are structure preserving (whenever a model r is well-formed, the transformed models $op_X(r) = r'$ and $op_Y(r) = r''$ are also well-formed):

$$\forall r \in D(\Upsilon, C) \Rightarrow (op_X(r) \in D(\Upsilon, C)), (op_Y(r) \in D(\Upsilon, C)) \qquad (14.43)$$

   The operators are orthogonal to the behavioral semantics X and Y, respectively, if

$$\forall r \in D(\Upsilon, C)) \Rightarrow (\llbracket r \rrbracket_X = \llbracket op_X(r) \rrbracket_X), (\llbracket r \rrbracket_Y = \llbracket op_Y(r) \rrbracket_Y) \qquad (14.44)$$

3. **Structural/behavioral invariants:** The design and verification tasks can be significantly simplified if behavioral properties can be guaranteed

by structural characteristics. Discovering and testing these invariants requires the precise representation of structural and behavioral semantics.

In this section, we describe a method for formalizing behavioral semantics based on *transformational interpretation*. As we discussed before, an *interpretation* is a mapping from the model realizations of one domain to the model realizations of another domain:

$$[\![]\!] : R_\Upsilon \mapsto R_{\Upsilon'} \tag{14.45}$$

Assuming that the $R_{\Upsilon'}$ target domain has well-defined behavioral semantics, the specification of the mapping assigns semantics to R_Υ if the mapping is well-defined. In other words, the behavioral interpretation of models in a DSML L requires two distinct components:

- A mathematical domain and a formal language for specifying behaviors
- A formal language for specifying transformation between domains

Selection of the first component largely depends on the goal of the formal specification. If the models represent behaviors that need to be simulated by computers or implemented on computers, selection of *operational semantics* is the right choice. In programming languages, operational semantics describe how a syntactically correct program is interpreted as sequences of abstract computational steps [17]. These sequences form the meaning of the program. Adopting this definition for the behavioral semantics of modeling languages, operational semantics of a DSML is defined by a model interpretation process formed by a sequence of abstract computational steps [10,12]. The model interpreter generates behavior from models. While primitive behaviors are described as sequences of computational steps, complex behaviors need to be defined compositionally as the interaction of behaviors of primitive components. Detailed treatment of this topic is out of scope for this chapter, we refer interested readers to the literature [5,11]. Another relevant method for specifying behavior is *denotational semantics*. Denotational semantics of programming languages provide mathematical objects for representing what programs do (as opposed to how they do it) [35]. For example, the Tagged Signal Model (TSM) provides a denotational framework representing sets of possible behaviors as a collection of events [29]. The primary goal of TSM is comparing models of computation and not simulation or code generation.

The second component is the specification of transformations between domains. Since these transformations are syntactic operations, their formal specifications can be expressed using a logic-based structural semantics foundation [23] or graph transformations [26]. We choose graph transformation specifications because of the availability of high-performance tools [27].

In this chapter, we describe one formalization of behavioral semantics based on Abstract State Machines (ASM) [7] and model transformations. We use ASMs for the formal specification of model interpreters and graph transformations to map models specified in various DSMLs into their ASM representation. This process, called *semantic anchoring*, is supported by an integrated tool suite including the Model Integrated Computing (MIC) tools [27] and the ASML tool suite [18]. We will use a simple example for demonstrating the method.

14.4.1   Overview of Semantic Anchoring

The outline above suggests that given a DSML $L = (\Upsilon, R_\Upsilon, C, \llbracket \rrbracket_{\text{struc}}, (\llbracket \rrbracket_i)_{i \in I})$ we need to specify for each $i \in I$ behavioral interpretation the following:

1. The $\llbracket \rrbracket : R_\Upsilon \mapsto R_{\Upsilon_{ASML}}$ mapping between the R_Υ domain of L and the $R_{\Upsilon_{ASML}}$ domain of an ASM language $ASML$
2. A model interpreter in ASML

While this is certainly possible, the required effort would be significant and in direct conflict with the ultimate goal of DSMLs: rapid formulation and evolution of semantically precise, highly domain specific abstractions. The difficulties are further exacerbated by the fact that specification of the mapping between the two domains (DSMLs and ASML) requires model transformations that use different formalisms on each side.

To mitigate these problems, we have developed a more practical approach to semantic anchoring that enables the reuse of specifications. The approach is based on the recognition that although DSMLs use many different modeling abstractions, model composition principles, and notations for accommodating needs of domains and user communities, the fundamental types of behaviors are more limited. Broad categories of component behaviors can be represented by behavioral abstractions, such as FSM, Timed Automaton, and Hybrid Automaton. This observation led us to propose a semantic anchoring infrastructure that includes the following elements [12]:

1. Specification of a $\left\{ L_j^{SU} \right\}_{j \in J}$ set of modeling languages for the basic behavioral abstractions. We use the term *semantic units* to describe these languages. The role of the semantic units is to provide a common behavioral foundation for a variety of DSMLs (or aspects of DSMLs) that are syntactically different, but semantically equivalent. For example, a number of modeling languages such as IF [37], UPPAAL [4], and Kronos [14] (and many others) have Timed Automata semantics. By specifying a Timed Automata Semantic Unit (TASU) using ASML, we can define semantics for IF, UPPAAL, and Kronos by specifying the transformation between them and TASU (see details in [9]). The clear advantage of this method is that the common semantic domain enables the semantically precise comparison of the different models.
2. Specification of the transformational interpretation $\llbracket \rrbracket^T : R_\Upsilon \to R_{\Upsilon_{ASML}}$ between the domains of a DSML L and a selected semantic unit $\left(L_j^{SU} \right)_{j \in J}$.

The T transformation "anchors" the semantics of L to the semantic unit $\left(L_j^{SU}\right)$.

3. Since the domain of $\left(L_j^{SU}\right)$ is a subset of the domain of ASML, we can exploit this for a further significant simplification. As we discussed in Section 14.2, a domain is defined by a metamodel expressed in a L_{meta} metamodeling language. Accordingly, the domain of $\left(L_j^{SU}\right)$ can be defined in two alternative forms: (a) using the metamodeling language that we use for defining domains for DSMLs, or (b) as an *Abstract Data Model* expressed using the type language of ASML. The conversion between the two representation forms is a simple syntactic transformation. This approach allows us to specify the model transformation T between the DSML and the selected semantic unit such that their domains are specified using the same formalism provided by L_{meta}.

Figure 14.11 shows our tool suite for semantic anchoring. It comprises (1) the ASM-based common semantic framework for specifying semantic units and (2) the MIC modeling and model transformation tool suites, the Generic Modeling Environment (GME), and Graph Rewriting and Transformation Tool (GReAT), respectively, which support the specification of the transformation between the DSML metamodels and the Abstract Data Models (ADM) of the semantic units.

As we discussed above, we selected ASMs, formerly called Evolving Algebras [17], as a formal framework for the specification of semantic units. The ASML tool suite [18] provides a specification language, simulator, test-case generation, and model checking tools for ASMs. GME [27] is employed for defining the metamodels for DSMLs using the Unified Modeling Language (UML)/OCL-based MetaGME metamodeling language [32,33]. The

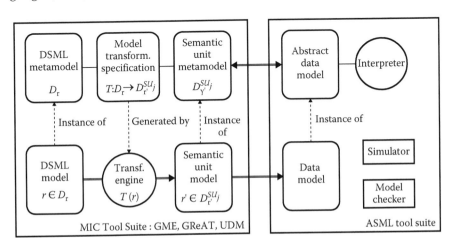

FIGURE 14.11
Tool suite for semantic anchoring.

semantic anchoring is defined by model transformation rules expressed in the UMT (Unified Model Transformation) language of the GReAT tool suite [26]. In UMT, model transformations are expressed as graph transformations that can be executed (in interpreted and/or compiled form) by the GReAT tool. In summary, semantic anchoring specifies DSML behavioral semantics by the operational semantics of selected semantic units (defined in ASML) and by the transformation rules (defined in UMT). The integrated tool suite enables the simulation of domain models defined in a DSML according to their "reference semantics" by automatically translating them into ASML data models using the transformation rules.

In the rest of this section, we show the process of semantic anchoring using a simple example. Readers can refer for detailed discussion about the use of the semantic anchoring tool suite in specifying a TASU in [9], and for specifying semantics for a complex DSML by composing several semantic units in [11].

14.4.2   Semantic Anchoring Example: Timed Automata

Timed Automata [2,3] model the behavior of real-time systems over the progression of time. This formalism extends the definition of state-based transition systems to include a finite set of real-valued clocks that synchronously progress time. The enabling of transitions is further restricted using constraints over the clock valuations, and transitions can specify a set of clocks to reset to zero upon firing.

Timed automata have previously been defined within the semantic anchoring framework [9]; however, the semantic unit included many features to provide semantic equivalences to other versions of timed automata modeling languages (e.g., UPPAAL [4] and IF [37]). The proposed TASU specified here is intended to be simple while capturing all the basic facilities of the original timed automata model. In the AsmL semantic definition, the structures mimic the abstract constructs of the behavioral formalism. The first two sections provide the Abstract Data Model (ADM) and operational semantics that govern a single automaton. The ADM describes the data structures of the semantic unit, and the operational semantics provide a model interpretation over an instance of the ADM, the data model. The succeeding sections will explain how automata are composed to form more complex systems and then define the modeling language of the semantic unit, L_s, required for semantic anchoring.

14.4.2.1   Timed Automata Overview

Structurally, a timed automaton is a mathematical 5-tuple $< L, l_0, C, E, Pr_e >$ with an event alphabet Σ:

1. L is a finite set of locations.
2. l_0 is an initial location, $l_0 \in L$.

3. C is a finite set of clock variables.
4. $E \subseteq L \times \Sigma \times 2^C \times L$ is a set of edges. An edge $< l, \alpha, \phi, \lambda, \omega, l' >$ is a transition from l to l' on a set of input events $\alpha \subset \Sigma$ and the satisfaction of the guard $\phi(v)$ over the valuation function v of the clocks $c \in C, v : c \to \mathbb{N}$. Upon the firing of a transition, the clocks $\lambda \subseteq C$ are reset to 0, and the set of output events $\omega \in \Sigma$ are generated.
5. $Pre : E \times \mathbb{N}^n \to \mathbb{N}$ is a map that assigns to each edge a nonnegative integer priority with respect to a given clock evaluation v, so that $Pr_e(e, v)$ is the priority of edge e at clock valuation v.

Transitions are assigned a priority to model the dynamics of many common real-time systems. This mechanism allows for dynamic priority assignment throughout the execution of a model. Following the model of [9], the progression of time is modeled as a transition; therefore, it too must be assigned a priority, the lowest priority in the specification. This supports the notion of urgent transitions; however, unlike [9], this semantic unit does not provide mechanisms for blocking time or most urgent transitions (i.e., the time transition is always enabled).

The semantics for a timed automaton fitting the structure above defines the state of a timed automaton as a pair of (l, v) where l is a location in L and v is a valuation of the clocks C. The possible state changes are enabled over the union of location-switching transitions and the progression of time defined respectively as

$$(l, v) \xrightarrow{\alpha} (l', v[\lambda := 0]) \Leftrightarrow \exists e =< l, \alpha, \phi, \lambda, \omega, l' > \in E$$

$$such\ that$$

$$\phi(v) = true\ and \tag{14.46}$$

$$\forall e' =< l, \alpha, \phi', \lambda', \omega', l'' > \in E \wedge \phi(v) = true$$

$$\Rightarrow Pr_e(e, v) \geqslant Pr_e(e', v)$$

$$(l, v) \xrightarrow{t} (l', v + \varepsilon) \Rightarrow Pr_e(e, v + \varepsilon) = 0 \tag{14.47}$$

In the following, we will use AsmL syntax in the specifications. The advantage of this approach is that these specifications are directly executable using the freely available AsmL tools [18] that can be used with the latest Microsoft Visual Studio distributions.

14.4.2.2 Semantic Unit Abstract Data Model

In ASML, *Events* and *Clocks* (shown in Figure 14.12) are enumerated stores for the defined events ($\in \Sigma$) and clock variables (C) of the automaton. The *TimedAutomaton* class captures the mathematical model described above for the semantic unit. The *id* field, present in all following class definitions, provides a naming construct to identify objects, but it has no semantic implications. The *initial* field holds the initial location of the automaton, and it must

```
1. enum Clocks
2. enum Events
3. class TimedAutomaton
4.     const id as String
5.     const locations as Set of Location
6.     const initial as Location
7.     const transitions as Set of Transition
8.     const local_clocks as Set of Clockss
9.     var v as Set of Clocks of Clocks to Integer = {- >}
10.    var cur as (Location, Set of Events) = ( null, {})
```

FIGURE 14.12
Structures: Clocks, events, and timed automata.

be a predefined location of the system, member of the field *locations*, or an error will occur. The *transitions* field holds the set of all defined transitions between the locations of this automaton. The variable v is the valuation of the clock variables in *local_clocks*. The valuation is a partial function specified as a mapping from the clock domain, C, to the domain of natural numbers, $\mathbb{N}$. Natural numbers were chosen for discrete time steps for clarity. Domain models from a DSML that uses variable discrete time steps can be scaled as required, complicating the operational code. The variable field *cur* is a 2-tuple indicating the location and set of active events in the current executing step of the simulation.

Location and *Transition* are defined as first-class types (Figure 14.13). Locations contain only the unique identifier, *id*. Transition defines a move from a location *src* to a location *dst*, given the appropriate input events, trigger, and the satisfaction of the time guard condition, $\phi(v)$. The variable *time_guard* is a partial function that maps clock valuations to a Boolean. The time guard must be defined as a variable since the valuation of the clocks $v(C)$ is variable over the progression of time. Upon taking a transition, the event set *outputs* is added to the set of currently active events. The *resets* field holds the clocks to

```
1. class Location
2.     const id as String

4. class Transition
5.     const id as String
6.     const src as Location
7.     const dst as Location
8.     const trigger as Set of Events
9.     const output as Set of Events
10.    const resets as Set of Clocks
11.    var time_guard as Map of (Map of Clocks to Integer) to Boolean

13. const time = new Transition("time", null, null, {}, {time_ev},
    {}, {->})
```

FIGURE 14.13
Structures: Locations, transitions, and time.

be reset to zero upon taking the transition. According to the original model, the progression of time is enabled in every state of a timed automaton; therefore, the *time* transition is defined as a constant. If the time transition is taken, all clocks in the timed automaton will be incremented by some value $\epsilon \in \mathbb{N}$ that must be defined for the model. Note that the time transition has an output event, *time_ev*. This event must be included in the data model definition.

14.4.2.3 Operational Semantics

As is the case for all MIC semantic units, the operational semantics of the timed automaton semantic unit are specified as a set of ASML rules that execute over models instantiated using the ADM data structures.

A global function *InputEvents* provides an interface to continue the execution of a model within the AsmL tools, that is, it provides a means to drive the simulation. This method receives an input parameter set of TimedAutomaton objects, and should return a set of Events to serve as input to the current simulation step. It must be provided by the consumer of the semantic unit (i.e., the simulation environment).

The *InitializeTA* method (shown in Figure 14.16) first initializes the *cur* variable to the initial location and an empty active event set, and then sets the valuation of all local clock variables to zero. The set of currently enabled transitions returned by the *EnabledTransitionsTA* method is the *time* transition added to the set of transitions that meet the enabling conditions: the source location of the transition is the current location, the triggering events of the transition are a subset or equal to the set of currently active input events, *et.trigger <= cur.Second*, and *time_guard(v)* evaluates to true.

The *EvolveTA* method (Figure 14.14) fires the transition passed to it, *tr*, by updating the current configuration to be the destination location of the transition and the generated output events of the transition and resets all clocks specified in the resets field of tr to zero. The partial update to the mapping *v* also maintains the current valuations of all other clocks not in resets, *local_clocks - (local_clocks intersect resets)*. Even though the time transition is always returned as an enabled transition, its effect it not yielded by passing it to the *EvolveTA* method. Instead, the *TimeProgressTA* method must be invoked to increment all local clock valuations by the defined integer value ϵ

```
1. class TimedAutomaton
2.    EvolveTA(tr as Transition)
3.       require tr in transitions
4.       cur := (tr.dst, tr.output)
5.       v := {clki  -> 0 | clki in (tr.resets intersect local_clocks) }
6.             union {clki  -> v(clki) |
7.                in local_clocks -(tr.resets intersect local_clocks) }
```

FIGURE 14.14
Execution: Stepping a timed automaton.

and must set the *time_ev* event as the only active event for the given automaton. The *UpdateEvents* and *GetEvents* methods are self-explanatory. The *PriorityTA* and *UpdateTimeGaurdTA* methods are not predefined functions as they are model dependent. Each must be specified when simulating an instance model of a DSML.

The *PriorityTA* method returns a nonnegative integer value for each transition in the system with the base priority being 0, that of the time transition. Returning identical priorities for multiple transitions allows for nondeterminism in the data models.

The *UpdateTimeGuardTA* method reevaluates the guard condition of every transition given the current valuation of the clocks. The time transition's time guard is never included in this method since it is always an enabled transition.

14.4.2.4 Composition of Timed Automata

The timed automaton semantic unit presented thus far describes the behavioral semantics and data structures of a single automaton. To model larger, more complex systems and behaviors, the semantic unit needs to be extended to model concurrently executing automata. Here, we show the synchronous parallel execution semantics for multiple automata and provide the appropriate metamodel for the TASU.

The modeling of concurrent automata was previously approached in [9]; however, the resulting semantic unit appeared overly complex and insufficiently expressive. These issues motivated this new specification that is intended to be simpler and extensible to a wider variety of execution semantics.

The set of globally scoped clocks, *global_clocks*, is a set of clocks that all automata can read and possibly reset. Conversely, the set of clocks in the *TimedAutomaton* class, *local_clocks*, is scoped exclusively to its respective automaton. The variable g_v is the partial function that maps the global clock variables to their valuations. Figure 14.15 contains the ASML structures for a global system definition.

The class *Comp_System* (Figure 14.15) contains the set of concurrently executing TimedAutomaton objects and a variable *E*. *E* represents the set of all

```
1. const global_clocks as Set of Clocks
2. var g_v as Map of Clocks to Integer = {clki -> 0 | clki in
   global_clocks}
4. class Comp_System
5.   const TA_components as Set of TimedAutomaton
6.   var E as Set of Events = {}
```

FIGURE 14.15
Execution: ASML structures supporting concurrent composition of independent timed automata.

```
1. class Comp_System
2.    InitializeCS_TA()
3.      forall c in TA_components
4.        require {t} intersect {tr.resets | tr in c.transitions} = {}
5.        require c.local_clocks intersect global_clocks = {}
6.        require c.local_clocks intersect {h.local_clocks | h in
          TA_components
7.                                          where h <> c} = {}
8.        c.InitializeTA()

10.   UpdateEventsCS_TA()
11.     let e = E union InputEvents(TA_components)
12.     E := e
13.     forall c in TA_components
14.       c.UpdateEvents(e)

16.   UpdateTimeGuardCS_TA()
17.     forall c in my_sys.TA_components
18.       c.UpdateTimeGuardTA()
```

FIGURE 14.16
Execution: Initialization and update for noninteracting concurrent timed automata.

active events in the composed system. This composition does not restrict or scope events that an automaton can see; therefore, events generated from one automaton will be visible for all other automata in the composed system as well.

The *InitializeCS_TA* method invokes the *InitializeTA* method for each TimedAutomaton object in the composed system after it checks constraints that are placed on the clock variables (Figure 14.16). First, in the composed system, the clock variable t is the system clock and is not allowed to be reset. The next two constraints ensure that no clock is defined both globally and locally and that no clock is defined as local to multiple automata. The *UpdateEventsCS_TA* method must correctly update the global set of active events for all TimedAutomaton objects in the system. Remember, all of the automata see the same set of active events (i.e., there is no scoping of local events versus global events). Still, this could be extended to allow event scoping in the semantics of other variations of a composed system. The *UpdateTimeGuardCS_TA* method simply calls the *UpdateTimeGuardTA* method for each *TimedAutomaton*.

The *TimeProgressCS_TA* and *EvolveCS_TA* methods (shown in Figure 14.17) are responsible for changing the state of the system for every execution step. Along with calling the *TimeProgressTA* method for each automaton, *TimeProgressCS_TA* must increment all global clock valuations by the defined constant ϵ. This method is only called if all automata in the system have elected to take the time transition in this step. The *EvolveCS_TA* method takes as input each timed automaton in the system with its respective highest priority currently enabled transition. *ev* and *nev* are initialized to

```
1.  class Comp_System
2.    TimeProgressCS_TA()
3.       g_v := {clki -> g_v (clki) + epsilon | clki in global_clocks}
4.       forall c in TA_components
5.         c.TimeProgressTA()

7.    EvolveCS_TA(cT as Set of (TimedAutomaton, Transition))
8.       let ev = {k | k in cT where k.Second <> time}
9.       let nev = {l | l in cT where l.Second = time}
10.      let gr = BigUnion({ h.Second.resets | h in ev})

12.      g_v := {clki -> 0 | clki in ( gr intersect global_clocks) }
13.            union { clki -> g_v (clki) | clki in global_clocks
14.                                      - ( gr intersect global_clocks)}

16.      forall m in ev
17.        m.First.EvolveTA(m.Second)
18.      forall n in nev
19.        n.FirstUpdateEvents({})
```

FIGURE 14.17

Execution: Progress for global time and individual automata.

the automata that will take a non-time transition in this step and those that will not, respectively. The automata in *nev* indicate that their highest priority transition is the time transition; however, all other automata (specified in *ev*) do not agree so the time transition will not be taken. If a transition is not taken by an automaton, the set of active output events for that automaton are cleared. Given the automata that will be taking a non-time transition, the method must also reset global clocks specified in the resets field of the enabled transitions. The *EvolveTA* method of each automaton will reset the local clocks. Accordingly, the *EvolveTA* method is called for each automaton in the set *ev*.

The *RunCS_TA* method provides a single execution loop for the composed system (Figure 14.18). First, each automaton of the composed system must be initialized. Notice that the variable *count* restricts the number of execution steps of the loop and has no semantic meaning for the system. During each iteration of the execution loop, the active event set is updated (*UpdateEventsCS_TA* method) to include the events generated from the last set of transitions, initially held in *E*, joined to the events returned from the *InputEvents* method. Also, all time guards must be reevaluated given the resulting system state from the last execution step.

Following these updates, the set of enabled transitions for all automata can be determined. First, the set of all enabled transitions for each automaton are stored in *eT* in the anonymous 2-tuple of type *<TimedAutomaton, Set of Transitions>* . Next, the set of enabled transitions for each automaton is reduced to a single transition that has the highest priority of all enabled transitions. This reduction is nondeterministic if multiple enabled transitions have the highest priority value. Time will progress only if all automata

```
1.  class Comp_System
2.     RunCS_TA()
3.        Var_ count = 1
4.        step my_sys.InitializeCS_TA()
5.        step while count < 35
6.           step
7.              UpdateEventsCS_TA()
8.              UpdateTimeGuardCS_TA()
9.           step
10.             let eT = {(ta, ta.EnabledTransitionsTA()) | ta in
                   TA_components}
11.             let eT2 = {(b.First, (any h | 1 in b.Second where
12.                b.First PriorityTA(h) = (max b.First.PriorityTA(tp)
13.                | tp in b.Second))) | b in eT}
14.             if {b2.Second | b2 in eT2} = {time} then
15.                TimeProgressCS_TA()
16.             else
17.                EvolveCS_TA(eT2)
18.          step
19.             E := BigUnion({ c.GetEvents()| c in  TA_Components})
20.             count := count + 1
21.
```

FIGURE 14.18
Execution: Global coordination code.

indicate that the time transition should be taken; otherwise, *eT2* is passed to *EvolveCS_TA*. Following either action (time progress or taking other transitions) all generated events from each automaton in the composed system are collected and stored in the variable field *E*.

14.4.2.5 TASU Modeling Language

Within the MIC semantic anchoring framework, model transformations between metamodels of a given DSML and the TASU specifies the anchoring of the DSML to the TASU. In order to connect the ASML and the MIC tool suites, we must represent the TASU ADM as a MIC metamodel (expressed in MetaGME) and must implement a translator that translates MIC model instances into ADM model instances. Figure 14.19 shows the metamodel of the TASU specified in the MetaGME metamodeling language [22] of the MIC tool GME.

The metamodel in Figure 14.19 captures the Abstract Data Model of the TASU. Note that the metamodel includes *Constant* and *Variable* objects for defining model-dependent data. The remaining piece of the TASU is a model translator that generates the ASML of data models from timed automata models created in the MIC tool suite. Since these time automata models are instances of the TASU Metamodel as well as the TASU ADM on the ASML side, the translator is a simple XML parser.

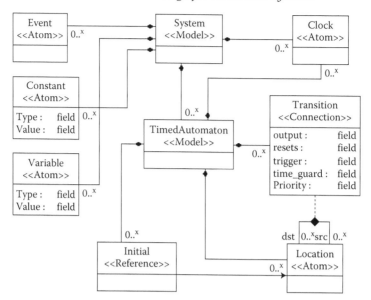

FIGURE 14.19
Metamodel for our timed automata abstract data model.

14.4.2.6 Semantic Anchoring Example: The Timing Definition Language

The timing definition language (TDL) [16,36] provides a high-level programming abstraction that allows distributed real-time system designers to describe the temporal behavior of components apart from their functional code and deployment/configuration concerns. This timing model is based on logical execution time (LET) introduced by the Giotto modeling language [20]. TDL preserves many other artifacts of Giotto, but some syntactic and semantic differences remain.

This case study will capture the timing behavior of a TDL application by anchoring a TDL modeling language to the TASU. For simplicity, in this example, all TDL communication artifacts will be excluded from the mapping since they have no effect on the temporal behavior of a TDL application.

In Giotto [20] a system is defined over a set of nonconcurrent modes, each of which contains periodically executing activities and a communication topology description. The periodic activities, task invocations under LET semantics, actuator updates, and mode switches, all conditionally execute a finite number of times, their frequency, within the mode's cyclic execution period. A task's timed execution cycle is initiated by a release for scheduled execution and finished by a completion event at time after the release. Actuator updates and mode switches are considered synchronous activities; therefore, they take zero logical time to execute.

In TDL [16] the notion of a mode is preserved; however, the model is extended to include a module, a TDL component. A module encapsulates what is characterized as a Giotto system, and a TDL system is defined as a set of modules and communication networks between them. Like Giotto, the modes within a single module do not execute concurrently; however, modules of a TDL system execute in parallel.

14.4.2.7 Anchoring the TDL Modeling Language to the TASU

Figure 14.20 shows an abbreviated metamodel for the TDL modeling language. The full metamodel contains data handling mechanisms, such as ports and drivers; however, these omitted artifacts do not affect the timing behavior of a TDL application. Notice that the ModeSwitch class inherits from the abstract class Periodics. The attribute frequency that ModeSwitch now contains will be ignored in the transformation since nonharmonic mode switches are not allowed in TDL unlike Giotto [20]. The TaskReference class is used to copy a previously defined Task into a new mode that may define a different frequency value for the TaskReferemce versus the original Task object's frequency.

The model transformation that provides the anchoring across the metamodels is specified in the UMT language of the MIC tool GReAT [26]. The transformation is given by a set of rules that define the mapping between the TDL metamodel and the TASU metamodel. The transformation takes a TDL

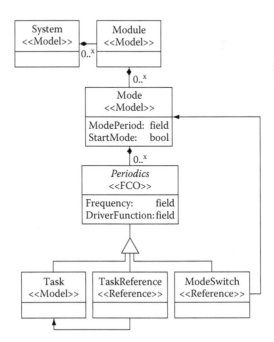

FIGURE 14.20
TDL MetaGME Metamodel (abbreviated).

domain model as the input and produces a corresponding TASU model as output.

A transformation in GReAT is formed by a sequence of rules, each of which contains a pattern graph [26]. The pattern graph is constructed from the objects defined in the TDL and TASU metamodels. Each pattern graph object is assigned an action to perform on matched objects upon execution of the transformation: *bind*, *createnew*, or *delete*.

The execution of a transformation rule proceeds by finding matches between the pattern graph objects annotated with the *bind* or *delete* action and instances in the input and output domain models. Upon the initial discovery of a nonempty set of matches, a boolean expression called a *Guard* can be included in the pattern graph to compare the attributes of the returned matches. Any matches that do not satisfy the Guard condition will not be included in the returned matches. Finally, the remaining matched objects marked as *delete* are removed and objects marked with the *createnew* action are created in their respective domain models. Before completion of the rule, the attribute fields of objects in the updated graph can be modified using an *AttributeMapping* object, for example, to correctly instantiate fields of new objects.

The model transformation between the TDL and TASU metamodels provides the mapping from the TDL objects that characterize the temporal behavior of the system to the corresponding TASU structures. The transformation steps are described using the following syntax:

$$TDL o_1; o_2; \ldots; o_n \rightarrow TASU o_1; o_2; \ldots; o_m \qquad (14.48)$$

Each matched set of n objects from the TDL domain found in the input graph is mapped to newly created instances of a set of m objects from the TASU domain. TASU object types in bold font are newly created in the current step. Each object can also be appended with relevant attribute values, for example, *clock.id()= "clock1."* The line following the step number gives a brief description of the object types involved in the transformation step, and parentheses following an object type indicate their respective container object. We shorten TimedAutomaton to TA for brevity.

1. Systems: (Top-Level Objects, only one per model):
   {System} → {System; clock.id()=" t";}

   ```
   event.id()="time_ev";
   constant.id()="epsilon"
   ```

2. Modules (System) to TA (System):
   {Module} → {TA}

3. Modes (Modules) to Locations (TA):
   {Mode} → {Location; clock.id()=Mode.id() +"_clk"}
   where the clock created is local to the automaton of the location. The value of this clock is used in all transitions concerning the activities of

the mode, and it will be reset by a transition once its counts up to the mode period value.

4. Transitions and Initial Configuration:

   (a) Start Mode (Module) to Initial Location (TA):
   ```
   {Mode.StartMode()==true} → {initial.reference()
      = Location.id()}
   {Where Location.id() == Mode.id()}
   ```

   (b) Mode Period (Mode) to Transition (TA):
   ```
   {Mode} → {Transition.time_guard()
      = Mode.ModePeriod(),
      resets() = self.id()+"_clk"}
   ```

   (c) Mode Switch (Mode) to Transitions (TA):
   ```
   {ModeSwitch} → {Transition.time_guard() = 0,
      src() = Mode.id(),
      dst() = ModeSwitch.reference(),
      resets() = ModeSwitch.reference() + "_clk"}
   ```

   (d) Task (Mode) to Transitions (TA):
   ```
   {Task}  n*{Transition.id()="R", time_guard()
      = n*dur;
      Transition.id() = "F", time_guard() = (n+1)*dur}
   ```
   Where $dur = Mode.ModePeriod/Task.Frequency()$ and $n = Task.Frequency()$. The transition with $id = "R"$ denotes a task release for execution, and the other transition with $id = "F"$ denotes a task finishing execution. Notice that we create $2n$ transitions for each task. A unique transition for release and finish must be created for each execution cycle in a given mode period.

   (e) Task Reference(Mode) to Transitions (TA):
   Repeat step 4.d but use *TaskReference*.

Figure 14.21 shows the GReAT transformation responsible for anchoring TDL to the TASU. The first three rules, steps 1–3 above, must be performed in sequence; however, the remaining steps, creating all of the transitions assigning the initial locations for each automaton are performed in parallel.

Figure 14.22 shows the GReAT transformation rule responsible for creating the mode switch transitions in the TASU. It is an individual element of the overall transformation, shown to illustrate the form of individual rules. The two port objects on the left side of the diagram are bound to Mode and TimedAutomaton objects indicating these are the input context for the rule. Bound objects in a GReAT transformation are black and only contain the small arrow icon in the lower left-hand corner. Objects designated to be newly created are colored in blue and contain a check mark in the lower right-hand corner, the Transition object in the diagram. The rule execution proceeds according to the pseudo-code description in Figure 14.23.

An example TDL model and its corresponding TASU ASML data model and execution trace are given in Figures 14.24 through 14.26. The TDL model

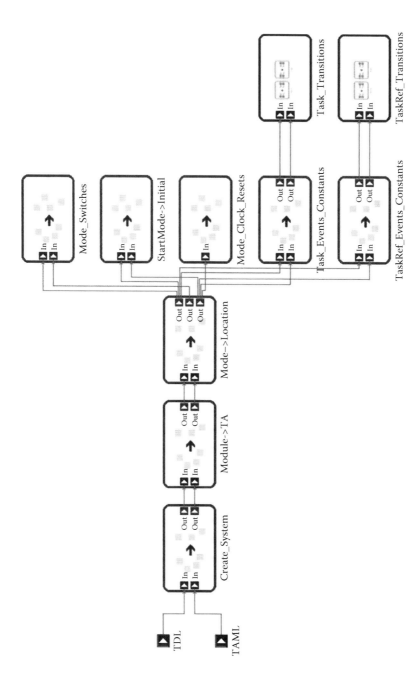

FIGURE 14.21
GReAT Transformation from TDL to TASU (top-level rules).

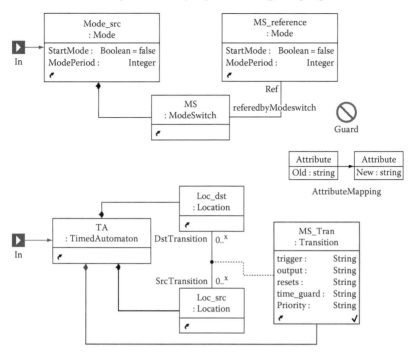

FIGURE 14.22

GReAT Transformation from TDL to TASU (rule for creating mode switch transitions).

(Figure 14.24) contains one module, "Count," two modes, "Up" and "Down," with period 4, two tasks per mode, "inc" and "dec," with a frequency given by the field "f," and each mode contains a Mode Switch specified as the diamond shape labeled with the target mode of the switch. The tasks "inc" and "dec" increment and decrement some system datum by 1 during each execution cycle. The mode "Up" will execute the task "inc" more often than "dec"; therefore, the datum will be increasing over time. The converse is true in mode "Down." The ASML data model representation (Figure 14.25) is generated by an XML parser that traverses the TASU model created by the semantic anchoring transformation. Simulation of this data model using the behavioral code previously described and the ASML interpreter yields an execution trace (Figure 14.26).

In the ASML data model code, the transition instantiations are not given to save space; however, their identifiers are included in the TimedAutomaton constructor, (e.g., "TUp_inc_r_0", "TUp_Down_sw", etc. ...). The format of the transition identifier indicates what activity the transition corresponds to: "TUp_inc_r_0" means in mode "Up," the release of task "inc" at mode time 0, "TDown_dec_f_2" means in mode "Down," the finishing of execution of task "dec" at mode time 2, and "TUp_Down_sw" means the

```
1.  Mode_Switches(Mode_src as Mode, TA as TimedAutomaton)
2.  // Find TDL matches
3.  let match_TDL = {(Mode_src, MS, MS.reference()) | MS in
    Mode_src.children()
4.                       where MS.type() == ModeSwitch}
5.  // Find TASU matches
6.  let match_TASU = {(TA, Loc_src, Loc_dst) | Loc_src, Loc_dst in
    TA.children()
7.                          where Loc_src.type() == Location
8.                          and Loc_dst.type() == Location}
9.     // Apply Guard to reduce matches
10.    let match_Total = {(m_TDL, m_TASU) | m_TDL in match_TDL and
    m_TASU in match_TASU
11.                      where m_TDL.First.id() == m_TASU.Second.id()
12.                      and m_TDL.Third.id() == m_TASU.Third.id()
13.  // Create new transitions and perform attribute mapping
14.  forall m in match_Total
15.  let MS_Tran = new Transition()
16.  MS_Tran.id() = "T" + m.Second.Second.id() + "_" +
    m.Second.Third.id()
17.  MS_Tran.src() = m.Second.Second
18.  MS_Tran.dst() = m.Second.Third
19.  MS_Tran.time_guard() = m.First.Second.id() + "_clk" == 0
```

FIGURE 14.23
Pseudo-code for mode switch rule.

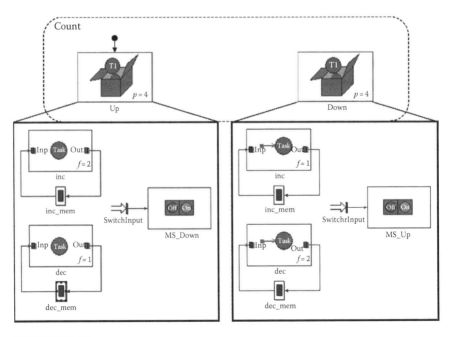

FIGURE 14.24
TDL sample model.

```
1. enum Clocks
2.    t
3.    Up_clk
4.    Down_clk

6. global_clocks as Set of Clocks = {t}
7. var g_v as Map of Clocks to Integer =
8.                    {clki -> 0 | clki in global_clocks}

10. Up = new Location("Up")
11. Down = new Location("Down")
12. Count = new TimedAutomaton("Count", {Up, Down}, Up,
13.          {TUp_inc_r_0, TUp_inc_r_2, TUp_inc_f_2, TUp_inc_f_4,
14.           TUp_dec_r_0, TUp_dec_f_4, TUp_Down_sw, TUp_clk,
15.           TDown_inc_r_0, TDown_inc_f_4, TDown_dec_r_0,
16.           TDown_dec_r_2, TDown_dec_f_2, TDown_dec_f_4,
17.           TDown_Up_sw, TDown_clk}, {Up_clk, Down_clk}}
18. my_sys = new Comp_System({Count})
```

FIGURE 14.25
TDL sample code.

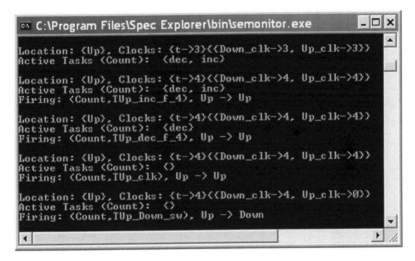

FIGURE 14.26
TDL sample execution trace.

mode switch transition from mode "Up" to mode "Down." The execution trace shows the simulation output for the following event trace:

1. $g\_v(t)$, $v(m1\_clk, m2\_clk) = 3 \rightarrow \{\}$
2. $g\_v(t)$, $v(m1\_clk, m2\_clk) = 4 \rightarrow \{Finish \ "inc"\}$
3. $g\_v(t)$, $v(m1\_clk, m2\_clk) = 4 \rightarrow \{Finish \ "dec"\}$
4. $g\_v(t)$, $v(m1\_clk, m2\_clk) = 4 \rightarrow \{Reset \ "m1\_clk"\}$
5. $g\_v(t)$, $v(m2\_clk) = 4; \ v(m1\_clk) = 0 \rightarrow$
   $\{Switch \ to \ mode \ "Down"\}$

14.4.3   Conclusion

The effective application of DSMLs for embedded software design requires developers to have an unambiguous specification of the semantics of modeling languages. Two key aspects of this problem, specification of structural and behavioral semantics have been explored in this chapter. Our first conclusion is that structural semantics is a deep and hugely important concept that is frequently neglected. Formal investigations of structural semantics are the foundation for composing modeling languages, investigating structurally dynamic adaptive systems, and developing model transformations as well as methods for constructive modeling.

In our proposed semantic anchoring framework, behavioral semantics are defined using model transformations. Much of the ongoing work concerning the semantic anchoring framework involves topics encountered here and throughout our previous work: the concrete specification of more fundamental models of computation for behavior and interaction, the construction and derivation of more complex behavioral models through the composition of multiple semantic units, investigation of mechanisms to model various categories of execution semantics of concurrent systems, and integration of the formalisms throughout the MIC tools to capture the end-to-end process of specifying and using domain-specific modeling. Concurrently, we are investigating approaches to determine the behavioral equivalences between the "reference traces" generated from an ASML data model of a semantic unit and the implementation generated from instance models of a DSML previously anchored to the semantic unit. Our preliminary investigations have led us to consider integrating a framework for constructing concurrency models and their corresponding event traces using POMSETs without knowledge of the initial state of the system. The observed traces are formed from an interleaving of the time intervals between observed events and the event labels. Coupling methods derived from such an approach with the event structures of POMSETs may produce novel techniques for determining correct behavior of concurrent system components in deployed real-time embedded systems.

Acknowledgments

This work is sponsored in part by the National Science Foundation (grants NSF-CCR-0225610 and NSF-CCF-0820088) and by the Air Force Office of Scientific Research, USAF (grants/contract numbers FA9550-06-0312 and FA9550-06-1-0267). The views and conclusions contained herein are those of the authors and should not be interpreted as necessarily representing the official policies or endorsements, either expressed or implied, of the Air Force Office of Scientific Research or the U.S. Government.

References

1. H. Aït-Kaci. *Warren's Abstract Machine: A Tutorial Reconstruction.* MIT Press, Cambridge, MA, 1991.

2. R. Alur. Timed automata. In N. Halbwachs and D. Peled, editors, *Proceedings of the 11th International Conference on Computer Aided Verification*, London, *LNCS* 1633, pp. 8–22, 1999. Springer-Verlag.

3. R. Alur and D. Dill. A theory of timed automata. *Theoretical Computer Science*, 126(2):183–235, April 1994.

4. Basic Research in Computer Science (Aalborg Univ.) / Dept. of Information Technology (Uppsala Univ.). Uppaal. http://www.uppaal.com/. Integrated tool environment for modeling, validation and verification of real-time systems.

5. A. Basu, M. Bozga, and J. Sifakis. Modeling heterogeneous real-time components in BIP. In *SEFM '06: Proceedings of the 4th IEEE International Conference on Software Engineering and Formal Methods*, Washington, DC, pp. 3–12, 2006. IEEE Computer Society.

6. Marc V. Benveniste. Writing operational semantics in z: A structural approach. *Springer LNCS v. 551: Proceedings of the 4th International Symposium of VDM Europe on Formal Software Development*, 1:164–188, 1991.

7. E. Börger and R. Stärk. *Abstract State Machines: A Method for High-Level System Design and Analysis.* Springer-Verlag, Berlin, 2003.

8. L. P. Carloni, F. De Bernardinis, C. Pinello, A. L. Sangiovanni-Vincentelli, and M. Sgroi. Platform-based design for embedded systems. In R. Zurawski, editor, *The Embedded Systems Handbook*. CRC Press, Boca Raton, FL, 2005.

9. K. Chen, J. Sztipanovits, and S. Abdelwahed. A semantic unit for timed automata based modeling languages. In *Proceedings of RTAS'06*, San Jose, CA, pp. 347–360, 2006.

10. K. Chen, J. Sztipanovits, S. Abdelwahed, and E. Jackson. Semantic anchoring with model transformations. In *Proceedings of European Conference on Model Driven Architecture-Foundations and Applications (ECMDA-FA)*, Nuremberg, Germany, 3748 *LNCS*, pp. 115–129, November 2005. Springer-Verlag.

11. K. Chen, J. Sztipanovits, and S. Neema. Compositional specification of behavioral semantics. In *Design, Automation, and Test in Europe: The Most*

Influential Papers of 10 Years DATE, part III, the Netherlands, pp. 253–265, 2008. Springer.

12. K. Chen, J. Sztipanovits, S. Neema, M. Emerson, and S. Abdelwahed. Toward a semantic anchoring infrastructure for domain-specific modeling languages. In *Proceedings of the Fifth ACM International Conference on Embedded Software (EMSOFT'05)*, Jersey City, NJ, pp. 35–44, September 2005.

13. E. Dantsin, T. Eiter, G. Gottlob, and A. Voronkov. Complexity and expressive power of logic programming. *ACM Computing*, 33(3):374–425, 2001.

14. C. Daws, A. Olivero, S. Tripakis, and S. Yovine. The tool kronos. In *Proceedings of "Hybrid Systems III, Verification and Control"*. *LNCS* 1066, pp. 208–219. Springer-Verlag, New York, 1996.

15. M. Emerson, S. Neema, and J. Sztipanovits. Metamodeling languages and metaprogrammable tools. In I. Lee, J. Leung, and S. H. Son, editors, *Handbook of Real-Time and Embedded Systems*. CRC Press, Boca Raton, FL, 2006.

16. E. Farcas, C. Farcas, W. Pree, and J. Templ. Transparent distribution of real-time components based on logical execution time. In *Proceedings of the 2005 ACM SIGPLAN/SIGBED Conference on Languages, Compilers, and Tools for Embedded Systems (LCTES '05)*, New York, pp. 31–39, June 2005. ACM Press.

17. Y. Gurevich. Evolving algebra 1993: Lipari guide. In *Specification and Validation Methods*, pp. 9–36. Oxford University Press, New York, 1995.

18. Y. Gurevich, W. Schulte, and M. Veanes. Toward industrial strength abstract state machines. Technical Report MSR-TR-2001-98, Microsoft Research, October 2001. Tools available: http://www.codeplex.com/AsmL.

19. D. Harel and B. Rumpe. Meaningful modeling: What's the semantics of "semantics"? *IEEE Computer*, 37(10):64–72, 2004.

20. T. A. Henzinger, C. M. Kirsch, M. A. A. Sanvido, and W. Pree. From control models to real-time code using giotto. *Control Systems Magazine*, 2(1):50–64, 2003.

21. The MathWorks Inc. Simulink/Stateflow Tools. http://www.mathworks.com.

22. Vanderbilt University ISIS. Generic Modeling Environment. http://www.isis.vanderbilt.edu/projects/gme.

23. E. Jackson and J. Sztipanovits. Formalizing the structural semantics of domain-specific modeling languages. *Journal of Software and Systems Modeling*, 4:1–28, 2009.

24. E. K. Jackson and J. Sztipanovits. Towards a formal foundation for domain specific modeling languages. *Proceedings of the Sixth ACM International Conference on Embedded Software (EMSOFT'06)*, Seoul, South Korea, pp. 53–62, October 2006.

25. E. Jackson, W. Schulte, and J. Sztipanovits. The power of rich syntax for model-based development. Technical Report MSR-TR-2008-86, Microsoft Research, June 2008.

26. G. Karsai, A. Agrawal, and F. Shi. On the use of graph transformations for the formal specification of model interpreters. *Journal of Universal Computer Science*, 9(11):1296–1321, November 2003.

27. G. Karsai, A. Ledeczi, S. Neema, and J. Sztipanovits. The model-integrated computing toolsuite: Metaprogrammable tools for embedded control system design. In *Proceedings of the IEEE Joint Conference CCA, ISIC and CACSD*, Munich, Germany, September 2006.

28. G. Karsai, J. Sztipanovits, A. Ledeczi, and T. Bapty. Model-integrated development of embedded software. *Proceedings of the IEEE*, 91(1):145–164, January 2003.

29. Edward A. Lee and Alberto L. Sangiovanni-Vincentelli. A denotational framework for comparing models of computation. Technical Report UCB/ERL M97/11, EECS Department, University of California, Berkeley, CA, 1997.

30. S. McConnell. *Rapid Development: Taming Wild Software Schedules*. Microsoft Press, Redmond, WA, 1996.

31. Object Management Group. Mda guide version 1.0.1. Technical Report omg/2003-06-01, 2003.

32. Object Management Group. Unified modeling language: Superstructure version 2.0, 3rd revised submission to omg rfp. Technical report, 2003.

33. Object Management Group. Object constraint language v2.0. Technical report, 2006.

34. K. Ogata. *Modern Control Engineering*. Prentice-Hall, Inc., Englewood Cliffs, NJ, 1970.

35. D. S. Scott. Outline of a mathematical theory of computation. Technical Report Technical Monograph PRGÜ2, Programming Research Group, Oxford University, 1970.

36. J. Templ. Tdl specification and report. Technical Report C059, Dept. of Computer Science, Univ. of Salzburg, 2004. http://www.cs.uni-salzburg.at/pubs/reports/T001.pdf.

37. Verimag. If verification tool. http://www-verimag.imag.fr/ async/IF/ index.htm.

15

Multi-Viewpoint State Machines for Rich Component Models

Albert Benveniste, Benoît Caillaud, and Roberto Passerone

CONTENTS

15.1 Introduction and Requirements

This chapter presents the modeling effort that sustains the workrelated to the IP-SPEEDS heterogeneous rich component (HRC) metamodel, its associated multiple viewpoint contract formalism, and the underlying mathematical model of machines supporting such contracts. We put the emphasis on combining different viewpoints and providing a simple and elegant notion of parallel composition.

The motivations behind this work are the drastic organizational changes that several industrial sectors involving complex embedded systems have experienced—aerospace and automotive being typical examples. Initially organized around large, vertically integrated companies supporting most of the design in house, these sectors were restructured in the 1980s because of the emergence of sizeable competitive suppliers. Original equipment manufacturers (OEM) performed system design and integration by importing entire subsystems from suppliers. This, however, shifted a significant portion of the value to the suppliers, and eventually contributed to late errors that caused delays and excessive additional cost during the system integration phase. In the past decade, these industrial sectors went through a profound reorganization in an attempt by OEMs to recover value from the supply chain, by focusing on those parts of the design at the core of their competitive advantage. The rest of the system was instead centered around standard platforms that could be developed and shared by otherwise competitors. Examples of this trend are AUTOSAR in the automotive industry [10] and integrated modular avionics (IMA) in aerospace [7]. This new organization requires extensive virtual prototyping and design space exploration, where component or subsystem specification and integration occur at different phases of the design, including at the early ones [19].

Component-based development has emerged as the technology of choice to address the challenges that result from this paradigm shift. Our objective is to develop a component-based model that is tailored to the specific requirement of system development with a highly distributed OEM/supplier chain. This raises the novel issue of dividing and distributing responsibilities between the different actors of the OEM/supplier chain. The OEM wants to define and know precisely what a given supplier is responsible for. Since components or subsystems interact, this implies that the responsibility for each entity in the area of interaction must be precisely assigned to a given supplier, and must remain unaffected by others. Thus, each supplier is assigned a design task in the form of a goal, which we call "guarantee or promise" that involves only entities for which the supplier is responsible. Other entities entering the subsystem for design are not under the responsibility of this supplier. Nonetheless, they may be subject to constraints assigned to the other suppliers, that can therefore be offered to this supplier as "assumptions." Assumptions are under the responsibility of other actors of the OEM/supplier chain but can be used by this supplier to simplify the task of achieving its own promises. This mechanism of assumptions and promises is structured into "contracts" [9], which form the essence of distributed system development involving complex OEM/supplier chains.

In addition to contracts, supporting an effective concurrent system development requires the correct modeling of both interfaces and open systems, as well as the ability to talk about partial designs and the use of abstraction mechanisms. This is especially true in the context of safety critical embedded systems. In this case, the need for high-quality, zero-defect software calls for techniques in which component specification and integration are supported by clean mathematics that encompass both static and "dynamic" semantics—this means that the behavior of components and their composition, and not just their port and type interface, must be mathematically defined. Furthermore, system design includes various aspects—functional, timeliness, safety and fault tolerance, etc.—involving different teams with different skills using heterogeneous techniques and tools. We call each of these different aspects a "viewpoint" of the component or of the system. Our technology of contracts is based on a mathematical foundation consisting of a model of system that is rich enough to support the different viewpoints of system design, and at the same time clean and simple enough to allow for the development of mathematically sound techniques. We build on these foundations to construct a more descriptive state-based model, called the HRC model, that describes the relationships between the parts of a component in an executable fashion. It is the objective of this chapter to present this higher level model. Nonetheless, we also provide a quick overview of the contract model it is intended to support—readers interested in details regarding this contract framework are referred to [5,6].

Our notion of contract builds on similar formalisms developed in related fields. For example, a contract-based specification was applied by Meyer in the context of the programming language Eiffel [17]. In his work, Meyer uses "preconditions" and "postconditions" as state predicates for the methods of a class, and "invariants" for the class itself. Similar ideas were already present in seminal work by Dijkstra [12] and Lamport [16] on "weakest preconditions" and "predicate transformers" for sequential and concurrent programs, and in more recent work by Back and von Wright, who introduce contracts [4] in the "refinement calculus" [3]. In this formalism, processes are described with guarded commands operating on shared variables. This formalism is best suited to reason about discrete, untimed process behavior.

More recently, De Alfaro and Henzinger have proposed interface automata as a way of expressing constraints on the environment in the case of synchronous models [11]. The authors have also extended the approach to other kinds of behaviors, including resources and asynchronous behaviors [8,15]. Our contribution here consists in developing a particular formalism for hybrid continuous-time and discrete state machines where composition is naturally expressed as intersection. We show how to translate our model to the more traditional hybrid automata model [14]. In addition, we identify specialized categories of automata for the cases that do not need the full generality of the model, and introduce probabilities as a way of representing failures.

The chapter is structured as follows. We will first review the concepts of component and contract from a semantic point of view in Section 15.2. We then describe the extended state machine (ESM) model in Section 15.3 and compare it to a more traditional hybrid model in Section 15.4. The syntax and the expressive power used for expressions in the transitions of the state-based model is reviewed in Section 15.5, followed, in Section 15.6, by the specialization of the model into different categories to support alternative viewpoints. Several examples complement the formalism throughout the chapter.

15.2 Components and Contracts

Our model is based on the concept of "component." A component is a hierarchical entity that represents a unit of design. Components are connected together to form a system by sharing and agreeing on the values of certain ports and variables. A component may include both "implementations" and "contracts." An implementation M is an instantiation of a component and consists of a set P of ports and variables (in the following, for simplicity, we will refer only to ports) and of a set of behaviors, or runs, also denoted by M, which assign a history of "values" to ports. Because implementations and contracts may refer to different viewpoints, as we shall see, we refer to the components in our model as HRC.

We build the notion of a contract for a component as a pair of assertions, which express its assumptions and promises. An assertion E is a property that may or may not be satisfied by a behavior. Thus, assertions can again be modeled as a set of behaviors over ports, precisely as the set of behaviors that satisfy it. An implementation M satisfies an assertion E whenever they are defined over the same set of ports and all the behaviors of M satisfy the assertion, that is, when $M \subseteq E$.

A contract is an assertion on the behaviors of a component (the promise) subject to certain assumptions. We therefore represent a contract C as a pair (A, G), where A corresponds to the assumption, and G to the promise. An implementation of a component satisfies a contract whenever it satisfies its promise, subject to the assumption. Formally, $M \cap A \subseteq G$, where M and C have the same ports. We write $M \models C$ when M satisfies a contract C. There exists a unique maximal (by behavior containment) implementation satisfying a contract C, namely, $M_C = G \cup \neg A$. One can interpret M_C as the implication $A \Rightarrow G$. Clearly, $M \models (A, G)$ if and only if $M \models (A, M_C)$, if and only if $M \subseteq M_C$. Because of this property, we can restrict our attention to contracts of the form $C = (A, M_C)$, which we say are in "canonical form," without losing expressiveness. The operation of computing the canonical form, that is, replacing G with $G \cup \neg A$, is well defined, since the maximal implementation is unique and idempotent. Working with canonical forms simplifies the definition of our operators and relations, and provides a unique representation for equivalent contracts.

The combination of contracts associated to different components can be obtained through the operation of parallel composition. If $C_1 = (A_1, G_1)$ and $C_2 = (A_2, G_2)$ are contracts (possibly over different sets of ports), the composite must satisfy the guarantees of both, implying an operation of intersection. The situation is more subtle for assumptions. Suppose first that the two contracts have disjoint sets of ports. Intuitively, the assumptions of the composite should be simply the conjunction of the assumptions of each contract, since the environment should satisfy all the assumptions. In general, however, part of the assumptions A_1 will be already satisfied by composing C_1 with C_2, acting as a partial environment for C_1. Therefore, G_2 can contribute to relaxing the assumptions A_1. And vice versa. The assumption and the promise of the composite contract $C = (A, G)$ can therefore be computed as follows:

$$A = (A_1 \cap A_2) \cup \neg(G_1 \cap G_2), \tag{15.1}$$

$$G = G_1 \cap G_2, \tag{15.2}$$

which is consistent with similar definitions in other contexts [11,13,18]. C_1 and C_2 may have different ports. In that case, we must extend the behaviors to a common set of ports before applying Equations 15.1 and 15.2. This can be achieved by an operation of inverse projection. Projection, or elimination, in contracts requires handling assumptions and promises differently, in order to preserve their semantics. For a contract $C = (A, G)$ and a port p, the "elimination of p in C" is given by

$$[C]_p = (\forall p\, A, \exists p\, G) \tag{15.3}$$

where A and G are seen as predicates. Elimination trivially extends to finite sets of ports, denoted by $[C]_P$, where P is the considered set of ports. For inverse elimination in parallel composition, the set of ports P to be considered is the union of the ports P_1 and P_2 of the individual contracts.

Parallel composition can be used to construct complex contracts out of simpler ones, and to combine contracts of different components. Despite having to be satisfied simultaneously, however, multiple viewpoints "associated to the same component" do not generally compose by parallel composition. We would like, instead, to compute the conjunction $\sqcap$ of the contracts, so that if $M \models C_f \sqcap C_t$, then $M \models C_f$ and $M \models C_t$. This can best be achieved by first defining a partial order on contracts, which formalizes a notion of substitutability, or refinement. We say that $C = (A, G)$ "dominates" $C' = (A', G')$, written $C \preceq C'$, if and only if $A \supseteq A'$ and $G \subseteq G'$, and the contracts have the same ports. Dominance amounts to relaxing assumptions and reinforcing promises, therefore strengthening the contract. Clearly, if $M \models C$ and $C \preceq C'$, then $M \models C'$.

Given the ordering of contracts, we can compute greatest lower bounds and least upper bounds, which correspond to taking the conjunction and disjunction of contracts, respectively. For contracts $C_1 = (A_1, G_1)$ and $C_2 = (A_2, G_2)$ (in canonical form), we have

$$C_1 \sqcap C_2 = (A_1 \cup A_2, G_1 \cap G_2), \qquad (15.4)$$

$$C_1 \sqcup C_2 = (A_1 \cap A_2, G_1 \cup G_2). \qquad (15.5)$$

The resulting contracts are in canonical form. Conjunction of contracts amounts to taking the union of the assumptions, as required, and can therefore be used to compute the overall contract for a component starting from the contracts related to multiple viewpoints. The following example illustrates the need for two different composition operators.

Example 15.1 (Viewpoint Synchronization) *We discuss here an example of viewpoint synchronization. Assume two contracts $C_i, i = 1, 2$ modeling two different viewpoints attached to a same rich component $\mathbf{C}$. For example, let $C_1 = (A_1, G_1)$ be a viewpoint in the functional category and $C_2 = (A_2, G_2)$ be a viewpoint of the timed category.*

Assumption A_1 specifies allowed data pattern for the environment, whereas A_2 sets timing requirements for it. Since contracts are in canonical forms, the promise G_1 itself says that, if the environment offers the due data pattern, then a certain behavioral property can be guaranteed. Similarly, G_2 says that, if the environment meets the timing requirements, then outputs will be scheduled as wanted and deadlines will be met. Thus, both $G_i, i = 1, 2$ are implications.

The greatest lower bound $C_1 \sqcap C_2$ can accept environments that satisfy either the functional assumptions, or the timing assumptions, or both. The promiseof $C_1 \sqcap C_2$ is the conjunction of the two implications: If the environment offers the due data pattern, then a certain behavioral property can be guaranteed, and, if the environment meets the timing requirements, then outputs will be scheduled as wanted and deadlines will be met. When both the environment offers the due data pattern and the environment meets the timing requirements, remark that both a certain behavioral property can be guaranteed and outputs will be scheduled as wanted and deadlines will be met.

To have a closer look at the problem, assume first that the two viewpoints are orthogonal or unrelated, meaning that the first viewpoint, which belongs to the functional category, does not depend on dates, while the second viewpoint does not depend on the functional behavior (e.g., we have a dataflow network of computations that is fixed regardless of any value at any port). Let these two respective viewpoints state as follows:

- If the environment alternates the values T, F, T, ... on port b, then the value carried by port x of component $\mathbf{C}$ never exceeds 5.
- If the environment provides at least one data per second on port b, then component $\mathbf{C}$ can issue at least one data every 2 s on port x.

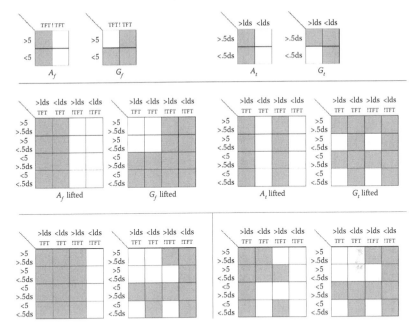

FIGURE 15.1

Truth tables for the synchronization of categories.

These two viewpoints relate to the same rich component. Still, having the two contracts $(A_i, G_i), i = funct, timed$ for **C** should mean that if the environment satisfies the functional assumption, then **C** satisfies the functional guarantees. Also, if the environment satisfies the timing assumption, then **C** satisfies the timing guarantees. Figure 15.1 illustrates the greatest lower bound of the viewpoints belonging to two different categories, and compares it with their parallel composition, introduced in Section 15.2. For this case, the right definition for viewpoint synchronization is the greatest lower bound.

The four diagrams on the top are the truth tables of the functional category C_f and its assumption A_f and promise G_f, and similarly for the timed category C_t. Note that these two contracts are in canonical form. In the middle, we show the same contracts lifted to the same set of variables b, d_b, x, and d_x, combining function and timing. On the bottom, the two tables on the left are the truth tables of the greatest lower bound $C_f \sqcap C_t$. For comparison, we show on the right the truth tables of the parallel composition $C_1 \parallel C_2$, revealing that the assumption is too restrictive and not the one expected.

So far we discussed the case of noninteracting viewpoints. But in general, viewpoints may interact as explained in the following variation of the same example. Assume that the viewpoints (the first one belonging to

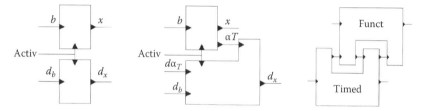

FIGURE 15.2
Illustrating the synchronization of viewpoints.

the functional category, while the other one belongs to the timed category)
interact as follows:

- If the environment alternates the values T, F, T, ... on port b, then the
  value carried by port x of **C** never exceeds 5; if x outputs the value 0,
  then an exception is raised and a handling task T is executed.
- If the environment provides at least one data per second on port b, then
  C can issue at least one data every 2 s on port x; when executed, task T
  takes 5 s for its execution.

For this case, the activation port α_T of task T is an output port of the functional view, and an input port of the timed view. This activation port is
boolean; it is output every time the component is activated and is true when
an exception is raised. Then, the timed viewpoint will involve α_T and d_{α_T} as
inputs, and will output the date d_T of completion of the task T according to
the following formula: $d_T = (d_{\alpha_T} + 5)$ when $(\alpha_T = \text{T})$. Note that d_{α_T} has no
meaning when $\alpha_T = \text{F}$.

Here we had an example of connecting an output of a functional viewpoint to an input of a timed viewpoint. Note that the converse can also occur.
Figure 15.2 illustrates the possible interaction architectures for a synchronization viewpoint.

Discussion. So far we have defined contracts and implementations in
terms of abstract assertions, that is, sets of runs. In Sections 15.3 and 15.4,
we describe in more precise terms the mathematical nature of these abstract
assertions.

To provide intuition for our design choices, we start by comparing two
alternative views of system runs, as illustrated in Figure 15.3. In the classical
approach, shown in Figure 15.3a, transitions take no time; time and continuous dynamics progress within states; they are specified by state invariants
and guarded. The alternative approach is dual: states are snapshot valuations of variables and take no time; time and continuous dynamics progress
within "thick" transitions that are guarded.

The two approaches have advantages and disadvantages. The classical
approach is preferred for abstractions based on regions, which are valid for
certain classes of models. The alternative approach makes it much easier to

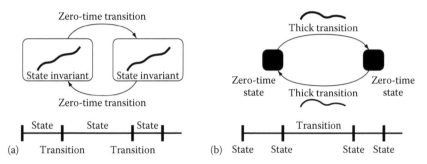

FIGURE 15.3
Runs. (a) Classical approach. (b) Alternative approach. State invariants on the left, or thick transitions on the right, involve the progress of time and continuous dynamics such as differential equations.

deal with composition and is able to capture open systems, as we shall see. Clearly, the two approaches are dual and can be exchanged without harm.

We shall develop the two approaches and relate them throughout this chapter.

15.3 Extended State Machines

ESMs follow the second approach illustrated in Figure 15.3. They are our preferred model, because of the simplicity of its associated parallel composition.

15.3.1 Variables and Ports, Events and Interactions, Continuous Dynamics

Interaction between viewpoints and components is achieved by synchronizing events and variables involved in both discrete and continuous dynamics. Synchronization events take place on ports. Dynamic creation or deletion of ports or variables is not supported by the model.

Values are taken in a unique domain D that encompasses all usual data types (booleans, enumerated types, integers, real numbers, etc.). We shall distinguish a subdomain $D_c \subset D$ in which variables involved in continuous evolutions take their values; D_c collects all needed Euclidean spaces to deal with differential equations or inclusions. Other type consistency issues are dealt within the static semantics definition of HRC and are disregarded in the sequel.

We are given a finite set V of "variables"; the set of variables is partitioned into two subsets $V = V_d \uplus V_c$: the variables belonging to V_d are used exclusively in the discrete evolutions, and those belonging to V_c can be used in both continuous and discrete evolutions. "States" correspond to the assignment of a value to each variable: $s : V \to D$.

A finite set of *ports* $\mathcal{P}$ is then considered. "Events" correspond to the assignment of a value to a port; therefore an event is a pair $(p, d) \in \mathcal{P} \times D$. "Interactions," also called "labels" in the sequel, are sets of events. The only restriction is that a given port may yield at most one event in an interaction. Hence interactions are partial mappings $\lambda : \mathcal{P} \rightharpoonup D$. The set of all interactions is denoted by Λ ($= \mathcal{P} \rightharpoonup D$). The empty interaction $\varepsilon_\mathcal{P}$ over ports $\mathcal{P}$ is the unique mapping $\varepsilon_\mathcal{P} : \mathcal{P} \rightharpoonup D$ that is undefined for any $p \in \mathcal{P}$.

Regarding continuous dynamics, we restrict ourselves to the case where a unique global physical time is available, denoted generically by the symbols t or τ and called the "universal time." Other time bases can be used, but need to be related to this universal time as part of the assertion specification. Investigating the consequences of relaxing this restriction is part of our future work.

Similarly, for $V_c \subseteq \mathcal{V}_c$, the "domain of continuous evolutions on V_c," denoted by $\mathcal{C}(V_c)$, is the set of all functions

$$\mathcal{C}(V_c) =_{\text{def}} \{ \varphi \mid \varphi : \mathbb{R}_+ \rightharpoonup V_c \to D_c \} \tag{15.6}$$

such that (we write $\varphi(t, v)$ instead of $\varphi(t)(v)$):

1. $\text{dom}(\varphi) = [0, t|$ for some $t > 0$, where symbol $|$ denotes either $]$ or $)$; call t the "duration" of φ and denote it generically by t_φ.
2. For every $v \in V_c$, $\tau \to \varphi(\tau, v)$ is smooth enough (typically at least differentiable on $(0, t)$) and possesses a left limit $Exit(\varphi) \in D^{V_c}$ defined by

$$Exit(\varphi, v) =_{\text{def}} \lim_{\tau \nearrow t} \varphi(\tau, v) \tag{15.7}$$

Each $\varphi \in \mathcal{C}(V_c)$ can be decomposed, for all $t \in (0, t_\varphi)$, as the concatenation

$$\varphi = \varphi_1 \cdot \varphi_2, \text{ where}$$
$$\varphi_1(\tau) = \varphi(\tau) \quad \text{for } 0 \leq \tau < t, \quad \text{dom}(\varphi_1) = [0, t)$$
$$\varphi_2(\tau) = \varphi(t + \tau) \quad \text{for } 0 \leq \tau < t_\varphi - t, \quad \text{dom}(\varphi_2) = [0, t_\varphi - t) \tag{15.8}$$

We denote these two evolutions by $\varphi_{<t}$ and $\varphi_{\geq t}$, respectively. We thus have the decomposition

$$\varphi = \varphi_{<t} \cdot \varphi_{\geq t} \tag{15.9}$$

15.3.2 ESM Definition

Having defined variables, ports, labels and interactions, it is possible to introduce ESMs as a syntactic means of defining assertions in HRC components.

Definition 15.1 (ESM) An ESM is a tuple with the following components:

$E = (V, P, \rho, \delta, I, F)$, where

$P \subseteq \mathcal{P}, V = V_d \uplus V_c, V_d \subseteq \mathcal{V}_d, V_c \subseteq \mathcal{V}_c$

$S =_{\text{def}} D^V$ is the set of states, projecting to

$S_d =_{\text{def}} D^{V_d}$ the set of discrete states, and

$S_c =_{\text{def}} D^{V_c}$ the set of continuous states,

$\rho \subseteq S \times \Lambda \times S$, where $\Lambda =_{\text{def}} (P \rightharpoonup D)$, is the discrete transition relation,

$\delta \subseteq S \times \mathcal{C}(V_c) \times S$ is the continuous transition relation,

$I \subseteq S$ is the set of initial states,

$F \subseteq S$ is the set of final states,

where we require that δ does not modify discrete states:

$$\forall (s, \varphi, s') \in \delta, \forall v \in V_d \Rightarrow s'(v) = s(v). \tag{15.10}$$

For convenience, we shall denote the disjoint union of sets of ports and variables by $W =_{\text{def}} P \uplus V$.

Runs. The runs recognized by an ESM are arbitrary finite interleavings of discrete and continuous evolutions, separated by snapshot states:

$$\sigma =_{\text{def}} s_0, w_1, s_1, w_2, s_2, \ldots, s_{k-1}, w_k, s_k, \ldots \tag{15.11}$$

where

$$\forall k > 0 : \begin{cases} s_0 & \in & I \\ \text{either} & w_k & = & (s_{k-1}, \lambda_k, s_k) & \in & \rho \\ \text{or} & w_k & = & (s_{k-1}, \varphi_k, s_k) & \in & \delta \end{cases}$$

Infinite runs are captured by considering their finite approximations. "Accepted runs" are finite runs ending in F. To capture nonterminating computations, just take $F = S$. In run σ, time progresses as follows: Discrete transitions take no time and continuous evolutions are concatenated. Formally

- State s_k is reached at time $\sum_{i=1}^{k} t_{w_i}$, where t_w denotes the duration of w; by convention, t_w is equal to t_φ (the duration of φ) if $w = (s, \varphi, s')$, and is equal to zero if $w = (s, \lambda, s')$.
- At time t, the number of transitions completed is $\max\{k \mid \sum_{i=1}^{k} t_{w_i} \leq t\}$.

Projection. For $W = P \oplus V$ a set of ports and variables, ρ a discrete transition relation defined over W, δ a continuous transition relation defined over W, and $W' \subseteq W$, let $\mathbf{proj}_{W,W'}(\rho)$ and $\mathbf{proj}_{W,W'}(\delta)$, respectively denote the projections of ρ and δ over W', obtained by existential elimination of ports or variables not belonging to W'. The results are discrete and continuous transition relations defined over W', respectively. Corresponding inverse projections are denoted by $\mathbf{proj}_{W,W'}^{-1}(\ldots)$.

Product. The composition of ESM is by intersection; interaction can occur via both variables and ports:

$$E_1 \times E_2 = (V, P, \rho, \delta, I, F), \text{ where}$$
$$V_d = V_{d,1} \cup V_{d,2} \text{ discrete variables can be shared}$$
$$V_c = V_{c,1} \cup V_{c,2} \text{ continuous variables can be shared}$$
$$P = P_1 \cup P_2 \text{ ports can be shared}$$
$$\rho =_{def} \mathbf{proj}_{W,W_1}^{-1}(\rho_1) \cap \mathbf{proj}_{W,W_2}^{-1}(\rho_2)$$
$$\delta =_{def} \mathbf{proj}_{W,W_1}^{-1}(\delta_1) \cap \mathbf{proj}_{W,W_2}^{-1}(\delta_2)$$
$$I =_{def} \mathbf{proj}_{W,W_1}^{-1}(I_1) \cap \mathbf{proj}_{W,W_2}^{-1}(I_2)$$
$$F =_{def} \mathbf{proj}_{W,W_1}^{-1}(F_1) \cap \mathbf{proj}_{W,W_2}^{-1}(F_2)$$

where we recall that $W = P \uplus V$. ESMs synchronize on discrete transitions thanks to shared ports and variables. Continuous evolutions synchronize only via shared variables. If $W = W_1 = W_2$, then $\rho = \rho_1 \cap \rho_2$ and $\delta = \delta_1 \cap \delta_2$, whence the name of "composition by intersection." When specialized to continuous dynamics made of differential equations, this boils down to systems of differential equations like in undergraduate mathematics.

Our interpretation of runs with snapshot states and thick transitions (see Figure 15.3) is instrumental in allowing for the above simple and elegant definition of parallel composition "by intersection." With thick states and zero-time transitions, it is more difficult to define composition, because synchronization takes place both on states and transitions.

Union or disjunction. The union of two sets of runs can be obtained from two ESMs by taking the union of their variables, and by adding a distinguished variable $\# \in V_c$ that indicates the particular state space in which we are operating ($\# = 0$ for the first ESM, $\# = 1$ for the second). Then, we simply take the union of the transition relations after inverse projection. Formally, for i indexing the set of components involved in the considered union, let

$$\rho\Big|_{\#=i}^{V} =_{def} \{(s, \lambda, s') \in S \times \Lambda \times S \mid s(\#) = i \text{ and } s'(\#) = i\}$$
$$\delta\Big|_{\#=i}^{V} =_{def} \{(s, \varphi, s') \in S \times C(V_c) \times S \mid s(\#) = i \text{ and } s'(\#) = i\}$$

be the transition relation that is true everywhere variable # is evaluated to i. Then

$$E_1 \cup E_2 = (V, P, \rho, \delta, I, F)$$
$$V_d = V_{d,1} \cup V_{d,2} \uplus \{\#\}$$
$$V_c = V_{c,1} \cup V_{c,2} \uplus \{\#\}$$
$$P = P_1 \cup P_2$$
$$\rho =_{\text{def}} \left(\mathbf{proj}^{-1}_{W,W_1}(\rho_1) \cap \rho|^V_{\#=1}\right) \cup \left(\mathbf{proj}^{-1}_{W,W_2}(\rho_2) \cap \rho|^V_{\#=2}\right)$$
$$\delta =_{\text{def}} \left(\mathbf{proj}^{-1}_{W,W_1}(\delta_1) \cap \rho|^V_{\#=1}\right) \cup \left(\mathbf{proj}^{-1}_{W,W_2}(\delta_2) \cap \rho|^V_{\#=2}\right)$$
$$I =_{\text{def}} \{s \in S \mid s_{|W_1} \in I_1 \wedge s(\#) = 1\} \cup \{s \in S \mid s_{|W_2} \in I_2 \wedge s(\#) = 2\}$$
$$F =_{\text{def}} \{s \in S \mid s_{|W_1} \in F_1 \wedge s(\#) = 1\} \cup \{s \in S \mid s_{|W_2} \in F_2 \wedge s(\#) = 2\}$$

Inputs and outputs. Whenever needed we can distinguish inputs and outputs, which we also call "uncontrolled" and "controlled" ports. In this paragraph, we define the corresponding algebra. Ports and variables are partitioned into inputs and outputs:

$$P = P^I \uplus P^O$$
$$V = V^I \uplus V^O$$

Regarding products, the set of ports of a product is again the union of the set of ports of each component. However, outputs cannot be shared.* That is, the product of two ESMs E_1 and E_2 is defined if and only if

$$P^O_1 \cap P^O_2 = \emptyset$$
$$V^O_1 \cap V^O_2 = \emptyset \tag{15.12}$$

In that case

$$P^I = (P^I_1 \cup P^I_2) - (P^O_1 \cup P^O_2)$$
$$P^O = P^O_1 \cup P^O_2 \tag{15.13}$$

with the corresponding rules for variables.

Receptiveness. For E an ESM, and $P' \subseteq P, V' \subseteq V$ a subset of its ports and variables, E is said to be (P', V')-*receptive* if and only if for all runs σ' restricted to ports and variables belonging to (P', V'), there exists a run in σ of E such that σ' and σ coincide over $P' \uplus V'$.

---

*We could allow sharing of outputs, and declare a failure whenever two components set a different value on an output port.

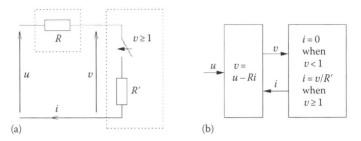

FIGURE 15.4
Nonreceptive composition of two receptive ESMs. (a) Electric circuit with two components. (b) Modeling of the circuit with two receptive ESMs.

Receptiveness is a semantic concept. It is often implicitly meant that an ESM should be receptive with respect to its inputs. However, the example in Figure 15.4 shows that receptiveness is generally not preserved by composition, even when Condition 15.12 is satisfied and Rule 15.13 is used for the composition. This example aims at modeling an electric circuit with two components (Figure 15.4a), a resistor R and a voltage sensitive switch that is opened when $v < 1$ and has resistance R' when $v \geq 1$. The ESM for resistor R (Figure 15.4b) inputs voltage u and current i and outputs voltage v. The switch ESM inputs voltage v and outputs current i. Each ESM is receptive: $v = u - Ri$ is the output of the first ESM for every value of u and i. The second ESM outputs $i = v/R'$ when $v \geq 1$ and $i = 0$ otherwise. The composition of these two ESMs has u as only input and v and i as outputs. The system of equations admits a solution when $u < 1$, in which case $v = u$ and $i = 0$, and when $u \geq 1 + R/R'$, in which case $v = R'/(R + R')u$ and $i = u/(R + R')$. However, it has no solution when $u \in [1; 1 + R/R']$. Clearly, the composition of the two ESMs is not receptive.

Openness. The ability to handle open systems is an important feature of ESMs. This can be achieved by requiring that the following conditions hold for discrete and continuous transitions:

$$\{(s, \varepsilon_P, s) \mid s \in S\} \subseteq \rho \tag{15.14}$$

$$\left.\begin{array}{c} (s, \varphi, s') \in \delta \\ t < t_\varphi \end{array}\right\} \Rightarrow \left\{\begin{array}{l} (s, \varphi_{<t}, Exit\,(\varphi_{<t})) \in \delta \\ (Exit\,(\varphi_{<t}), \varphi_{\geq t}, s') \in \delta \end{array}\right. \tag{15.15}$$

where, by abuse of notation, we extend $Exit\,(\varphi_{<t})$ to the set of discrete variables by copying the value they had in state S.

Condition 15.14 on discrete evolutions is the usual stuttering invariance condition for discrete transition systems. It requires that it is always possible, for an ESM, to perform a discrete stuttering transition that emits no event and leaves states unchanged. This leaves room for other components to perform discrete transitions.

Condition 15.15 on continuous evolutions expresses that it is always possible to interrupt a continuous evolution and resume it immediately. The reason for doing this is to allow other components to perform a discrete transition (which takes no time) in the middle of a continuous evolution of the considered component.

Observe that conditions for openness imply that any finite run can be extended to an infinite one; whence our definition for accepted runs.

Locations or macrostates. Certain aggregations of states are useful for use in actual syntax. For example, hybrid systems "locations" contain the continuous evolutions. Also, *macro-states* are considered when gluing states together. Locations or macro-states are obtained in our framework by

1. Selecting a subset $V'_d \subset V_d$ of discrete variables
2. Grouping together all states s having the same valuation for all $w \in V'_d$.

For example, one could have one particular discrete variable $w \in V_d$, of enumerated type, that indexes the locations; in this case we would take $V'_d = \{w\}$. Note that the description of the dynamics still requires the discrete and continuous transitions as above. This is further elaborated on in Section 15.4.

15.4 HRC State Machines

In this section, we introduce the model that corresponds to the first (classical) approach illustrated in Figure 15.3. Its interest is that it more closely fits the type of model in use when considering timed automata [1] or their generalization hybrid automata [14]. We call this model "HRC state machines." Then we show how to translate between HRC state machines and ESMs, thus providing a way to switch to the best framework depending on the considered activity (analysis or composition). To simplify our presentation, we consider only flat HRC state machines that do not include hierarchical or-states such as in statecharts. Extension to hierarchical or-states raises no difficulty.

Inspired by the definition of hybrid automata in Henzinger [14], we define:

Definition 15.2 (HRC State Machine) A *HRC state machine* is a tuple

$$\mathcal{H} = (V, P; \mathbf{G}, init, inv, flow, final; trans) \tag{15.16}$$

where

- $V = V_d \uplus V_c$ is a finite set of variables decomposed into discrete and continuous variables; set $S = D^V$, where D is the domain of values.
- P is a finite set of ports.

- **G** is a finite directed multigraph $\mathbf{G} = (\mathbf{L}, \mathbf{E})$, where **L** is a finite set of locations and **E** is a finite set of switches.
- Four vertex labeling functions *init, inv, flow,* and *final,* that assign to each location $\ell \in \mathbf{L}$ four predicates; $init(\ell), inv(\ell),$ and $final(\ell)$ are expressions of boolean type over V, and $flow(\ell) \subseteq \mathcal{C}(V_c)$, see (15.6).
- An edge labeling function *trans* that assigns to each switch $\mathbf{e} \in \mathbf{E}$ a relation $trans(\mathbf{e}) \subseteq S \times \Lambda \times S$, where $\Lambda =_{\text{def}} (P \rightharpoonup D)$.

HRC State Machine $\mathcal{H}$ can be re-expressed as the following equivalent ESM (in that they possess identical sets of runs):

$$E_{\mathcal{H}} = \left(V \uplus \{loc\}, P, \rho, \delta, I, F\right),$$

where

- V is as in Equation 15.16 and *loc* is an additional "location variable" taking values in the finite set **L**; a value for *loc* is therefore a location ℓ; the corresponding set of states is the set of all possible configurations of the tuple (V, loc); such states are generically written as pairs (s, ℓ).
- P is as in Equation 15.16.
- The discrete transition relation ρ is defined as follows:

$$\left((s, \ell), \lambda, (s', \ell')\right) \in \rho$$

if and only if there exists a switch **e** with source ℓ and target ℓ' such that $(s, \lambda, s') \in trans(\mathbf{e})$.
- The continuous transition relation δ is defined as follows:

$$\left((s, \ell), \varphi, (s', \ell')\right) \in \delta$$

if and only if $\ell' = \ell$ and continuous evolution φ satisfies both predicates $inv(\ell)$ and $flow(\ell)$.
- $(s_0, \ell_0) \in I$ if and only if $inv(\ell_0)(s_0) = \text{T}$ and $init(\ell_0)(s_0) = \text{T}$.
- $(s_f, \ell_f) \in F$ if and only if $inv(\ell_f)(s_f) = \text{T}$ and $final(\ell_f)(s_f) = \text{T}$.

Conversely, let $E = (V, P, \rho, \delta, I, F)$ be an ESM in which a subset $loc \subset V_d$ of discrete variables has been distinguished. Then, E can be represented as the following HRC state machine:

$$\mathcal{H}_E = (W, P; \mathbf{G}, init, inv, flow, final; trans) \tag{15.17}$$

where $W =_{\text{def}} V - loc$ and

- $\mathbf{G} = (\mathbf{L}, \mathbf{E})$, where $\mathbf{L} = D^{loc}$ and $\mathbf{e} = (\ell, \ell') \in \mathbf{E}$ if and only if there exists an event $\lambda \in \Lambda$ of E, such that $(\ell, \lambda, \ell') \in \mathbf{proj}_{V, loc}(\rho)$.
- For $\mathbf{e} = (\ell, \ell') \in \mathbf{E}$, $(s, \lambda, s') \in trans(\mathbf{e})$ if and only if $((s, \ell), \lambda, (s', \ell')) \in \rho$.
- For $\ell \in \mathbf{L}$, $inv(\ell)$ is satisfied by state s if and only if $((s, \ell), \lambda, (s', \ell')) \in \rho$, for some event λ, some switch $\mathbf{e} = (\ell, \ell') \in \mathbf{E}$, and some state $s' \in D^W$.

- Since, by Equation 15.10, continuous transition relation δ does not modify discrete states, it does not modify locations. Therefore, if $(s, \varphi, s') \in \delta$, then $s(loc) = s'(loc)$, we denote it by ℓ; then $flow(\ell)$ is the set of $\varphi \in \mathcal{C}(V_c)$ such that there exists a pair of states (s, s') with $\ell = s(loc) = s'(loc)$ and $(s, \varphi, s') \in \delta$.
- $init(\ell)$ is satisfied by state $s \in D^W$ if and only if the pair (ℓ, s) belongs to I.
- $final(\ell)$ is satisfied by state $s \in D^W$ if and only if the pair (ℓ, s) belongs to F.

The following are natural questions: how does $\mathcal{H}_{E_{\mathcal{H}}}$ relate to $\mathcal{H}$? and how does $E_{\mathcal{H}_E}$ relate to E? These are not strictly identical but "almost" so. More precisely

- $\mathcal{H}_{E_{\mathcal{H}}}$ is identical to $\mathcal{H}$.
- $E_{\mathcal{H}_E}$ identifies with E in which the subset $loc \subseteq V_d$ of discrete variables has been replaced by a single variable whose domain is the product of the domains of variables belonging to loc.

Having the translation of HRC state machines into ESMs allows them to inherit from the various operators associated with ESMs. In particular

$$\mathcal{H}_1 \times \mathcal{H}_2 = \mathcal{H}_{E_{\mathcal{H}_1} \times E_{\mathcal{H}_2}}$$

where, in defining $\mathcal{H}_{E_{\mathcal{H}_1} \times E_{\mathcal{H}_2}}$, we take $loc = loc_1 \uplus loc_2$. This is an indirect definition for the product—it can also be used to define other operators on HRC state machines. It involves the (somehow complex) process of translating HRC state machines to ESMs and vice versa. But one should remember that defining the product directly on HRC State Machines is complicated as well. Our technique has the advantage of highlighting the very nature of product, namely, by intersection.

15.5 Mathematical Syntax for the Labeling Functions of HRC State Machines

In this section, we refine the definition of the labeling functions occurring in Definition 15.2 of HRC state machines. Location or vertex labeling functions *init, inv, final,* and *flow* are specified by using expressions. Switch or edge labeling function *trans* will be specified via a pair (guard, action), where the guard is composed of a predicate over locations and variables and a set of triggering events on ports; the action consists in assigning the next state following the transition. Guards and actions will also be specified by means of expressions.

15.5.1　Expressions and Differential Expressions

We consider two distinct copies $\mathcal{V}$ and $\mathcal{V}'$ of the set of all variables, where each $V' \in \mathcal{V}'$ is the primed version of $V \in \mathcal{V}$.

Expressions.　We assume a family *Expr* of "expressions" over unprimed variables, primed variables, and ports. Thus all (partial) functions we introduce below are expressed in terms of *Expr*. Expressions are generically denoted by the symbol **E**. Whenever needed, we shall define subfamilies *Expr'* $\subset$ *Expr*. This mechanism will be useful when we need to make the mathematical syntax of special families of expressions precise.

Expressions over ports.　In particular, we shall distinguish $Expr_{pure}$, the family of "expressions over ports of type" **pure** (carrying no value) which involve the special operator "present" and the three combinators $\vee, \wedge, \ominus$:

$$\text{"present"}(p) \text{ is true iff } p \text{ occurs}$$

$$p_1 \vee p_2 \text{ occurs iff } p_1 \text{ occurs or } p_2 \text{ occurs}$$

$$p_1 \wedge p_2 \text{ occurs iff } p_1 \text{ occurs and } p_2 \text{ occurs}$$

$$p_1 \ominus p_2 \text{ occurs iff } p_1 \text{ occurs but } p_2 \text{ does not occur} \qquad (15.18)$$

where the expression "p occurs" means that p is given a value in the considered transition (see the last bullet in Definition 15.2).

Differential expressions.　Let

$$Expr_{|\mathcal{V}_c} \subset Expr$$

be the subfamily of expressions involving only variables belonging to $\mathcal{V}_c$. Let $Expr_{cont}$ be the set of "differential expressions," recursively defined as

$$Expr_{cont} \supseteq Expr_{|\mathcal{V}_c}$$

$$\forall \mathbf{E} \in Expr_{cont} \Rightarrow \frac{d}{dt}(\mathbf{E}) \in Expr_{cont} \qquad (15.19)$$

where $\frac{d}{dt}(\mathbf{E})$ denotes the time derivative $\frac{d\mathbf{E}}{dt}$ of the continuous evolution of the valuation of **E**. Thus, expressions such as $\mathbf{E} \in C$, where $\mathbf{E} \in Expr_{cont}$ and C is a subset of D_c, specify "differential inclusions" [2]. Continuous evolutions defined in Equation 15.6 are specified with the following syntax:

$$\mathbf{E} \in C \text{ where } \mathbf{E} \in Expr_{cont} \text{ and } C \subseteq D_c$$

For $\mathbf{E} \in Expr_{cont}$, let *Exit* (**E**) be the left limit of the valuation of **E** at the maximal instant t of its domain, as shown in (Equation 15.7).

15.5.2　Invariants

An "invariant" is the association to a location of a pair $(inv, flow)$ (see Definition 15.2). Invariants are generically denoted by symbol ι (the greek letter "iota"). Invariant *inv* is expressed by means ot expressions, whereas invariant "flow" uses differential expressions.

15.5.3 Mathematical Syntax for Transition Relation *trans*

Referring to the last bullet of Definition 15.2, the switch labeling function *trans* is specified as a pair (γ, α) of a guard γ and an action α so that

$$(s, \lambda, s') \in trans$$

iff

$$(s, \lambda) \models \gamma \text{ (the guard)} \ \wedge \ s' \in \alpha(s, \lambda) \text{ (the action)}$$

The pair (γ, α) must be such that

$$\mathrm{dom}(\alpha) \supseteq \left\{ (s, \lambda) \in \overline{S} \ \middle| \ (s, \lambda) \models \gamma \right\},$$

where $\overline{S}$, guards γ, and actions α, are defined next.

Guards. Guards consist of a predicate over (previous) states and a set of triggering events on ports. We group the two by introducing the notion of "extended states," which consist of states augmented with valuations of ports. Formally (see Definition 15.2):

$$\overline{S} =_{\mathrm{def}} D^V \uplus \Lambda$$

A "guard" is a predicate over extended states:

$$\gamma : \overline{S} \to \{F, T\}$$

We say that an extended state (s, l) satisfies γ, written $(s, l) \models \gamma$, if $\gamma(s) = T$. Guards can be defined as boolean-valued expressions involving (unprimed) state variables and ports. Expressions over ports introduced in Equation 15.18 are important for guards, in order to be able to refer to the presence/absence of certain ports in a given transition.

Actions. An "action" is a partial nondeterministic function over extended states:

$$\alpha : \overline{S} \rightharpoonup \wp(S')$$

where $\wp$ denotes power set. Actions assign values to primed variables, nondeterministically. It is allowed for actions to be nondeterministic in order to support partial designs.

Whereas guards generally need to use $Expr_{\mathrm{pure}}$ over ports, this is not needed for actions. Thus, the action language can be "classical," in that it does not need to involve $Expr_{\mathrm{pure}}$ over ports, that is, the presence/absence of selected ports in defining the considered action. Specifying this is the role of the guard, whereas the action that follows can be restricted to refer only to values carried by ports that are known to be present in the considered transition. Whenever needed, auxiliary ports of the form $p = p_1 \vee p_2$ or $p' = p_1 \ominus p_2$ can be introduced for that purpose, when defining the guard.

15.5.4 Products in Terms of Guards and Actions

We return now to our formalism of ESM, where products are naturally defined. The above mathematical syntax for HRC state machines induces a corresponding mathematical syntax for ESMs. Accordingly, the product of two ESMs $E = E_1 \times E_2$ is refined as follows:

$$\text{invariants: } \iota = \iota_1 \wedge \iota_2$$
$$\text{guards: } \gamma = \gamma_1 \wedge \gamma_2$$
$$\text{actions: } \alpha = \mathbf{proj}_{W,W_1}^{-1}(\alpha_1) \cap \mathbf{proj}_{W,W_2}^{-1}(\alpha_2) \qquad (15.20)$$

This formula has several interesting special cases:

- If $\gamma_i, i = 1, 2$ involves only ports of type "pure," then $\gamma_1 \wedge \gamma_2$ in Equation 15.20 expresses that the two ESMs must synchronize on their shared ports.
- If $\iota_i, i = 1, 2$ involves only flows, then $\iota_1 \wedge \iota_2$ in Equation 15.20 denotes the system consisting of the continuous evolutions for the two ESMs.
- If $\gamma_i, i = 1, 2$ involves only ports x, y, z, where y is shared, and has the form

$$\gamma_1 : y = f(x)$$
$$\gamma_2 : z = g(y)$$

then $\gamma_1 \wedge \gamma_2$ in Equation 15.20 denotes the conjunction of $y = f(x)$ and $z = g(y)$. This case captures the composition mechanism of dataflow formalisms; thus the composition mechanism of dataflow formalisms is supported by guards, not by actions. Note that the dependency of z on x through y is immediate, that is, involves no logical delay.
- If $\gamma_i, i = 1, 2$ has the form

$$\gamma_1 : y = f(x)$$
$$\gamma_2 : z = g(v_y)$$

where y is a port and v_y is a state variable storing the value of y at previous transition

$$\alpha_2 : v'_y := y$$

then $\gamma_1 \wedge \gamma_2$ introduces a "logical delay" in the composition of the two systems.

Thus, we see here a simple syntactic condition to ensure the existence of a logical delay from input ports to output ports while composing two ESMs.

15.6 Categories as Specialization of HRC State Machines

We now specialize our model of HRC state machine into several *categories* of assertions, or "viewpoints," generically denoted by the symbol Γ. This is achieved by

1. Restricting the subset of ports and variables that characterize a category; formally, we define subsets $\mathcal{P}_\Gamma \subseteq \mathcal{P}$ and $\mathcal{V}_\Gamma \subseteq \mathcal{V}$.
2. Specializing how the two transition relations ρ and δ restrict to these ports and variables.

We do not need to define the synchronization of different assertions/viewpoints, as this is just a particular case of product of HRC state machine. In fact, our HRC state machine model has built-in cross-category heterogeneity. In the next sections, we define basic categories considered within HRC.

Semantic atoms. For categories other than "discrete," we also provide the "semantic atoms," that is, the minimal set of building blocks that are sufficient for building any model belonging to the considered viewpoint. Semantic atoms must be combined with a suitable model that belongs to the discrete viewpoint. They will be defined in terms of the mathematical syntax of Section 15.5. (The paragraphs on atoms can be skipped for a first reading.)

15.6.1 Discrete Functional Category

In a pure discrete HRC state machines the continuous dynamics is trivial. Allowed ports and variables for this category are

$$\mathcal{P}_\Gamma = \mathcal{P}$$
$$\mathcal{V}_\Gamma = \mathcal{V}_d$$
$$\mathit{flow} = \mathbf{Triv}$$

Since $\mathcal{V}_d = \emptyset$, continuous evolutions $\varphi : \mathbb{R}_+ \rightharpoonup \emptyset \to \emptyset$ are all trivial: They just let time progress until their duration t_φ has elapsed and perform nothing else. We call **Triv** the set of all trivial continuous evolutions—note that these are entirely parameterized by their duration. Composing with **Triv** has no effect for continuous evolutions.

15.6.2 Timed Category

In a timed viewpoint, only clocks are considered in combination with enumerated state variables for the discrete part:

$$\mathcal{P}_\Gamma = \mathcal{P}$$
$$\mathcal{V}_\Gamma = \mathcal{V}$$
$$S_d : \text{finite set}$$
$$\forall \ell \in \mathbf{L}, \varphi \models flow(\ell) \Rightarrow \frac{d\varphi}{dt} \equiv 1 \text{ (corresponds to the clocks)}$$

Semantic atoms. Atoms for timed systems are simply "timers" with their activation guard. Thus timers are HRC state machines having two variables: the clock c (a continuous variable) and a trigger b_c, a discrete variable of boolean type. In addition, a continuous guard γ_c is provided as a constraint of the form $c \in C$, where C is some subset of the positive real line (typically, $c \leq c_{max}$, for some threshold c_{max}). Clock c is active whenever $\gamma_c \wedge [b_c = \mathtt{T}]$.

A timed system will be obtained by composing clocks with a discrete HRC state machine providing the b_c's as outputs, and taking the exit values of the clocks as inputs. The use of this category in expressing contracts is illustrated by the following example.

Example 15.2 (Timing Pattern, Figure 15.5) *Consider the timing pattern in Figure 15.5a. It aims at specifying a timed communication medium. Its intended (informal) meaning is that, whenever the delay between the two events s_b and t_b is less than τ_b, then it is guaranteed that the delay between the two events s_a and t_a is less than τ_a. Figure 15.5b shows two assertions: A, and ¬G. The pair (A, G) constitutes contract C. Ports of C are s_a, s_b, t_b, t_a, e. Among these ports, s_a and t_b are uncontrollable. The two clocks h_a and h_b are local variables of the contract; they*

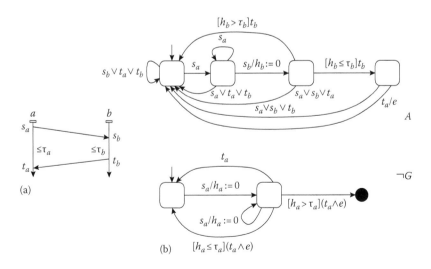

FIGURE 15.5
Assumption/promise. (a) Represented informally as a timing pattern. (b) Represented as the contract $C = (A, G)$ (the black circle is an accepting state).

satisfy the dynamics $\frac{dh}{dt} = 1$. Assertion A emits e whenever the desired pattern is completed with the due timing constraint on the pair s_b, t_b. Assertion G ensures that, whenever e is received, the timing constraint on the pair s_a, t_a is satisfied. This contract is not in canonical form. To make it in canonical form, simply replace ¬G by the product A × ¬G.

15.6.3 Hybrid Category

The hybrid category simply corresponds to the general case of HRC state machines.

Semantic atoms. Atoms for hybrid systems are "differential inclusions" with their guard. Differential inclusions are HRC State Machines having two sets of variables: a set $X = \{X_1, \ldots, X_n\}$ of continuous variables and the trigger b_X, a discrete variable of boolean type. In addition, a continuous guard γ_X is provided as a constraint of the form $\exp_c(X_1, \ldots, X_n) \in C$, where $\exp_c$ is some differential expression with values in $\mathbb{R}^p$ and C is some subset of $\mathbb{R}^p$. The differential inclusion is active whenever $\gamma_X \wedge b_X$ holds.

A hybrid system is obtained by composing clocks with a discrete HRC state machine providing the b_X's as outputs, and taking the exit values of the differential inclusions as inputs. Figure 15.6 gives such a decomposition for a variant of the electric circuit presented in Figure 15.4. The switch is modeled

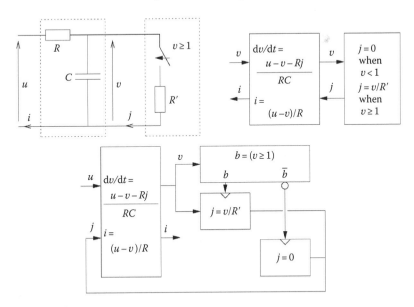

FIGURE 15.6
Use of clocked hybrid atoms. Top-left: Electric circuit. Top-right: Modeling of the circuit as a composition of ESMs. Bottom: Composite state machine with clocks to control hybrid atoms.

with three hybrid atoms: one for each state of the switch (opened and closed) and one for controlling the two former atoms. Consider the hybrid atom $j = v/R'$. When clock b is true, variable j is controlled by this atom, otherwise it is not constrained by this atom.

15.6.4 Safety or Probabilistic Category

Probabilistic ESMs specify what is considered random in a given ESM. Such a framework is useful when dealing with reliability models in which reliability properties interact with functional properties. For example, the risk for a component to fail may become zero in certain operating modes. In this category we provide means to specify such systems in a flexible yet simple way. More precisely, we assume that randomness will apply only to a specified subset **p** of ports. To be consistent with our approach, **p** must consist of ports that make the considered ESM receptive. The idea is that the environment will be the source of randomness for these ports. An element of the safety category thus consists of the following:

- An HRC state machine $\mathcal{H}$ with set of ports P.
- A subset $\mathbf{p} \subseteq P$ of "probabilistic ports," such that $\mathcal{H}$ is **p**-receptive. See Section 15.3.2 for the definition of receptiveness.
- For each $p \in \mathbf{p}$, an "activation port" $a_p \in P$ of pure type. Each event received on port a_p triggers the emission of an event on port p with a value drawn at random from some distribution μ_p. The different random trials are independent between different probabilistic ports.

Probabilistic ports are categorized into "time-related" and "value-related." If port p is time-related, then μ_p is a distribution on $\mathbb{R}_+$ or $\mathbb{N}_+$ and the value emitted by p is interpreted as a timing delay (e.g., for use in modeling the occurrence of failures). The probabilistic semantics is straightforward. Since $\mathcal{H}$ is **p**-receptive

1. One can draw at random the entire random sequence for each probabilistic port p (it need not be an independent sequence, it can be Markov, or even general).
2. These random sequences are stored for feeding the probabilistic ports of $\mathcal{H}$.
3. Each probabilistic sequence of data is then offered to $\mathcal{H}$ when activation port a_p requests these.

Comments.   Note that this is still compatible with nondeterminism. And other ways of modeling failure generation can be considered. For some applications, failures can be state-dependent. If there are only finite dependencies, then just provide one random source per different possible failure, and select the right one in a state-dependent way. If correlation between failures must be covered, this can be generally achieved by generating appropriate joint distributions by transforming joint distributions for independent random variables. Of course, all these tricks have a cost, and it will be the role of the use cases to check feasibility of this simple and pragmatic approach.

Semantic atoms. Semantic atoms for the safety category consist of an HRC state machine $\mathcal{H}_p$ having one probabilistic port p, the associated activation port a_p, associated distribution μ_p, and no variable.

Composing probabilistic ESMs. For $i = 1, 2$, let $\mathcal{P}_i = (\mathcal{H}_i, P_i, \mathbf{p}_i, (\mu_p^i)_{p \in \mathbf{p}_i})$ be two probabilistic ESMs. Their parallel composition $\mathcal{P}_1 \parallel \mathcal{P}_2$ is defined only if

$$\mathbf{p}_1 \text{ and } \mathbf{p}_2 \text{ are two disjoint sets of uncontrolled ports in } \mathcal{H}_1 \parallel \mathcal{H}_2. \quad (15.21)$$

Then

$$\mathcal{P}_1 \parallel \mathcal{P}_2 = (\mathcal{H}, P, \mathbf{p}, (\mu_p)_{p \in \mathbf{p}}) \text{ where } \begin{cases} \mathcal{H} = \mathcal{H}_1 \parallel \mathcal{H}_2 \\ \mathbf{p} = \mathbf{p}_1 \uplus \mathbf{p}_2 \\ \mu_p = \mu_p^i, \text{ where } i \text{ is such that } p \in \mathbf{p}_i \end{cases}$$

Comments regarding Condition 15.21 and a technique of wrappers. The reason for Condition 15.21 is to keep composition simple for probabilistic systems. If this condition does not hold, then indirect coupling between the probabilities may occur, due to constraints resulting from taking the product $\mathcal{H}_1 \parallel \mathcal{H}_2$. Condition 15.21 allows us to capture failure models, as well as random timing models for input signals.

The consequences of Condition 15.21 regarding compositionality are, however, nontrivial, as the following example shows. Consider a situation where we have a component having a port x which is either a source of failure, or is subject to failure propagation from another component. In the first case, the model of this component should look like $\mathcal{P} = (\mathcal{H}, P, \mathbf{p}, \mu)$, where $\mathbf{p} = \{x\}$ and port x is uncontrolled. The second case, on the other hand, may be obtained by composing $\mathcal{P}$ with another ESM in which x is an output and therefore controlled. This is ruled out by our Condition 15.21, however. Thus, it seems that this definition prevents us from capturing the above natural situation.

However, a simple mechanism of wrappers solves the problem as explained in the following text. Isolate the nonprobabilistic part $\mathcal{H}$ of our probabilistic ESM $\mathcal{P} = (\mathcal{H}, \mathbf{p}, \mu)$. Next, wrap $\mathcal{H}$ with the following small probabilistic ESM $\mathcal{P}_x$, which has one controlled port x and three uncontrolled ports: $x_{\text{source}}, x_{\text{herit}}$, and an additional port c taking values in the set {source, herit}. The only probabilistic port of $\mathcal{P}_x$ is x_{source}, we equip it with the original probability distribution μ. There is no assumption for ESM $\mathcal{P}_x$, and its guarantee is the following assertion

$$E =_{\text{def}} \quad x = x_{\text{source}} \text{ if } c = \text{source else } x = x_{\text{herit}} \text{ if } c = \text{herit},$$

which specifies a selector. Wrapping our original ESM in this way prepares it for the desired parallel composition in a valid way.

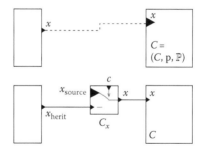

FIGURE 15.7
Illustrating the wrapper mechanism.

This is illustrated in Figure 15.7. In Figure 15.7, thick triangles denote probabilistic ports. The incorrect composition is shown at the top; it gives rise to a mismatch between thick and thin triangles. The corrected version, with its wrapper $\mathcal{P}_x$, is shown at the bottom. Probabilistic ESM $\mathcal{P}_x$ has one probabilistic port x_{source} with probability μ, and one uncontrolled port x_{herit}; uncontrolled boolean port c selects which input is propagated to the wrapped ESM $\mathcal{H}$. The design can be prepared for composition by this mechanism of wrapping. Wrapping must be performed manually, however.

15.6.5 Illustrating Multi-Viewpoint Composition

Our approach aims at supporting component-based development of heterogeneous embedded systems with multiple viewpoints, both functional and nonfunctional. The following simple example illustrates this for the case of functional, timed, and safety viewpoints. The overall system architecture is shown on Figure 15.8. It consists of a simple controller that can let the underlying plant to "start," "stop," or "work" (signals r, s, and w). The controller is subject to "failure" **f** of fail/stop type. The underlying plant has limited capacity and thus the controller should not accumulate in excess w messages during a certain period. This is ensured by the supervisor. The supervisor monitors the flow of w's. When they get too frequent, an "overloaded" message **o** is sent to the controller, which reduces the controller's pace. When appropriate, the human operator can decide to switch the controller back to

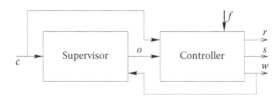

FIGURE 15.8
The overall system architecture.

its nominal mode, by sending the "cleaned" message **c** to the pair controller/ supervisor.

This system involves three viewpoints: functional, quality of service (QoS) of timed nature, and safety. In designing the system, the designer may follow three different methodologies. He she may consider each of the two components with its three viewpoints, implement each of them and then compose the result. Alternatively, he she may perform a first design by ignoring the safety viewpoint. The safety aspect is then added in a second stage. Finally, he she may consider all contracts for all components in a flat manner. The semantics of our framework has been designed to yield consistent results when following these three methods. For more details on this aspect, we refer the interested reader to [5].

The different contracts. Figure 15.9 depicts the set of contracts associated to the controller. For each contract, we show its assumption (top) and promise

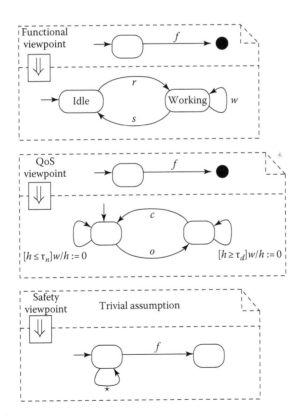

FIGURE 15.9

The three contracts C_{funct}, C_{QoS}, and C_{safety} specifying the three viewpoints of the controller. The assumption is put on top of the promise and both are separated by the implication symbol $\Downarrow$.

(bottom). The third contract has trivial, empty assumption. Assumptions are specified as observers, meaning that the corresponding diagrams define the negation of these assumptions. In these diagrams, the circles filled in black denote accepting states.

The first contract C_{funct} describes the functional aspect under the no failure assumption: The controller is activated by commands r ("run") and s ("stop"), and it can let the controlled system (not shown) work, by performing action w. This contract holds in absence of a failure, as shown by its assumption.

Contract C_{QoS} indicates that, under the no failure assumption, there exist two modes: nominal and degraded. Event **o** (for "getting overloaded") is not controlled by this component; in turn, when in overloaded mode, the human operator (not shown) can decide to perform "cleaning," corresponding to input event **c** to the system. This contract holds in absence of a failure, as shown by its assumption. Contract C_{QoS} relates to timing. When in nominal mode, the controller performs its task (whose termination is abstracted with the action w) in at most τ_n milliseconds. When in degraded mode, the controller performs its task in at least τ_d milliseconds, with $\tau_d > \tau_n$.

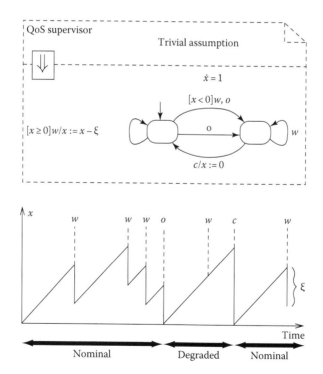

FIGURE 15.10
Contract C_s of the supervisor and its behavior.

Finally, contract C_{safety} specifies the safety aspect, which under no assumption states that a fault can occur at any time.

Figure 15.10 depicts the QoS contract for the supervisor in charge of avoiding system collapse by turning it to degraded mode. The assumption is trivial since the supervisor is not subject to failure. The promise is specified in terms of a hybrid automaton of the timed category. This hybrid automaton uses a timer x bound to physical time, thus satisfying the differential equation $\dot{x} = 1$ (x increases with constant speed 1). The behavior of this timer is depicted on the second diagram. When action w occurs too frequently in the long range, timer x starts decreasing and eventually reaches zero, which causes the emission of message **o** and switches the mode to "overloaded," where latency is at least τ_d. At some point, the cleaning message **c** is input by the operator, which resets the timer to 0 and brings the system back to its nominal mode.

15.7 Conclusion

We have briefly presented a framework for multiple viewpoint contracts. This framework is supported by the HRC metamodel, for which we have presented an underlying mathematical model of machine. We have emphasized how to support the combination of different viewpoints and have provided a simple and elegant notion of parallel composition.

In order to support partial designs, we have favored a constrained, non-functional, style for our model. Also, we have considered that our systems are open, that is, are subject to further combination with other, yet to be defined, subsystems. Our mathematical model is stratified in that it is progressively refined by detailing more and more its mathematical objects—from abstract transitions to combinations of guards and assignments.

One important feature of this model is that it has two equally important (and equivalent) versions. In the first version, states are snapshots whereas transitions are "thick"—transitions support continuous progress and invariants. For this version, parallel composition is by intersection, which is particularly simple and elegant. In the second version, transitions are snapshots whereas states are "thick"—states support continuous progress and invariants. This second version conforms to region-based models of systems, which are preferred by model checking tools. We have shown how the two versions can be intertranslated. Since the notions of state and transition are in fact interchanged between the two versions, it was essential not to constrain the way systems can interact. We have thus chosen to support both common state variables and common ports as vehicles for interaction.

Finally, we have characterized "categories," that is, subclasses of systems focusing on a particular aspect or viewpoint. One particular category required specific attention in dealing with parallel composition, namely, the probabilistic one.

The resulting mathematical model is the basis for a precise behavioral semantics for the HRC metamodel and provides a precise semantic for component composition, an often neglected issue in the design of frameworks for component-based design.

Acknowledgment

This research has been developed in the framework of the European IP-SPEEDS project number 033471.

References

1. R. Alur and D. L. Dill. A theory of timed automata. *Theoretical Computer Science*, 126(2):183–235, 1994.

2. J.-P. Aubin and A. Cellina. *Differential Inclusions, Set-Valued Maps and Viability Theory, Grundl. der Math. Wiss.*, vol. 264, Springer, Berlin/Heidelberg, 1984.

3. R.-J. Back and J. von Wright. *Refinement Calculus: A systematic Introduction*. Graduate Texts in Computer Science. Springer-Verlag, New York 1998.

4. R.-J. Back and J. von Wright. Contracts, games, and refinement. *Information and Computation*, 156:25–45, 2000.

5. A. Benveniste, B. Caillaud, A. Ferrari, L. Mangeruca, R. Passerone, and C. Sofronis. Multiple viewpoint contract-based specification and design. In *Proceedings of the Software Technology Concertation on Formal Methods for Components and Objects (FMCO07)*, Revised Lectures, *Lecture Notes in Computer Science*, Amsterdam, the Netherlands, October 24–26, 2007.

6. A. Benveniste, B. Caillaud, and R. Passerone. A generic model of contracts for embedded systems. Rapport de recherche 6214, Institut National de Recherche en Informatique et en Automatique, June 2007.

7. H. Butz. The Airbus approach to open Integrated Modular Avionics (IMA): Technology, functions, industrial processes and future development road map. In *International Workshop on Aircraft System Technologies*, Hamburg, Germany, March 2007.

8. A. Chakrabarti, L. de Alfaro, T. A. Henzinger, and M. Stoelinga. Resource interfaces. In *Proceedings of the Third Annual Conference on Embedded*

Software (EMSOFT03), Lecture Notes in Computer Science, Philadelphia, PA, 2855:117–133, 2003. Springer, Berlin/Heidelberg.

9. W. Damm. Controlling speculative design processes using rich component models. In *Fifth International Conference on Application of Concurrency to System Design (ACSD 2005)*, St. Malo, France, pp. 118–119, June 6–9, 2005.

10. W. Damm. Embedded system development for automotive applications: Trends and challenges. In *Proceedings of the Sixth ACM & IEEE International Conference on Embedded Software (EMSOFT06)*, Seoul, Korea, October 22–25, 2006.

11. L. de Alfaro and T. A. Henzinger. Interface automata. In *Proceedings of the Ninth Annual Symposium on Foundations of Software Engineering*, Vienna, Austria, pp. 109–120, 2001, ACM Press, New York.

12. E.W. Dijkstra. Guarded commands, nondeterminacy and formal derivation of programs. *Communications of the ACM*, 18(8):453–457, August 1975.

13. D. L. Dill. *Trace Theory for Automatic Hierarchical Verification of Speed-Independent Circuits*. ACM distinguished dissertations. MIT Press, Cambridge, MA, 1989.

14. T. A. Henzinger. The theory of hybrid automata. In *LICS*, New Brunswick, NJ, p. 278–292, 1996, IEEE Computer Society Press.

15. T. A. Henzinger, R. Jhala, and R. Majumdar. Permissive interfaces. In *Proceedings of the 13th Annual Symposium on Foundations of Software Engineering (FSE05)*, Lisbon, Portugal, pp. 31–40, 2005, ACM Press, New York.

16. Lamport, L. Win and sin: Predicate transformers for concurrency. *ACM Transactions on Programming Languages and Systems*, 12(3):396–428, July 1990.

17. B. Meyer. Applying "design by contract." *IEEE Computer*, 25(10):40–51, October 1992.

18. R. Negulescu. Process spaces. In *CONCUR, Lecture Notes in Computer Science*, University Park, PA, 1877, 2000. Springer-Verlag, Berlin/Heidelberg.

19. A. Sangiovanni-Vincentelli. Reasoning about the trends and challenges of system level design. *Proceedings of the IEEE*, 95(3):467–506, 2007.

16

Generic Methodology for the Design of Continuous/Discrete Co-Simulation Tools

Luiza Gheorghe, Gabriela Nicolescu, and Hanifa Boucheneb

CONTENTS

16.1   Introduction

The past decade witnessed the shrinking of the chips' size simultaneously with the expansion of a number of components, heterogeneous architectures, and systems specific to different application domains, for example, electronic, mechanics, optics, and radio frequency (RF) integrated on the same chip [16]. These heterogeneous systems enable cost-efficient solutions, an advantageous time-to-market, and high productivity. However, one will notice the increase of the variability of design related parameters. Given their application in various domains such as defense, medical, communication, and automotive, the continuous/discrete (C/D) systems emerge as important heterogeneous systems. This chapter focuses on these systems, their modeling and simulation.

Because of the complexity of these systems, their global design specification and validation are extremely challenging. The heterogeneity of these systems makes the elaboration of an executable model for the overall simulation more difficult. Such a model is very complex; it includes the execution of different components, the interpretation of interconnects, as well as the adaptation of the components. Their design requires tools with different models of computation and paradigms. The most important concepts manipulated by the discrete and the continuous components are

- In discrete models, *time* represents a global notion for the overall system and advances discretely when passing by time stamps of events, while in continuous models, the time is a global variable involved in data computation and it advances by integration steps that may be variable.
- In discrete models, *processes* are sensitive to events while in continuous models processes are executed at each integration step [12].
- Each model has to be able to *detect*, *locate* in time, and react to *events* sent by the other model.

The International Technology Roadmap for Semiconductors (ITRS) emphasizes that "a more structured approach to verification demands an effort towards the formalization of a design specification" and that "in the long term, formal techniques will be needed to verify the issues at the boundary of analog and digital, treating them as hybrid systems" [16].

Generally, in the design of embedded systems, the technique favored for the systems validation is co-simulation. Co-simulation allows for the joint simulation of heterogeneous components with different execution models. One of the advantages of this technique is the reusability of the models already developed in a well-known language and using already existing powerful tools (i.e., Simulink® [24] for the continuous domain and VHDL [33], Verilog [31], or SystemC [30] for the discrete domain). Thus, the development time, the time-to-market, and the cost are reduced. Moreover, this technique allows the designer to use the best tool for each domain and to provide capabilities to validate the overall model. This methodology requires the elaboration of a global simulation model.

The global validation of continuous/discrete systems requires co-simulation interfaces providing synchronization models for the accommodation of the heterogeneous. The interfaces play also an important role in the accuracy and the performance of the global simulation. This implies a complex behavior for the simulation interfaces, their design being time consuming and an important source of error. Therefore, their automatic generation is very desirable. An efficient tool for the automatic generation of the simulation interfaces must rely on the formal representation of the co-simulation interfaces [29].

This chapter presents a generic methodology, independent of simulation language, for the design of continuous/discrete co-simulation tools. This chapter is organized in nine sections. Section 16.2 gives several previous approaches to the modeling of continuous/discrete systems. The execution models for the continuous and the discrete domains are presented in Section 16.3. Section 16.4 details the methodology while Section 16.5 proposes a continuous/discrete synchronization model. Section 16.6 exemplifies the application of the methodology described in Section 16.4. Section 16.7 presents the formalization and the verification of the simulation interfaces. An example of a tool implemented with respect to the presented methodology is shown in Section 16.8. Finally, Section 16.9 gives our conclusions.

16.2   Related Work

The existing work on the validation of continuous/discrete heterogeneous systems can be classified into a few categories. They mostly include two approaches: simulation-based approach and formal representation-based approach.

The simulation-based approaches can be divided into two groups that use different techniques to obtain the global execution model:

1. The extension of existing tools and languages. Most of the tools created using this approach started from classical hardware description languages (HDLs) and new concepts specific to other domains such as analog mixed signal (AMS) or synchronous data flow (SDF) kernel were added (VHDL-AMS) [15], Verilog-AMS [10], SystemC–AMS

[32] or SystemC [27] extended with SDF kernel. These extensions are usually designed from scratch and by consequence their libraries are not as strong as the well established tools for this field (i.e., Simulink).

2. The definition of new models and tools. The systems are designed by assembling different components [23,28]. HyVisual [21] is a systems modeler based on Ptolemy [28] that supports the construction of hierarchal systems for continuous-time dynamical systems (see Chapter 15 and [21]). However, the different subsystems and components need to be developed in the same environment in order to be compatible and therefore they do not solve the problem of IP reuse in system design. Moreover, Ptolemy is based on formal representation, but the formal verification of the simulation models is not considered.

In the formal representation-based approaches, the integration is addressed as a composition of models of computation. These approaches propose a single main formalism to represent different models and the main concern is building interfaces between different models of computation (MoC). These approaches bring a deep conceptual understanding of each MoC. In other work [22], a framework of tagged signal models is proposed for comparison of various MoCs. The framework was used to compare certain features of various MoCs such as dataflow, sequential processes, concurrent, sequential processes with rendezvous, Petri nets, and discrete-event systems. The role of computation in abstracting functionalities of complex heterogeneous systems was presented in [17]. In [18] the author proposes the formalization of the heterogeneous systems by separating the communication and the computation aspects; however the interfaces between domains were not taken into consideration.

In [34], the authors introduce an abstract simulation mechanism that enables event-based, distributed simulation (discrete event system specifications—DEVS), where time advances using a continuous time base. DEVS is a formal approach to build the models, using a hierarchical and modular approach and more recently it integrates object-oriented programming techniques. Based on this formalism, [8] has proposed a tool for the modeling and simulation of hybrid systems using Modelica and DEVS. The models are "created using Modelica standard notation and a translator converts them into DEVS models" [8]. In [20] the authors propose a heterogeneous simulation framework using DEVS BUS. NonDEVS-compliant models are converted through a conversion protocol into DEVS-compliant models. CD++ is a general toolkit written in C++ that allows the definition of DEVS and Cell-DEVS models. DEVS-coupled models and Cell-DEVS models can be defined using a high-level specification language [35]. PythonDEVS is a tool for constructing DEVS models and generating Python code. A model is described by deriving coupled and/or atomic DEVS descriptive classes from this architecture, and arranging them in a hierarchical manner through composition [4]. DEVSim++ is an environment for object-oriented modeling of discrete event systems [19].

16.3 Execution Models

This section presents the global execution models of continuous/discrete heterogeneous systems. The execution model can be viewed as the interpretation of a computation model. Discrete and continuous systems are characterized by different physical properties and modeling paradigms.

16.3.1 Global Execution Model

The global execution model of a heterogeneous system is the realization of the system's functionality. A C/D system and its corresponding global execution model are illustrated in Figure 16.1. There are three types of basic elements that compose the model [26]:

- The *execution models* of the different components constituting the heterogeneous system (corresponding to Component 1 and Component 2 in Figure 16.1)
- The *co-simulation bus*
- The *co-simulation interfaces*

The *co-simulation bus* is in charge of interpreting the interconnections between the different components of the system.

The *co-simulation interfaces* enable the communication of different components through the simulation bus. They are in charge of the adaptation of different simulators to the co-simulation bus in order to guarantee the transmission of information between simulators executing the different

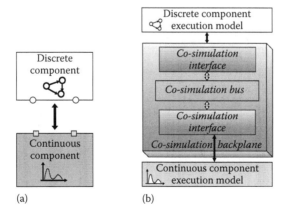

(a) (b)

FIGURE 16.1
Continuous/discrete (a) heterogeneous system and its corresponding (b) execution model.

components of the heterogeneous systems. They also have to provide efficient synchronization models for the modules adaptation.

The *co-simulation backplane* is the element of the global execution model that guarantees the synchronization and the communication between the different components of the system. It is composed of the above mentioned simulation interfaces and the simulation bus.

The implementation and the simulation of an execution model in a given context is called *co-simulation instance*. Several instances may correspond to the same execution model and these instances may use different simulators and may present different characteristics (e.g., accuracy and performances).

16.3.2 Discrete Execution Model

The execution model for a discrete system is a model where changes in the state of the system occur at discrete points in the execution time.

The discrete system can be described by the state–space equations [6]:

$$\begin{cases} x_d(t_{k+1}) = f(x_d(t_k), u(t_k), t_k) & \text{with} \quad x(t_0) = x_0 \\ y(t_k) = g(x_d(t_k), u(t_k), t_k) \end{cases} \tag{16.1}$$

where
  f and g are transformations
  x_d is the discrete state vector
  u is the input signal vector
  y is the output signal vector

For the linear discrete systems, Equation 16.1 becomes

$$\begin{cases} x_d(t_{k+1}) = A_d x_d(t_k) + B_d u(t_k) \\ y(t_k) = C_d x_d(t_k) + D_d u(t_k) \end{cases} \tag{16.2}$$

where A_d, B_d, C_d, and D_d are matrices that can be time varying and describe the dynamics of the system [6].

A discrete event system execution concentrates on processing events, each event having assigned a time stamp. Each event computation can modify the state variables, schedule new events or retract existing events. The unprocessed events are stored in a pending events list. The events are processed in the order of their time stamp. Figure 16.2 shows a possible update event schema. At each simulation cycle, the first event with the smallest time stamp is processed and the processes sensitive to this event are executed [34].

If several processes are sensitive to one or several events (with the same time occurrence) then these processes have to be executed in parallel. Executions often occur on sequential machines that can only execute one instruction at a time (therefore, one process). The consequence is that this execution

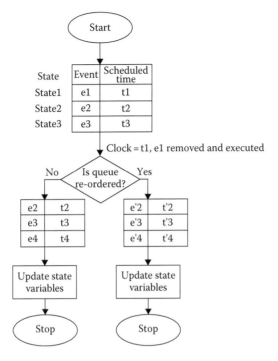

FIGURE 16.2
Event update schema.

cannot parallelize the processes. The solution consists in emulating the parallelism, where the processes are executed as if the parallelism is real and the environment does not change while executing all the processes. Once all events with discrete time stamp equal to the current time have been treated, the simulator advances the time to the nearest scheduled discrete event.

16.3.3 Continuous Execution Model

The continuous time system is described by the state–space equations:

$$
\begin{cases}
\dot{x}_c(t) = A_c x_c(t) + B_c u(t) \\
y(t) = C_c x_c(t) + D_c u(t)
\end{cases}
\tag{16.3}
$$

where
x_c is the state vector
u is the input signal vector
y is the output signal vector
A_c, B_c, C_c, and D_c are constant matrices that describe the dynamic of the system

The execution of continuous model, described by differential and algebraic equations, requires solving these equations numerically. A widely used class of algorithms discretizes the continuous time line into an increasing set of discrete time instants, and numerically computes values of state variables at these ordered time instants. The next state of derivative systems cannot be specified directly but the derivative functions are used to specify the rate of change of state variables [34].

The execution of a continuous system raises problems because given a state q_k and a vector x for a time t_k, the derivative offers information only for dq_k/dt but not the system's behavior over time. For a nonzero interval $[t_k, t_{k+1}]$ the computation has to be realized without knowing the behavior in the interval (t_k, t_{k+1}). This problem can be solved using numerical integration methods. Some of the most commonly used methods are

- *Euler method* that consists in signal integration:

$$\frac{dq(t)}{dt} = \lim_{h \to \infty} \frac{q(t+h) - q(t)}{h}.$$

For an h small enough (in order to obtain accurate results), the following approximation can be used:

$$q(t+h) = q(t) + h * \frac{dq(t)}{d(t)}$$

This solution has low efficiency and does not have stability problems for small enough h and it is very robust [34].

- *Causal methods* that are a linear combination of states and derivative values at time instants with coefficients chosen to minimize errors from the computed estimate to the real value [34].

This solution has high efficiency but it has stability and robustness problems.

- *Noncausal methods* that use "future" values of states, derivative, and inputs. In order to do that the model is executed past the needed time and the values that are necessary are stored to estimate the present values [34].

16.4 Methodology

This section introduces a methodology for the design of continuous/discrete co-simulation tools (as shown in Figure 16.3). To enable the design of co-simulation tools, this methodology presents several steps that are independent of the simulation tools used for the continuous and discrete

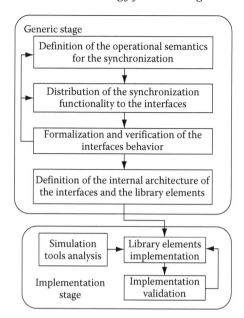

FIGURE 16.3
A generic methodology for the design of C/D co-simulation tools.

components of the system. During these generic steps, the co-simulation interfaces are defined in a conceptual framework; their functionality and the internal structure of simulation interfaces are expressed using existing formalisms and temporal logic. After the rigorous definition of the required functionality for simulation interfaces, the designer will start the steps related to the implementation.

The main steps of the proposed methodology (illustrated in Figure 16.3) can be divided into two stages:

1. A generic stage with the following actions:

   - Definition of the operational semantics for the synchronization in continuous/discrete global execution models.
   - Distribution of the synchronization functionality to the simulation interfaces.
   - Formalization and verification of the simulation interfaces behavior.
   - Definition of the library elements and the internal architecture of the simulation interfaces.

2. An implementation stage with the following actions:

   - The analysis of the simulation tools for the integration in the co-simulation framework.
   - The implementation of the library elements specific to different simulation tools.

This section focuses on the generic stage and its steps will be detailed in the next subsections. A possible implementation stage will be detailed further in Section 16.8.

16.4.1 Definition of the Operational Semantics for the Synchronization in Continuous/Discrete Global Execution Models

The first step of the methodology for co-simulation tools design is the definition of the operational semantics for the synchronization in continuous/discrete global execution models. An operational semantics gives a detailed description of the system's behavior in mathematical terms. This model serves as a basis for analysis and verification. The description provides a clear language independent model that can serve as a reference for different implementations.

The operational semantics for continuous/discrete systems requires the rigorous representation of the relation between the simulators (communication/synchronization and data exchanged between the continuous and the discrete simulators) as well as their high level and dynamic representations.

16.4.2 Distribution of the Synchronization Functionality to the Simulation Interfaces

Based on the operational semantics, we can now define the synchronization functionality between the continuous and the discrete simulators. This functionality is insured by the interfaces that are the link between the different execution models and the co-simulation bus (see Figure 16.1). They are each in charge with a part of the synchronization between the two models. To ensure system's flexibility, the synchronization functionality has to be distributed to the simulation interfaces. Moreover, each computation step has to be thoroughly specified.

16.4.3 Formalization and Verification of the Simulation Interfaces Behavior

The formalization and verification of the simulation interfaces behavior stage can be roughly divided in three steps: formalization (that can be the formal specification of the heterogeneous system), the validation by model simulation, and the formal verification. The two main techniques that can be used for the formal verification of the interfaces are [36]:

- *Model checking*, where the system descriptions are given as automata, the specification formulas are given as temporal logic formulas, and the checking consists of the verification which ensures that all models of a given system description satisfy a given specification formula. It focuses mainly on automatic verification. *Completeness* and *termination guarantee* of model checking are some features of this technique, as well as it enables the tool to guarantee the correctness of a given property, or produce a counterexample otherwise.

- *Theorem proving,* where the verification plan is manually designed and the correctness of the steps in the plan is verified using theorem provers. Completely automatic decision procedures are impossible because the input language (the model and the specification) is of higher order logic and that eliminates the decidability. Moreover, everything has to be translated in higher order logic, and, therefore, the structure of the system may be lost and its representation can become large and difficult to work with.

Considering that the system is dynamic, it is necessary to use a formalism that allows the expression of dynamic properties (the state of a system changes and by consequence the properties of the state also change). The temporal logic handles formalization where the properties evolve over time and in general uses:

- Propositions that describe the states (i.e., elementary formulas and logical connectors)
- Temporal operators that allow the expression of the properties of the states successions (called executions)

The differences between the logics are in terms of temporal operators and objects on which they are interpreted (such as sequences or state trees) [25].

The most commonly used logics are Linear Temporal Logic (LTL), Computation Tree Logic (CTL* and CTL, both of them untimed temporal logics) and their timed extensions TCTL and Metric Interval Temporal Logic (MITL).

- CTL* allows the use of all temporal and branching operators but the property verification is very complex. For this reason, most of the tools actually used allow the verification of fragments of CTL*.
- LTL is a fragment of CTL* that excludes the trajectory quantifiers. In this case only the trajectory predicates are considered. LTL does not provide a means for considering the existence of different possible behaviors starting from a given state (sequential) 0.
- CTL is also a fragment of CTL* and it is obtained when every occurrence of a temporal operator is immediately preceded by a branching operator. In the case of CTL we have state trees.
- TCTL is a timed temporal logic that is an extension of CTL obtained by subscribing the modalities with time intervals specifying time restrictions on formulas.

For our formal model, the properties that need to be checked are branching properties that are expressed using CTL or TCTL logics.

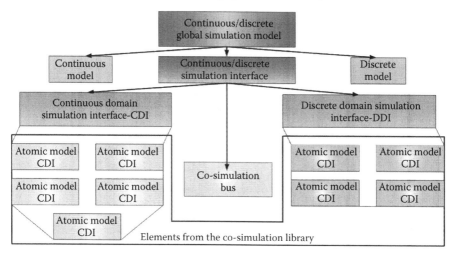

FIGURE 16.4
Hierarchical representation of the generic architecture of the co-simulation model.

16.4.4 Definition of the Internal Architecture of the Simulation Interfaces

The formalization of the simulation interfaces behavior step is naturally followed by the definition of their internal architecture. This definition eases the automatic generation of the simulation interfaces. We present in Figure 16.4 the hierarchical representation of the global simulation model used in our approach.

At the top hierarchical level, the global model is composed of the continuous and discrete models and of the C/D simulation interface required for the global simulation [12].

The second hierarchical level of the global simulation model includes the domain specific simulation interfaces and the co-simulation bus in charge of the data transfer between these interfaces.

The bottom hierarchical level includes the elements from the co-simulation library that are the atomic modules of the domain specific simulation interface. These atomic components implement basic functionalities of the synchronization model.

16.4.5 Analysis of the Simulation Tools for the Integration in the Co-Simulation Framework

The considerations presented in the previous steps of the methodology show that specific functionalities are required for the co-simulation of continuous and discrete modes. Therefore, the integration of a simulation tool in the co-simulation environment requires their analysis. Thus, in the case of continuous simulator integration in the co-simulation tool, this simulator has to provide application programming interfaces (APIs) enabling the following controls:

- Detection and location of state events
- Setting break points during differential equation solving
- Online update of the breakpoints settings
- Sending processing results and information for synchronization (i.e., the time step of the state event) to the discrete simulator. This generally implies the possibility to integrate C-code and Inter-Process Communications (IPC).

For the integration of a discrete simulator in the co-simulation tool, the simulator has to enable the addition of the following functionalities:

- Detection for the end simulation cycle
- Insertion or retraction of new events (*state events*) in the scheduler's queue. This must be done before the advancement of the simulator time
- Sending processing results and information for synchronization to the continuous simulator (i.e., the *time stamp* of its next discrete event).

16.4.6 Implementation of the Library Elements Specific to Different Simulation Tools

The last step of the methodology for the design of co-simulation tools for continuous/discrete systems is the implementation of the library elements that are specific to different simulation tools. This step depends highly on the simulation tools chosen in the previous step, the analysis of the simulation tools.

16.5 Continuous/Discrete Synchronization Model

The methodology focuses mostly on the co-simulation interfaces. One of the most important functions of the interfaces is providing the synchronization that is defined as coordination with respect to time. Thus, for a better understanding of the methodology it is very important for one to comprehend the synchronization model. The synchronization between the continuous-time domain and the discrete-event domain is realized using a canonical algorithm as it was presented in [11]. For a rigorous synchronization the discrete kernel has to detect the events generated by the analog (continuous) solver and the continuous solver must detect the scheduled events from the discrete kernel.

The events exchanged between the discrete and the continuous simulators are [7]:

- *Occurred/scheduled events* that are timed events scheduled by the discrete simulator.
- *State events* that are unpredictable events generated by the continuous simulator. Their time stamp depends on the values of state variables (e.g., a zero-passing or a threshold crossing).

For the discrete event processes, time does not advance during the execution. The next execution time is the time of the next event in the event queue. The execution of the analog solver advances the simulation time. Let t_k be the synchronization time for the discrete kernel and the analog solver. The analog solver advances to the next synchronization time t_{k+1}, known in advance from the digital kernel. At this point the analog solver suspends while the digital kernel resumes and the events in t_{k+1} are executed. If a state event occurs in the time interval $[t_k, t_{k+1}]$, the analog solver suspends to allow the digital kernel to take this event into consideration. This way the analog solver and the digital kernel are synchronized again.

Figure 16.5a presents the synchronization model in the C/D cosimulation interface without taking into consideration the state event occurrence. The state event is taken into consideration in Figure 16.5b.

At a given time, the discrete simulator is in the state (x_{dk}, t_k). At this point, the discrete simulator has executed all the processes sensitive to the event with the time stamp t_k and sends the time of the next event t_{k+1} and the data to the continuous simulator and the context is switched from the discrete to the continuous simulator before advancing the time.

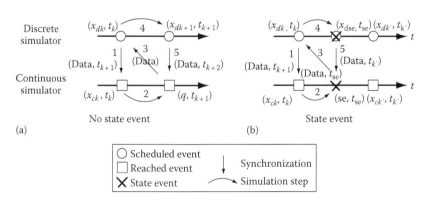

FIGURE 16.5
The synchronization model in the C/D simulation interface without state event (a) or with state event (b).

The state of the continuous simulator is (x_{ck}, t_k) and the advance in time of the simulator cannot be further than t_{k+1}, the time sent by the discrete simulator. Consequently, the continuous simulator computes signals by resolving the system's equations until it reaches with accuracy the time t_{dk+1}, updates signals with the values calculated at this time its new state being (q, t_{k+1}), sends the new data, and the context is switched to the discrete model (arrow 3). The discrete simulator will also advance to the time t_{k+1}(arrow 4) and the cycle restarts as shown in Figure 16.5a. The continuous model may generate a state event. In this case, it updates signals with the values calculated at this time, t_{se} in Figure 16.5b its new state being, (se, t_{se}), it indicates its presence by sending the state event's time stamp (t_{se}) and the corresponding data to the discrete model before switching the simulation context. The discrete model has to be able to detect this event, by advancing the local time to the time stamp, and to execute the processes that are sensitive to it and arrive at the state (x_{dse}, t_{se}).

16.6 Application of the Methodology

This section proposes a possible application of the methodology that was proposed in Section 16.4. First, the basic concepts that are used in our specific methodology are introduced: DEVS and timed automata [1].

16.6.1 Discrete Event System Specifications

DEVS is a formalism supporting a full range of dynamic system representation, with hierarchical and modular model development. The abstraction separates modeling from simulation and provides atomic models that can be used to build complex models that allow the integration of continuous and discrete-event models [34]. It also provides all the mechanisms for the definition of an operational semantics for the continuous/discrete synchronization model, the high level representation of the global formal model.

A DEVS is defined as a structure [34].

DEVS = $< X, S, Y, \delta_{int}, \delta_{ext}, \lambda, t_a >$ where

$X = \{(p_d, v_d) | p_d \in InPorts, v_d \in Xp_d \}$ set of *input* ports and their values in the discrete event domain,

S—set of *sequential states*

$Y = \{(p_d, v_d) | p_d, \in OutPorts, v_d \in Yp_d\}$ set of *output* ports and their values in the discrete event domain.

δ_{int}: $S \rightarrow S$ the *internal transition* function

δ_{ext}: $QxX \rightarrow S$ the *external transition* function, where:

$Q = \{(s, e) | s \in S, 0 \le e \le ta(s)\}$ set of *total state*,

e is the *time elapsed* since the last transition

$\lambda : S \to Y$ output function

$t_a: S \to R_{0,\infty}^+$ set of positive reals with 0 and ∞.

The system's state at any time is s. There are two possible situations:

- Case 1—where we assume that no external events occur. In this case the system stays in this state s for the time $t_a(s)$. When the elapsed time e equals $t_a(s)$ (that is the time allocated for the system to stay in state s), the system outputs the value $\lambda(s)$. The state s changes to the state $s\prime$ as a result of the transition $\delta_{int}(s)$. We emphasize here that the output is possible only before the internal transitions. We propose the definition of this type of transition using the following rule of the form $\dfrac{\text{Premises}}{\text{Conclusions}}: \dfrac{e=t_a(s) \wedge s'=\delta_{int}(s)}{(s,e) \overset{!\lambda(s)}{\to} (s',0)}$, where "!" represents the send operator.

- Case 2—where there is an external event x before the expiration time, $t_a(s)$ (the system is in state (s,e), with $e \leq t_a(s)$), the system's state changes to state s' as a result of the transition $\delta_{ext}(s,e,x)$. For the definition of this type of transition, we propose the following rule: $\dfrac{e \leq t_a(s) \wedge s' = \delta_{ext}(s,e,x)}{(s,e) \overset{?x}{\to} (s',0)}$, where "?" represents the receive operator.

Thus, the internal transition function dictates the system's new state when no external events occurred since the last transition while the external transition function dictates the system's new state when an external event occurs—this state is determined by the input x, the current state s and the time period during which the system has been in this state, e. In both cases the system is then in some new state s' with some new expiration time $t_a(s')$.

We also give here DEVS coupled models as defined by the same formalism. For the case where we have ports, the specification includes external interfaces with input and output ports and values and coupling relations:

- $N = (X,Y,D, \{M_d|d \in D\}, EIC,EOC,IC)$ where:
- $X = \{(p, v)|p \in InPorts, v \in Xp\}$ set of *input* ports and values
- $Y = \{(p, v)|p \in OutPorts, v \in Yp\}$ set of *output* ports and values
- D = set of components names
- $M_d = (X_d, S, Y_d, \delta_{int}, \delta_{ext}, \lambda, t_a)$ is a DEVS with X_d, Y_d the set of input/output ports and values
- EIC (External Input Coupling) = the coupling between the input in the coupled model and the external environment
- EOC (External Output Coupling) = the coupling between the output from the coupled model and the external environment
- IC (Internal Coupling) = the coupling between the modules that compose the coupled module [12]

In our work we used the parallel DEVS coupled formalism. Each module composing the interface performs a different task according to the synchronization model specified in Section 16.5.

16.6.2 Timed Automata

In this section we briefly introduce timed automata. Chapter 14 gives a more detailed presentation of this formalism. A timed automaton [1] is the formalism for modeling and verification of real time systems. It can be seen as classical finite state automata with clock variables and logical formulas on the clock (temporal constraints) [3]. The constraints on the clock variables are used to restrict the behavior of the automaton. The logical clocks in the system are initialized to zero when the system is started and then increase at the uniform rate counting time with respect to a fixed global time frame. Each clock can be separately reset to zero. The clocks keep track of the time elapsed since the last reset [1]. There are two types of clock constraints: constraints associated with transitions and constraints associated with locations. A transition can be taken when the clocks' values satisfy the *guard* labeled on it. Figure 16.6 illustrates an example of a timed automaton. The constraints associated with locations are called *invariants* and they specify the amount of time that may be spent in a location. The invariant "true" for a location means there are no constraints for the time spent in the location.

The process shown in Figure 16.6 starts at the location p with all its clocks (x and y) initialized to 0. The values of the clocks increase synchronously with time at the location q.

At any time, the process can change the location following a transition $p \xrightarrow{g;a;r} q$ if the current values of the clocks satisfy the enabling condition g (guard). A guard is a Boolean combination of integer bounds on clocks and

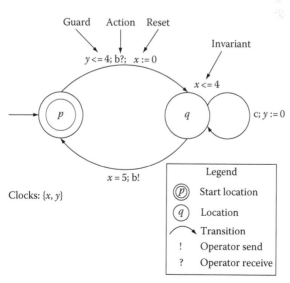

FIGURE 16.6
Example of a timed automaton.

clock differences. With this transition, the variables are updated by r (reset), which is an action performed on clocks. The actions are used for synchronization and are expressed by a (`action`) [3]. A synchronization label is of the form *Expression?* or *Expression!* where ! represents the operator `send` and ? represents the operator `receive`.

The semantics for a time automaton are defined as "a transition system where a state or configuration consists of the current location and the current values of clocks" [3]. Thus, the state is represented by the tuple: (l, v) where l is the `location` and v is the `clock` valuation (a function that associates a real positive value, including zero, to each clock). Given the system, we can have two types of transitions between locations: a delay transition when the automaton may delay for some time or an action transition when the transition follows an enabled transition.

The transition showing the time passing is $(l, v) \xrightarrow{t} (l', v')$ if and only if:

$$\begin{cases} v' = v + t \\ \forall t' \in [0, t], (v + t') \text{ verifies } \text{Inv}(l) \end{cases} \tag{16.4}$$

where $Inv(l)$ is the *invariant* in the location $l, l = l', v' = v + t$ showing that for all clocks $x, v'(x) = v(x) + t$.

For the discrete transitions $(p, v) \xrightarrow{g;a;r} (q, v')$ v' has to satisfy the *invariant* of q. v' is obtained from v by resetting the clocks indicated by the *reset r*.

Timed automata have the following characteristics that make them desirable for our formal model:

- The ease and the flexibility of systems' modeling
- The existence of a whole range of powerful tools that are already implemented and that allow different verification techniques
- The adequate expressivity in order to model time constrained concurrent systems

Our formal model needs to support concurrency between continuous/discrete systems and thus it was represented as a parallel composition of several timed automata with no constraints regarding the time spent in the locations.

16.6.3 Definition of the Operational Semantics for the Synchronization in C/D Global Execution Models

The operational semantics for the C/D synchronization model is given by the set of rules presented in Table 16.1. DEVS, as defined in Section 16.7.1 allows for the definition of the operational semantics of the behavior of the simulation interfaces with respect to the synchronization model presented in Section 16.5.

In Table 16.1, *DataToBus* is the output function from the discrete domain interface $\lambda(s_d)$, and *DataFromBus* is the output function from the continuous

TABLE 16.1

Operational Semantics for the C/D Synchronization Model

Rule	Arrows in Figure 16.5
$$\dfrac{synch = 1 \wedge flag = 1 \wedge q = \delta_{ext}(q)}{(s_d, e_d) \xrightarrow{!(DataFromBus,\, t_a(s_d));\, flag:=0} (s_d, e_d);\, q \xrightarrow{?(DataFromBus,\, t_a(s_d));\, synch:=0} q}$$	Arrow 1 Figure 16.5a and b
$$\dfrac{flag = 0 \wedge \neg stateevent(t) \wedge q' = \delta_{int}(q)}{q \xrightarrow{\delta_{int}} q' \xrightarrow{!DataToBus;\, flag:=1} q'}$$	Arrow 2 and 3 in Figure 16.5a
$$\dfrac{synch = 0 \wedge flag = 1\, \neg stateevent \wedge s'_d = \delta_{ext}(s_d)}{(s_d, e_d) \xrightarrow{t_a(s_d) - e_d} (s_d, t_a(s_d)) \xrightarrow{?DataToBus;\, \delta_{int}(s'_d);\, \lambda(s'_d);\, synch:=1} (s'_d, 0)}$$	Arrow 4 in Figure 16.5a
$$\dfrac{flag = 1 \wedge stateevent \wedge q' = \delta_{int}(q)}{q \xrightarrow{!DataToBus} q' \xrightarrow{!DataToBus;\, t_{se};\, flag:=1} q'}$$	Arrow 2 and 3 in Figure 16.5b
$$\dfrac{synch = 0 \wedge flag = 1 \wedge stateevent \wedge s'_d = \delta_{ext}(s_d, t)}{(s_d, e_d) \xrightarrow{?t_{se}} (s_d, t_{se}) \xrightarrow{?DataToBus;\, \delta_{int}(s'_d);\, \lambda(s'_d);\, synch:=1} (s'_d, 0)}$$	Arrow 4 in Figure 16.5b

Source: Gheorghe, L. et al., Formal definition of simulation interfaces in a continuous/discrete co-simulation tool, *Proceedings of the 17th IEEE International Workshop on RSP*, Chania, Crete, Greece, pp. 186–192, 2006. With permission.

domain interface $\lambda(s_d)$. The semantics of the global variable "flag" is related to the context switch between the continuous and discrete simulators. When "flag" is set to "1," the discrete simulator is executed. When it is "0," the continuous simulator is executed. The global variable "synch" is used to impose the order of the different operations expressed by the rules.

For a better explanation, we present the first rule in more detail here, corresponding to arrow 1 in Figure 16.5a and b. The premises of this rule are: the "synch" variable has value "1," the "flag" variable has value "1," and we have an external transition function (δ_{ext}) for the continuous model. The discrete model is initially in the total state (s_d, e_d), this means it has been in the state s_d for the time e_d. In this state, the discrete simulator performs the following actions:

send the data and the value of its next time stamp (this action is expressed by !($DataFromBus, t_a(s_d)$))

- Switch the simulation context to the continuous model (this action is expressed by $flag = 0$).

For the same rule, the continuous model is in state q and performs the following actions:

- Receive the data and the value of the time stamp from the discrete simulator (expressed by?(($DataFromBus, t_a(s_d)$)).
- Set the global variable synch to "0" (action expressed by $synch = 0$) in order to respect the premise of the rule corresponding to the arrow 4.

The actions expressed by this rule will be executed by the discrete simulator when the context will be switched to it [12].

16.6.4 Distribution of the Synchronization Functionality to the Simulation Interfaces

The second step of the methodology consists in the distribution of the synchronization functionality to the simulation interfaces. The synchronization functionality was presented in Section 16.4.2.

The behavior of the discrete domain interface can be described by a few processing steps detailed in Figure 16.7.

The interface is in charge of:

- Exchanging *data* between the simulators (send/receive)
- Sending the *time stamps* of the next events
- Considering the *state events*
- The *context switch* to the continuous interface

The behavior of the continuous domain interface can also be described by a few processing steps detailed in Figure 16.8. This interface handles:

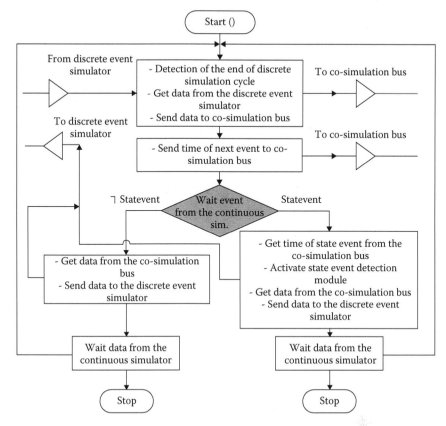

FIGURE 16.7
Flowchart for the discrete domain interface. (From Gheorghe, L. et al., Formal definition of simulation interfaces in a continuous/discrete co-simulation tool, *Proceedings of the 17th IEEE International Workshop on RSP*, Chania, Crete, Greece, pp. 186–192, 2006. With permission.)

- Exchanging *data* between the simulators (send/receive)
- Sending the *time stamps* of the following events
- The indication (to the discrete interface) of the occurrence of a *state event*
- The *context switch* to the discrete interface

16.6.5 Formalization and Verification of the Simulation Interfaces Behavior

This section presents details of the formalization and the formal verification of the behavior of the simulation interfaces using, as explained in Section 16.4, timed automata and TCTL. In order to model, validate, and check our model we used UPPAAL [2]. The main advantage of UPPAAL is

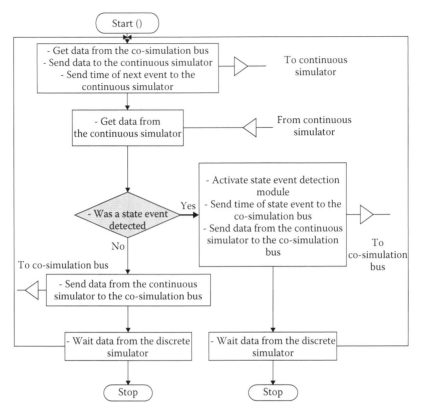

FIGURE 16.8
Flowchart for the continuous domain interface. (From Gheorghe, L. et al., Formal definition of simulation interfaces in a continuous/discrete co-simulation tool, *Proceedings of the 17th IEEE International Workshop on RSP*, Chania, Crete, Greece, pp. 186–192, 2006. With permission.)

that the product automaton is computed on-the-fly during verification. This reduces the computation time and the required memory space. It also allows interleaving of actions as well as hand-shake synchronization. The tool provides a user-friendly graphical interface and a simulator.

1. *Formalization of the simulation interfaces:* In [14] the authors demonstrate the equivalence between a DEVS model and the timed automata. The timed-automata model completes the DEVS graph with the addition of the timing evolution notions. Figure 16.9 shows the formal model for the discrete domain interface using timed automata. The model has only one initial location (marked in Figure 16.9 by a double circle) *Start*.

The discrete interface will change location from *Start* to *NextTimeGot* following the transition *Start* $\xrightarrow{\textit{DataFromDisc}?}$ *NextTimeGot*. This is an external

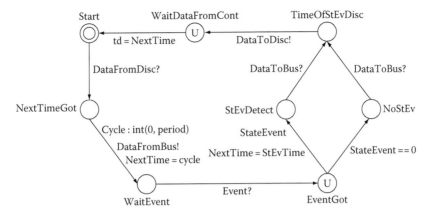

FIGURE 16.9
The discrete domain interface model (IDiscrete).

transition realized in zero time and it is triggered by receiving the data (that is also synchronization between the discrete simulator and the interface) from the discrete simulator (`DataFromDisc?`). Here the interface receives the data from the discrete simulator and the time of the next event in the discrete domain.

The location changes to *WaitEvent* following the transition:

$$NextTimeGot \xrightarrow{\textit{DataFromBus!, NextTime}=\textit{cycle,cycle:int}[0,\textit{period}]} WaitEvent$$

In order to change the location, the continuous interface sends the time of the next event (occurred/scheduled event) in discrete (the synchronization `DataFromBus!`) to the discrete interface. The variable `NextTime` is the time of the next event in the discrete domain. This variable takes, in this mode, the value `cycle`. The theory normally assumes equidistant sampling intervals. This assumption is not usually achieved in practice. For an accurate simulation we assume that cycle takes random values in an interval defined here as [0, *period*]. In *WaitEvent* location, the context is switched from the discrete to the continuous simulator.

When the context is switched back to the discrete simulator, the location is changed to *EventGot* following the synchronization transition:

$$WaitEvent \xrightarrow{\textit{Event?}} EventGot.$$ During this transition the discrete interface receives from the continuous interface the synchronization `Event?`. In this location the occurrence of a state event in the continuous domain is considered. *EventGot* is an urgent location (as defined in Section 16.6.2). This will not allow the discrete model to miss a state event generated by the continuous model. Two cases are possible:

- When no state event was generated by the continuous domain, the location changes from *EventGot* to *NoStEv*. The transition

$EventGot \xrightarrow{StateEvent == 0} NoStEv$ is annotated in this case only with the guard `StateEvent==0`.

- When a state event was generated by the continuous domain the location changes from *EventGot* to *StEvDetect* following the transition:

$$EventGot \xrightarrow{StateEvent,NextTime=StEvTime} StEvDetect.$$

This transition is annotated with a guard (`StateEvent`) and the update of the `NextTime` in the discrete domain as the time when the state event occurred in the continuous domain `StEvTime` (for a rigorous synchronization, the discrete domain has to consume this event and stop at the time when it was generated by the continuous domain interface). This is the time of the next event that is going to be sent to the continuous simulator. From both locations *StEvDetect* and *NoStEv*, the system can reach the next location: *TimeOfStEvDisc*. In both cases the model performs synchronization (`DataToBus?`). At this point the discrete interface will synchronize and send data to the discrete simulator (`DataToDisc`) and changes the location to *WaitDataFromCont*. The next location is *Start*, the discrete time variables is initialized on this channel (`td=NextTime`)and the cycle restarts.

Figure 16.10 shows the formal model (using timed automata) for the continuous domain interface. The model also has only one initial location (marked in Figure 16.10 by a double circle) *Start*.

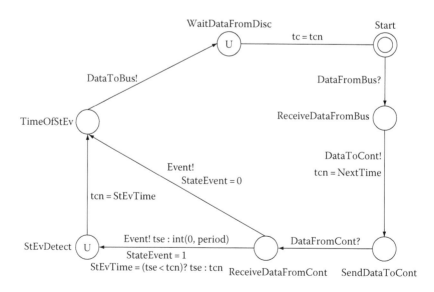

FIGURE 16.10

The continuous domain interface model (IContinu).

The continuous interface will leave the initial location *Start* following the transition:

Start $\xrightarrow{DataFromBus?}$ *ReceiveDataFromBus*. This is also an external transition realized with zero time and it is triggered by the reception of the data from the discrete interface (DataFromBus?) that is also the first synchronization point between the discrete interface and the continuous interface. The interface receives the data from the discrete simulator and the time of the next event in the discrete model. From the *ReceiveDataFromBus* location the process moves to the next location *SendDataToCont* following the transition

$$\textbf{ReceiveDataFromBus} \xrightarrow{DataToCont!,\ ten\,=\,NextTime} \textbf{SendDataToCont}$$

The value NextTime, the time of the next event (occurred/scheduled event) in the discrete simulator is assigned to tcn, the next time in the continuous simulator. In our model, the synchronization on this transition is between IContinu and SimCont (where SimCont is the continuous domain simulator), the interface sends data received from IDiscrete and the time of the next event in the discrete domain to the simulator.

The system changes the location from *SendDataToCont* to *Receive-DataFromCont* following the synchronization transition:

SendDataToCont $\xrightarrow{DataFromCont?}$ *ReceiveDataFromCont*. During this transition the continuous interface receives data from the continuous simulator and, if a state event occurred, the time of the state event. In the *ReceiveDataFromCont* location, the continuous interface evaluates if a state event was generated. Two cases are possible:

- When no state event is generated, the location changes from Receive-DataFromCont to TimeOfStEv following the transition *ReceiveDataFromCont* $\xrightarrow{Event!StateEvent\,=\,0}$ *TimeOfStEv*. The transition is annotated in this case by the synchronization Event! and with the update StateEvent = 0.
- When a state event is generated, the location changes from *Receive-DataFromCont* to *StEvDetect* following the transition:

$$\textbf{ReceiveDataFromCont} \xrightarrow{Event!\ StateEvent\,=\,1,StEvTime\,=\,(tse<tcn)?tse:tcn,tse:int[0,period]} \textbf{StEvDetect}$$

This transition is annotated with a synchronization (Event!) and three variable updates: StateEvent=1 (for the detection of a state event), StEvTime=(tse<tcn)? tse:tcn, tse:int[0,period] (for the time of the state event that occurs during the time interval [0, *period*]; this time will be sent to the discrete simulator). *StEvDetect* is an urgent location. The location *StEvDetect* changes to *TimeOfStEv* following the transition *StEvDetect* $\xrightarrow{tcn\,=\,StEvTime}$ *TimeOfStEv*.

At this point there is no synchronization, only an update of the time in the continuous domain having assigned the time of the state event StEvTime: tcn=StEvTime.

TimeOfStEv location is common for both cases, StateEvent=0 or StateEvent=1. This location changes to *WaitDataFromDisc*. The system performs synchronization (DataToBus!) between the continuous interface and the continuous simulator. The next location is *Start*, the continuous time variables is initialized on this channel (tc=tcn) and the cycle restarts.

2. *Formal model simulation:* The UPPAAL tool allows the validation of the system's expected behavior regarding functionality: synchronization, conflicts, and communication. We simulated all the possible dynamic executions of our model.

Figure 16.11 shows a screenshot with the simulator. We observe that the left panel is the simulation control window. It highlights the enabled transition as well as the symbolic traces. The middle panel shows the variables. It displays the values of the data and clock variables in the current location or transition selected in the trace of the simulation control panel (the symbolic traces). The right panel allows the visualization of the message sequence chart (also known as simulator).

The vertical lines in the simulator window in Figure 16.11 represent the transitions between the locations while the horizontal lines are the synchronization points. In this figure, the communication between the interfaces as well as the communication between the simulators and the domain specific interfaces is represented by the same horizontal lines. As shown here, the simulation was stopped by the user after the detection of a state event in the continuous domain. The state event was indicated to the discrete simulator

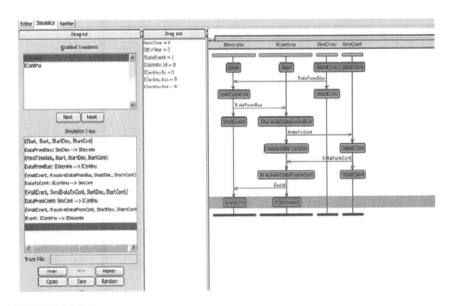

FIGURE 16.11
Formal model simulation screen capture.

and the time of the state event (StEvTime) is to be sent from the continuous to the discrete interface. The variable panel shows that the variable StateEvent=1, the time of the state event StEvTime=2, and the NextTime=10. The discrete simulator, instead of advancing the time to 10, will advance only to StEvTime.

3. *Formal verification*: The formal verification consists of checking properties of the system for a broad class of inputs [25]. In our work, we checked properties that fall into three classes:

   - Safety properties—the system does not reach an undesirable configuration, (e.g., deadlock) [9].
   - Liveness properties—some desired configuration will be visited eventually or infinitely (e.g., expected response to an input) [9].
   - Reachability properties—the system always has the possibility of reaching a given situation (some particular situation can be reached) [25].

The properties verified in order to validate the synchronization model are described below.

P0 Absence of deadlock (safety property)
Deadlock exists among a set of processes if every process waits for an event that can be caused only by another process in the set. In UPPAAL, deadlock is expressed by a formula using the keyword `deadlock`. A state is a deadlock state if there are no outgoing action transitions either from the state itself or any of its delay successors [36].

P1 State event detected by the discrete domain (liveness property)
The indication of a state event by the continuous interface and its detection by the discrete interface is very important for continuous/discrete heterogeneous systems. We defined a liveness property in order to check this behavior that is stated as follows:
Definition: A state event detected in the continuous domain `leads` to a state event detected in the discrete.

P2 No state event in discrete if no state event in continuous domain (safety property)
In order to avoid false responses from the discrete simulators, we defined a safety property to verify if the system will "detect" a state event in the discrete simulator when it was not generated (and indicated) by the continuous domain:
Definition: `Invariantly` a state event detected in the discrete domain imply state event in the continuous.

P3 Synchronization between the interfaces (reachability property)
One of the most important properties characterizing the interaction between the continuous and the discrete domains is the communication and

implicitly the synchronization. This property verifies that after a cycle executed by each model, both are at the same time stamp (and by consequence are synchronized)

Definition: Invariantly both processes in the *Start* location (initial state) imply the time in the continuous domain t_c is equal with the time in the discrete domain t_d.

P4 Causality principle (liveness property)
The causality can be defined as a cause and effect relationship. The causality of two events describes to what extent one event is caused by the other. The causality is already verified by *P3* for scheduled events. However, when a state event is generated by the continuous domain, the discrete domain has to detect this event at the same precise time (the cause precedes or equals the effect time) and not some other possible event existing at a different time in the continuous domain.

Definition: Invariantly both processes in the *StEvDetect* location (detection of state event) imply the time in the continuous t_c is equal with the time in the discrete t_d.

16.6.6　Definition of the Internal Architecture of the Simulation Interfaces

The overall continuous/discrete simulation interface is formally defined using the DEVS formalism. As shown in Figure 16.4, the interface is described as a set of coupled models: the continuous domain interface (CDI), the discrete domain interface (DDI), and the co-simulation bus. Figure 16.12 shows the atomic modules composing the interface used in our implementation.

The specific functionalities of the interfaces were presented in Section 16.4.2. In terms of internal architecture, the blocks assuring these features are

For the Continuous Model Simulation Interface

- The *State Event Indication and Time Sending* block (SETS)
- The *Signal Conversion and Data Exchange* block (SCDE)
- The *Event Detection* block (DED)
- The *Context Switch* block (CS)

For the Discrete Model Simulation Interface

- The *End of Discrete Simulation Cycle Detection and Time Sending* block (DDTS)
- The *Data Exchange* block (DE)
- The *Event Detection* block (DEC)
- The *Context Switch* block (CS)

These atomic modules are forming the co-simulation library and the co-simulation tools enable their parameterization and their assembly in order to generate a new co-simulation instance.

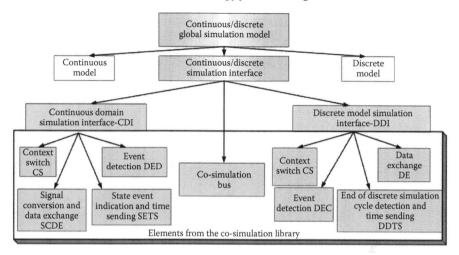

FIGURE 16.12
The hierarchical representation of the generic architecture of the co-simulation model with the elements of the co-simulation library defined. (From Gheorghe, L. et al., Formal definition of simulation interfaces in a continuous/discrete co-simulation tool, *Proceedings of the 17th IEEE International Workshop on RSP*, Chania, Crete, Greece, pp. 186–192, 2006. With permission.)

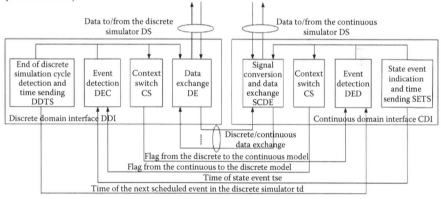

FIGURE 16.13
Internal architecture of the continuous/discrete simulation interface. (From Gheorghe, L. et al., Formal definition of simulation interfaces in a continuous/discrete co-simulation tool, *Proceedings of the 17th IEEE International Workshop on RSP*, Chania, Crete, Greece, pp. 186–192, 2006. With permission.)

Figure 16.13 presents the atomic modules interconnection in each domain specific simulation interface as well as the signals and interactions between the interfaces.

The internal architecture is defined as a set of coupled modules that respect the coupled modules DEVS formalism as presented in Section 16.6.1:

$-N_{interface} = (X, Y, D, \{M_d | d \in D\}, EIC, EOC, IC)$

$-X = \{(p_d, v_d) | p_d \in InPorts, v_d \in Xp_d\}$

$-Y = \{(p_d, v_d) | p_d \in OutPorts, v_d \in Yp_d\}$

$-Xp_d, Yp_d =$ values for input ports, respectively output ports

$-InPorts = P_{in,c} \cup P_{in,d} \cup P_{in,td} \cup P_{in,tse} \cup P_{in,flag}$ where

$P_{in,c}-$ set of ports receiving data from the continuous model; $P_{in,d}-$ set of ports receiving data from the discrete model (via the co-simulation bus);

$P_{td}-$ port receiving the timestamp of the next discrete event

$P_{in,flag}-$ port receiving the command for the context switch

$- OutPorts = P_{out,c} \cup P_{out,d} \cup P_{out,td} \cup P_{out,tse} \cup P_{out,flag}$

$P_{out,flag}, P_{out,c}, P_{out,d}$ are defined similarly to $P_{in,flag}, P_{in,c}$ and $P_{in,d}$

$- D = \{$"Continuous Domain Interface" (with associated model $N_{interfaceCDI}$), "Discrete Domain Interface" (with associated model $N_{interfaceDDI}$), "co-simulation bus" (with associated model M_{cosim})$\}$

$$-M_d = (N_{interfaceCDI}, M_{interfaceDDI}, M_{cosim})$$

$$-EIC = \{((N_{interface}, \text{"in}''_{c,1}), (N_{interfaceCDI}, \text{"in}''_{c,1})); ...;$$

$$((N_{interface}, \text{"in}''_{c,n}), (N_{interfaceCDI}, \text{"in}''_{c,n}));$$

$$((N_{interface}, \text{"in}''_{d,0}), (N_{interfaceDDI}, \text{"in}''_{d,0})); ...;$$

$$((N_{interface}, \text{"in}''_{d,m}), (N_{interfaceDDI}, \text{"in}''_{d,m}))\}$$

$$-EOC = \{((N_{interfaceCDI}, \text{"out}''_{c,1}), (N_{interface}, \text{"out}''_{c,1})); ...;$$

$$((N_{interfaceCDI}, \text{"out}''_{c,p}), (N_{interface}, \text{"out}''_{c,p}));$$

$$((N_{interfaceDDI}, \text{"out}''_{d,1}), (N_{interface}, \text{"out}''_{d,1})); ...;$$

$$((N_{interfaceDDI}, \text{"out}''_{d,q}), (N_{interface}, \text{"out}''_{d,q}))\}$$

$$-IC = \{((N_{interfaceCDI}, op_{CDI}), (M_{cosim}, ip_{cosim})) | N_{interfaceCDI},$$

$$- M_{cosim} \in D, op_{CDI} \in OutPorts_{CDI}, ip_{cosim} \in InPorts_{cosim}\} \cup$$

$$\{((N_{interfaceDDI}, op_{DDI}), (M_{cosim}, ip_{cosim})) | N_{interfaceDDI},$$

$$M_{cosim} \in D, op_{DDI} \in OutPorts_{DDI}, ip_{cosim} \in InPorts_{cosim}\} \cup$$

$$\{((M_{cosim}, op_{cosim}), (N_{interfaceCDI}, ip_{CDI})) | N_{interfaceCDI},$$

$$M_{cosim} \in D, op_{cosim} \in OutPorts_{cosim}, ip_{CDI} \in InPorts_{CDI},\} \cup$$

$$\{((M_{cosim}, op_{cosim}), (N_{interfaceDDI}, ip_{DDI})) | N_{interfaceDDI},$$

$$- M_{cosim} \in D, op_{cosim} \in OutPorts_{cosim}, ip_{DDI} \in InPorts_{DDI},\} [12]$$

We show here the atomic module **co-simulation bus** that can be formally defined as follows:

$- X. = \{(p_d, v_d) | p_d \in InPorts, v_d \in Xp_d\}$

$- Y = \{(p_d, v_d) | p_d \in OutPorts, v_d \in Yp_d\}$

- $InPorts = P_{in,c} \cup P_{in,d} \cup P_{in,td} \cup rmP_{in,tse} \cup P_{in,flag}$
- $OutPorts = P_{out,c} \cup P_{out,d} \cup P_{out,td} \cup P_{out,tse} \cup P_{out,flag}$

where $P_{in,c}$, $P_{in,d}$ $P_{in,td}$, $P_{in,tse}$, $P_{in,flag}$ $P_{out,c}$, $P_{out,d}$, $P_{out,td}$, $P_{out,tse}$, $P_{out,flag}$ as well as Xp_d, Yp_d were previously defined.

States triplet S: (phase * σ * job) where:

phase: ("passive," "active")

$\sigma : \Re_0^+$ advance time

job: ("store," "respond")

$S = \{$"passive," "active"$\} * \Re_0^+ * \{$"store", "respond"$\}$

δ_{ext} (("passive" * σ * job), e, x))=

("passive," σ −e, x), if x = 0

("active", σ−e, job), if x! = 0

δ_{int} **(s)** = ("active", σ, job)

λ ("active," σ, job) = {"store", "respond" }

t_a(phase, σ, job) = σ

The architecture of the discrete domain interface and the continuous domain interface are also formally defined as a set of coupled modules. Formal descriptions for DDI and CDI respect the coupled module DEVS formalism. Each element of the structure follows the concepts presented in Section 16.6.1 and that were applied for the overall continuous/discrete simulation interface.

16.6.7 Analysis of the Simulation Tools for the Integration in the Co-Simulation Framework

The previous steps that describe the gradual formal definition of the simulation interfaces and the required library elements are independent of the different simulation tools and specification languages used generally for the specification/execution of the continuous and discrete subsystems. After the analysis of the existing tools we found that Simulink is an illustrative example of a continuous simulator enabling the control functionalities that were presented in Section 16.4.5. These functionalities can be added in generic library blocks and a given Simulink model may be prepared for the co-simulation by parameterization and addition of these blocks.

Several discrete simulators present the characteristics detailed in Section 16.4.5. SystemC is an illustrative example. Since it is an open source, SystemC enables the addition of the presented functionalities in an efficient way—the scheduler can be modified and adapted for co-simulation. In this way, the co-simulation overhead may be minimized. However, the addition of simulation interfaces is more difficult than in Simulink because the specifications in SystemC are textual and a code generator is required to facilitate the addition of simulation interfaces. The automatic generation of the co-simulation interfaces is very suitable, since their design is time consuming and an important source of errors. The strategy currently used is based on the configuration of the components and their assembly. These components are selected from a co-simulation library.

16.6.8　Implementation of the Library Elements Specific to Different Simulation Tools

The implementation for the validation of continuous/discrete systems was realized using SystemC for the discrete simulation models and Simulink for the continuous simulation models.

For Simulink, the interfaces are functional blocks programmed in C++ using S-Functions [24]. These blocks are manipulated like all other components of the Simulink library. They contain input/output ports compatible with all model ports that can be connected directly using Simulink signals. The user starts by dragging the interfaces from the interface components library into the model's window, then parameterizes them, and finally connects them to the inputs and the outputs of his model.

For SystemC, in order to increase the simulation performance, part of the synchronization functionality has been implemented at the scheduler's level, which is a part of the state event management and the end of the discrete cycle detection (detects that there are no more delta cycles at the current time). For the generation of the co-simulation interfaces for SystemC, the implementation of a code generator was necessary. This script has as input user-defined parameters such as sampling periods, number and type of ports, and synchronization ports.

16.7　Formalization and Verification of the Interfaces

One of the key issues for the automatic generation of co-simulation interfaces is the rigorous definition of the behavior and architecture of simulation interfaces and this can be achieved by their formalization and formal verification in terms of behavior. Formal definitions can be used to develop easy-to-verify designs [16]. On this mathematical foundation one can define the criteria that allows for the automatic selection of the components from the co-simulation library and the automatic generation of the co-simulation interfaces. This section presents the formalization and the verification of the co-simulation interfaces.

16.7.1　Discrete Simulator Interface

This section presents the operational semantics of the discrete simulation interfaces (DSI). The semantics were defined with respect to the synchronization model presented in Section 16.6.1, using the DEVS formalism. Table 16.2 presents a set of rules that show the transitions between states.

For all the rules, the semantics of the global variable *flag* is related to the context switch between the continuous and discrete simulators. When the flag is set to "1," the discrete simulator is executed. When it is "0," the continuous simulator is executed. The global variable *synch* is used to impose

TABLE 16.2

Operational Semantics for the DSI

Nr.	Rule
1.	$$synch = 1 \wedge flag = 1 \wedge (x_{dk}, t_{dk}) = \delta_{ext}((x_{dk}, t_{dk}), 0, x)$$ $$((x_{dk}, t_{dk}), \infty) \xrightarrow{?DataFromDisc} ((x_{dk}, t_{dk}), 0) \xrightarrow{!(data, t_{dk+1}((x_{dk},t_{dk})));flag:=0} ((x_{dk}, t_{dk}), t_{dk+1})$$
2.	$$synch = 0 \wedge flag = 1 \wedge \neg stateevent \wedge (x_{dk}, t_{dk}) = \delta_{ext}((x_{dk}, t_{dk}), 0, x)$$ $$((x_{dk}, t_{dk}), e_{dk}) \xrightarrow{?Event} ((x_{dk}, t_{dk}), 0) \xrightarrow{?data;synch:=1} ((x_{dk}, t_{dk}), 0) \xrightarrow{!DataToDisc} ((x_{dk+1}, t_{dk+1}), e_{dk+1})$$
3.	$$synch = 0 \wedge flag = 1 \wedge stateevent \wedge (x_{dk}, t_{dk}) = \delta_{ext}((x_{dk}, t_{dk}), 0, x)$$ $$((x_{dk}, t_{dk}), e_{dk}) \xrightarrow{?Event} ((x_{dk}, t_{dk}), 0) \xrightarrow{?(data, t_{se});synch:=1} ((x_{dk}, t_{dk}), 0) \xrightarrow{!(DataToDisc, t_{se})} ((x_{se}, t_{se}), e_{se})$$

the order of the different operations expressed by the rules. The first rule covers arrow 1 in Figure 16.5a and b. The second and third rules correspond to arrows 3 (on the receiving part) and 4 in Figure 16.5a and b.

In order to clarify, we present here the first rule in detail. The premises of this rule are: the *synch* variable has value "1," the *flag* variable has value "1," and we have an external transition function (δ_{ext}) for the DSI.

This rule expresses the following actions of the discrete simulator interface (DSI):

- Receiving data from the discrete model. This is an external transition (δ_{ext}) expressed by ?(*DataFromDisc*).
- Sending data to the continuous simulator interface (CSI) (!*DataToCSI*). The data sent to the CSI is the output function $\lambda(x_{dk}, t_{dk})$ and it is possible, in accordance with DEVS formalism, only as a consequence of an internal transition (δ_{int}). In our case, the output is represented by !(*data,$t_{dk+1}(x_{dk},t_{dk})$*). This transition corresponds to arrow 1 in Figure 16.5a and b.
- Switching the simulation context from the discrete to the continuous domain (action expressed by *flag*: = 0).

All the other rules presented in this table follow the same format [13].

TABLE 16.3

Operational Semantics for the CSI

Nr.	Rule
1.	$$synch = 1 \wedge flag = 1 \wedge q_k = \delta_{ext}((x_{ck}, t_{ck}), 0, x)$$ $$(x_{ck}, \infty) \xrightarrow{?(data, t_{dk+1});synch:=0} (x_{ck}, t_{ck}) \xrightarrow{!(DataToCont, t_a(x_{dk}, t_{dk}))} (x_{ck}, t_{ck})$$
2.	$$synch = 0 \wedge flag = 0 \wedge \neg stateevent \wedge q_{k+1} = \delta_{int}((x_{ck}, t_{ck}))$$ $$(x_{ck}, t_{ck}) \xrightarrow{?(DataFromCont)} (x_{ck}, t_{ck}) \xrightarrow{!(data);flag:=1} (q, t_{d,k+1})$$
3.	$$synch = 0 \wedge flag = 0 \wedge stateevent \wedge q_{k+1} = \delta_{int}(q_k)$$ $$(x_{ck}, t_{ck}) \xrightarrow{?(DataFromCont)} (x_{ck}, t_{ck}) \xrightarrow{!(data, t_{se});flag:=1} (se, t_{se})$$

16.7.2   Continuous Simulator Interface

The operational semantics for the CSI is given by the set of rules presented in Table 16.3. In these rules, the *Data* notation refers to the data exchanged between the DSI and the discrete simulator. All the rules presented in Table 16.3 can be explained analogously to the one already illustrated in Section 16.7.1.

16.8   Implementation Stage: CODIS a C/D Co-Simulation Framework

In Section 16.4, we proposed a generic methodology, in two stages, for the efficient design of C/D co-simulation tools. These stages are: a generic stage and an implementation stage. The generic stage, divided in four steps, was presented in detail and applied in Sections 16.4.1 through 16.4.4 and Sections 16.6.3 through 16.6.6.

This section presents, CODIS [5], a tool for the generation of simulation models in more detail. This tool can automatically produce the global simulation model instances for C/D systems simulation using SystemC and Simulink simulators. This is done by generating and providing the interfaces that implement the simulation model layers and building the co-simulation bus. In the development of this framework, the generic steps of the methodology that we presented in detail in this chapter are "hidden" in different stages of the simulation flow. The "definition of the library elements and the internal architecture of the co-simulation interfaces" step represents the foundation for the generation of the co-simulation library and implicitly for the generation of the co-simulation interfaces. The internal architecture of the interfaces respects the definition of the operational semantics, the distribution of the synchronization functionality, as well as the synchronization model.

Figure 16.14 gives an overview of the flow of the instance generation in the case of CODIS. The inputs in the flow are the continuous model in Simulink and the discrete model in SystemC which are schematic and textual models, respectively. The output of the flow is the global simulation model (co-simulation model) instance.

For Simulink, the interfaces can be parameterized starting with their dialog box. The user starts by dragging the interfaces from the interface components library into the model's window, then parameterizes them, and finally connects them to the inputs and the outputs of the model. Before the simulation, the functionalities of these blocks are loaded by Simulink from the .dll libraries. The parameters of the interfaces are the number of input and output ports, their type, and the number of state events.

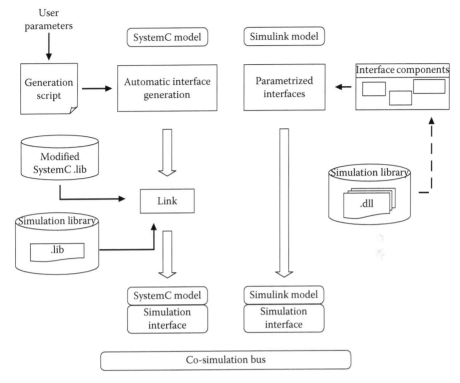

FIGURE 16.14
Automatic generation of global simulation model instances. (From Gheorghe, L. et al., Formal definition of simulation interfaces in a continuous/discrete co-simulation tool, *Proceedings of the 17th IEEE International Workshop on RSP*, Chania, Crete, Greece, pp. 186–192, 2006. With permission.)

For SystemC, the interface parameters are: the names, the number and the data type of the discrete model inputs ports and the sampling periods. The interfaces are automatically generated by a script generator that has as input the user-defined parameters. The tool also generates the function sc_main (or modifies the existing sc_main) that connects the interfaces to the user model. The model is compiled and the link editor calls the library from SystemC and a static library (the simulation library in Figure 16.14).

16.9 Conclusion

This chapter presented a generic methodology for the design of efficient continuous/discrete co-simulation tools. The methodology can be divided into two main stages: (1) a generic stage, defining simulation interface functionality in a conceptual framework when formal methods for the specification and validation are used, and (2) a stage that provides the implementation for the rigorously defined functionality. Given the importance of the co-simulation interfaces, the methodology concentrates on the co-simulation interfaces, their behavior, as well as the synchronization they assure in the global validation of a C/D heterogeneous system.

We also present here a formal representation of the functionality of each interface independent of the other.

We illustrate the methodology with CODIS, a tool for C/D systems co-simulation. In the development of CODIS, all the steps of the methodology we proposed in this paper are "hidden" in different stages of the simulation flow. The "definition of the library elements and the internal architecture of the co-simulation interfaces" step represents the foundation for the generation of the co-simulation library and implicitly for the co-simulation interfaces generation. The definition of the operational semantics and the distribution of the synchronization functionality as well as their behavior play an important role at the output flow with the behavior of the co-simulation interfaces and the synchronization model. The "analysis of the simulation tools for the integration in the co-simulation framework" helped choosing the tools that were used for the modeling of the continuous and the discrete simulators while the "implementation of the library elements specific to different simulation tools" constitutes the final implementation of the libraries.

References

1. R. Alur and D. Dill, Automata for modeling real-time systems, in *Proceedings of the Seventeenth International Colloquium on Automata, Languages and Programming*, Warwick University, England, 1990, Vol. 443, pp. 322–335.

2. G. Behrmann, A. David, and K. Larsen, A tutorial on UPPAAL, *Real-Time Systems Symposium*, Miami, FL, 2005.

3. J. Bengtsson and W. Yi, Timed automata: Semantics, algorithms and tools, Uppsala University, Uppsala, Denmark, 1996.

4. J.-S. Bolduc and H. Vangheluwe, The modelling and simulation package PythonDEVS for classical hierarchical DEVS. MSDL technical report MSDL-TR-2001-01, McGill University, Montreal, Quebec, Canada, June 2001.

5. F. Bouchhima, G. Nicolescu, M. Aboulhamid, and M. Abid, Generic discrete–continuous simulation model for accurate validation in heterogeneous systems design, *Microelectronics Journal*, 38, 2007, 805–815.

6. C. G. Cassandras, *Discrete Event Systems: Modeling and Performance Analysis*, Richard Irwin, New York, 1993.

7. F. E. Cellier, Combined continuous/discrete system simulation languages—usefulness, experiences and future development, *Methodology in Systems Modelling and Simulation*. North-Holland, Amsterdam, the Netherlands, 1979, pp. 201–220.

8. M. D'Abreu and G. Wainer, M/CD++: Modeling continuous systems using Modelica and DEVS, in *Proceedings of the IEEE International Symposium of MASCOTS'05*, Atlanta, GA, 2005, pp. 229–238.

9. S. Edwards, L. Lavagno, E. Lee, and A. L. Sangiovanni-Vincentelli, Design of embedded systems: Formal models, validation, and synthesis, *Proceedings of the IEEE*, 85, 1997, 366–390.

10. P. Frey and D. O'Riordan, Verilog-AMS: Mixed-signal simulation and cross domain connect modules, in *Proceedings BMAS International Workshop*, Orlando, FL, 2000, pp. 103–108.

11. H. R. Ghasemi and Z. Navabi, An effective VHDL-AMS simulation algorithm with event, in *International Conference on VLSI Design*, Kolkata, India, 2005, pp. 762–767.

12. L. Gheorghe, F. Bouchhima, G. Nicolescu, and H. Boucheneb, Formal definition of simulation interfaces in a continuous/discrete co-simulation tool, in *Proceedings of the 17th IEEE International Workshop on RSP*, Chania, Crete, Greece, 2006, pp. 186–192.

13. L. Gheorghe, F. Bouchhima, G. Nicolescu, and H. Boucheneb, Semantics for model-based validation of continuous/discrete systems, in *Proceedings of the DATE*, Munich, Germany, 2008, pp. 498–503.

14. N. Giambiasi, J.-L. Paillet, and F. Chane, From timed automata to DEVS, in *Proceedings of the 2003 Winter Simulation Conference*, New Orleans, LA, 2003.

15. IEEE Standard VHDL Analog and Mixed-Signal Extensions (1999), IEEE Std 1076.1-1999

16. International Technology Roadmap for Semiconductor Design. [Online]. Available at: http://public.itrs.net/

17. A. Jantsch and I. Sander, Models of computation and languages for embedded system design, *IEE Proceedings Computers and Digital Techniques*, 152, 2005, 114–129.

18. A. Jantsch, *Modeling Embedded Systems and SoCs—Concurrency and Time in Models of Computation. Systems on Silicon.* Morgan Kaufmann Publishers, San Francisco, CA, June 2003.

19. T. G. Kim, *DEVSim++ User's Manual*, SMSLab, Dept. of EECS, KAIST, Taejon, Korea, 1994, http://smslab.kaist.ac.kr

20. Y. J. Kim, J. H. Kim, and T. G. Kim, Heterogeneous simulation framework using DEVS-BUS, in *Simulation, the Society for Modeling and Simulation International*, 79, 2003, 3–18.

21. E. A. Lee and H. Zheng, Operational semantics of hybrid systems, in *Hybrid Systems: Computation and Control: 8th International Workshop, HSCC*, Zurich, Switzerland, 2005, pp. 25–53.

22. E. A. Lee and A. L. Sangiovanni-Vincentelli, Comparing models of computation, in *IEEE Proceedings of the International Conference on Computer-Aided Design (ICCAD)*, San Jose, CA, 1996, pp. 234–241.

23. S. Levitan, J. Martinez, T. Kurzveg, P. Marchand, and D. Chiarulli, Multi technology system-level simulation, *Analog Integrated Circuits and Signal Processing*, 29, 2001, 127–149.

24. MATLAB-Simulink [Online]. Available at: www.mathworks.com

25. J.-F. Monin, *Understanding Formal Methods*, Springer, Berlin, 2003.

26. G. Nicolescu et al., Validation in a component-based design flow for multicore SoCs, in *Proceedings of ISSS*, Kyoto, Japan, 2002, pp. 162–167.

27. D. H. Patel and S. K. Shukla, *SystemC Kernel—Extensions for Heterogeneous System Modeling*, Kluwer Academic Publishers, Dordrecht, the Netherlands, 2004.

28. Ptolemy project [Online]. Available at: http://ptolemy.eecs.berkeley.edu/

29. S. Romitti, C. Santoni, and P. François, A design methodology and a prototyping tool dedicated to adaptive interface generation, in *Proceedings of the 3rd ERCIM Workshop on "User Interfaces for All"*, Obernai, France, 1997.

30. SystemC LRM [Online]. Available at: www.systemc.org

31. SystemVerilog [Online]. Available at: www.systemverilog.org

32. A. Vachoux, C. Grimm, and K. Einwich, Analog and mixed signal modeling with SystemC-AMS, in *Proceedings International on Symposium Circuits System*, Bangkok, Thailand, 2003, pp. 914–917.

33. VHDL [Online]. Available at: www.vhdl.org

34. B. P. Zeigler, H. Praehofer, and T. G. Kim, *Modeling and Simulation—Integrating Discrete Event and Continuous Complex Dynamic Systems*, Academic Press, San Diego, CA, 2000.

35. G. Wainer, Modeling and simulation of complex systems with cell-DEVS, in *Winter Simulation Conference*, Washington, DC, 2004, pp. 49–60.

36. F. Wang, Formal verification of times systems: A survey and perspective, *Proceedings of the IEEE*, 92, 2004, 1283–1305.

17

Modeling and Simulation of Mixed Continuous and Discrete Systems

Edward A. Lee and Haiyang Zheng

CONTENTS

17.1 Introduction

An embedded system mixes digital controllers realized in hardware and software with the continuous dynamics of physical systems [30]. Such systems are semantically heterogeneous, combining continuous dynamics, periodic timed actions, and asynchronous event reactions. Modeling and design of such heterogeneous systems is challenging. A number of researchers have defined concurrent **models of computation** (MoCs) that support modeling, specification, and design of such systems [11,22,26,28,34].

A variety of approaches have been tried for dealing with the intrinsic heterogeneity of embedded systems. This chapter describes a particularly useful combination of semantics, providing a disciplined and rigorous mixture of synchronous/reactive (SR) systems [4], discrete-event (DE) systems [13,19,29,49], and continuous-time (CT) dynamics [20,35,42,46]. Our approach embraces heterogeneity, in that subsystems can be modeled using any of the three semantics, and these subsystem models can be combined hierarchically to form a whole system. We leverage the idea of an **actor abstract semantics** [33] to provide a coherent and rigorous meaning for

559

the heterogeneous system. Our approach also provides improvements to conventional DE and CT semantics by leveraging the principles of SR languages. These improvements facilitate the heterogeneous combination of the three distinct modeling styles.

17.2   Related Work

A number of authors advocate heterogeneous combinations of semantics. Ptolemy Classic [11] introduced the concept, showing useful combinations of asynchronous models based on variants of dataflow and timed DE models. The concept was picked up for hardware design in SystemC (version 2.0 and higher) [45], on which some researchers have specifically built heterogeneous design frameworks [26]. Metropolis [22] introduced communication refinement as a mechanism for specializing a general MoC in domain-specific ways, and also introduced **quantity managers** that provide a unified approach to resource management in heterogeneous systems.

Our approach in this chapter is closest in spirit to SML-Sys [41], which builds on Standard ML to provide for mixtures of MoCs. SML-Sys combines asynchronous models (dataflow models) with synchronous models (which the authors call "timed"). Our approach, in contrast, combines only timed models, including both DE and CT dynamics.

A particular form of heterogeneous systems, hybrid systems provide for joint modeling of continuous and discrete dynamics. A few software tools have been built to provide simulation of hybrid systems, including Charon [2], Hysdel [47], HyVisual [10], Modelica^{TM}[46], Scicos [16], Shift [15], and Simulink$^{®}$/Stateflow^{TM}(from The MathWorks). An excellent analysis and comparison of these tools is given by Carloni et al. [12]. We have previously extensively studied the semantics of hybrid systems as heterogenous combinations of finite state machines (FSM) and continuous dynamics [35]. Our approach in this chapter extends this to include SR and DE models. We focus here on the interactions between SR, DE, and CT, because the interactions between FSM and SR, DE, and CT have already been extensively studied [21,35].

Several authors advocate unified MoCs as a binding agent for heterogeneous models [3,9,23]. Heterogeneous designs are expressed in terms of a common semantics. Some software systems, such as Simulink from The MathWorks, take the approach of supporting a general MoC (CT systems in the case of Simulink) within which more specialized behaviors (like periodic discrete-time) can be simulated. The specialized behaviors amount to a **design style** or **design pattern** within a single unified semantics. Conformance to design styles within this unified semantics can result in models from which effective embedded software can be synthesized, using, for example, Real-Time Workshop^{TM} or TargetLink^{TM} from dSpace.

Our approach in this chapter is different in that the binding agent is an **abstract semantics**. By itself, it is not sufficiently complete to specify system designs. Its role is exclusively as a binding agent between diverse concrete MoCs, each of which is expressive enough to define system behavior (each in a different way).

We are heavily inspired here by the fixed-point semantics of synchronous languages [4], particularly Lustre [25], Esterel [8], and Signal [24]. SCADE [7] (Safety Critical Application Development Environment), a commercial product of Esterel Technologies, builds on the synchronous language, Lustre [25], providing a graphical programming framework with Lustre semantics. All the synchronous languages have strong formal properties that yield quite effectively to formal verification techniques. Our approach, however, is to use the principles of synchronous languages in the style of a **coordination language** rather than a programming language. This coordination language approach has been realized in Ptolemy [17] and ForSyDe [44]. It allows for "primitives" in a model to be complex components rather than built-in language primitives. This approach will allow for heterogeneous combinations of MoCs, since the complex components may themselves be given as compositions of further subcomponents under some other MoCs.

A number of researchers have combined synchronous languages with asynchronous interactions using a principle called **globally asynchronous, locally synchronous** or GALS (see for example [5]). In our case, all the MoCs we consider are timed, so there is a measure of synchrony throughout. Instead, we focus on combinations of heterogeneous timing properties, including abstracted discrete sequences (SR), time-stamped events (DE), and continuous-time dynamics (CT).

17.3   Actor-Oriented Models

Our approach here closely follows the principles of **actor-oriented design** [33], a component methodology where components called *actors* execute and communicate with other actors in a *model*, as illustrated in Figure 17.1. Figure 17.1 shows an actor with a single output port that produces a sequence of data values that constitute a sine wave. Internally, the actor is realized as a network of actors. It has three parameters, *frequency*, *phase*, and *samplingFrequency*, with default values shown.

Actors have a well-defined component interface. This interface abstracts the internal state and behavior of an actor, and restricts how an actor interacts with its environment. The interface includes *ports* that represent points of communication for an actor, and *parameters* that are used to configure the operation of an actor. Often, parameter values are part of the *a priori* configuration of an actor and do not change when a model is

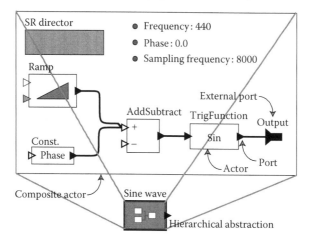

FIGURE 17.1
Illustration of a composite actor (above) and its hierarchical abstraction (below).

executed. The configuration of a model also contains explicit communication *channels* that pass data from one port to another. The use of channels to mediate communication implies that actors interact only with the channels that they are connected to and not directly with other actors.

Like actors, which have a well-defined external interface, a composition of actors may also define an external interface, which we call its *hierarchical abstraction*. This interface consists of *external ports* and *external parameters*, which are distinct from the ports and parameters of the individual actors in the composite. The external ports of a composite can be connected (on the inside) by channels to other external ports of the composite or to the ports of actors within the composite. External parameters of a composite can be used to determine the values of the parameters of actors inside the composite. Actors that are not composite actors are called **atomic actors**. We assume that the behavior of atomic actors is given in a **host language** (in Ptolemy II, Java, or C).

Taken together, the concepts of actors, ports, parameters, and channels describe the **abstract syntax** of actor-oriented design. This syntax can be represented concretely in several ways, such as graphically, as in Figure 17.1, in XML [32], or in a program designed to a specific API (as in SystemC). Ptolemy II [18] offers all three alternatives. In some systems, such as SML-Sys, the syntax of the host language specifies the interconnection of actors.

It is important to realize that the syntactic structure of an actor-oriented design says little about the semantics. The semantics is largely orthogonal to the syntax, and is determined by an MoC. The model of computation might give operational rules for executing a model. These rules determine when actors perform internal computation, update their internal state, and perform external communication. The model of computation also defines the nature of communication between components (buffered, rendezvous, etc.).

Our notion of actor-oriented modeling is related to the term *actor* as introduced in the 1970s by Carl Hewitt of MIT to describe the concept of autonomous reasoning agents [27]. The term evolved through the work of Agha and others to describe a formalized model of concurrency [1]. Agha's actors each have an independent thread of control and communicate via asynchronous message passing. We are further developing the term to embrace a larger family of models of concurrency that are often more constrained than general message passing. Our actors are still conceptually concurrent, but unlike Agha's actors, they need not have their own thread of control. Moreover, although communication is still through some form of message passing, it need not be strictly asynchronous.

In this chapter, we will consider three MoCs, namely, SR, DE, and CT. We carefully define the semantics of DE and CT so that the three MoCs are related in an interesting way. Specifically, SR is a special case of DE, and DE is a special case of CT. This does not mean that we should automatically use the most general MoC, CT, because execution efficiency, modeling convenience, and synthesizability all may be compromised. In fact, there are good reasons to use all three MoCs.

Most interestingly, we will show that these three MoCs can be combined hierarchically in arbitrary order. That is, in a hierarchical model like that in Figure 17.1, the higher level of the hierarchy and the lower level need not use the same MoC. In fact, all combinations of SR, DE, and CT are supported by our framework. We describe a prototype of this framework constructed in Ptolemy II.

17.4   Actor Abstract Semantics

In order to preserve the specialization of MoC while also building general models overall, we concentrate on the hierarchical composition of heterogenous MoC. The composition of arbitrary MoC is made tractable by an *abstract semantics*, which abstracts how communication and flow of control work. The abstract semantics is (loosely speaking) not the union of interesting semantics, but rather the intersection. It is abstract in the sense that it represents the common features of MoC as opposed to their collection of features.

A familiar example of an abstract semantics is represented by the Simulink S-function interface. Although not formally described as such, it in fact functions as such. In fact, simulink works with stateflow to accomplish a limited form of hierarchical heterogeneity through this S-function interface. We will describe an abstract semantics that is similar to that of Simulink, but simpler. It is the one realized in the Ptolemy II framework for actor-oriented design.

In Ptolemy II models, a **director** realizes the model of computation. A director is placed in a model by the model builder to indicate the model of computation for the model. For example, an SR director is shown visually as the uppermost icon in Figure 17.1. The director manages the execution of the model, defining the flow of control, and also defines the communication semantics.

When a director is placed in a composite actor, as in Figure 17.1, the composite actor becomes an **opaque composite actor**. To the outside environment, it appears to be an atomic actor. But inside, it is a composite, executing under the semantics defined by the local director. Obviously, there has to be some coordination between the execution on the outside and the execution on the inside. That coordination is defined by the abstract semantics.

The flow of control and communication semantics are abstracted in Ptolemy II by the *executable* and *receiver* interfaces, respectively. These interfaces define a suite of methods, the semantics of which are the actor abstract semantics of Ptolemy II. A receiver is supplied for each channel in a model by the director; this ensures that the communication semantics and flow of control work in concert to implement the model of computation.

In the Ptolemy II abstract semantics, actors execute in three phases, *setup*, a sequence of iterations, and *wrapup*. An **iteration** is a sequence of operations that read input data, produce output data, and update the state, but in a particular, structured way. The operations of an iteration consist of one or more invocations of the following pseudocode:

```
if (prefire()) {
    fire();
}
```

If *fire* is invoked at least once in the iteration, then the iteration concludes with exactly one invocation of *postfire*.

These operations and their significance constitute the executable interface and are summarized in Figure 17.2. The first part of an iteration is the invocation of *prefire*, which tests preconditions for firing. The actor thus determines whether its conditions for firing are satisfied. If it

setup	Initialize the actor.
prefire	Test preconditions for firing.
fire	Read inputs and produce outputs.
postfire	Update the state.
wrapup	End execution of the actor.

FIGURE 17.2
The key flow of control operations in the Ptolemy II abstract semantics. These are methods of the executable interface.

indicates that they are (by a return value of true), then the iteration proceeds by invoking *fire*. This may be repeated an arbitrary number of times. The contract with the actor is that *prefire* and *fire* do not change the state of the actor. Hence, multiple invocations with the same input values in a given iteration will produce the same results. This contract is essential to guarantee convergence to a fixed point.

If *prefire* indicates that preconditions are satisfied, then most actors guarantee that invocations of *fire* and *postfire* will complete in a finite amount of time. Such actors are said to realize a *precise reaction* [37]. A director that tests these preconditions prior to invoking the actor, and fires the actor only if the preconditions are satisfied, is said to realize a *responsible framework* [37]. Responsible frameworks coupled with precise reactions are key to hierarchical heterogeneity.

The abstract semantics also provides the set of primitive communication operations as shown in Figure 17.3. These operations allow an actor to query the state of communication channels, and subsequently retrieve information from the channels or send information to the channels. These operations are invoked in *prefire* and *fire*. Actors are also permitted to read inputs in *postfire*, but they are not permitted to produce outputs (by invoking *put*). Violations of this contract can lead to nondeterminism.

These operations are abstract, in the sense that the mechanics of the communication channel is not defined. It is determined by the model of computation. A domain-polymorphic actor is not concerned with how these operations are implemented. The actor is designed assuming only the abstract semantics, not the specific realization.

A hierarchically heterogeneous model is supported by this abstract semantics as follows. Figure 17.1 shows an opaque composite actor. It is opaque because it contains a director. That director gives the composite a behavior like that of an atomic actor viewed from the outside. A director implements the executable interface, and thus provides the operations of Figure 17.2.

Suppose that in Figure 17.1 the hierarchical abstraction of the sine wave component is used in a model of computation different from SR. Then from the outside, this model will appear to be a domain-polymorphic actor. When its *prefire* method is invoked, for example, the inside director must determine whether the preconditions are satisfied for the model to execute (in the SR case, they always are), and return true or false accordingly. When *fire* is

get	Retrieve a data token via the port.
put	Produce a data token via the port.
hasToken(k)	Test whether *get* will succeed k times.
hasRoom(k)	Test whether *put* will succeed k times.

FIGURE 17.3
Communication operations in Ptolemy II. These are methods of the receiver interface.

invoked, the director must manage the execution of the inside model so that input data (if any) are read, and output data are produced. When *postfire* is invoked, the director must update the state of the inside actors by invoking their *postfire* methods. Obviously, directors must be carefully designed to obey the actor abstract semantics contract. By obeying it, they gain the ability to be nested arbitrarily with other directors that also obey the contract.

The communication across the hierarchical boundary will likely end up heterogeneous. In Figure 17.1, the connection between the TrigFunction actor and the external port will be a channel obeying SR semantics (that is, it will be realized as a simple buffer with length one, support for *unknown* state, and enforcement of monotonicity constraints). The connection between the external port and some other port on the outside will obey the semantics of whatever director is provided on the outside. This need not be the same as the SR semantics.

In this chapter, we will focus on the use of three directors in Ptolemy II implementing SR, DE, and CT MoCs.

17.5   Synchronous/Reactive Models

We begin with the principle of synchronous languages, but used in the style of a coordination language rather than a programming language, as done in Ptolemy [17] and ForSyDe [44]. We will show that by adding time to this, we get a clean semantics for DE systems, and that by adding continuous dynamics, we get a clean semantics for CT, hybrid, and mixed signal systems (CT models).

The principle behind synchronous languages is simple, although the consequences are profound [4]. Execution follows "ticks" of a global "clock." At each tick, each variable (represented visually by the wires that connect the blocks) may have a value (it can also be absent, having no value). Its value (or absence of value) is defined by functions associated with each block. That is, at each tick, each block is a function from input values to output values (the function can vary from tick to tick). In Figure 17.4, the variables x and y at a particular tick are related by

$$x = f(y), \text{ and } y = g(x).$$

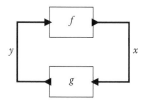

FIGURE 17.4
A simple feedback system.

The task of the compiler is to synthesize a program that, at each tick, solves these equations. We assume a flat order on the values that x and y can take on at each tick, with $\perp$ being the bottom of the order, representing "unknown." It is well known that if the functions f and g are monotonic in this order, then a unique least fixed point solution can be found in finite time. See [17] for a readable exposition of this semantics.

In most synchronous languages, there is no metric associated with the time between ticks (ForSyDe is an exception, as it defines a fixed constant "distance" between events of a signal). This means that programs can be designed to simply react to events, whenever they occur. This contrasts with Simulink, which has temporal semantics.

In our case, SR models by default have no notion of the passage of time, and only a notion of sequences of ticks. The semantics can be easily extended by associating a fixed interval of time between ticks, or even by associating a variable interval (somehow determined by the environment). These extensions have no effect on the SR semantics, and hence just amount to interpretations of an execution.

Unlike most work with synchronous languages, we assume that the SR model is being used to coordinate components of arbitrary complexity, rather than to coordinate known language primitives. For example, a Lustre compiler "knows" that *pre** breaks data precedences, whereas in our framework, we may not be able to know whether a component breaks data precedences. We can give an execution model that is very simple, but rather inefficient. In particular, at each tick of the clock, we can start with all signal values unknown, $\perp$, and invoke *prefire* and *fire* on each actor, in arbitrary order, until signal values stabilize at a fixed point. Here, *prefire* will specify whether there is sufficient information about the inputs for *fire* to execute. Once a fixed point is reached, then *postfire* is invoked on each actor exactly once, allowing the actor to update its state.

This execution strategy requires no knowledge of the internals of the components, except that they conform with the actor abstract semantics and implement monotonic functions. But it may result in rather inefficient execution. Following [17], we rely on causality interface information for actors (see [36]) to develop more efficient execution strategies. A key property of SR models is that the results of execution do not depend on the execution strategy. Only the efficiency of execution does.

To give a denotational semantics for such systems, we follow the approach in [38], which specializes the tagged signal model [34] to timed systems. Denotationally, a *signal* is the entire history of a communication between an output port and an input port. Specifically, let $T = \mathbb{N}$ be the tag set, where $\mathbb{N}$ is the set of natural numbers with the usual numerical order. Let V be an arbitrary family of values (a data type, or alphabet). Let

$$V_\varepsilon = V \cup \{\varepsilon\}$$

be the set of values plus "absent." Then a signal s is a partial function:

$$s \colon T \rightharpoonup V_\varepsilon$$

---

*In Lustre, *pre* is an operator that on each tick of the clock outputs the value of the input seen on the previous tick of the clock.

defined on an initial segment of T. Execution of an SR model begins with all signals being empty (defined on the empty initial segment $\emptyset \subset \mathbb{N}$). The first step of execution extends the signals so that they are all defined on the initial segment $\{0\} \subset \mathbb{N}$. If this is not possible (some signals remain undefined, or equivalently, unknown ($\perp$), at tag 0), then we declare the model to be flawed (it has a "causality loop").* The second step of execution extends the signals so that they are defined on the initial segment $\{0, 1\} \subset \mathbb{N}$. The third step defines them on $\{0, 1, 2\} \subset \mathbb{N}$, and so on.

For hierarchical models, like those in Figure 17.1, we can exploit the actor abstract semantics to get multiclock SR models. Specifically, consider the example in Figure 17.5. This shows a classic example from the early literature on synchronous languages, the "guarded count." The heart of the model is the *CountDown* actor, which given an input nonnegative integer, counts down to zero from that integer on each tick of the global clock. When the output reaches zero, it requests another input from the *CountDown* actor.

The interesting part of the model is that the *CountDown* actor has its own SR director, and hence is an opaque composite actor. The outside director cannot distinguish between it and an atomic actor, like the When actor at the bottom. This particular composite actor has a specialized behavior. When its *fire* method is invoked, it examines the *enable* input port. If the signal value at the current tick is known and present at this port, then it fires the inside actors until it reaches a fixed point. This fixed point may define output values. When the *postfire* method of the composite actor is invoked, then it invokes *postfire* on all the contained actors that were fired in the iteration. Thus, the clock of the inside model is a subclock of the outside model. The composite actor with a director itself behaves just like an atomic actor, conforming to the actor abstract semantics. It is precisely this compositional property that will enable us to put SR models within DE or CT models and vice versa.

Note that the composite actor at the bottom of Figure 17.5 has no director, and hence is said to be **transparent**. The actors within it are controlled by the top-level director as if there were no hierarchy. Readers familiar with synchronous language operators such as "when" and "default" should be able to read this model and understand how it implements the guarded count.

17.6 Discrete-Event Models

There is a long history of DE modeling techniques, and a number of widely used simulation systems (e.g., Opnet, NS-2) and programming languages (e.g., VHDL, Verilog) with DE semantics. The approach we take, however, is a bit different from all the DE systems we know of. In particular, we define

---

*This conforms to the *constructive semantics* of Berry [6].

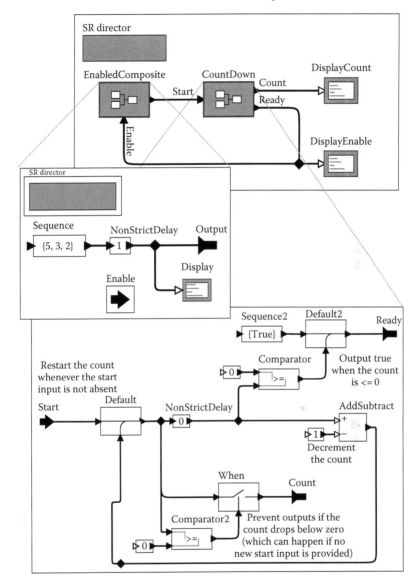

FIGURE 17.5
Example of a multiclock SR model.

DE to be an extension of SR where there is a measurable time between ticks of the clock. Specifically, a DE model becomes an SR model if we ignore the passage of time and focus instead only on the sequence of times at which events occur (each such time is a **tick** of the SR clock). The SR model is an abstraction of the DE model.

Since for DE we want to measure the passage of time, the tag set $T = \mathbb{N}$ that we used for SR becomes inadequate. A first attempt at a denotational semantics would simply use real numbers for time increments between ticks of the clock. Specifically, we would replace the tag set T with $T = \mathbb{R}_+$, the nonnegative real numbers, instead of $T = \mathbb{N}$. A signal s becomes a partial function,

$$s\colon \mathbb{R}_+ \rightharpoonup V_\varepsilon$$

defined on an initial segment of $\mathbb{R}_+$. If we let $I \subseteq \mathbb{R}_+$ be the initial segment where the signal is defined, then for s to be a DE signal, we require that $s(\tau) = \varepsilon$ for all $\tau \in I$ except for a discrete subset $D \subset \mathbb{R}_+$.*

This semantic model, however, is not sufficiently rich. In particular, signals cannot have multiple values at the same value of time. This makes the model awkward for models of software sequences that are abstracted as instantaneous (the perfect synchrony hypothesis [4]), transient states in modal models [35], and batch arrivals in network systems, to name a few examples.

Following [35,39,40], we solve these problems by using **super-dense time**. Let $T = \mathbb{R}_+ \times \mathbb{N}$ be a set of tags, and give a signal s as a partial function:

$$s\colon T \rightharpoonup V_\varepsilon$$

defined on an initial segment of T, assuming a lexical ordering on T:

$$(t_1, n_1) \leq (t_2, n_2) \iff t_1 < t_2, \text{ or } t_1 = t_2 \text{ and } n_1 \leq n_2 .$$

For a particular tag $t = (\tau, n) \in T, \tau \in \mathbb{R}_+$ represents physical time, whereas $n \in \mathbb{N}$ represents the ordering of events that occur at that physical time. We again require $s(\tau) = \varepsilon$ for all tags τ in the initial segment on which s is defined except a discrete subset.

Execution again starts with all signals being empty (i.e., nowhere defined, or unknown at all tags), and finds a fixed point at the first tag $(0,0) \in T$, just as in SR. To proceed to the next step, however, it would not be sufficient to simply go to the second tag $(0,1) \in T$. With such an execution policy, we would have an infinite sequence of steps to resolve signal values at all tags of the form $(0,n)$, where $n \in \mathbb{N}$. Physical time would not advance. We thus need to augment the actor abstract semantics.

The augmentation of the actor abstract semantics that is realized in Ptolemy II reflects a preference for absent values, ε over present values.[†] Specifically, we assume that if at a tag $t \in T$ an actor is presented with all input signals that have value ε at that tag, then all its outputs will be ε unless it has specifically declared otherwise. Operationally, each time the

---

*A discrete subset is one that is order isomorphic to a subset of the natural numbers. This property ensures that every DE model has an SR abstraction.

[†]This observation is due to Eleftherios Matsikoudis.

setup phase or *postfire* method of an actor is invoked, it has the option of requesting a firing at some future tag t, in which case it will be fired at that tag even if all its inputs are ε. In Ptolemy II, the actor does this by calling a Director method called *fireAt*, passing it a future time value at which the actor wishes to be fired. Otherwise, it has no expectation of being fired at any tag when its inputs are all absent.

Thus, when a fixed point is reached at any tag $t = (\tau, n)$ (e.g., on the first step, a fixed point is reached at $t = (0, 0)$), then when *postfire* is invoked on all the actors, some of them may call *fireAt*. If no actor called *fireAt*, then implicitly all actors have declared that they will not produce non-absent events given absent inputs any time in the future, and we can infer that all signals are therefore absent for all future tags. Now all signals are completely defined on the tag set T and the execution is complete. More commonly, some actors will have called *fireAt* specifying a time value. We find the minimum such time value τ_{min}. If $\tau_{min} = \tau$, the current time value, then we proceed with the next iteration at tag $t = (\tau, n + 1)$. Otherwise, we proceed with the next iteration at tag $t = (\tau_{min}, 0)$.

In our operational semantics *postfire* is called exactly once at each tag for any actor that was fired. Thus, τ_{min} is uniquely determined for each tag τ at which actors are fired. The first τ at which to fire actors is uniquely determined by giving each actor an opportunity to call *fireAt* in its setup phase. Thus, composition of actors remains associative, as it is with SR.

Of course, this execution strategy does not guarantee that time will advance to infiinty. In particular, by invoking *fireAt* with sufficiently small time increments, any actor may block the progress of time, preventing it from exceeding some limit. Such a situation is known as a Zeno condition. Zeno conditions can be prevented with appropriate constraints on the behavior of actors (see for example [14,38,39].

When we proceed from tag to tag, we must keep track of all previous *fireAt* requests that have not yet been satisfied. This task is, essentially, what a typical DE simulator uses an **event queue** for. In our model, an event queue can be as simple as a set of tags in the future. A more efficient implementation (but semantically equivalent) would keep the set ordered (using, for example, a priority queue). A still more efficient implementation would keep track, for each event, of which actor(s) requested a firing at the tag. Then the execution engine would not need to fire actors that have already (implicitly) declared that their outputs will be absent.

Although these implementations are more efficient, they are semantically equivalent to a very simple execution strategy. Start at tag $t = (0, 0)$ and fire actors to find a fixed point. If a fixed point results in some signals being undefined ($\perp$) at $t = (0, 0)$, then declare the model to be flawed (causality loop). Otherwise, postfire all actors that were fired. Then select the smallest tag in the event queue, increment t, and repeat. Semantically, this is exactly SR, except that now there is a measurable time between ticks of the clock, and the measure of the time increment is determined by the actor invocations of *fireAt*.

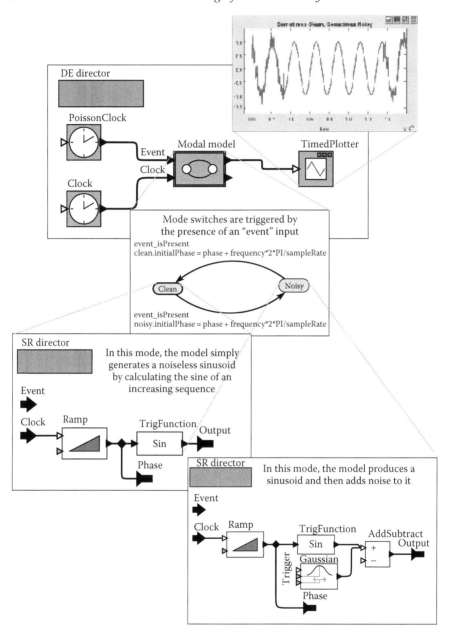

FIGURE 17.6
Example of mixed DE and SR model.

Consider the example in Figure 17.6, which represents a fairly typical scenario. The top-level model is a DE model of a system that is clocked at a regular rate (by the Clock actor), but is also affected by irregular events generated by the PoissonClock actor. This example produces a sinusoidal signal, generating one sample for each event from the Clock actor. When the PoissonClock actor produces an event, however, the model switches from generating a clean sinusoid to a noisy sinusoid, or vice versa. The switching is modeled one level down in the hierarchy by a **modal model** [21] with two modes, labeled "clean" and "noisy." The sinusoids themselves are generated one level further down in the hierarchy by SR opaque composite actors, as shown. The SR composites themselves have no timed semantics. They simply produce the next sample of the sinusoid on each tick. When these ticks occur is controlled above these composites in the hierarchy.

In this example, the ability to put SR composites in a modal model within a DE model is a direct consequence of the fact that each of these opaque composite actors conforms to the actor abstract semantics.

The converse containment is also possible, where DE is put within SR. However, if SR is at the top level, then the top-level SR director needs to regulate the passage of time, effectively acquiring DE semantics. That is, it cannot abstract away the passage of time if its submodels have to be concrete about the passage of time. For this reason, the SR director in Ptolemy II includes a *period* parameter, that if set, defines a fixed time increment between ticks of the SR model. If used, this turns an SR model into a simple DE model with a fixed time increment between ticks.

17.7 Continuous-Time Models

CT models include continuous dynamics, typically given as ordinary differential equations (ODEs). Figure 17.7 shows a block diagram representing a simple third-order nonlinear differential equation. An ODE can be represented by a set of first-order differential equations on a vector-valued state,

$$\dot{x}(t) = g(x(t), u(t), t),$$
$$y(t) = f(x(t), u(t), t),$$

where $x : \mathbb{R} \to \mathbb{R}^n$, $y : \mathbb{R} \to \mathbb{R}^m$, and $u : \mathbb{R} \to \mathbb{R}^l$ are state, output, and input signals. The functions $g : \mathbb{R}^n \times \mathbb{R}^l \times \mathbb{R} \to \mathbb{R}^n$ and $f : \mathbb{R}^n \times \mathbb{R}^l \times \mathbb{R} \to \mathbb{R}^m$ are state functions and output functions, respectively. The state function g is represented collectively by the actors in the gray area (the gray box is purely decorative and devoid of semantics) on the left in Figure 17.7.

In [35], we have shown how to extend the semantics of such differential equation systems to superdense time, allowing us to use the same model of time that we used with DE. This is useful for modeling hybrid systems, which combine continuous dynamics given in Figure 17.7 with discrete mode transitions given in Figure 17.6. In fact, for the tags at which these discrete mode

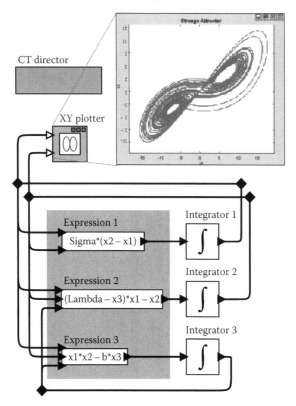

FIGURE 17.7
Continuous-time model.

transitions occur, the model behaves exactly like a DE model. The interesting extension with CT is that between these tags, instead of all signals being absent, some (or all) signals have continuously evolving values, as governed by an ODE.

Of course, in a digital computer, values do not evolve continuously. The ODE must be approximated by a **solver** [43]. The solver will provide samples of the continuous evolution, and typically, to maintain adequate accuracy, must control the time steps between such samples. The actor abstract semantics, it turns out, supports the inclusion of such solvers as part of the execution of models.

One family of solvers use *Runge-Kutta* (RK) methods, which perform interpolation at each integration step to approximate the derivative at a discrete subset of time points. An explicit k stage RK method has the form

$$x(t_n) = x(t_{n-1}) + \sum_{i=0}^{k-1} c_i K_i, \qquad (17.1)$$

where

$$K_0 = h_n g(x(t_{n-1}), u(t_{n-1}), t_{n-1}),$$

$$K_i = h_n g(x(t_{n-1}) + \sum_{j=0}^{i-1} A_{i,j} K_j, u(t_{n-1} + hb_i),$$

$$t_{n-1} + hb_i), \quad i \in \{1, \ldots, k-1\}$$

and $A_{i,j}$, b_i, and c_i are algorithm parameters calculated by comparing the form of a Taylor expansion of x with Equation 17.1.

The first order RK method, also called the *forward Euler* method, has the (much simpler) form

$$x(t_n) = x(t_{n-1}) + h_n \dot{x}(t_{n-1}). \tag{17.2}$$

This method is conceptually important but not recommended for practical usage for simulating physical applications. More practical RK methods have $k = 3$ or 4, and also control the step size for each integration step. An RK method implemented in the MATLAB® ODE suite is a $k = 3$ stage method and given by

$$x(t_n) = x(t_{n-1}) + \frac{2}{9}K_0 + \frac{3}{9}K_1 + \frac{4}{9}K_2, \tag{17.3}$$

where

$$K_0 = h_n g(x(t_{n-1}), u(t_{n-1}), t_n - 1), \tag{17.4}$$
$$K_1 = h_n g(x(t_{n-1}) + 0.5K_0, u(t_{n-1} + 0.5h_n),$$
$$t_{n-1} + 0.5h_n), \tag{17.5}$$
$$K_2 = h_n g(x(t_{n-1}) + 0.75K_1, u(t_{n-1} + 0.75h_n),$$
$$t_{n-1} + 0.75h_n). \tag{17.6}$$

Notice that in order to complete one integration step, this method requires evaluation of the function g at intermediate times $t_{n-1} + 0.5h_n$ and $t_{n-1} + 0.75h_n$, in addition to the times t_{n-1}, where h_n is the step size. This fact has significant consequences for compositionality of this method. In fact, any method that requires intermediate evaluations of the state functions g, such as the classical fourth-order RK method, the linear multistep methods (LMS), and Burlirsch-Store methods, will have to face the same issue during composition. An additional complication is that validity of a step size h_n is not known until the full integration step has been completed.

To show how solving these problems is facilitated by the actor abstract semantics, consider the model shown in Figure 17.7. In the gray area on the left is a collection of actors that collectively implement the state function g. We assume these actors to be black boxes that conform with the actor abstract semantics. That is, they could be internally implemented as composite actors with SR or DE directors, for example, or as modal models [21]. To evaluate

g at $t_{n-1} + 0.5h_n$ and $t_{n-1} + 0.75h_n$, we must *fire* but not *postfire* these actors. Postfiring the actors would erroneously commit them to state updates before we know whether the step size h_n is valid. Thus, in effect, the solver must provide them with tentative inputs at each tag (one tag for each of these time values), as shown in Equations 17.5 and 17.6, and find a fixed point at that tag. But it must not commit the actors to any state changes until it is sure of the step size. Avoiding invocation of the *postfire* method successfully avoids these state changes, as long as all actors conform to the actor abstract semantics. This mechanism is similar to that used in Simulink, where the *model_update* method is not invoked until a simulation step is concluded.

We can now see that CT operates similar to DE models, with the only real difference being that in addition to using an event queue to determine the advancement of time, we must also consult an ODE solver. The same *fireAt* mechanism that we used in DE would be adequate, but for efficiency we have chosen to use a different mechanism that polls relevant actors for their constraints on the advancement of time and aggregates the results. In our implementation, any actor can assert that it wishes to exert some influence on the passage of time by implementing a *ContinuousStepSizeController* interface. All such actors will be consulted before time is advanced. The *Integrator* actors implement this interface and serve as proxies for the solver. But given this general mechanism, there are other useful actors that also implement this interface. For example, the *LevelCrossingDetector* actor implements this interface. Given a CT input signal, it looks for tags at which the value of the signal crosses some threshold given as a parameter. If a step size results in a crossing of the threshold, the actor will exert control over the step size, reducing it until the time of the crossing is identified to some specified precision.

Since the CT director only assumes that component actors conform to the actor abstract semantics, these actors can be opaque composite actors that internally contain SR or DE models. Moreover, a CT model can now form an opaque composite actor that exports the actor abstract semantics, and hence CT models can be included within SR or DE models and vice versa (subject again to the constraint that if SR is at the top level, then it must be explicit about time).

A simple example is shown in Figure 17.8. The top-level model is DE representing a sequence of discrete jobs with increasing service requirements. For each job, a random (exponential) service rate is generated. The inside model uses a single integrator to model the (continuous) servicing of the job and a level-crossing detector to detect completion of the job.

17.8 Software Implementation

A prototype of the techniques described here in Ptolemy II is available in an open-source form (BSD-style license) at http://ptolemy.org. We started with

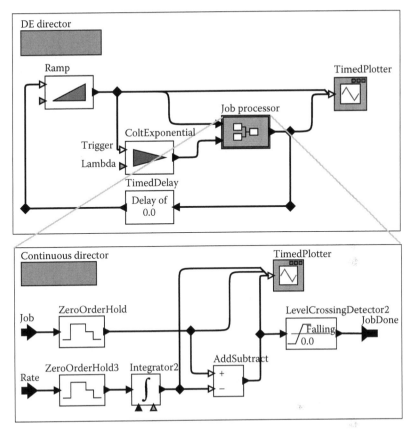

FIGURE 17.8
CT opaque composite actor within a DE model.

the *SRDirector* created by Whitaker [48], which was based on an SR director in Ptolemy classic created by Edwards and Lee [17]. We then used this director as a base class for a new *ContinuousDirector*. Unlike the predecessor *CTDirector* created by Liu [37], this new director realizes a fixed point semantics at each discrete time point. The discrete time points are selected from the time continuum, as explained above, in response to actors that invoke *fireAt* and actors that implement *ContinuousStepSizeController*. The latter include integrator actors, which use an underlying ODE solver with variable step size control.

We modified *SRDirector* and implemented *ContinuousDirector* so that both now rigorously export the actor abstract semantics. That is, when the *fire* method of either director is invoked, the director does not commit to any state changes, and it does not invoke *postfire* on any actors contained in its composite. Thus, if those actors conform to the actor abstract semantics, then so does the opaque composite actor containing the director.

These improvements led to significant improvements in simplicity and usability. Before we had a menagerie of distinct versions of *CTDirector*, but now we only need one. Previously, in order to compose CT models with other MoCs (such as DE for mixed signal models and FSM for modal models and hybrid systems), we needed to implement specialized cross-hierarchy operations to coordinate the speculative execution of the ODE solver with the environment. This resulted in distinct directors for use inside opaque composite actors and inside modal models.

We also acquired the ability to put SR inside CT models. This is extremely convenient, because SR can be used to efficiently specify numeric computations and complex decision logic, where the continuous dynamics of CT is irrelevant and distracting. Note that it would be much more difficult to use dataflow models, such as SDF [31] inside CT models. This is because in dataflow models, communication between actors is queued. In order to support the speculative executions that an ODE solver performs, we would have to be able to backtrack the state of the queues. This would add considerable complexity. SR has no such difficulty.

Since the CT MoC is a generalization of the SR, in principle, SR becomes unnecessary. However, SR is much simpler, not requiring the baggage of support for ODE solvers, and hence is more amenable to formal analysis, optimization, and code generation.

17.9 Conclusions

In this chapter, we explain an operational semantics that supports mixtures of SR, DE, and CT MoC, and outline a corresponding denotational semantics. Dialects of DE and CT are developed that generalize SR, but provide complementary modeling and design capabilities. We show that the three MoCs can be combined hierarchically in arbitrary order.

Acknowledgments

We thank to Jie Liu, Xiaojun Liu, Eleftherios Matsikoudis, and Reinhard von Hanxleden for their major contributions to our understanding of this topic, to the software on which we base this chapter, and to the contents of this chapter.

The work described in this chapter was supported in part by the Center for Hybrid and Embedded Software Systems (CHESS) at UC Berkeley, which receives support from the National Science Foundation (NSF awards #0720882 (CSR-EHS: PRET), #0647591 (CSR-SGER), and #0720841 (CSR-CPS)), the U. S. Army Research Office (ARO #W911NF-07-2-0019), the U. S. Air Force Office of Scientific Research (MURI #FA9550-06-0312 and AF-TRUST #FA9550-06-1-0244), the Air Force Research Lab (AFRL), the State of California Micro Program, and the following companies: Agilent, Bosch, HSBC, Lockheed-Martin, National Instruments, and Toyota.

References

1. G. A. Agha, I. A. Mason, S. F. Smith, and C. L. Talcott. A foundation for actor computation. *Journal of Functional Programming*, 7(1):1–72, 1997.

2. R. Alur, T. Dang, J. Esposito, Y. Hur, F. Ivancic, V. Kumar, I. Lee, P. Mishra, G. J. Pappas, and O. Sokolsky. Hierarchical modeling and analysis of embedded systems. *Proceedings of the IEEE*, 91(1):11–28, 2003.

3. A. Basu, M. Bozga, and J. Sifakis. Modeling heterogeneous real-time components in BIP. In *International Conference on Software Engineering and Formal Methods (SEFM)*, pp. 3–12, Pune, India, September 11–15, 2006.

4. A. Benveniste and G. Berry. The synchronous approach to reactive and real-time systems. *Proceedings of the IEEE*, 79(9):1270–1282, 1991.

5. A. Benveniste, L. Carloni, P. Caspi, and A. Sangiovanni-Vincentelli. Heterogeneous reactive systems modeling and correct-by-construction deployment. In *EMSOFT*, Philadelphia, PA, 2003, Springer.

6. G. Berry. *The Constructive Semantics of Pure Esterel*. Book Draft, 1996. http://www-sop.inria.fr/meije/esterel/doc/main-papers.html

7. G. Berry. The effectiveness of synchronous languages for the development of safety-critical systems. White paper, Esterel Technologies, 2003.

8. G. Berry and G. Gonthier. The Esterel synchronous programming language: Design, semantics, implementation. *Science of Computer Programming*, 19(2):87–152, 1992.

9. R. Boute. Integrating formal methods by unifying abstractions. In E. Boiten, J. Derrick, and G. Smith (editors,) *Fourth International Conference on Integrated Formal Methods (IFM)*, 2999: Canterbury, Kent, U.K., April 4–7, 2004, *LNCS*, pp. 441–460, Springer-Verlag.

10. C. Brooks, A. Cataldo, C. Hylands, E. A. Lee, J. Liu, X. Liu, S. Neuendorffer, and H. Zheng. HyVisual: A hybrid system visual modeler. Technical

report UCB/ERL M03/30, University of California, Berkeley, CA, July 17, 2003.

11. J. T. Buck, S. Ha, E. A. Lee, and D. G. Messerschmitt. Ptolemy: A framework for simulating and prototyping heterogeneous systems. *International Journal of Computer Simulation*, Special issue on *"Simulation Software Development,"* 4:155–182, 1994.

12. L. P. Carloni, M. D. DiBenedetto, A. Pinto, and A. Sangiovanni-Vincentelli. Modeling techniques, programming languages, and design toolsets for hybrid systems. Technical Report IST-2001-38314 WPHS, Columbus Project, June 2004.

13. C. G. Cassandras. *Discrete Event Systems, Modeling and Performance Analysis*. Irwin, Boston, MA, 1993.

14. A. Cataldo, E. A. Lee, X. Liu, E. Matsikoudis, and H. Zheng. A constructive fixed-point theorem and the feedback semantics of timed systems. In *Workshop on Discrete Event Systems (WODES)*, Ann Arbor, MI, July 10–12, 2006.

15. A. Deshpande, A. Gollu, and P. Varaiya. The Shift programming language for dynamic networks of hybrid automata. *IEEE Transactions on Automatic Control*, 43(4):584–587, 1998.

16. R. Djenidi, C. Lavarenne, R. Nikoukhah, Y. Sorel, and S. Steer. From hybrid simulation to real-time implementation. In *11th European Simulation Symposium and Exhibition (ESS99)*, pp. 74–78, Erlangen-Nuremberg, Germany, October 1999.

17. S. A. Edwards and E. A. Lee. The semantics and execution of a synchronous block-diagram language. *Science of Computer Programming*, 48(1):21–42, 2003.

18. J. Eker, J. W. Janneck, E. A. Lee, J. Liu, X. Liu, J. Ludvig, S. Neuendorffer, S. Sachs, and Y. Xiong. Taming heterogeneity—The Ptolemy approach. *Proceedings of the IEEE*, 91(2):127–144, 2003.

19. G. S. Fishman. *Discrete-Event Simulation: Modeling, Programming, and Analysis*. Springer-Verlag, New York, 2001.

20. P. Fritzson. *Principles of Object-Oriented Modeling and Simulation with Modelica 2.1*. Wiley, New York, 2003.

21. A. Girault, B. Lee, and E. A. Lee. Hierarchical finite state machines with multiple concurrency models. *IEEE Transactions On Computer-Aided Design of Integrated Circuits and Systems*, 18(6):742–760, 1999.

22. G. Goessler and A. Sangiovanni-Vincentelli. Compositional modeling in Metropolis. In *Second International Workshop on Embedded Software (EMSOFT)*, Grenoble, France, October 7–9, 2002, Springer-Verlag.

23. G. Goessler and J. Sifakis. Composition for component-based modeling. *Science of Computer Programming*, 55:161–183, 2005.

24. P. L. Guernic, T. Gauthier, M. L. Borgne, and C. L. Maire. Programming real-time applications with SIGNAL. *Proceedings of the IEEE*, 79(9):1321–1336, 1991.

25. N. Halbwachs, P. Caspi, P. Raymond, and D. Pilaud. The synchronous data flow programming language LUSTRE. *Proceedings of the IEEE*, 79(9):1305–1319, 1991.

26. F. Herrera and E. Villar. A framework for embedded system specification under different models of computation in SystemC. In *Design Automation Conference (DAC)*, San Francisco, CA, July 2006. ACM.

27. C. Hewitt. Viewing control structures as patterns of passing messages. *Journal of Artifical Intelligence*, 8(3):323–363, 1977.

28. A. Jantsch. *Modeling Embedded Systems and SoCs—Concurrency and Time in Models of Computation*. Morgan Kaufmann, San Francisco, CA, 2003.

29. E. A. Lee. Modeling concurrent real-time processes using discrete events. *Annals of Software Engineering*, 7:25–45, 1999.

30. E. A. Lee. Embedded software. In M. Zelkowitz (editor), *Advances in Computers*, vol. 56. Academic Press, London, U.K., 2002.

31. E. A. Lee and D. G. Messerschmitt. Synchronous data flow. *Proceedings of the IEEE*, 75(9):1235–1245, 1987.

32. E. A. Lee and S. Neuendorffer. MoML—A modeling markup language in XML. Technical report UCB/ERL M00/12, UC Berkeley, Berkeley, CA, March 14, 2000.

33. E. A. Lee, S. Neuendorffer, and M. J. Wirthlin. Actor-oriented design of embedded hardware and software systems. *Journal of Circuits, Systems, and Computers*, 12(3):231–260, 2003.

34. E. A. Lee and A. Sangiovanni-Vincentelli. A framework for comparing models of computation. *IEEE Transactions on Computer-Aided Design of Circuits and Systems*, 17(12):1217–1229, 1998.

35. E. A. Lee and H. Zheng. Operational semantics of hybrid systems. In M. Morari and L. Thiele (editors), *Hybrid Systems: Computation and Control*

(HSCC), Zurich, Switzerland, March 9–11, 2005. *LNCS*, 3414: pp.25–53, Springer-Verlag.

36. E. A. Lee, H. Zheng, and Y. Zhou. Causality interfaces and compositional causality analysis. In *Foundations of Interface Technologies (FIT), Satellite to CONCUR*, San Francisco, CA, August 21, 2005.

37. J. Liu. Responsible frameworks for heterogeneous modeling and design of embedded systems. PhD thesis Technical Memorandum UCB/ERL M01/41, December 20, 2001.

38. X. Liu and E. A. Lee. CPO semantics of timed interactive actor networks. Technical report EECS-2006-67, UC Berkeley, Berkeley, CA, May 18, 2006.

39. X. Liu, E. Matsikoudis, and E. A. Lee. Modeling timed concurrent systems. In *CONCUR 2006—Concurrency Theory*, Bonn, Germany, August 27–30, 2006. *LNCS*, 4137: Springer.

40. Z. Manna and A. Pnueli. Verifying hybrid systems. *Hybrid Systems, LNCS*, 736: pp. 4–35, 1992, Springer.

41. D. A. Mathaikutty, H. D. Patel, and S. K. Shukla. A functional programming framework of heterogeneous model of computation for system design. In *Forum on Design and Specification Languages (FDL)*, Lille, France, September 2004.

42. P. Mosterman. An overview of hybrid simulation phenomena and their support by simulation packages. In F. Varager and J. H. v. Schuppen (editors), *Hybrid Systems: Computation and Control (HSCC)*, Bergen Dal, the Netherlands, 1999, *LNCS*, 1569: pp. 165–177, Springer-Verlag.

43. W. H. Press, S. Teukolsky, W. T. Vetterling, and B. P. Flannery. *Numerical Recipes in C: The Art of Scientific Computing*. Cambridge University Press, Cambridge, MA, 1992.

44. I. Sander and A. Jantsch. System modeling and transformational design refinement in forsyde. *IEEE Transactions on Computer-Aided Design of Circuits and Systems*, 23(1):17–32, 2004.

45. S. Swan. An introduction to system level modeling in SystemC 2.0. Technical report, Open SystemC Initiative, May 2001.

46. M. M. Tiller. *Introduction to Physical Modeling with Modelica*. Kluwer Academic Publishers, Norwell, MA, 2001.

47. F. D. Torrisi, A. Bemporad, G. Bertini, P. Hertach, D. Jost, and D. Mignone. Hysdel 2.0.5—User manual. Technical report, ETH, 2002.

48. P. Whitaker. The simulation of synchronous reactive systems in Ptolemy II Master's Report Memorandum UCB/ERL M01/20, Electronics Research Laboratory, University of California, California, CA, May 2001.

49. B. P. Zeigler, H. Praehofer, and T. G. Kim. *Theory of Modeling and Simulation*. 2nd edition, Academic Press, Orlando, FL, 2000.

18

Design Refinement of Embedded Mixed-Signal Systems

Jan Haase, Markus Damm, and Christoph Grimm

CONTENTS

18.1 Introduction

There is a growing trend for closer interaction between embedded hardware/software (HW/SW) systems and their analog physical environment. This leads to systems in which digital HW/SW is functionally interwoven with analog and mixed-signal blocks such as radio-frequency (RF) interfaces, power electronics, and sensors and actuators, as shown, for example, by the communication system in Figure 18.1. We call such systems "embedded analog/mixed-signal (E-AMS) systems." Examples of E-AMS systems are cognitive radios, sensor networks, and systems for image sensing. A challenge for the development of E-AMS systems is to understand and consider the interaction between HW/SW and the analog and mixed-signal subsystems at architecture level.

Complexity of modern systems often requires methodologies that hide complexity and allow designers an incremental, interactive approach that

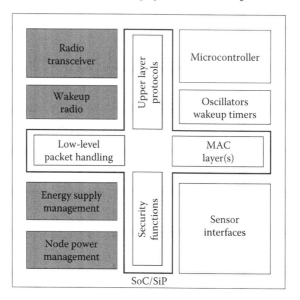

FIGURE 18.1
A node of a sensor network serving as an example of an E-AMS architecture.

step-by-step leads to an implementation. In this approach, it is crucial to obtain very early feedback on the impact of nonideal properties onto overall system performance, which requires considering interaction of HW/SW and AMS subsystems.

In the SW engineering community, extreme programming [17] uses a stepwise approach that starts with code fragments that are successively "refined" by SW engineers. Refinement of SW systems has been known for a long time (e.g., [15]). However, the SW-oriented approaches are restricted to pure SW systems and do not deal with specific problems in the design of E-AMS systems. In the realm of formal "property refinement" of embedded systems, a (formal) property that is present and proved in a system specification is maintained by proved design steps (e.g., [16]). In this chapter, we describe a design refinement approach for E-AMS systems. Similar to extreme programming, and in the same vein of "property refinement," it is an incremental approach. Compared to extreme programming, however, it is more specifically tailored to E-AMS system design, whereas compared to property refinement we do not intend to provide a formal proof.

18.1.1   Previous Work

SystemC [1] supports the refinement of HW/SW systems down to RTL by providing a discrete-event (DE) simulation framework. Design refinement of HW/SW systems starts with a functional, untimed specification that is successively augmented with timing information, power consumption, and more accurate modeling of communication and simulation of potential HW/SW architectures. However, the properties of E-AMS systems are much more diverse than only timing, performance, and power consumption. A key issue is the accuracy that is determined by noise, sampling frequencies, quantization, and very complex, nonlinear dependencies between subsystems: Often, accuracy of the AMS part is improved by digital signal processing (DSP) software in the HW/SW part. SystemC offers neither an appropriate methodology for refinement of E-AMS nor support for the modeling and simulation of analog, continuous-time systems.

Support for modeling and simulation of E-AMS systems is offered by tools such as Simulink® [3] and Ptolemy II [4]. While their support for system-level design also facilitates capturing continuous-time behavior, these tools lack appropriate support for the design of HW/SW (sub)systems at the architecture level in a manner that, for example, SystemC does.

Hardware description languages (HDLs) dedicated to the design of AMS systems such as VHDL-AMS [5] and Verilog-AMS [6] target the design of mixed-signal subsystems close to implementation level such as analog/digital (A/D) converters, but modeling HW/SW systems based on HDLs is cumbersome. To support HW/SW system design, cosimulation solution frameworks mix SystemC and Verilog/VHDL-AMS. However, although the resulting heterogeneous framework allows designers the modeling of mixed HW/SW and AMS architectures, it does not support interactive evaluation of different potential architectures in a seamless design refinement flow.

18.1.2 Design Refinement of E-AMS Systems with OSCI AMS Extensions

An earlier work by the open SystemC initiative (OSCI) [12] presented an AMS extension that augments SystemC with the ability to model and simulate AMS subsystems at functional and architectural level [7,8]. Furthermore, this work specifically intends to support design refinement of E-AMS. Design refinement of E-AMS starts with a functional description that is used as an "executable specification." In "architecture exploration," properties of different subsystems such as

- Noise, distortions, and limitation effects
- Quantization and sampling frequencies
- Partitioning of (A/D/SW)

are added to the functional specification. The impact of these properties on the overall system performance (accuracy, power consumption, etc.) is determined by modeling and simulation.

In this chapter, we assume that the reader is familiar with SystemC 2.0. We first present a brief overview of the SystemC AMS extensions. A more detailed overview of the AMS extensions is provided in [9,12]. Then, we describe typical use cases of SystemC AMS extensions to focus on architecture exploration and to classify different levels of refinement and refinement activities.

18.2   OSCI SystemC-AMS Extensions

The SystemC AMS extensions provide support for signal flow, data flow, and electrical networks, as shown in Figure 18.2. Electrical networks and signal-flow models use a linear differential and algebraic equation (DAE) solver that solves the equation system and that is synchronized with the SystemC kernel. The use of a linear DAE solver restricts networks and signal-flow components to linear models in order to provide high simulation performance. Data-flow simulation is accelerated using a static schedule that is computed before simulation starts. This schedule is activated in discrete time steps, where synchronization with the SystemC kernel introduces timed semantics. It is therefore called "timed" data flow (TDF).

The SystemC AMS extensions define new language constructs identified by the prefix `sca_`. They are declared in dedicated namespaces `sca_tdf` (TDF), `sca_eln` (electrical linear networks (ELN)), and `sca_lsf` (linear signal flow (LSF)) according to the underlying semantics. By using namespaces, similar primitives as in SystemC are defined to denote ports, interfaces, signals, and modules. For example, a TDF input port is an object of class `sca_tdf::sca_in<type>`.

LSF and linear electrical networks (LEN) are specified by instantiating components of the AMS extensions library such as resistors, capacitors, and

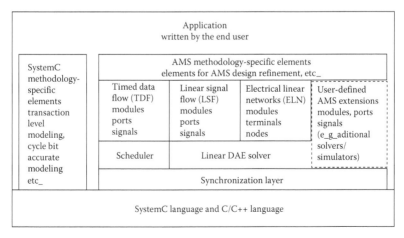

FIGURE 18.2
AMS extensions for the SystemC language standard.

(controlled) current sources for LEN or integrators and adders for LSF. TDF requires some new syntactic elements and is crucial for understanding the SystemC AMS extensions. In the following, we will concentrate on TDF models.

TDF models consist of TDF modules that are connected via TDF signals using TDF ports. Connected TDF modules form a contiguous graph structure called TDF cluster. Clusters must not have cycles without delays, and each TDF signal must have one source. A cluster is activated in discrete time steps. The behavior of a TDF module is specified by overloading the predefined methods set_attributes(), initialize(), and processing():

- The method set_attributes() is used to specify attributes such as rates, delays, and time steps of TDF ports and modules.
- The method initialize() is used to specify initial conditions. It is executed once before the simulation starts.
- The method processing() describes time–domain behavior of the module. It is executed with each activation of the TDF module during the simulation.

It is expected that there is at least one definition of the time step value and, in the case of cycles, one definition of a delay value per cycle. TDF ports are single-rate by default. It is the task of the elaboration phase to compute and propagate consistent values for the time steps to all TDF ports and modules. Before simulation, the scheduler determines a schedule that defines the order of activation of the TDF modules, taking into account the rates, delays, and time steps. During simulation, the processing() methods are executed at discrete time steps. Example 18.1 shows the TDF model of a mixer. The processing() method will be executed every 1μs.

```
SCA_TDF_MODULE(mixer) // TDF primitive module definition
  {
  sca_tdf::sca_in<double> rf_in, lo_in; // TDF in ports
  sca_tdf::sca_out<double> if_out; // TDF out ports
  void set_attributes()
  {
  set_timestep(1.0, SC_US); // time between activations
  }
  void processing() // executed at each activation
  {
  if_out.write( rf_in.read() * lo_in.read() );
  }
  SCA_CTOR(mixer) {}
  };
```

Example 18.1 *TDF model of a mixer. Predefined converter ports (sca_tdf:: sc_out or sca_tdf::sc_in) can establish a connection to a SystemC DE channel, for instance, sc_signal<T>, to read or write values during the first* *delta cycle of the current SystemC time step. Example 18.2 illustrates the use of such a converter port in a TDF module modeling a simple A/D converter with an output port to which a SystemC DE channel can be bound. This A/D converter also scales the input values based on a given input range to fit the range of the output data type, while clipping the inputs out of range.*

```
SCA_TDF_MODULE(ad_converter) // simple AD converter
{
sca_tdf::sca_in<double> in_tdf; // TDF port
sca_tdf::sc_out<sc_dt::sc_uint<12> > out_de;
                        // converter port to DE domain
double in_range_min, in_range_max; // expected range of
                                   input values
double scaleFactor; // scaling factor due to input range
void processing()
{
double val;
if(in_tdf.read() > in_range_max) val = pow(2,12)-1;
                        // clip if
else if(in_tdf.read() < in_range_min) val = 0;
                        // necessary
else val = (in_tdf.read() - in_range_min) * scaleFactor;
                        // scale otherwise
out_de.write( static_cast<sc_dt::sc_uint<12> >(val) );
{
ad_converter(sc_module_name n, double _in_range_min,
                        double _in_range_max)
{
in_range_min = _in_range_min;
in_range_max = _in_range_max;
scaleFactor = (pow(2,12)-1)/(in_range_max - in_range_min);
{
};
```

Example 18.2 *TDF model of a simple A/D converter using a converter port. The SystemC AMS simulation kernel uses its own simulation time t_{TDF} that usually differs from the SystemC simulation time t_{DE}. If a pure TDF model is used in a simulation, the SystemC AMS simulation kernel blocks the DE kernel, and so the DE simulation time does not proceed at all. That is, in general we have $t_{TDF} \geq t_{DE}$. In a mixed TDF-DE model, interconnected by converter ports, there is of course the need for synchronization. If there is an access to a converter port within the processing() method of a TDF module, the SystemC AMS simulation kernel interrupts the execution of the static schedule of TDF modules and yields control to the SystemC DE simulation kernel, such that the DE part of the model can now execute, effectively proceeding t_{DE} until it is equal to t_{TDF}. Now, the DE modules reading from signals driven by TDF*

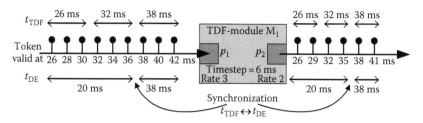

FIGURE 18.3
Synchronization between $t_{TDF} \leftrightarrow t_{DE}$.

modules can read their new values at the correct time, and TDF modules reading from signals driven by DE modules can read their correct current values.

Figure 18.3 shows an example using a TDF module M_1 with input rate 3, output rate 2, and a time step of 6 ms. The data tokens consumed are on the left-hand axis, and those produced are on the right-hand axis. The numbers below the tokens denote the time (in ms) at which the respective token is valid. The time spans above the tokens indicate the values of t_{TDF} when the respective tokens are consumed and produced, respectively. The time spans below indicate the according values for t_{DE}. At the beginning of the example, $t_{TDF} > t_{DE}$ already holds until $t_{TDF} = 38$ ms. Then the SystemC-AMS simulation kernel initiates synchronization, for example because M_1 contains a converter port that it accesses at this time or because another TDF module within the same TDF cluster accesses its converter port.

It has to be underlined that SystemC AMS extensions are recently available as beta version and are under public review. This means that language features may change. In the current draft standard, for example, name spaces are introduced and the keywords are slightly changed.

18.3 Design Refinement of Embedded Analog/Digital Systems

To increase design productivity, it is not sufficient to just use the AMS extensions in the same way as, for example, VHDL-AMS was used. The most important thing is to use the new language in a design methodology that enables designers to fully take advantage of the new features of SystemC AMS extension.

18.3.1 Use Cases of SystemC AMS Extensions

System AMS extensions are extensively applicable to all design issues of E-AMS systems at the architecture level: executable specification, architecture exploration, virtual prototyping, and integration validation as discussed in [12]. For executable specification, a functional model is used that most notably uses the TDF model of computation and (continuous-time) transfer functions that are embedded in single TDF nodes. For architecture exploration, for each node of the executable specification an implementation (analog, digital HW, DSP + SW) is assumed and evaluated using modeling and simulation. Virtual prototyping allows SW developers to develop and analyze software that, for instance, compensates nonideal properties of the analog parts. After implementation, integration validation provides confidence in the functional correctness of each designed component in the context of the overall system.

Figure 18.5 depicts an example of a simple executable specification captured by using the TDF MoC by a signal processing chain using functional blocks. For architecture exploration, partitions of the TDF model are mapped to components from different implementation domain (analog, DSP-SW) at the architecture level. To analyze a different architecture, filter 1 in Figure 18.5 can, for example, be shifted from the analog domain to a DSP implementation.

18.3.2 Design Refinement Methodology

The intent of the design refinement methodology is to "successively" augment and integrate properties of an implementation into a functional model, and to instantaneously analyze their impact by modeling and simulation.

We assume that the refinement starts with a functional model given as an executable specification using the TDF model of computation and embedded continuous transfer functions as shown by Figure 18.4.

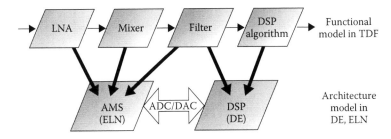

FIGURE 18.4
An architecture exploration by mapping a functional model to architecture-level processors, and repartitioning of an E-AMS system by moving a component from the analog domain to a DSP implementation.

In order to incrementally transform this functional model into models that allows designers evaluation of different architectures, we use the following refinement activities:

- *Refinement of computation* replaces algorithms, models, and data types of the functional model by algorithms, models, and data types that more accurately match the intended architecture. However, the model of computation and the overall structure of the executable specification remain unchanged. For example, refinement of computation can add noise and distortions to a mixer. A model is called a "computation accurate model," if it yields similar results as the implementation, where similar results means that the expected properties of the implementation and the properties of the computation-accurate model are—within tolerances—similar. The purpose of the refinement of computation is to analyze the impact of nonideal behavior of an architecture using a simple functional model.
- *Refinement of structure (repartitioning)* maps partitions of the functional model to components at architecture level that can realize the same (or similar in the same sense as above) behavior using other methods, operating principles, or models of computation (MoC). For example, refinement of structure can map a continuous-time transfer function to a DSP method implemented using a DSP processor. We furthermore distinguish between

  Top-down refinement by creative design of new structures.

  Bottom-up refinement by integration (reuse) of existing models, code fragments, and designs.

  After application of all steps of refinement to a model, we refer to it as a "structure accurate model." The purpose of the refinement of the structure is to introduce the architecture's structure and the architecture's MoC in all components.
- *Refinement of interfaces* replaces functional means for modeling communication (e.g., an interface specification) and synchronization with physical models of communication and synchronization (e.g., a pin-accurate bus or electrical nodes). We call the result of (full) refinement of interfaces an "interface-accurate or pin-accurate model." The purpose of refinement of interfaces is to enable circuit level design which is based on pin-accurate models, and to enable the validation of system integration.

Because we consider an incremental methodology, refinement activities that only refine a partition of a system are supported as well. If only a subset of all refinement activities has been performed, we call the model partially (computation/structure/interface) accurate.

Although many refinement activities can be performed in an independent manner, there is a "natural" or "logical" order to the refinement activities:

First algorithms, (functional) models and methods should be defined by a refinement of computation—this to consider the expected or estimated inaccuracies from analog parts by a refinement of computation. This step is done first because the impact of noise and inaccuracies in AMS systems is often mission-critical. Usually a refinement of structure is implicitly introduced by defining (DSP) methods and (functional) models because this assumes analog or digital implementations, but without distinguishing between HW or SW implementations.

Second, the algorithms must be mapped to processors by refinement of structure. This most notably adds accuracy of timing, provides more accurate computation results compared with the "implementation level," and provides more accurate information on the use of other resources (e.g., power and CPU time). The changed operation principle often includes the specification of new methods. This is especially the case if we refine a block from the TDF MoC to a pure digital implementation. In this case, we also have to provide new methods for communication and synchronization that ensure the function of the block to be executed in the same order as before.

Finally, the interfaces of the components are refined. This is usually the final step, because it does not contribute to the functional behavior but significantly decreases the simulation performance. Nevertheless, interface-accurate models are necessary as a starting point for ASIC/FPGA design, for integration verification, and for virtual prototyping.

It is important to note that for integration validation and virtual prototyping the models should be computation accurate and only be interface accurate where needed in order to obtain high simulation performance.

Figure 18.5 shows a typical schedule of the use cases and refinement activities. A major example is that the refinement activities and use cases can be overlapping in time: ASIC/FPGA/SW design of a single component can start after refinement of structure, but before all interfaces are completely defined, and integration validation can start immediately when an interface-accurate model becomes available.

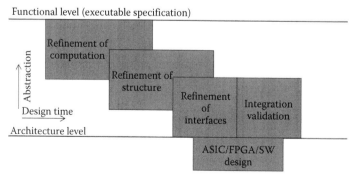

FIGURE 18.5
Use cases and refinement activities.

18.3.3 Methodology-Specific Support in SystemC AMS Extensions

While many modeling languages and tools can be used to specify the models necessary for design refinement as described above, the corresponding effort required by the designer may vary significantly. In the following we identify means to efficiently perform the refinement activities. SystemC AMS extensions already offer some support for refinement that are likely to make it easier for designers to follow a seamless flow as described in this chapter:

Refinement of computation uses the functional model from the executable specification. Therefore, the required changes can be made in the `processing()` section of the TDF modules. For example, the noise of an assumed analog implementation of a mixer can be modeled and analyzed by adding just a single line of code to a mixer modeled as multiplier in the executable specification:

```
void processing() // Mixer refined with distortions
                       and noise
{
double rf = in1.read();
double lo = in2.read();
double rf_dist = (alpha - gamma * rf * rf ) * rf;
                  // added line
double mix_dist = rf_dist * lo;
if_out.write( mix_dist + my_noise() );
}
```

This "refinement of computation" basically yields similar behavior as the original model, and allows a designer to very quickly evaluate the impact of noise and distortions in a mixer to, for example, the bit error rate of a transceiver system while taking advantage of the simulation performance of a functional model.

Refinement of structure/repartitioning activities most notably change the model of computation in which a component is used. This unfortunately requires some effort from the designer. The SystemC AMS extension supports this by providing name spaces for all MoCs in which the same identifiers are used for the modeling infrastructure such as ports, signals, etc. Although a complete reuse of models in another MoC is not possible because this requires translating the semantics, for example, from a DAE to a discrete model, at least ports, signals, and other basic, rather syntactic parts can be maintained, and changes are restricted to the lines that really are important to express the semantics. Often a reuse of existing models of components is possible—no design really proceeds in a strict top-down manner. Therefore, the case of "bottom-up refinement of structure" is quite often. In this case, an existing structure (e.g., a circuit level implementation) is integrated into a functional or architecture level model. SystemC AMS supports these refinement activities by a wide range of converters between the available MoC. Where possible, converter ports are provided (TDF). When MoC do not provide means to describe new primitive modules, and thus no converter ports can be used, converter modules are provided (ELN, LSF). The motivation for providing a basic set of predefined converters is that replacing single blocks of a TDF cluster with structure or even interface-accurate models leads to a heterogeneous structure that combines

- TDF (and embedded transfer functions) for the executable specification
- LSF signals (or ELN nodes) for analog implementation
- SystemC DE signals for digital implementation
- TDF with bit-true data type for DSP SW implementation

The potential use of SystemC AMS modules (and thus converter ports, converter modules) in hierarchical models as well as in hierarchical channels motivates a more sophisticated support for the design refinement of structure.

18.3.4 Methodology-Specific Support in a Methodology-Specific Library

Although SystemC AMS provides a basic set of converters, a richer and more flexible set of converters is useful for design refinement, especially in the case of structure refinement, where a functional (or computation accurate) block is replaced with an interface and structure accurate block. In this case, the MoC often differ from the originally used TDF model of computation. Although in all cases the basic set of converters allows the design engineer to set up a valid model, some additional effort and changes to the overall system structure are typically needed. However, these changes may violate principles of refinement that require changes to be made in a local sense, and with low effort This violation may occur in particular in case of structural refinement by integrating circuit level models into functional models: In this case, a refinement step would require the design engineer to modify the overall model by introducing converter modules or converter ports that translate the functional semantics (e.g., data flow, signal flow) into physical sizes of electrical nodes.

To relieve the design engineer of this awkward and error prone conversion tasks, the concept of polymorphic signals has been developed [10], and has been recently extended to a facility called "converter channel" [11]. Converter channels can automatically connect SystemC modules that are modeled using different MoCs, such as the TDF MoC of the SystemC AMS extensions, or the DE MoC that is native to SystemC, while also performing data-type conversion. The idea of the converter channel is that the designer declares the MoC and the data type of the module that writes to the converter channel, as well as up to two additional data types for the modules that read from this channel. The MoC of the reading modules can be determined automatically by the interface that is implemented by the respective port.

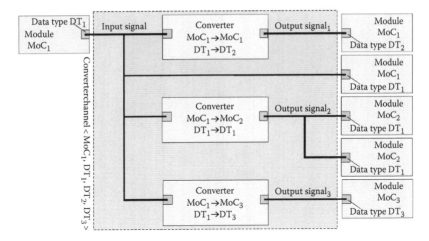

FIGURE 18.6
Internal structure of a converter channel connected to multiple reading modules.

The converter channel then instantiates the appropriate signal(s) for the writing module and the reading module(s), together with converter modules that do the actual conversion work (see Figure 18.6). In the case of TDF→DE conversion, for example, the respective converter module will utilize the converter port described above. If the data types are different, the converter module will also perform the data-type conversion. As an option, the converter channel can also scale the input data, for example when converting double values to data types such as sc_uint<n>. In this case, the user passes the expected input range to the converter channel, such that this range is scaled to the natural range of the target data type, while clipping every input data out of bounds. Therefore, the converter channel can completely replace the simple A/D converter from Example 18.2.

Besides the conversion work, the converter channel can also be used for "integration validation." It can check, for example, if a DE input is under sampled when converting to TDF, which can occur if the DE input changes more frequently compared to the sample rate of the TDF reader. Also, the converter channel can check and handle some corner cases in process network (PN) ↔ TDF conversion (see [11]). Finally, data type conversion issues such as overflow are detected.

18.4 Simple Example for a Refinement Step Using Converter Channels

In this section, we present an example that illustrates a single refinement step in the design of a software defined radio (SDR). An overview

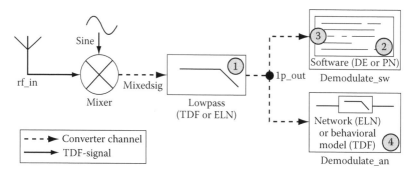

FIGURE 18.7
A refinement example: SW-defined radio.

of a simple SDR system is shown in Figure 18.7. The RF input signal is mixed with a sine wave of the same frequency as the carrier signal and the result is processed by a lowpass filter. After that, the demodulation follows. For demodulation, different realizations are possible: The most power efficient implementation is pure hardware (A/D) and the most flexible one considering support for different standards is the pure software realization. Usually, a compromise between these two realizations has to be found considering flexibility, accuracy/bit width/bit error rate (BER), power consumption, and required computing performance (costs). The evaluation of the architecture variants necessitates the study of an overall system. Starting with a pure functional model, a refinement step could, for example, replace the TDF modulator with a model using the DE MoC. More general, the A/D/SW partitioning could be altered successively and the results could be compared until evaluation yields acceptable results. These refinement steps can be modeled efficiently using converter channels because they automatically adopt data types and model of computation.

As a simple example for some refinement steps, we use the simple SDR concept as illustrated by Figure 18.7. The input of the software demodulator is an integer with fixed bit width. Example 18.3 shows the corresponding top-level SystemC code using two converter channels.

```
bitwidth=8;
sca_tdf::sca_signal<double> rf_in; // incoming RF-signal
sca_tdf::sca_signal <double> sine; // sine waverf_in
converterchannel<TDF, double> mixedsig;
                 // RF-signal multiplied with sine-wave
converterchannel<TDF, double, sc_int<bitwidth> > lp_out;
                 // output of lp-filter
lp_out.setRangeScaling( -1. , 1. ); // assuming the
                          value range within [-1,1]
```

```
mixer mix("mix"); // mixes the two input signals
mix.in1(rf_in);
mix.in2(sine);
mix.out(mixedsig);
lowpass_behavioural lp("lp"); // lowpass filter, either
                                        T-SDF-module
lp.in(mixedsig); // or electrical network
lp.out(lp_out);
demodulate_sw dem_sw("dem_sw", bitwidth); // software
                                        demodulator with
dem_sw.in(lp_out); // sc_int<bitwidth> input
demodulate_an dem_an("dem_an"); // analogue demodulator
dem_an.in(lp_out);
```

Example 18.3 *SW-defined radio with converter channels Regarding design space exploration and mixed level simulation, this example gives rise to the following tasks (the numbers refer to those in Figure 18.7):*

1. *Realizing the lowpass filter either as a (behavioral) TDF-module or as an electrical network.*
2. *Realizing the software demodulator either as a PN or as a DE module.*
3. *Varying the bit width of the input of the software demodulator.*
4. *Realizing the analog demodulator either as a (behavioral) TDF-module or as an electrical network.*

To further complicate matters, any subset of these tasks can be performed in parallel. It is clear that the effort for manually inserting (and adapting) the appropriate converters would be significant. For example, assume an initial model with the lowpass filter and the analog demodulator modeled as TDF modules and the software demodulator modeled within the PN MoC, taking sc_int<bitwidth> inputs. We then would need a TDF→PN converter from the output of the lowpass filter to the software modulator, which would also convert double values to sc_int<bitwidth> values. Now, executing the tasks above would require the following manual conversion steps (in addition to the design of ELN modules itself):

1. Realizing the lowpass filter as an electrical network:
   I. Insert a TDF→ELN converter between the mixer and the lowpass filter.
   II. Replace the initial TDF→DE converter by an ELN→DE converter, which also converts physical sizes in double precision to sc_int<bitwidth>.
   III. Instantiate appropriate signals to connect the TDF→ELN converter with the ELN→KPN converter.
2. Realizing the software demodulator as a DE module: Replace the initial TDF→DE converter by a TDF→DE converter, which also converts

double to sc_int<bitwidth>. Note, that using a SystemC AMS converter port makes no sense here since the analog demodulator is still in the TDF domain. Instantiate appropriate signals to connect them.

3. Varying the bit width of the input of the software demodulator: Change the output of the TDF→DE converter and the type of the signal that connects it to the software demodulator to sc_int<new_bitwidth>. The data-type conversion algorithm of the converter must also be altered slightly. Note, that a converter that takes the bitwidth as parameter would alleviate the design effort.

4. Realizing the analog demodulator as an electrical network: Insert a TDF→ELN converter between the lowpass filter and the analog demodulator. Instantiate an appropriate signal to connect them.

Obviously, steps 1–4 are a significant effort for a design engineer and therefore are against the principle of interactive, iterative refinement. By using converter channels, however, the code for each of the possible variants would be very similar to the code in Example 18.3. The value of the variable bitwidth would change as well as the class names of the respective modules (e.g., changing lowpass_behavioral to lowpass_electrical). This simple example shows the convenience that converter channels offer to the design engineer.

18.5 Conclusion and Outlook

With the AMS extensions, SystemC becomes amenable to modeling HW/SW systems and—at functional and architecture level—analog and mixed-signal subsystems. The intended use cases include executable specification, architecture exploration, virtual prototyping, and integration validation. We have described a methodology that efficiently uses the AMS extensions together with the newly introduced converter channels. To support it the concept of a converter channel has been introduced.

It is desirable to incorporate an even more abstract view into the methodology, namely *Transaction Level Modeling* (TLM) [2], which allows design engineers to perform abstract modeling, simulation, and design of HW/SW system architectures. The idea of TLM is to abstract away the low-level events occurring in bus communication into a single data object called "transaction," that is passed from process to process by method calls. TLM not only enables early software development, but also enhances simulation performance. Recently, the OSCI released the TLM 2.0 standard [13], and an extension library with facilities for TLM (mainly method interfaces and a standard transaction object, the "generic payload"). Parts of the standard are coding guidelines for different coding styles. Especially the "loosely timed coding" style contains state of the art simulation techniques for fast simulation.

By allowing processes to run ahead of global simulation time (temporal decoupling) locally, context switches can be reduced, which increases simulation performance, possibly for the price of a lower simulation accuracy.

In [14], an early approach was presented to couple loosely timed TLM models with models using the AMS extensions. It was shown that the loosely timed modeling style of TLM 2.0 can be exploited efficiently to fit with the AMS extensions TDF MoC, preserving the high simulation performance of both the simulation approaches. The key idea there emerged from the observation that TDF processes also run ahead of the SystemC simulation time (see Figure 18.4). Therefore, it is possible to set up converters that incorporate both temporal decoupling effects as well as trigger synchronization only when needed. With an efficient TLM-AMS extensions coupling available, the first step is implemented for an integrated E-AMS systems refinement approach using TLM and AMS extensions.

References

1. IEEE Std. 1666–2005. IEEE Press, New York.

2. F. Ghenassia (editor), *Transaction Level Modeling with SystemC*, Springer, Dordrecht, the Netherlands, 2005.

3. The MathWorks® Simulink®, http://www.mathworks.com/products/simulink

4. J. Eker, J. W. Janneck, E. A. Lee, J. Liu, X. Liu, J. Ludvig, S. Neuendorffer, S. Sachs, and Y. Xiong, Taming heterogeneity—the Ptolemy approach, *Proceedings of the IEEE*, v.91, No. 1, pp. 127–144, January 2003.

5. IEEE Std. 1076.1-2007. IEEE Press, New York.

6. Accellera: *Verilog-AMS Language Reference Manual* Version 2.2, 2004. Analog & Mixed-Signal Extensions to Verilog HDL; http://www.verilog.org/verilog-ams/

7. OSCI AMS Working Group, *SystemC AMS extensions Requirements Specification*, 2007.

8. A. Vachoux, C. Grimm, and K. Einwich, SystemC extensions for heterogeneous and mixed discrete/continuous systems. In: *International Symposium on Circuits and Systems 2005 (ISCAS '05)*, Kobe, Japan, May 2005. IEEE Press, New York.

9. C. Grimm, Modeling and refinement of mixed signal systems with SystemC. In: *SystemC: Methodologies and Applications*, Kluwer Academic Publisher (KAP), Norwell, MA, June 2003.

10. R. Schroll, Design komplexer heterogener Systeme mit Polymorphen Signalen. PhD thesis, Institut für Informatik, Universität Frankfurt, Germany, 2007.

11. M. Damm, F. Herrera, J. Haase, E. Villar, and C. Grimm, Using converter channels within a top-down design flow in SystemC. In: *Proceedings of the Austrochip 2007*, Graz, Austria, 2007.

12. C. Grimm, M. Barnasconi, A. Vachoux, and K. Einwich, An introduction to modeling embedded analog/mixed-signal systems using SystemC AMS extensions. OSCI, June 2008. www.systemc.org

13. Open SystemC Initiative. OSCI TLM2.0 standard, June 2008. http://www.systemc.org

14. M. Damm, C. Grimm, J. Haase, A. Herrholz, and W. Nebel, Connecting Systemc-AMS models with OSCI TLM 2.0 models using temporal decoupling. In: *Proceedings of the Forum on Specification and Design Languages (FDL)*, Stuttgart, Germany, 2008.

15. N. Wirth, Program development by stepwise refinement. *Communications of the ACM*, 14 (1971), S. 221–227.

16. J. Romberg and C. Grimm, Refinement of hybrid systems from formal models to design languages. In C. Grimm (editor), *Languages for System Specification*, Kluwer Academic Publisher: Dordrecht, Boston, New York, London, 2004.

17. M. Fowler, *Refactoring: Improving the Design of Existing Code*, Addison-Wesley, London.

19

Platform for Model-Based Design of Integrated Multi-Technology Systems

Ian O'Connor

CONTENTS

19.1 Rationale for Multi-Technology Design Exploration

Until recently, the single recognized vector enabling the improvement of the silicon system performance was perceived to be the progressive reduction (scaling) in CMOS device dimensions. Popularly known as Moore's law, this trend will lead to the emergence of transistors of physical gate lengths of 10–18 nm in the 2010–2015 timeframe, according to the predictions of the ITRS.* Consequently, the complexity of systems on chip (SoC) will reach unprecedented levels (high-performance microprocessors already contain over 10^9 transistors). However, the pursuit of Moore's law through scaling will meet significant future, intrinsic device hurdles (such as leakage,

---

*International Technology Roadmap for Semiconductors (http://www.itrs.net/).

interconnect, static power, quantum effects) to the capability of realizing system architectures using CMOS transistors with the performance levels required by future applications. It is recognized that these limitations, as much fundamental as economic, require the semiconductor industry to explore the use of novel devices able to complement or even replace the CMOS transistor in SoC within the next decade.

Hence, from the miniaturization of existing systems (position sensors, labs on chip, etc.) to the creation of specific integrated functions (memory, radio frequency (RF) tuning, energy, etc.), nanoscale and nonelectronic devices are being integrated to create nanoelectronic and heterogeneous systems in package (SiP), SoC, and 3D integrated circuits. This approach for future systems will have significant impact on several economic sectors, and is driven by

- The need for the miniaturization of existing systems to benefit from technological advances and improve performance at lower overall cost
- The potential replacement of specific functions in SoC/SiP with nanoscale or nonelectronic devices (nanoswitches, optical interconnect, magnetic memory, etc.)
- The advent of high-performance user interfaces (virtual surgical operations, games consoles, etc.)
- The rise of low-power mobile systems (communications and mobile computing) and wireless sensor networks for the measurement of phenomena inaccessible to single-sensor systems

In fact, the maturity and necessary diversification of integration technologies mean that three research directions are now open, as shown in Figure 19.1, an extended version of ENIAC*'s vision: "More Moore" (continued scaling); "More than Moore" (diversification); "Beyond Moore" (alternative devices). While "More Moore" focuses on the pursuit of traditional scaling for computation, "More than Moore" enables interaction with the real world and also system performance improvement through "equivalent" scaling with unconventional devices. It is clear that future SoC will be based on increasingly complex and diversified integration technologies in order to achieve unprecedented levels of functionality.

While the general benefits of heterogeneous integration appear to be clear, this evolution represents a strong paradigm shift for the semiconductor industry. This shift toward diversification and away from the scaling trend that has lasted over 40 years is possible because the integration technology (or at least the individual technological steps) exists to do so [ROO2005]. However, the capacity to translate system drivers into technology requirements (and consequently guidance for investment) to exploit such diversification is severely lacking. Such a role can only be fulfilled by a radical shift in design technology to address the new and vast problem of heterogeneous system design while remaining compatible with standard "More

---

*European Nanoelectronics Initiative Advisory Council (http://www.cordis.lu/ist/eniac).

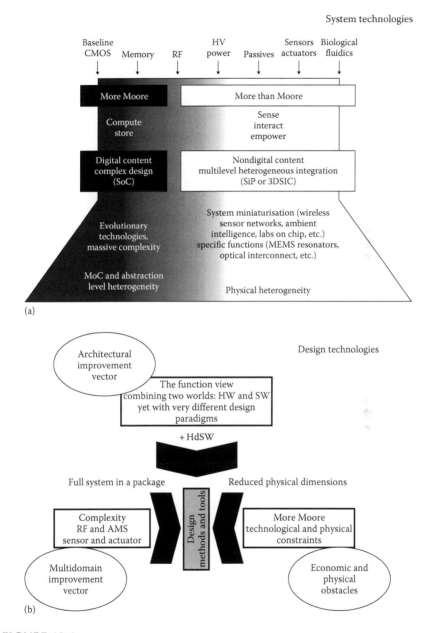

FIGURE 19.1
(a) ENIAC's vision of future integration technology diversification in the semiconductor industry. (b) The impact of technology diversification impact on design technology. (Adapted from ENIAC (European Nanoelectronics Initiative Advisory Council) SRA2007, http://www.eniac.eu.)

Moore" flows, that is, capable of simultaneously handling both "silicon complexity" and "system complexity." Designing in the context of increased silicon complexity (i.e., the number of individual elements) is managed through the development of methods capable of handling multiple abstraction levels and models of computation; while increased system complexity (i.e., number of different domains or concepts) requires that methods integrating other physical domains be developed. The urgency of this functionality for current SoC/SiP design flows is only too apparent from the data available from the ITRS (see Table 19.1), where it is clear that the earliest bottlenecks stem from the integration of heterogeneous content.

The field of design methods, in general terms, is a vibrant field of research, and is often applied to the management of design, production, logistics, and maintenance processes for complex systems in the aeronautic, transport, and civil engineering sectors, to name but a few. The microelectronics industry, over the years and with its spectacular and unique evolution, has built its own specific design methods while focusing mainly on the management of complexity through the establishment of abstraction levels. Today, the emergence of device heterogeneity requires a new approach, and no existing tool has the necessary architecture to enable the satisfactory design of physically heterogeneous embedded systems. The development of such software tools is a critical step to enable the widespread deployment of such systems.

The main objective of such an evolution is to reduce the design time in order to meet time to volume constraints. It is widely recognized that for complex systems at advanced technology nodes, a radical evolution in design tools and methods is required to reduce the "design productivity gap." Production capacity increases annually by around 50%, while design capacity increases annually by a rate of only 20%–25% [ITR2007]. All ITRS Roadmaps (and intermediate updates) since 2003 clearly state that "Cost [of design] is the greatest threat to continuation of the semiconductor roadmap. ... Today, many design technology gaps are crises," and identify this topic as one of the three main challenges to system design in the current post–45 nm era. It is clearly expressed that these "New technologies ... will require new modeling approaches, new types of creation and integration guidelines and rules, and, depending on the numbers of such design starts, may foster whole new toolsets." The issues pertaining to heterogeneous systems design methods and associated tools thus form part of the spectrum of highly relevant and long-term research topics. The European Commission also stresses the importance of design technology for nanoelectronic architectures of the future*: "There will be a need for new design approaches that make it possible to reuse designs easily when new generations or families of products appear. These approaches should be coupled with automatic translation of the resulting high-level designs into device manufacture."

---

*"Vision 2020—Nanoelectronics at the centre of change" http://cordis.europa.eu/nanotechnology.

TABLE 19.1
Selected Design Technology Bottleneck Predictions

Year of Production	2007	2008	2009	2010	2011	2012	2013	2014	2015	2016	2017	2018	2019	2020	2021	2022
DRAM half-pitch (nm)	68	59	52	45	40	36	32	28	25	22.5	20	17.9	15.9	14.2	12.6	11.3
Design block reuse (% to all logic size)	35	36	38	40	41	42	44	46	48	49	51	52	54	55	57	58
Accuracy of high-level estimates (% vs. measurements)	60	63	66	70	73	76	80	83	86	90	92	94	95	97	99	100
SoC reconfigurability (% of SoC functionality reconfigurable)	28	28	30	35	38	40	42	45	48	50	53	56	60	62	65	68
AMS automation (% vs. digital automation)	17	17	24	24	27	30	32	35	38	40	43	46	50	52	55	58
AMS modeling methodology (% vs. digital methodology)	58	60	62	65	67	70	73	76	78	80	83	86	90	92	95	98
Parameter uncertainty (% effect (on signoff delay))	6	8	10	11	11	12	14	15	18	20	20	20	22	25	26	28
Simultaneous analysis objectives (# objectives during optimization)	4	5	6	6	6	6	7	8	8	8	8	8	8	8	8	8
Synthesized AMS content (% of total design analog content)	15	16	17	18	19	20	23	25	28	30	35	40	45	50	55	60

Manufacturable solutions exist, and are being optimized.

Manufacturable solutions are known.

Interim solutions are known.

Manufacturable solutions are not known.

Source: ITRS, Sematech, 2007, http://www.itrs.net.

Without the introduction of new design technology, design cost becomes prohibitive and leads to weak integration of high added value devices (such as sensors and RF circuits) for the various application sectors (automotive/-transport, biomedical, telecommunications, etc.). A high-level vision of the maturity of existing abstraction levels for various physical domains is given in Table 19.2, with examples of adequate modeling languages or simulation engines where solutions exist.

To achieve design technology capable of fully exploiting the potential of heterogeneous SoC/SiP in a holistic approach, high-level modeling techniques capable of covering more physical domains should be developed, and cosimulation/cosynthesis methods and tools should aim to cover more abstraction levels. It is consequently clear that the impact of heterogeneity on design flows is or will be high, and necessary to facilitate heterogeneous device integration in SoC/SiP.

This chapter is structured as follows. We first describe the architecture and philosophy of the RUNE[II] platform for the development of predictive design methods and tools for heterogeneous SoC, in a "More than Moore" context. We focus specifically on the design process, on the use of specific abstraction levels in the process and how design information can be captured in a model for synthesizable analog and mixed-signal (AMS)/multi-technology (MT) intellectual property (IP), implemented in a high-level Unified Modeling Language (UML)/Extensible Markup Language (XML) framework. This design technology is applied to the exploration of an MT example: The elaboration of novel integrated optical interconnect schemes in the context of heterogeneous 3D integration. We focus on the use of a photonic interconnect layer enabled by 3D integration, and on the quantitative exploration of how such an approach can improve performance metrics of on-chip communication systems. We cover the establishment of functional and structural models for the simulation and synthesis of an optical link, and develop a method for optical point-to-point link synthesis. The investigation program, defined with respect to a set of performance metrics such as gate area, delay, and power, is shown to give uniquely detailed results for this new technology. Finally, some ideas will be given for the future evolution of integrated SoC and for the requirements for design technology to accompany this evolution.

19.2 Rune[II] Platform

The ongoing Rune[II] project* aims at researching novel design methods capable of contributing to the management of the increasing complexity of the

---

*http://sourceforge.net/projects/runeii.

TABLE 19.2
Abstraction Levels for Various Physical Domains and Related Modeling Languages

Level of Abstraction	Domain							
	Software	Digital	Analog	Radiofrequency	Mechanical	Optical	Fluidic	Chemical
Transaction	SystemC/UML SystemVerilog		SystemC					
Macro-architecture	SystemC SystemVerilog		Ptolemy/Matlab/SystemC-AMS			SystemC Matlab		
Micro-architecture	SystemC/VHDL SystemVerilog							
Block			VHDL	RF simulation / VHDL-AMS	VHDL-AMS	VHDL-AMS	VHDL-AMS	
Circuit		Electrical simulation		RF simulation				
Physical			Finite elements methods			Finite difference	FEMLab	Analytical

Manufacturable solutions exist, and are being optimized.
Manufacturable solutions are known.
Manufacturable solutions are NOT known.

SoC/SiP design process because of growth in both silicon complexity and in system complexity. As indicated above, current design technology is at its limits. It is in particular incapable of allowing any exploration of high- and low-level design tradeoffs in systems comprising digital hardware/software components and multiphysics devices (e.g., instruction line or software arguments against sensor or device characteristics). Such functionality is required to design (for example) systems in which power consumption is critical.

The ultimate overall goals of the platform include

- The definition and development of a coherent design process for heterogeneous SoC/SiP capable of effectively managing the whole of the heterogeneous design stages—through multiple domains and abstraction levels. A primary objective is to make clear definitions of the levels of abstraction, the associated design and modeling languages, and the properties of the objects at each level, whatever their natures (software components, digital/AMS/RF/multiphysics hardware). We consider it to be necessary to clearly define the scheduling of the design stages (which one can also regard as transformation actions applied to the various components) in an approach of "V-cycles" or "spiral" type, as well as the rules necessary for the validation of each stage. This makes it possible to establish the logistics of the design process, in particular for actions that could be carried out in parallel, and to take a first step toward a truly holistic design flow including economic and contextual constraints.

- The heterogeneous specification of the system by high-level modeling and cosimulation approaches; as well as the establishment of methods for executable high-level specifications of SoC/SiP including AMS and multiphysics components. The rationale for this is to allow the analysis of design criteria early in the design cycle.

- The extension of current hardware/software partitioning processes to nondigital hardware. Methods to formalize power, noise, silicon real estate, and uncertainty estimation in AMS and multiphysics components need to be developed, thus allowing the estimation of feasibility as critical information for the partitioning process. Although this information is intrinsically related to the implementation technology, efforts need to be made to render the formulation of information as qualitative as possible (thus circumventing the need to handle, in the early stages of the design process, the necessary numerical transposition to the technology). This formulation is employed to enrich the high-level models in the system.

- The development of hierarchical synthesis and top-down exploration methods, coherent with the design process model mentioned above, for SoC/SiP comprising multiple levels of abstraction and physical domains. Synthesis information for AMS/MT components is formalized and added to behavioral models as a basis for synthesizable AMS/MT IP, and developed tools exploit this information and are intended to guarantee the transformation of the system specifications into a feasible set of components specified at a lower (more physical) hierarchical level. Since multiple levels of abstraction are implied in the process, it is necessary to clearly specify bridges between the levels (through performance-based partitioning and synthesis). Here again, technology independence is a key point for the establishment of a generic approach, and makes it possible to generate predictive information when the approach is coupled with device models at future technology nodes.

- The validation of design choices using model extraction and cosimulation techniques. This relates to a bottom-up design process and requires model order reduction techniques for the modeling of non-electronic components (including the management of process and environmental variability), as well as the abstraction of time at the system level. This opens the way to the development of formal verification methods for AMS/MT to supplement the design flow for "More than Moore" systems.

These concepts contribute to our vision of a high-level design flow embodied in an experimental design platform for heterogeneous SoC/SiP, shown in Figure 19.2. The ultimate goal is to enable the concurrent handling of hardware/software and AMS/MT components in architectural exploration. As shown in Figure 19.2, there is a clear bridge between system-level and physical-level (or domain-specific) phases of design—in our view this bridge

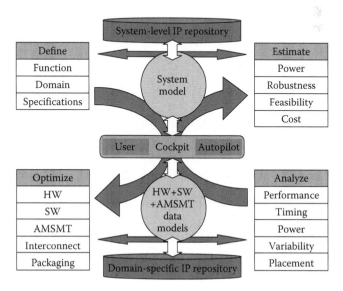

FIGURE 19.2
Target SoC/SiP design flow.

is critical to setting up a continuum of abstraction levels between system and physical design, enabling detailed analysis of cross-domain tradeoffs and the correct balancing of constraints over the various domains, and hence the achievement of optimally designed heterogeneous systems. The main impact would be to combat the current inefficiency in design processes between (1) the *a priori* generation of component specification sets at the system level in the presence of uncertainty concerning the feasibility of these sets in the target technology and (2) the *a posteriori* evaluation of the differences between specified and real component performance levels, generated at the physical level in the presence of uncertainty of their impact and potential degrees of freedom available at the system level. Learning systems can of course capitalize on the repeated use of estimation, optimization, and analysis to refine the accuracy of predictions.

A typical example of where such exploration would be required is in the optical interconnect demonstrator: based on software application constraints, system optimization requires the analysis of tradeoffs between (for example) (1) the number of cores (and parallel software tasks) that can be linked efficiently by optical interconnect to reduce the power contribution of the data processing part of the application and (2) the technology characteristics leading to a specific data rate/power ratio and the power contribution of the data communication part of the application. Hence such design technology can also be viewed as a high-level guide for design management.

In this section, we will first focus on a preliminary definition of abstraction levels and how they fit into a model for the design process. We will then consider the various elements pertaining to heterogeneous components required for the design process at various abstraction levels, and formalize this in a model for synthesizable AMS/MT IP, which we then show how to implement in UML/XML.

19.2.1   Abstraction Levels

The concept of abstraction levels is one that must be addressed for heterogeneous SoC/SiP. Valid abstraction (i.e., when there is a clear and explicit path to simplify representation at a higher level for all considered objects) is difficult to achieve when tightly coupled physical phenomena are present in the system—this is the case even for digital electronics at nanometric technology nodes, and the rise in AMS, RF, and heterogeneous content to address future application requirements compounds this problem. Efficient ways must be found to incorporate nondigital objects into design flows in order to ultimately achieve AMS/RF/heterogeneous/digital hardware/software codesign.

While hierarchy in the digital domain is based on the levels of signal abstraction, AMS/MT hierarchy is typically based on structural decomposition. It is necessary to combine and represent both types of hierarchy for

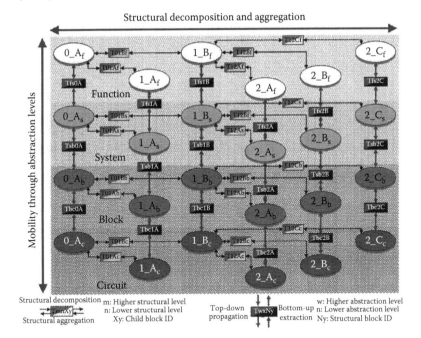

FIGURE 19.3
Modeling abstraction and structural hierarchies.

heterogeneous synthesis processes: hierarchy based on levels of abstraction, and hierarchy based on structural decomposition. Figure 19.3 shows a Petri net style diagram [GIR2002] where the ovals (places) represent IP blocks with various levels of abstraction (F, functional/mathematical; S, system; B, block; C, circuit/component). A loose association between these levels and existing modeling languages can be established: MATLAB®*/Simulink®† for the top two levels; Verilog-AMS for the block level; and SPICE/Spectre for the circuit/component level. In the diagram, the boxes situated on arcs between places at different abstraction levels represent transitions, and indicate the processes used to move between abstraction levels while preserving the properties of each block's functionality.

Structural decomposition can be represented by a set of transitions from one block to several other blocks (usually at the same abstraction level) and is also represented in Figure 19.3. For example, 0_A is the overall system to be designed and is comprised of components 1_A and 1_B. 1_B can be further decomposed into 2_A, 2_B, and 2_C. As can be seen from the diagram, some places (with dotted lines) are not accessible: at the functional level, this concerns 2_{A,B,C}_f and illustrates the nonrepresentativity of strong physical

---

*http://www.mathworks.com/products/matlab.
†http://www.mathworks.com/products/simulink.

coupling between IP blocks at this abstraction level; and at the circuit level, it concerns 0_A_c, illustrating the nonsimulability of the overall system at circuit level.

It is worth noting that while single direction transitions are the usual representation in this type of diagram, we have for simplicity here merged both structural transitions (decomposition and aggregation) and abstraction level transitions (top-down propagation, or "refinement," and bottom-up extraction, or "abstraction").

This approach enables clarification of the available/necessary steps in the design process. It is quite clear that several routes exist to achieve complete top-down synthesis of each individual component in the system, and conversely several routes enable the bottom-up validation of the whole (the top-down and bottom-up routes do not necessarily pass through the same places).

19.2.1.1 Model for Synthesizable AMS/MT IP

Most analog and RF circuits are still designed manually today, resulting in long design cycles and increasingly apparent bottlenecks in the overall design process [GIE2005]. This explains the growing awareness in industry that the advent of AMS/MT synthesis and optimization tools is a necessary step to increase design productivity by assisting or even automating the AMS/MT design process. The fundamental goal of AMS/MT synthesis is to generate quickly a first-time-correct sized circuit schematic from a set of circuit specifications. This is critical since the AMS/MT design problem is typically under-constrained with many degrees of freedom and with many interdependent (and often conflicting) performance requirements to be taken into account.

Synthesizable (soft) AMS IP [HAM2003] extends the concept of digital and software IP to the analog domain. It is difficult to achieve because the IP hardening process (moving from a technology-independent, structure-independent specification to a qualified layout of an AMS/MT block) relies to a large extent on the quality of the tools being used to do this. It is our belief that a clear definition of AMS/MT IP is an inevitable requirement to provide a route to system-level synthesis incorporating AMS/MT components. Table 19.3 summarizes the main facets necessary to AMS/MT IP, where each constituent element of design information is identified and the role of each is described.

Figure 19.4 shows how these various facets of AMS/MT IP should be brought together in an iterative single-level synthesis loop. This represents "structural" decomposition transitions. First, the set S of the performance criteria, originating from the higher hierarchical structural level $n + 1$ in Figure 19.4, is used to quantify how the IP block should carry out the defined function. Performance criteria are composed of functional specifications and performance specifications: for example in an amplifier, S will

TABLE 19.3

AMS/MT IP Block Facets

Property	Short Description
Function definition	Class of functions to which the IP block belongs.
Performance criteria set S	Quantities necessary to specify and to evaluate the IP block.
Terminals	Input/output links to which other IP blocks can connect.
Structure	Internal component-based structure of the IP block.
Design variable set V	List of independent design variables to be used by a design method or optimization algorithm.
Physical parameter set P	List of physical parameters associated with the internal components.
Evaluation method $*e$	Code defining how to evaluate the IP block, that is, transform physical parameter values to performance criteria values. Can be equation or simulation based (the latter requires a parameter extraction method).
Parameter extraction method	Code defining how to extract performance criteria values from simulation results (simulation-based evaluation methods only).
Synthesis method $*m$	Code defining how to synthesize the IP block, that is, transform performance criteria requirements to design variable values. Can be procedure- or optimization based.
Constraint distribution method $*c$	Code defining how to transform IP block parameters to specifications at a lower hierarchical level.

contain gain (the single functional specification), bandwidth, power supply rejection ratio, offset, etc. They have two distinct roles, related to the state of the IP block in the design process.

- As block parameters when the IP block is a component of a larger block, higher up in the hierarchy, in the process of being designed. In this case it can be varied and must be constrained to evolve within a given design space, i.e., $s_{low\_i} < s_i < s_{high\_i}$.
- As specifications when the IP block is the block in the process of being designed (such as here). In this case the value s_i is a fixed target and will be used to drive the design process through comparison with real performance values s_{ri}. Several possibilities exist to construct an error function ε, the most common being a sum of n (the size of S) weighted

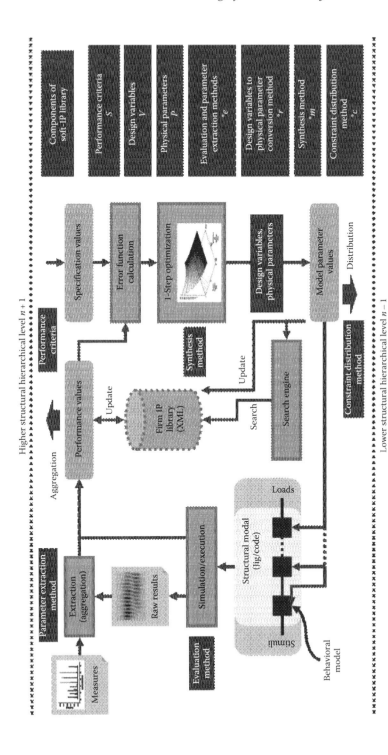

FIGURE 19.4
Single-level AMS/MT synthesis loop showing context of AMS/MT IP facet use.

($w_i \forall i \in \{0,n-1\}$) and normalized squared differences subject to specification type (constraint, cost, condition, etc.):

$$\varepsilon = \sum_{i=0}^{i=n-1} w_i \left(\frac{s_i - s_{ri}}{s_i} \right)^2$$

This comparison between specified and real performance criteria values, the error function in Figure 19.4, drives $*m$, the synthesis method, which describes the route to determine design variable values. It is possible to achieve this in two main ways:

- Through a direct procedure definition, if the design problem has sufficient constraints to enable the definition of an explicit solution.
- Through an iterative optimization algorithm. If the optimization process cannot, as is usually the case, be described directly in the language used to describe the IP block, then a communication model must be set up between the optimizer and the evaluation method. A direct communication model gives complete control to the optimization process, while an inverse communication model uses an external process to control data flow and synchronization between optimization and evaluation. The latter model is less efficient but makes it easier to retain tight control over the synthesis process [MAS1991].

The synthesis method generates a new set V of combinations of design variables as exploratory points in the design space according to $*m:S \rightarrow V$. The number of design variables defines the number of dimensions of the design space. The design variables must be independent of each other, and as such represent a subset of IP block parameters (i.e., performance criteria, described above) in a structure definition. For example, a differential amplifier design variable subset could be reduced to a single gate length, bias voltage, and three transistor widths for the current source, matched amplifier transistors, and matched current mirror transistors. Physical variables (in the set P, which outputs to the model parameter values in Figure 19.4) are directly related to design variables according to a mapping method $*r$ such that $*r:V \rightarrow P$, and serve to parameterize all components in the structure definition during the IP block evaluation process. In the above example, the design variable subset would be expanded to explicitly define all component parameters.

The evaluation method $*e$, at the left of Figure 19.4, describes the route from the physical variable values to the performance criteria values such that $*e:P \rightarrow S$, and thus completes the iterative single-level optimization loop. Evaluation can be achieved in two main ways:

- Through direct code evaluation, such as for active surface area calculations.
- Through simulation (including behavioral simulation) for accurate performance evaluation (gain, bandwidth, distortion, etc.). If the IP

block is not described in a modeling language that can be understood by a simulator, then this requires a gateway to a specific simulator and to a jig (a set of files to describe the environment surrounding the IP block and the stimuli to be applied in order to extract meaningful indicators of performance) corresponding to the IP block itself. For the simulator, this requires definition of how the simulation process will be controlled (part of the aforementioned communication model). For the jig, this requires the transmission of physical variables as parameters, and the extraction of performance criteria from the simulator-specific results file. The latter describes the role of the parameter extraction method, which is necessary to define how the design process moves up the hierarchical levels during bottom-up verification phases.

Once the single-level loop has converged, the constraint distribution method $*c$ defines how the design process moves down the hierarchical levels during top-down design phases, and defines the specifications for the lower hierarchical structural level $n - 1$ in Figure 19.4. At the end of the synthesis process at a given hierarchical level, an IP block will be defined by a set of physical variable values, some of which are parameters of an IP subblock. To continue the design process, the IP subblock will become an IP block to be designed and it is necessary to transform the block parameters into specifications according to $*c : P_k \rightarrow S_{k+1}$ (where k represents the structural hierarchy level). This requires a definition of how each specification will contribute to the error function ε for the synthesis method in the new block.

19.2.2 UML/XML Implementation

We have incorporated these concepts into an existing in-house AMS/MT synthesis framework, Rune[II]. A schematic showing the various inputs and data files is given in Figure 19.5. From the user's point of view, there are two main phases to AMS/MT synthesis: AMS/MT soft-IP definition, which can be done via UML, XML, or through a specific GUI (all inputs are interoperable—the internal database format is XML); and AMS/MT firm-IP synthesis, which can be run from the GUI or from scenarios.

At the system level, in order to enable the satisfactory partitioning of system-level performance constraints among the various digital, software, and AMS/MT blocks in the system architecture, top-down synthesis functionality needs to be added to AMS/MT blocks. The goal of this approach is to enable accurate prediction of analog/RF architectural specification values for block-level synthesis in an optimal top-down approach by making reasoned architectural choices about the structure to be designed. For general compatibility with system-level design flows, we chose to represent this aspect with UML. UML 2.0, adopted as a standard by Object Management

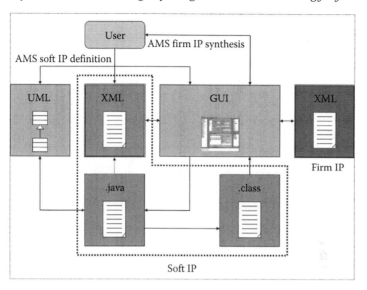

FIGURE 19.5
UML/XML/GUI use flow in Rune[II].

Group (OMG) in 2005, consists of graphical* languages enabling the expression of system requirements, architecture, and design, and is mainly used in industry for software and high-level system modeling. The use of UML for high-level SoC design in general appears possible and has generated interest in several research and industrial groups [RIC2005]. For AMS/MT systems, [CAR2004] demonstrated the feasibility of describing AMS/MT blocks in UML and then translating them to VHDL-AMS, building on other approaches to use a generic description to target various design languages [CHA2004]. Considerable effort is also being put into the development of "AMS/MT-aware" object-oriented design languages such as SystemC-AMS [VAC2003] and SysML [VAN2005]. These languages can be linked to a UML approach (SysML is directly derived from UML, and SystemC as an object-oriented language can be represented in UML also), and as such it should be possible to map UML-based work to these derived or related languages.

In order to develop a UML-based approach to hierarchical AMS/MT synthesis, it is necessary to map the AMS/MT IP element requirements given in Table 19.3 to UML concepts. UML has many types of diagrams, and many concepts that can be expressed in each—many more, in fact, than are actually needed for the specific AMS/MT IP problem. Concerning the types of diagram, two broad categories are available:

---

*A language for textual representation of UML diagrams also exists (OCL—Object Constraint Language. http://www.omg.org/).

- Structural diagram, to express the static relationship between the building blocks of the system. We used a class diagram to describe the properties of the AMS/MT IP blocks and the intrinsic relations between them. The tenets of this approach and how to generate UML-based synthesizable AMS/MT IP will be described in this section.
- Behavioral diagram, showing the evolution of the system overtime through response to requests, or through interaction between the system components. An activity diagram can be used to describe the AMS/MT synthesis process [OCO2006].

To describe class relationships for AMS/MT IP blocks, it is first necessary to establish a clear separation of a single function definition (entity and functional-level model for top-down flows) from n related structural models (for single-level optimization and bottom-up verification). Each structural model contains lower-level components, which should be described by another function definition. It is also necessary to establish functionality and requirements common to all structural models whatever their function. By representing all this in a single diagram, shown in Figure 19.6a, we are in fact modeling a library of system components; not the actual system to be designed itself.

A class diagram constitutes a static representation of the system. It allows the definition of classes among several fundamental types, the class attributes and operations, and the time-invariant relationships between the various classes. From the above analysis, we require

- A single, noninstantiable (abstract) class representing common functionality and requirements, in a separate publicly accessible package. We called this class topology.
- A single class representing the function definition, which inherits from topology. An alternative solution would be to separate "evaluatable" functionality and "synthesizable" functionality through the use of interfaces. This is certainly a debatable point, but our view is that it would tend to overcomplicate the description process. Another point is that one can also be tempted to separate the entity aspect from the behavioral model aspect, which would then allow the entity class to become abstract. Again, this also appears to be somewhat overcomplicated to carry out.
- A number of classes representing the structural models, which all inherit from the function definition class. Each structural variant is composed of a number of components at a lower hierarchical level, represented by a single function definition class for each component with different functionality. As the structural variant cannot exist if the component classes do not exist, this composition relationship is strengthened to an aggregation relationship.

Having established how to separate particular functionality between common, functional, and structural parts of an AMS/MT hierarchical model,

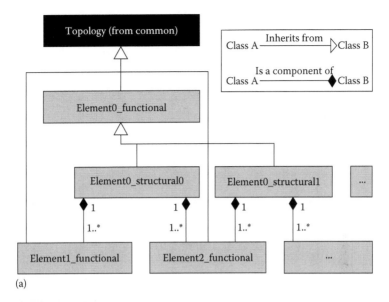

(a)

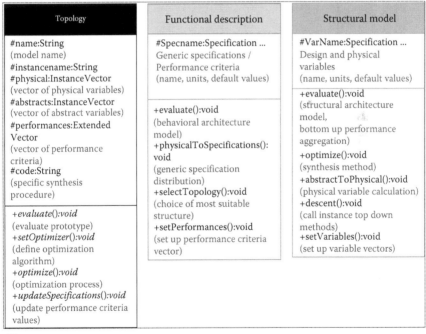

(b)

FIGURE 19.6

(a) UML class diagram showing representation of hierarchical dependencies between AMS/MT IP blocks. (b) UML class definitions for hierarchically dependent AMS/MT IP blocks.

each facet of the AMS/MT IP requirements set out in Table 19.3 can be included in the various model types, as shown in Figure 19.6b. It is worth noting that the performance criteria and variables are defined with type "specification." This is a specific data type, which plays an important role in the definition of AMS/MT IP, as will be seen in the following sections on the example optical interconnect application.

19.3   Multi-Technology Design Exploration Application: Integrated Optical Interconnect

Rune[II] has been extensively used in the exploration of integrated optical interconnect tradeoffs, both (1) to automatically size interface circuits according to link specifications and technology characteristics, thus enabling complete sizing of the optical link; and (2) to explore the impact of optical or photonic device characteristics on link performance and thus extract data leading to the identification of bottlenecks at the device level. Because of the very diverse nature of the exploration space variables, and the level of detail required in the investigations and analyses, this work could only be carried out using an automated and predictive simulation-based synthesis approach.

In this section, we will cover the establishment of models required for the simulation and synthesis of an integrated optical link; the development and implementation of the specification- and technology-driven optical point-to-point link synthesis method; and the definition of the performance metrics and specification sets to be used in the investigation program.

19.3.1   Models for the Simulation and Synthesis of an Optical Link

In order to extract meaningful physical data from analyses where advanced CMOS technologies are involved and accurate device models are key to the relevance of investigation, it is essential to work toward design technology including the simulation of a complete optical point-to-point link in an EDA framework. A direct consequence of this is that it is necessary to develop behavioral models for the optoelectronics devices and passive waveguides, for concurrent simulation with the transistor-level interface circuit schematics. For all behavioral models, the choice of an appropriate level of description is a prerequisite to developing and using the models in the required context. Essentially, the description level falls into one of two categories: functional modeling or structural modeling (this is a particular case of model types mentioned in Section 19.2.3). A functional model will describe the behavior of a device according to its specifications and behavioral equations, without defining the structure of the device. A structural model will describe the behavior of a device according to its internal structure and physical parameters, without necessarily satisfying the specification criteria (which do not have to be formalized in this approach).

Ideally, both functional and structural models should exist for all devices considered. However, this is not absolutely necessary, and careful consideration was given to choosing the appropriate description level for each device. Since the source behavior is arguably the most complex and likely to exhibit nonlinear behavior (thermal roll off, temperature changes) important to complete link simulation, it was decided to model this element at a structural level. The nonlinear behavior of the microsource laser was modeled (enabling the visualization of physical limits) and converges systematically in Spectre. The waveguide and detector were modeled at a functional level. The organization of the interface circuit and active optoelectronic device model libraries, complying to the UML modeling rules set out in Section 19.2.3, are shown in Figure 19.7a and b respectively.

The models were all implemented in the OVI-96 Verilog-A subset of Verilog-AMS, an extension of the IEEE 1364-1995 Verilog hardware description language (VHDL). This extension is an industry standard for analog simulation model description and can be simulated with a number of general-purpose circuit simulators (we used Spectre). This way, the optical and photonic devices can be simulated together with the interface circuitry and with the rest of the optical link given adequate simulation models. This enables interesting optimization strategies (e.g., joint power optimization) and the analysis of link performance sensitivity to various parameter variations as well as temperature changes.

Detailed description of these models is outside the scope of this chapter, but the device parameters for optical interconnect varied in this analysis are shown in Table 19.4, with minimum and maximum values defining the limits of the parameter variation. These limits are based on discussions with experts in the field and on data available in the literature on sources, waveguides, and detectors [BIN2005,FU2004,FUJ2000,ROE2006,SAK2001,VAN2007]. The values in bold italics represent the (pessimistic) nominal values.

Achieving complete link simulation was a necessary step to enable subsequent simulation-based link synthesis (using interface circuit design variables) over a range of target technologies and specification sets to extract link performance data. The iterative optimization step is facilitated by the low simulation time required for the complete link (a few seconds for 10 data bits on a 1.3 GHz processor).

19.3.2 Optical Point-to-Point Link Synthesis

The objective of our work was to carry out transistor-level sizing of the receiver and of the driver circuits according to complete link specifications. The optical link under consideration is shown in Figure 19.8. This includes, as shown in Figure 19.8a, a photonic communication layer integrated above a standard CMOS circuit above the metallic interconnect layers. CMOS

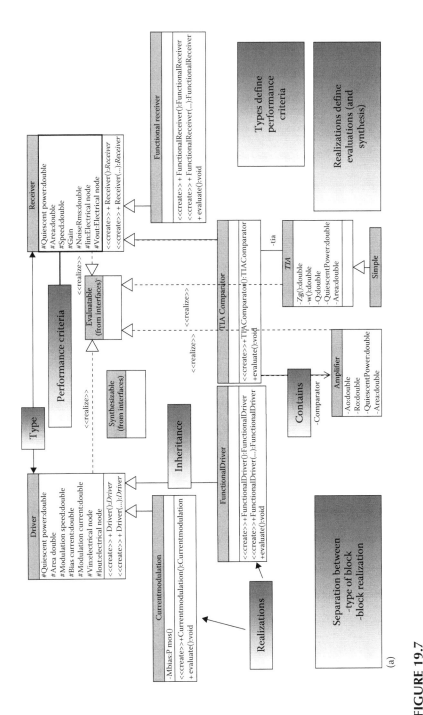

FIGURE 19.7
UML class diagram showing representation of hierarchical dependencies between AMS/MT IP blocks (a) CMOS interface circuit model library (b) active optoelectronics device model library.

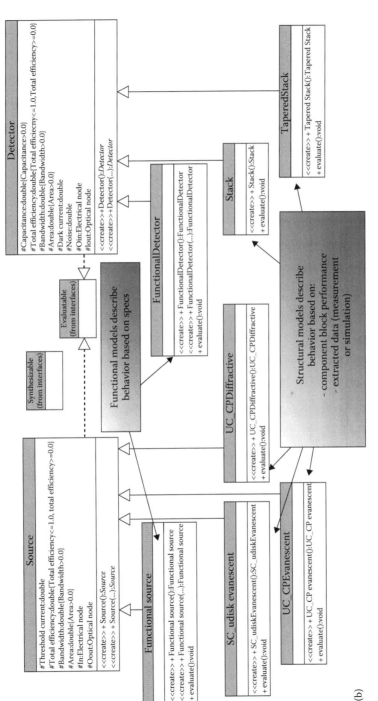

FIGURE 19.7 (continued)

(b)

TABLE 19.4

Optical Device Parameters

Source	Parameter	Min. Value	Max. Value	Units
	Total efficiency	**0.1**	0.3	mW/mA
	Area	10×10	100×100	μm^2
	Threshold current	1.5e-5	**1.5e-3**	A
	Bandwidth	10		GHz
Detector				
	Responsivity	*0.5*	0.7	mA/mW
	Area	10×10	100×100	μm^2
	Capacitance	1e-15	**1e-13**	F
	Dark current	1e-18		A
	Noise current	1e-15		A
	Bandwidth	20		GHz
Waveguide		**SOI**	**Si$_3$N$_4$**	
	Guide index	3.45	2.0	
	Cladding index	1.46	1.46	
	Height	0.22	0.8	μm
	Width	0.5	0.4	μm
	Pitch	1.1e-6	**4e-6**	m
	Loss	**2.7**	1.5	dB/cm
	Delay	*13.3*	6.7	ps/mm
	Excess bend loss	0.027	**1.52**	dB/90°
	Bend radius	2	**10**	μm

transmitter and receiver circuits modulate laser current and transform detector current respectively, through a via stack as represented by R_{via} in Figure 19.8b. Light emitted by an InP laser is transported through a passive Si/SiO$_2$ structure at wavelengths around 1.5 μm to an InP or SiGe detector. In this design approach, no architectural variants are considered (i.e., the CMOS topologies used at the transistor level are fixed in terms of their structure— the variables for the design problem consist essentially of bias currents and transistor sizes).

The synthesis approach implemented consists of creating "scenarios" allowing the specification of each model via a generic class containing all structures to be optimized, as well as necessary evaluation and design methods (Figure 19.9). Communication between the different blocks is ensured using the synthesizable AMS/MT IP blocks, and the actual synthesis and evaluation scenario relies upon the instantiation of the generic top-level object in a testbench structure. A device library containing the synthesizable models of each device in the optical link based on the UML language was developed to allow the modeling of this hierarchical synthesis problem, as was already shown in Figure 19.7.

The procedure used to automatically synthesize an optical point-to-point link, and implemented as a synthesis scenario, is shown in Figure 19.10. The

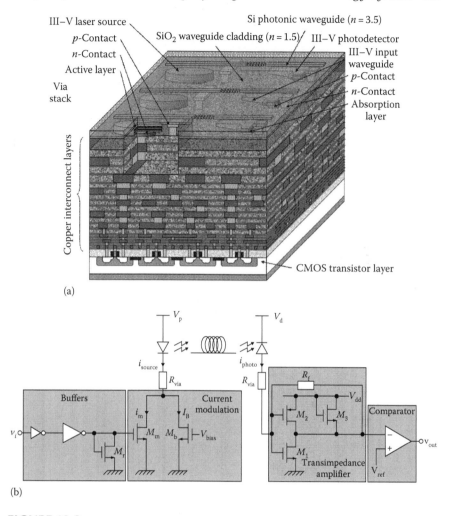

FIGURE 19.8
Integrated optical link (a) physical implementation (b) schematic.

process starts by defining the photodetector characteristics and the required data rate. Using a transistor-level synthesis method described in [OCO2003], the transistor-level schematic for the transimpedance amplifier is automatically generated and linked to a manually scaled comparator circuit.

The value of the rms noise power, i_n, is extracted from the simulation of the schematic, and updated for each synthesis loop, using the Morikuni formula [MOR1994] in the transimpedance amplifier noise calculations:

$$i_N^2 = \left(2q\left(I_{gate} + I_{dark}\right) + \frac{4kT}{R_f}\right)\frac{C}{4D} + 4kT\Gamma\frac{C^2}{16\pi^2 DE}\frac{(2\pi C_r)^2}{g_m}$$

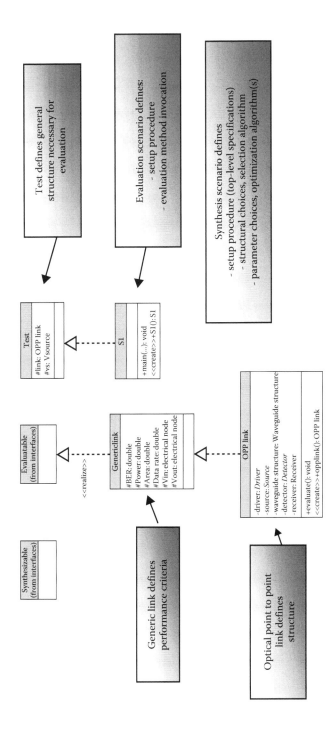

FIGURE 19.9

Definition of classes in UML for test benches, evaluation and synthesis scenarios for the integrated optical link.

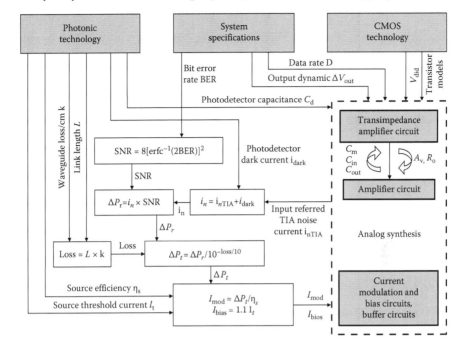

FIGURE 19.10
Optical link sizing method.

where

$$C = 1 + g_m R_f$$
$$D = R_o \left(C_x + C_y \right) + R_f \left(C_x + C_m \right) + g_m R_f R_o C_m$$
$$E = R_f R_o \left[\left(C_x + C_y \right) C_m + C_x C_y \right]$$

For a given bit error rate (BER) specification and noise signal associated with the photodiode and transimpedance circuit, we can then calculate the minimum optical signal power, ΔP_s, required by the receiver to operate at the given error probability:

$$\Delta P_s = i_n \times \text{SNR}$$

where

$$\text{SNR} = \left(\text{erfc}^{-1} \left(2\,\text{BER} \right) \right)$$

Here, SNR represents the linear signal to noise ratio (absolute value, not in dB). BER, defined as the rate of error occurrences, is one of the main criteria in evaluating the performance of digital transmission systems. In our analyses we fixed BER at 10^{-18} bits^{-1} (this corresponds to 1 error / 3.17 years for a single link at 10 Gbps communication, or 1 error/18 days for a 64-bit data bus at the same data rate.)

The value of the power that needs to be emitted by the laser source is evaluated from the calculated value of the minimum optical power at the receiver, and from the power losses induced by the geometry of the waveg-uide structure (length and intrinsic loss, number of bends, and loss/90° bend) and coupling. These figures depend to a large extent on the materials used and are defined in Table 19.4.

The final sizing step is to calculate the driver and associated bias and buffer circuits using the emitted power value and the source characteristics in conjunction with the method shown in Figure 19.10. This then enables the simulation of the complete optical link, using transistor-level schematics for the interface circuits and the developed behavioral models for the pho-tonic devices. From the simulation results, the performance criteria can be extracted.

Using this approach, the synthesis problem is considered to be complete, such that no constraint partitioning is required. In fact, the constraints are derived directly from system specifications, and thus constraint exploration is achieved directly by the user.

19.3.3 Performance Metrics and Specification Sets

In order to be able to evaluate and optimize link performance criteria cor-rectly, a clear definition of the performance metrics is required. First, the aim is to establish the overall power dissipation for an optical link at a given data rate and BER. The calculation is essentially conditioned by the receiver as explained above, since the BER defines the lower limit for the received optical power. This lower limit can then be used to calculate the required power coupled into waveguides by optical sources, the required detector efficiency (including optical coupling), and acceptable transmission losses. Power can then be estimated from source bias current and photoreceiver front-end design methodologies.

For interconnect density aspects, source and detector sizes must be taken into account, while the width, pitch, and required bend radius of waveguides is fundamental to estimating the size of the photonic layer. On the circuit layer, the additional surface due to optical interconnect is in the driver and receiver circuits, as well as the depassivated link to the photonic layer. The circuit layout problem is compounded by the necessity of using clean supply lines (i.e., separate from digital supplies) to reduce noise (for BER).

The data rate is essentially governed by the bandwidth of the photore-ceiver: high modulation speed at the source is generally more easily attain-able than similar detection speed at the receiver. This is largely due to the photodiode parasitic capacitance at the input of the transimpedance amplifier.

The limitations of this analysis as carried out here are that

- Predictive technology model (PTM)* models do not take noise into account particularly well, which means that no real noise analysis can

---

*Predictive Technology Model (http://www.eas.asu.edu/~ptm/).

be carried out. However in the sizing process this problem was circumvented by using the Morikuni formula to estimate noise at block level, as previously described.

- No automatic layout generation tools were used, as it is not in general possible to achieve optimal layout for high-speed analog circuits. Parasitic capacitances were therefore extracted from layout estimations using lambda rules [OCO2007] rather than from actual layouts.

19.4 Integrated Optical Interconnect Investigation Program and Results

In this section, we cover the values of the performance metrics generated by the synthesis procedures described in the previous section. The analyses were carried out for

- Two sets of optical device parameters as described in Table 19.4, which will be denoted in the following analyses as S1 ("pessimistic" values) and S2 ("optimistic" values)
- Three predictive technologies (gate lengths of 65, 45, and 32 nm) using PTM models and frequencies defined by the ITRS (as local clock frequencies) for the corresponding technology nodes
- Various optical link lengths from 2.5 mm to the maximum chip side dimension (20 mm)

Table 19.5 shows the sets of specifications used for analysis and interface circuit sizing and to demonstrate the capacity of the platform and implemented method to synthesize optical links subject to technological specifications, both CMOS and optical. The generation of each data point requires approximately 5 min on a 1.3 GHz processor with 4 Gb memory.

TABLE 19.5

Link Specification Set

Parameter		Scenario		
		PTM65	**PTM45**	**PTM32**
BER	bit^{-1}	10^{-18}	10^{-18}	10^{-18}
ITRS max. frequency	bit/s	2.98×10^9	5.20×10^9	1.10×10^{10}
Link length	mm	{2.5,20}	{2.5,20}	{2.5,20}
Activity rate		1	1	1
Ambient temperature	°C	70	70	70
V_{dd} (CMOS)	V	1.2	1.1	1.0

The results obtained were compared to the performance of 1.1 μm pitch unshielded electrical interconnect, synthesized with another toolset and using the same specifications as drivers for the synthesis process [OCO2007]. The simulation conditions were

- Inverters (as buffers for each electrical interconnect segment) with a 2/1 PMOS-to-NMOS ratio were used.
- For each CMOS technology used, the maximal overall link input capacitance was restricted to that of a CMOS inverter with minimal gate length (defined as 2 λ) and 60λ and 30λ for the PMOS and NMOS widths, respectively.
- The minimal output drive strength was set to that of the same inverter.

The comparison results are presented in the form of reduction factors, calculated as P_e/P_o, where P_o represents the optical performance figure and P_e represents the electrical interconnect figure (where smaller performance figures for area, delay, and power mean improved performance).

19.4.1 Gate Area Analysis

The link sizing method described in Section 19.3.2 was applied according to the specifications for the PTM 65, 45, and 32 nm technologies. Figure 19.11 shows the results in terms of gate area (i.e., transistor channel dimensions only), extracted as the sum of all transistor gate channel areas $W \cdot L$. These results show that the gate area metric approximately verifies the scaling law $(A_{32nm} \approx A_{45nm} \cdot s^2 \approx A_{32nm} \cdot s^2 \cdot s^2)$, where s is equal to 0.7 (scaling factor between technology generations).

The specification set S_2 reduces total CMOS gate area by a factor of between 2 (BPT 32 nm) to 4 (BPT 65 nm) with respect to S_1. These figures demonstrate an impressive reduction in gate area in favor of optical interconnect (of the order of 60x–90x for link lengths above 1 cm and for the two most advanced technology nodes) with respect to electrical interconnect. Optical interconnect will thus introduce a significantly lower area penalty for data routing functions at transistor level. These figures are to be considered in the context of scenarios indicating the use of up to 25% of transistors on chip in electrical interconnect buffers. While only a part of this number is used for long links, it is still clear that the use of optical links will free up a large number of transistors for use in functions other than interconnect (data processing, memory, etc.).

19.4.2 Delay Analysis

The link sizing method described in Section 19.3.2 was applied according to the specifications for the PTM 65, 45, and 32 nm technologies. The 50%

Total CMOS gate area (μm²) vs. interconnect length—optical link

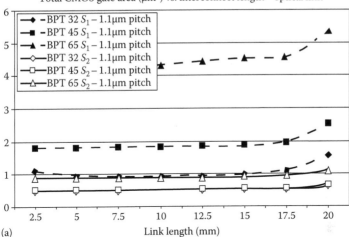

(a)

Link length (mm)

Gate area reduction factor vs. interconnect—length

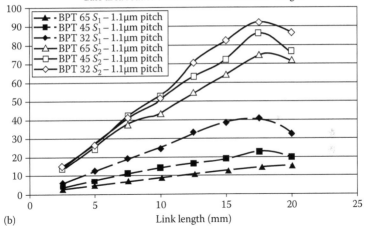

(b)

Link length (mm)

FIGURE 19.11
(a) Total CMOS gate area (μm²) for varying optical interconnect length and technologies (b) Gate area comparison for varying interconnect length and technologies at 1.1 μm pitch : reduction factor.

propagation delay was extracted from simulation as

$$\Delta t = t_{V_{out}=\frac{V_{pit_{max}}+V_{out_{min}}}{2},slope=+,5} - t_{V_{in}=\frac{V_{in_{max}}+V_{in_{min}}}{2},slope=+,5}$$

where the notation

$$t_{V=V_1,slope=\{+,-\},n}$$

signifies the time corresponding to the nth simulation point where the quantity V is equal to V_1 on a rising (+) or falling (−) slope. Also calculated as a point of reference was the intrinsic waveguide delay, using τ_{TOF}= 13.3 ps/mm for the Si/SiO$_2$ waveguides.* Figure 19.12 shows the delay results for varying link lengths.

It can be seen that (a) the circuit delay (i.e., the difference between the total delay and the intrinsic waveguide delay) decreases with smaller gate lengths, and (b) the same quantity also decreases with longer interconnect. This latter effect is because of higher driver modulation current, I_m, required to compensate higher overall waveguide loss, being able to drive the source capacitance faster and thus tends toward a small contribution to overall delay. As such the delay is dominated at these lengths by detector and source delay, waveguide delay, and comparator delay. Since these parameters do not change between S_1 and S_2, no significant improvement can be observed between the results for these two optical parameter sets. At best, a further 50 ps delay reduction can be achieved at the shorter link lengths (where optical links are less likely to be used).

With respect to electrical interconnect, it is shown that optical interconnect will have a slight advantage in terms of delay for long interconnect lengths (above 10 mm). The underlying reason for this is that delay for optical interconnect does not depend as strongly on interconnect length as electrical interconnect, because no additional circuit stages are added—the increase stems from higher intrinsic waveguide delay only. However, the advantage decreases for more advanced technology nodes and indeed does not achieve any delay reduction (actually the opposite) for unscaled interconnect at 32 nm gate length.

19.4.3 Power Analysis

The link sizing method described in Section 19.3.2 was applied according to the specifications for the PTM 65, 45, and 32 nm technologies. The average static power was extracted from transient simulations using

$$\bar{P} = \frac{I_{source_0} + I_{source_1}}{2} \cdot V_p + \frac{I_{det_0} + I_{det_1}}{2} \cdot V_d \frac{I_{cct_0} + I_{cct_1}}{2} \cdot V_{dd}$$

where I_{source}, I_{det}, and I_{cct} represent the currents flowing through the source, detector, and circuit voltage supplies of V_p, V_d, and V_{dd}, respectively. Figure 19.13a shows the average static power results for varying link lengths. Figure 19.13b shows the dynamic power results for varying link lengths, calculated from rising and falling edge transitions (the average switching energy extracted from simulations as the integral of supply currents in edge transitions).

Significant reductions in static power are observed between S_1 and S_2 (a minimum factor of 2 for BPT 45 and 3 for BPT 32 at low link lengths; and a

---

*TOF, time of flight.

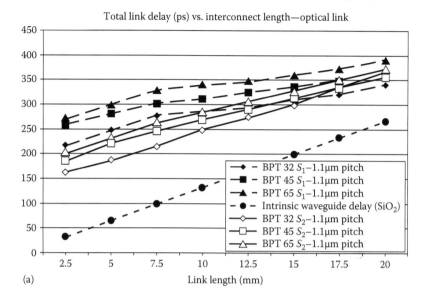

(a)

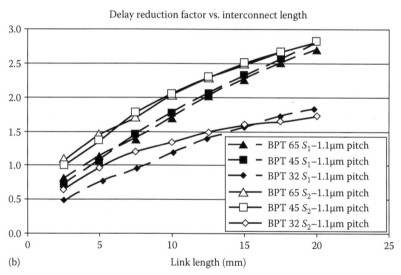

(b)

FIGURE 19.12
(a) Total link delay (ps) for varying optical interconnect length and technologies. (b) Delay comparison for varying interconnect length and technologies at 1.1 μm pitch : reduction factor.

maximum factor of 4 for BPT 65 and BPT 32 at high link lengths). This result is due mainly to the lower source threshold current, but also to higher source efficiency and detector responsivity leading to lower modulation currents,

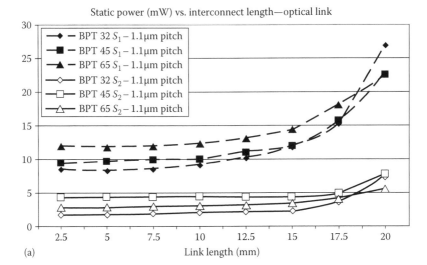

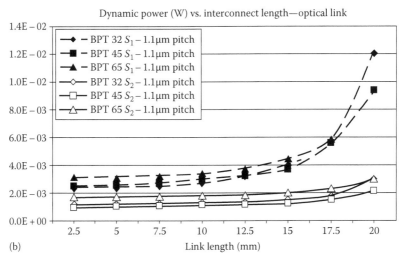

FIGURE 19.13
Power vs. interconnect length for BPT 65, 45, and 32 nm technologies (a) Average static power (in mW). (b) Average dynamic power (in W).

and to lower detector capacitance, which lowers the receiver circuit quiescent current.

Dynamic power reductions between S_1 and S_2, of the order of 2–4, are observed. This is attributed mainly to lower transistor capacitances because of lower bias current in the driver (due to the reduction in source threshold current) and because of lower modulation currents in the driver (due to the increase in source efficiency and detector responsivity).

Total power (mW) vs. interconnect length—optical link

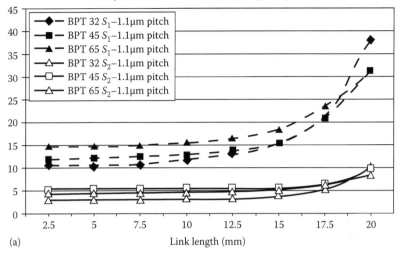

(a)

Power reduction factor vs interconnect length

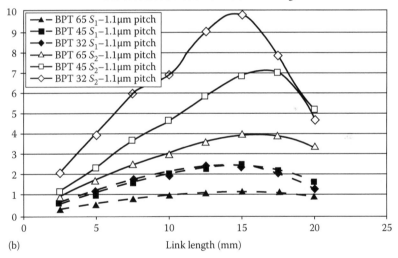

(b)

FIGURE 19.14
(a) Total power (in mW) vs. interconnect length for BPT 65, 45, and 32 nm technologies. (b) Average total power comparison for varying interconnect length and technologies at 1.1 μm pitch: reduction factor.

Using static and dynamic power information, the total energy can be calculated (Figure 19.14).

The overall power is shown to be reduced between S_1 and S_2 by factors of between 2 and 4. The greatest reduction is achieved at higher link lengths, which is the expected context for the use of such optical links. For

S_2, power reduction can be considered to be a major argument in favor of optical interconnect. For S_1, it is clear that the static power comparison is the weak point for optical interconnect, because of continuous biasing of the source (avoiding turn-on times to achieve the required bit rate) and of the receiver circuit (the circuit bandwidth is directly related to quiescent bias current). The reduction in source threshold current and in detector capacitance in S_2 has a significant impact on both these factors, to the extent that the static power in optical links, while still higher than that of electrical interconnect, is no longer dominant. Dynamic power in optical interconnect is further reduced with the smaller overall circuit transistors used in this analysis. The total power comparison shows power reduction factors between 5x and 10x for link lengths above 10 mm and for the two most advanced technology nodes. It is likely that comparisons using technologies with transistor gate lengths below 32 nm will further improve this comparison since on the electrical side the static power dissipation will increase with leakage current. We can thus consider that the optical device improvements constitute the main path to the solution to the following recommendation: "Power, and particularly static power, is a key performance metric to optimize during exploration of optical interconnect device specifications." This is an illustration of the type of feedback our approach can give to photonic device engineers.

19.5 Conclusions and Ideas for the Future

In this chapter, we have looked at several aspects of heterogeneous design methods in the context of increasing diversification of integration technologies. The rationale and analysis of the situation in terms of technological evolution and severe gaps in design technology show a clear need for advances in this domain. The experimental Rune[II] heterogeneous design platform addresses some of these needs—in particular we have shown how design processes can be formalized over multiple abstraction levels and multiple domains using a common model for design knowledge formulation called AMS/MT IP. To address the need to represent this knowledge at higher abstraction levels in order to retain compatibility with system-level design methods, we have demonstrated the feasibility of the use of UML and established parallels between the UML concepts and widely used concepts in AMS/MT descriptions. We have successfully used these concepts to build class diagrams for functional and structural models of integrated optical link component libraries, and implemented a synthesis scenario to explore, in a quite detailed way, the available design space over a number of very different dimension types. This design method and technology is particularly useful for the repetitive design of fixed optical link structures subject to varying design constraints, technology parameters, and performance requirements. We have illustrated the direct application of our approach for optical link synthesis and technology performance characterization by analyzing optical link performance for two sets of photonic component parameters and three CMOS technology generations. Importantly for technological development, the results of such analyses can generate useful feedback from system designers to component designers.

In our view, the next major step in technological evolution for SoC concerns 3D integration. This approach, exploiting the vertical dimension, provides an opportunity to continue to achieve the performance levels predicted by the extrapolation of Moore's law, but using a different technological approach. The main impact is to enable the construction of highly complex systems (e.g., multiple core data-processing systems with multiple memory banks in close spatial proximity; highly heterogeneous systems using multiple technologies for specific functional layers) while reducing the cost of communication by several orders of magnitude. The direct consequences of this approach are to (1) improve isochronous signal coverage and (2) allow evolution toward modular integrated systems. Such an approach should also in the long term enable novel vertical interconnect solutions (e.g., embedded repeaters, vertical switches, etc.). While by no means technologically simple, it does represent a major design paradigm shift away from the conventional approach of traditional Moore's law scaling toward "equivalent scaling" and "functional diversity" through unconventional approaches.

In this context, several tradeoff situations can be clearly shown to require heterogeneous design methods, which cannot as yet be processed by existing tools. Some examples are given here.

- Tradeoffs for data-processing systems using, for example, 45 nm CMOS for processor cores and a specific technology (Flash–DRAM) for memory. While the gain over conventional planar architectures is clear (high core-memory bandwidth and high memory capacity leading to reduction/elimination of cache structure, enabling orders of magnitude improvements in processing time for algorithms requiring high data-rate memory access), the organization of the memory and its spatial organization in relation to achievable through-silicon via (TSV) density and characteristics, and to the number and size of processor cores, has to be explored.
- Partitioning in transformations from a planar chip to a 3D architecture. It will, for example, be necessary to explore comparisons between a complete system on chip built from an aggressive planar CMOS technology and a technological partition between using the same aggressive CMOS technology for data processing, but a less aggressive, more stable CMOS technology for I/O functions, analog, power, voltage regulators to achieve interfacing and to compensate for variability, etc. The tradeoffs here concern cost, variability, performance, power, and process simplification.

- Tradeoffs for mobile computing nodes in a 3D context using, for example, an aggressive CMOS technology for processor cores, specific advanced technology for memory, mature technology for analog, and a specific technology for RF (e.g., passive RF MEMS). Here, the focus is on overall power minimization exploiting the reduced communication cost between the data-processing layer(s) (interpreting received symbols with more complex interpolation functions) and the RF layer (relaxed constraints on RF MEMS tunability and speed based on algorithm efficiency).

Acknowledgment

This work was partially funded by the European FP6 IST program under PICMOS FP6-2002-IST-1-002131.

References

[BIN2005] P. R. A. Binetti et al., A compact detector for use in photonic interconnections on CMOS ICs, in *Proceedings of the Symposium IEEE/LEOS* (Benelux Chapter), Mons, Belgium, 233–236, 2005.

[CAR2004] C. T. Carr, T. M. McGinnity, and L. J. McDaid, Integration of UML and VHDL-AMS for analogue system modeling, *BCS Formal Aspects of Computing*, 16(1), 80–94, 2004.

[CHA2004] V. Chaudhary, M. Francis, W. Zheng, A. Mantooth, and L. Lemaitre, Automatic generation of compact semiconductor device models using Paragon and ADMS, in *Proceedings of the IEEE International Behavioral Modeling and Simulation Conference (BMAS) 2004*, San Jose, CA, 107, 2004, IEEE, Piscataway, NJ.

[ENI2007] Strategic Research Agenda, European Nanoelectronics Initiative Advisory Council (ENIAC), 2007. http://www.eniac.eu

[FU2004] T. Fukazawa et al., Low loss intersection of Si photonic wire waveguides, *Japanese Journal of Applied Physics*, 43(3), 646–647, 2004.

[FUJ2000] R. Fujita, R. Ushigome, and T. Baba, Continuous wave lasing in GaInAsP microdisk injection laser with threshold current of 40 µA, *Electronics Letters*, 36(9), 790–790, 2000.

[GIE2005] G. Gielen et al., Analog and digital circuit design in 65 nm CMOS: End of the road? in *Proceedings of the Design Automation and Test in Europe*, Munich, Germany, 2005.

[GIR2002] G. Girault and R. Valk, *Petri Nets for Systems Engineering*, Springer, Berlin, Germany, 2002.

[HAM2003] M. Hamour, R. Saleh, S. Mirabbasi, and A. Ivanov, Analog IP design flow for SoC applications, *Proceedings of the International Symposium on Circuits and Systems* (ISCAS), Bangkok, Thailand, IV-676, 2003.

[ITR2007] The International Technology Roadmap for Semiconductors (ITRS), Sematech, 2007. http://www.itrs.net

[MAS1991] R. E. Massara, *Optimization Methods in Electronic Circuit Design*, Longman Scientific & Technical, Harlow, U.K., 1991.

[MOR1994] J. J. Morikuni et al., Improvements to the standard theory for photoreceiver noise, *IEEE Journal of Lightwave Technology*, 12, 1174, 1994.

[OCO2003] I. O'Connor, F. Mieyeville, F. Tissafi-Drissi, G. Tosik, and F. Gaffiot, Predictive design space exploration of maximum bandwidth CMOS photoreceiver preamplifiers, in *IEEE International Conference on Electronics, Circuits and Systems*, Sharjah, United Arab Emirates, 483–486, December 14–17, 2003.

[OCO2006] I. O'Connor, F. Tissafi-Drissi, G. Revy, and F. Gaffiot, UML/XML-based approach to hierarchical AMS synthesis, in *Advances in Specification and Design Languages for SoCs*, A. Vachoux (ed.), Kluwer Academic Publishers, Dordrecht, the Netherlands, 2006.

[OCO2007] I. O'Connor et al., Systematic simulation-based predictive synthesis of integrated optical interconnect, *IEEE Transactions on Very Large Scale Integration (VLSI) Systems*, 15(8), 927–940, August 2007.

[RIC2005] E. Riccobene, P. Scandurra, A. Rosti, and S. Bocchio, A SoC design methodology involving a UML 2.0 profile for SystemC, in *Proceedings of the Design Automation and Test in Europe (DATE) 2005*, Munich, Germany, 704–709, 2005, IEEE Computer Society, Washington, DC.

[ROE2006] G. Roelkens, D. Van Thourhout, R. Baets, R. Notzel, and M. Smit, Laser emission and photodetection in an InP/InGaAsP layer integrated on and coupled to a silicon-on-insulator waveguide circuit, *Optics Express*, 14(18), 8154–8159, 2006.

[ROO2005] F. Roozeboom et al., Passive and heterogeneous integration towards a Si-based system-in-package concept, *Thin Solid Films*, 504(1–2), 391–396, May 2006.

[SAK2001] A. Sakai, G. Hara, and T. Baba, Propagation characteristics of ultrahigh-Δ optical waveguide on silicon-on-insulator substrate, *Japanese Journal of Applied Physics—Pt. 2*, 40(383), L383–L385, 2001.

[VAC2003] A. Vachoux, C. Grimm, and K. Einwich, SystemC-AMS requirements, design objectives and rationale, in *Proceedings of the Design Automation and Test in Europe (DATE) 2003*, IEEE Computer Society, Munich, Germany, 388–393, 2003.

[VAN2005] Y. Vanderperren and W. Dehaene, UML 2 and SysML: An approach to deal with complexity in SoC/NoC design, in *Proceedings of the Design Automation and Test in Europe (DATE) 2005*, Munich, Germany, 716–717, 2005, IEEE Computer Society, Washington, DC.

[VAN2007] J. Van Campenhout et al., Electrically pumped InP-based microdisk lasers integrated with a nanophotonic silicon-on-insulator waveguide circuit, *Optics Express*, 15(11), 6744–6749, 2007.

20

CAD Tools for Multi-Domain Systems
on Chips

Steven P. Levitan, Donald M. Chiarulli, Timothy P. Kurzweg, Jose A.
Martinez, Samuel J. Dickerson, Michael M. Bails, David K. Reed, and
Jason M. Boles

CONTENTS

20.1 Introduction

In the last several years there has been much success in the realm of multi-domain, mixed-signal system on chip (SoC) technology. Devices ranging from heterogeneous multi-core processors to micro-electromechanical systems (MEMS) to labs-on-chips are becoming highly integrated into chip-scale packages. However, the complexity of this multi-technology integration increases the difficulty of verifying such systems. Since different technology domains, such as electrical (digital and analog), optical (free-space and fiber), and mechanical (micro and macro), coexist in one package, there has emerged a need for tools that can verify such heterogeneous systems. For these integrated micro-systems the goal is to model large numbers of both linear and nonlinear components with sufficient speed and accuracy to explore the design space at the system level. Beyond functional design, mixed-technology tools, working at the system level, must support the traditional models of performance (e.g., speed, power, and area) as well as the special needs of mixed-technology systems. This means being able to analyze such things as crosstalk, noise, and mechanical tolerance in an interactive environment, and leads to the requirement of a computationally efficient yet accurate mixed-technology simulation framework. These problems are exacerbated by the need to model the behavior of the controlling digital hardware and/or software and the feedback between these two worlds. Most importantly, the tools must be able to capture the interaction of these realms in order to support the designer in making both architectural and technological tradeoffs.

These requirements emphasize the need for high-level models for optical, electronic, and electromechanical components, accurate and computationally efficient analog simulation, and an interface to traditional digital simulation and embedded software development tools. To date, no single CAD tool has been able to completely model the complexity of these multi-domain systems on chips (MDSoCs). Current MDSoC design methodology is to use a variety of "point tools" for each domain present in the design, and then stitch together the results using an additional tool. This process is both time consuming and inefficient.

Therefore, the need to perform high-level system simulations in a single framework has driven both academia and industry to the development of "system simulation" environments. Since most of these support top-down design, the focus is on hardware and software codesign and verification. Some examples of commercially available products include Seamless from Mentor Graphics, Incisive Simulator Products from Cadence, and MultiSim from Electronics Workbench. Many academic tools have also been developed such as Ptolemy [1] from the University of California at Berkeley, Pia [2] from the University of Washington, and CoSim [3] from the TIMA Laboratory at the Institut National Polytechnique Grenoble and SGS-Thomson.

Most of these simulation environments target hardware–software co-simulation and rely on other simulators to perform tasks such as mechanical finite element analysis (FEA), optical propagation analysis (e.g., RSoft), and circuit-level simulation (e.g., SPICE). Some of the modern tools, such as System Vision from Mentor Graphics, allow for complete system modeling and simulation in mechanical and mixed-signal electrical domains, but do not support optical or fluidic systems.

In this chapter we introduce a tool that can simulate and thus verify the behavior of MDSoCs from the system architectural level down to the physical level of such technologies. This is accomplished by using the mixed-domain, mixed-signal simulation environment, Chatoyant [4], and a commercial mixed-language HDL simulator, ModelSim. The combination of these two simulators is accomplished using UNIX-style inter-process communication (IPC) as an implementation method for parallel discrete event (DE) simulation.

The methodology here is similar to the work presented in an earlier chapter by Lee and Zheng [5]. However, we have developed our models hierarchically such that lower-level "component" models support continuous time semantics, while composition of those models is done with discrete time semantics. Components pass complex messages among themselves under a global simulation framework. Similar to the ideas of Gheorghe, Nicolescu, and Boucheneb in this volume [6], message semantics are defined by common message classes. Additionally, conversion between these analog models and the multivalued digital models of a hardware description language (HDL) simulator is mediated by a set of predefined semantic rules.

The rest of this chapter is organized as follows: It begins with the investigation of methods for modeling digital free-space optoelectronic systems. These are systems that incorporate electronic digital and analog components, optoelectronic interface devices, such as laser and detector arrays, and free-space optical interconnects that are composed of passive and active optical elements. These models have been successfully incorporated into an optoelectronic system-level design tool called Chatoyant [4,7–9]. We present the features of Chatoyant that are useful in the modeling, simulation, and analysis of MDSoCs. Next, we introduce electrical, mechanical, and optical models that are used as building blocks in multi-domain system design. We then present the analog/digital co-simulation environment and discuss issues in synchronization and signal conversion between the analog domains, managed by Chatoyant, and the digital domain, managed by ModelSim. Finally, we show the utility of the co-simulation environment with several example systems.

20.2   Chatoyant Multi-Domain Simulation

Chatoyant is a mixed-domain, mixed-signal simulation environment developed at the University of Pittsburgh. It is capable of simulating MEMS and optical MEMS or MOEMS at a system and architectural level. This permits design space exploration by examining the effects of variations in component parameters on system performance and the interaction of these components across technology domains. For example, one can model the small adsorption of optical power in a MEMS mirror, and how that power, as heat, causes the mirror to deform. That deformation, in turn, could degrade the quality of the analog signal that is modulating the light beam used for chip-to-chip communications between a processor and L3 cache in a 3D optoelectronic package. Of course, the degraded signal could be recovered with good analog circuitry, but it could also have error correcting codes embedded in digital data. Some typical questions a system-level designer would ask in this case are these: Where should they invest more design, fabrication effort, and product expense? Should it be better mirrors, lower power optics with better analog signal processing, or more bits of ECC code? Design exploration and tradeoff analysis of this nature motivated the development of Chatoyant.

20.2.1   System Simulation in Chatoyant

The Chatoyant environment is a series of multipurpose libraries that are built upon the Ptolemy framework from the University of California, Berkeley. Ptolemy provides the basic infrastructure for different domains of simulation such as dynamic data flow (DDF), static data flow (SDF), and discrete event (DE). Chatoyant builds upon the simulation domains provided by Ptolemy by adding components that perform analog netlist simulation, optical modeling and analysis, and mechanical elemental analysis [10].

Chatoyant is based on a methodology of system-level architecture design. In this methodology, architectures are defined in terms of models for "modules," the "signals" that pass between them, and the "dynamics" of the system behavior. For electrical, mechanical, and optical systems, signals are represented as electronic waveforms, mechanical deformations, and modulated carriers, (i.e., beams of light). Using the characteristics of these signals, we define models for the system components in terms of the manner in which they transform the characteristic parameters of these signals. Chatoyant's component models are written in C++ with sets of user-defined parameters for the characteristics of each module instance.

Component models are based on three modeling techniques. The first is a "derived model" technique where analytic models are used based on an underlying physical model of the device. These can be very abstract "0th-order" models, or more complex models involving time varying functions, internal state, or memory. The second class of models is based on empirical measurements from fabricated devices. These models use measured data and interpolation techniques to directly map input signal values to output values. The third technique is reduced-order or response surface models. For these models, we use the results of low-level simulations, such as finite element solvers, or simulators, and generate a reduced-order model, that covers the range of operating points for the component. We have implemented this technique using a variety of methods from a polynomial curve fit, or simple interpolation over the range of operation, to nonlinear model order reduction [11,12].

We have successfully used all three of these methods to create four component libraries: The optoelectronic library, which includes devices such as vertical cavity surface emitting lasers (VCSELs), multiple quantum well (MQW) modulators, and p-i-n detectors. The optical library contains components such as refractive and diffractive lenses, lenslets, mirrors, and apertures. The electrical library includes CMOS drivers and transimpedance amplifiers, and the mechanical library contains beams, plates, and mechanical assemblies such as scratch drive actuators (SDA) and deformable mirrors.

Signal information is carried between modules using a C++ "message class." To maximize our modeling flexibility, the signals in Chatoyant are composite types, representing the attributes of position and orientation for both optical and mechanical signals, voltages and impedances for electronic signals, and wave front, phase, and intensity for optical signals. The composite type is extensible, allowing us to add new signal characteristics as needed. The advantage of using such a class is that one single message contains optical, electrical, or mechanical information, and each component type-checks the data, extracting the relevant information. The message class also carries time information for each message in the stream of data.

The DE simulation scheduler allows modeling of multi-dynamic systems where every component can alter the rate of consumed/produced data during simulation. The scheduler also provides the system with buffering capability. This allows the system to keep track of all the messages arriving at one module when multiple input streams of data are involved.

Before the discussion of individual signal models and to further understand the development of our system-level simulation tool, we first introduce our device and component modeling methodology.

20.2.2 Device and Component Models

In our methodology, we make a distinction between device-level and component-level modeling. Device-level models focus on explicitly modeling the processes within the physical geometry of a device such as fields, fluxes, stresses, and thermal gradients. For component-level models these distributed effects are characterized in terms of device parameters, and the models focus on the relationships between these parameters and state variables (e.g., optical intensity, phase, current, voltage, displacement, or temperature) as a set of linear or nonlinear differential equations. In the electronic domain these are often called "small signal models" or "circuit models."

Circuit-level (or more generally, component-level) modeling techniques can be used for optoelectronic device modeling, but, for most models, the degree of accuracy does not match that required for performance analysis in these types of devices. Fast transient phenomena, the dependencies on the physical geometry of the device, and large-signal operation are generally not well characterized by these kinds of models. Device-level simulation techniques offer the degree of accuracy required to model fast transients (e.g., chirp), fabrication geometry dependencies, and steady-state solutions in the optical device [13]; however, modeling these processes requires specialized techniques and large computational resources that produce results that are not compatible with simulators required for other domains. For instance, it is difficult to model the behavior of a laser in terms of carrier population densities, and at the same time, the emitted light in terms of electromagnetic field propagation.

There are two obvious techniques to deal with this problem of device simulation versus circuit simulation. The first is the use of two levels of simulation, a device-level simulation for each unique domain, coupled to a common circuit-level simulation that coordinates the results of each. However, for the case of device and circuit co-simulation, this technique has all the drawbacks previously mentioned for the device-level simulation with the additional computational resources required to coordinate analog simulators, which means not just in making time-stamps match but to force them to converge to a common point of operation [14,15].

Rather, our approach is to increase the accuracy of the circuit-level (component-level) models. That is, to incorporate the transient solution, along with other second order effects, of the device analysis within the circuit-level simulation. This is accomplished by creating circuit models for these higher order effects and incorporating them into the circuit model of the optoelectronic device. Different methodologies can be used to translate the device-level expressions, which characterize the semiconductor device operation (e.g., Poisson's, carrier current, and carrier continuity equation) into a set of temporal linear/nonlinear differential equations [13,16]. The advantage of having this representation is that we can simulate electronic and optoelectronic models in a single mixed-domain component-level simulator.

These enhanced component models can then fit in a DE simulation engine, since convergence of the analog models is compartmentalized in each device. The result is an abstract representation of the system consisting of a set of loosely coupled modules interchanging information as energy signals. However, this approach brings the challenge of choosing which circuit/component modeling techniques will be optimal for accurate and fast characterization of the different modules involved in this system.

20.2.3 Simulation Issues

Traditional circuit simulators based on numerical integration solvers offer the required accuracy to solve linear and nonlinear DE systems; however, they are too computationally expensive to consider for evaluating individual modules in a mixed-domain framework [17,18]. In the linear case, successful low order reduction techniques have been used to model high-density interconnection networks with excellent computational efficiency [19–22]. In the nonlinear case, however, the success is only partial. Work has been conducted to apply reduction techniques to obtain macro-models for the interconnection section and use them in circuit simulators, such as SPICE [23], as a way to simplify the computational task carried out by such solvers [18,22]. Merging both techniques maintains the accuracy offered by circuit simulators, but also the problems associated with them.

Two problems with this technique are the difficulties guaranteeing the convergence of the solution and the relatively high computational load. Pioneer nonlinear network modeling using piecewise models in a timing simulator RSIM [24] was conducted by Kao and Horowitz [25]. While well suited for delay estimation in dense nonlinear networks, the limited complexity of models and tree analysis technique used do not allow piecewise linear (PWL) timing analyzers to simulate higher order effects that are of significant importance in the modeling of typical optoelectronic devices.

The fact that the density of the network generated for modeling of our optoelectromechanical devices is moderate allows us to consider PWL modeling merged with linear numerical analysis as a way to achieve the desired accuracy with a lower computational demand. More importantly, the amount of feedback between active devices in such models is limited when compared with dense VLSI networks, which makes the scheduling task feasible even for increased numbers of regions of operation for each device.

For simulation, we perform a linear numerical analysis in order to solve the differential equation necessary to obtain an accurate solution, using piecewise modeling to overcome the iteration process encountered in the integration technique used in traditional circuit simulators for the nonlinear case. Linearizing the behavior of the nonlinear devices by regions of operation simplifies the computational task to solve the system. This also allows us to trade accuracy for speed. Most importantly, PWL models for these devices allow us to integrate mechanical, electrical, and optical components in the same simulation. We have successfully used this technique to model electric, optical, and mechanical components, and are currently expanding this same methodology to incorporate fluid models. These models will be discussed in the next section.

20.2.4 Electrical and Optoelectronic Models

Our optoelectronic modeling is accomplished as shown in Figure 20.1. Given a device, such as an optical transmitter, we perform linear and nonlinear sub-block decomposition of the circuit model of the device. This decomposes the design into a linear multiport subblock section and nonlinear subblocks. The linear multiport subblock can be thought of as characterizing the interconnection network or parasitics while the nonlinear subblocks characterize the active devices.

Then, modified nodal analysis (MNA) [26] is used to create a matrix representation for the device, as shown in Figure 20.2. In this figure, [S] is the storage element matrix, [G] is the conductance matrix, [x] is the vector of state variables, [b] is a connectivity matrix, [u] is the excitation vector, and [I] is the current vector.

The linear subblock elements can be directly matched to this representation, but the nonlinear elements need to first undergo a further transformation. We perform piecewise modeling of the active devices for each nonlinear subblock. When we form each nonlinear subblock, an MNA template is used for each device in the network. The use of piecewise models is based on the ability to change these models for the active devices depending on the changes in conditions in the circuit, and thus the regions of operation.

The templates generated can be integrated to the general MNA containing the linear components adding their matrix contents to their corresponding counterparts. This process is shown in Figure 20.2 for the S matrix. This same composition is done for the other matrices. The size of each of the

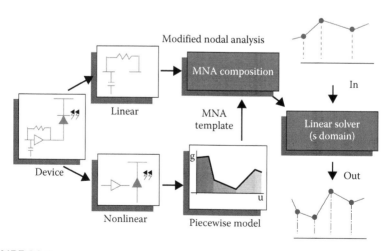

FIGURE 20.1
Piecewise modeling for electrical/optoelectrical devices. (From Kurzweg, T.P. et al., *J. Model. Simul. Micro-Syst.*, 2, 21, 2001. With permission.)

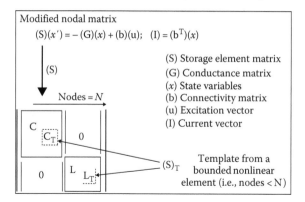

FIGURE 20.2
MNA representation and template integration. (From Kurzweg, T.P. et al., *J. Model. Simul. Micro-Syst.*, 2, 21, 2001. With permission.)

template matrices is bounded by the number of nodes connected to the nonlinear element.

Once the integrated MNA is formed, a linear analysis in the frequency domain can be performed to obtain the solution of the system. Constraining the signals in the system to be piecewise in nature allows us to use a simple transformation to the time domain without the use of costly numerical integration.

During each time step in the simulation, the state variables in the module will change and might cause the active devices to change their modes of operation. Therefore, we recompute and recharacterize the PWL solution caused by changes between piecewise models. Depending on the number of segments used in the PWL model, on average there will be a large number of time steps during which the system representation is unchanged, justifying the computational savings of this technique.

Understanding that the degree of accuracy of PWL models depends mainly on the step size chosen for the time base, we have incorporated an adaptive control method for the simulation time step [26]. A binary search over the time step interval is the basis for this dynamic algorithm. The algorithm discards nonsignificant samples, which do not appreciatively affect the output, and adds samples when the output change is greater than a user-defined tolerance. The inclusion of the samples during fast transitions or suppression of time-points during "steady-state" periods optimizes the number of events used in the simulation.

To support the interaction of electrical models, Chatoyant's message class contains parameters that represent general electrical signals that are passed between electrical devices. Three parameters that are in the message class for electrical signals are output potential, V, capacitance, C_{sb}, and conductance, g_{sb}. The last two fields define the output impedance of the signal, providing a model of loading between electrical devices. We next show an example of how we use our electrical technique in the modeling of CMOS circuits.

20.2.4.1 Example Modeling of CMOS Circuits

To illustrate our modeling of the active optoelectronic devices in modular networks, we focus on CMOS driver circuits based on the simple complementary inverter. Considering the classical nonlinear V–I equations for MOS transistors (Level II) as characterizing the behavior of every FET device, a linearization of drain-source current (I_{ds}) is presented using

$$\Delta I_{ds} = g_m(P)\Delta v_{gs} + g_{ds}(P)\Delta v_{ds} \tag{20.1}$$

where P represents the PWL region of operation for the device. Transconductance (g_m) and conductance (g_{ds}) are the parameters characterizing the device. In Figure 20.3a, the parasitic effects (C_{ds}, C_{gs}, and C_{ds}) are introduced. An MNA template is created from this representation and is shown in Figure 20.3b. This MNA formulation allows us to incorporate the FET as a three-port element into the MNA of a complete optoelectronic module. The nonlinear nature of the FET is modeled by piecewise changes in values of the parameters (g_{ds}, g_m, C_{ds}, C_{gs}, and C_{ds}) depending on the region of operation which are functions of v_g, v_d, and v_s.

To show the speed and accuracy of the PWL approach, we performed several experiments comparing our results to that of SPICE 3f4 (Level II). The test was a multistage amplifier with a significant number of drivers. PWL models were tested versus SPICE at 10 and 1000 MHz. Figure 20.4 shows that the speed-up achieved for the same number of time-points is at least two orders of magnitude compared to SPICE. Accuracy was less than 10% RMS error. These results show that PWL models are well suited to perform accurate and fast simulations for the typical multistage CMOS drivers and transimpedance amplifiers widely used in optoelectronic applications. In the next section, we show how this same procedure for modeling electronic signals can be extended for modeling mechanical structures.

20.2.5 Mechanical Models

The general module for solving sets of nonlinear differential equations using PWL can be used to integrate complex mechanical models in our design tool. The model for a mechanical device can be summarized in a set of differential equations that define its dynamics as a reaction to external forces. This model must then be converted to the form seen in the electrical case to be given to the PWL solver for evaluation.

In the field of MEM modeling, there has been an increasing amount of work that uses a set of ordinary differential equations (ODEs) to characterize MEM devices [27–29]. ODE modeling is used instead of techniques such as finite element analysis, to reduce the time and amount of computational

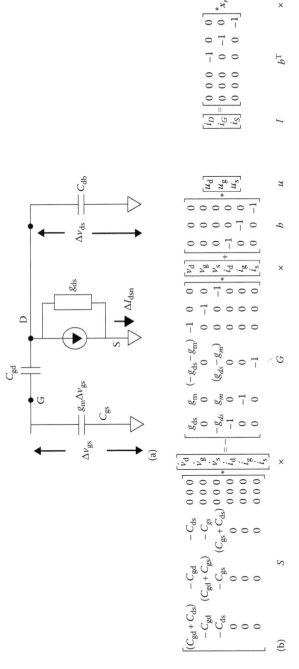

FIGURE 20.3

MOSFET (a) model, (b) template. (From Kurzweg, T.P. et al., *J. Model. Simul. Micro-Syst.*, 2, 21, 2001. With permission.)

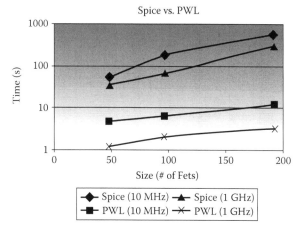

FIGURE 20.4
Spice versus PWL models in a system of multiple FETs.

resources necessary for simulation. The model uses nonlinear differential equations in multiple degrees of freedom and in mixed domains. The technique models a MEM device by characterizing its different basic components such as beams, plate-masses, joints, and electrostatic gaps, and by using local interactions between components.

Our approach to modeling mechanical elements is to reduce the mechanical ODE representation to a form matching the electronic counterpart, seen in the equation in Figure 20.2. This enables the use of the PWL technique previously discussed for simulating the dynamic behavior of electrical systems.

With damping forces proportional to the velocity, the motion equation of a mechanical structure with viscous damping effects is

$$F = [K]U + [B]V + [M]A \tag{20.2}$$

where
  $[K]$ is the stiffness matrix
  U is the displacement vector
  $[B]$ is the damping matrix
  V is the velocity vector
  $[M]$ is the mass matrix
  A is the acceleration vector
  F is the vector of external forces affecting the structure

Obviously, knowing that the velocity is the first derivative and the acceleration is the second derivative of the displacement, the above equation can be recast to

$$F = [K]U + [B]U' + [M]U'' \tag{20.3}$$

Similar to the electrical modeling, this equation represents a set of linear ODEs if the characteristic matrices $[K]$, $[B]$, and $[M]$ are static and independent of the dynamics in the body. If the matrixes are not static and independent (e.g., with aerodynamic load effects), they represent a set of nonlinear ODEs.

To reduce the above equation to a standard form, we use a modification of Duncan's reduction technique for vibration analysis in damped structural systems [30]. This modification allows for the general mechanical motion equation to be reduced to a standard first order form, similar to Equation 20.1, which allows for the complete characterization of a mechanical system.

$$\begin{bmatrix} 0 & M \\ M & B \end{bmatrix} \begin{bmatrix} U'' \\ U' \end{bmatrix} + \begin{bmatrix} -M & 0 \\ 0 & K \end{bmatrix} \begin{bmatrix} U' \\ U \end{bmatrix} = \begin{bmatrix} 0 \\ I \end{bmatrix} F \qquad (20.4)$$

Using substitutions, the equation is rewritten as

$$[Mb]\, X' + [Mk]\, X = [E]\, F \qquad (20.5)$$

where the new state variable vector $X = \begin{bmatrix} U' \\ U \end{bmatrix}$.

Each mechanical element (beam, plate, etc.) is characterized by a template consisting of the set of matrices $[Mb]$ and $[Mk]$, composed of matrices $[B]$, $[M]$, and $[K]$ in the specified form seen above. If the dimensional displacements are constrained to be small and the shear deformations are ignored, the derivation of $[Mb]$ and $[Mk]$ is simplified and independent of the state variables in the system. Additionally, the model for elements is formulated assuming a one-element idealization (e.g., two nodes for a beam). Consequently, only the static resonant mode is considered. Multiple-element idealization can be performed combining basic elements to characterize higher order modes.

As an example of our mechanical modeling methodology, we present the response of an anchored beam in a 2D plane with an external force applied on the free end. The template for the constrained beam is composed of the following matrices [31]:

$$K = \frac{EI_z}{l^3} \begin{bmatrix} \frac{Al^2}{I_z} & 0 & 0 \\ 0 & 12 & -6l \\ 0 & -6l & 4l^2 \end{bmatrix} ; \quad M = \frac{\rho Al}{420} \begin{bmatrix} 140 & 0 & 0 \\ 0 & 156 & -22l \\ 0 & -22l & 4l^2 \end{bmatrix} ;$$

$$B = \delta \begin{bmatrix} 1 & 0 & 0 \\ 0 & 1 & 0 \\ 0 & 0 & 0 \end{bmatrix} \qquad (20.6)$$

where
E is Young's modulus
I_z is the inertia momentum in z

A is the area of the beam
l is the length
ρ is the density of the material
δ is the viscosity factor in the system acting over x and y components

The analysis of this element is obtained using the PWL technique presented above. Constraining the input/output signals to PWL waveforms, the time domain response is completed in one step, without costly numerical integration.

To test our results, a comparison against NODAS [28] was performed. Figure 20.5 shows the frequency response and corresponding resonant frequencies for this constrained beam (183 μm length, 3.8g μm width, poly-Si) from both our PWL technique and NODAS. The transient response to 1.8 nN nonideal step (rise time of 10 μs) rotational torque is also simulated. The rotational deformation to this force is shown in Figure 20.6. The comparison between our results and those of NODAS are very close. NODAS uses SABER, a circuit analyzer performing numerical integration for every analyzed point, which results in costly computation time. Our linear piecewise solver is computational expensive during the eigenvalue search; however, this procedure is performed only one time, at the beginning of the simulation run. Overall, this results in a more computationally efficient simulation. However, as previously mentioned, the accuracy of the analysis depends on the granularity of the piecewise characterization for the signals used in the system, which can increase computation time.

Typically, this beam is only a part of a bigger device made from individual components that are characterized using similar expressions.

	Resonant frequencies	
	f_1	f_2
NODAS	154.59 KHz	1.52 MHz
PWL simulator	150.04 KHz	1.48 MHz

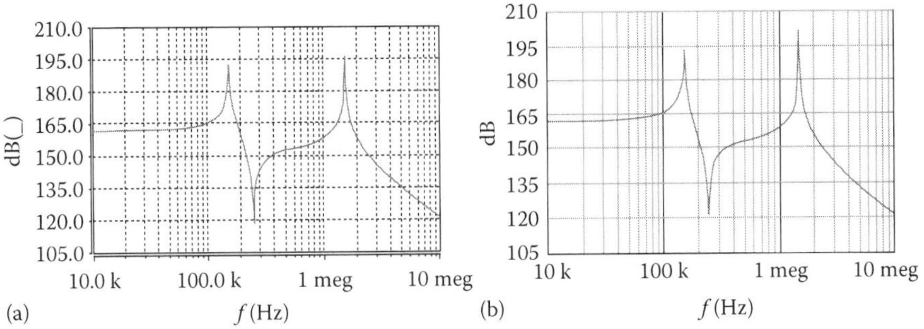

(a) f (Hz)   (b) f (Hz)

FIGURE 20.5
Frequency response of a beam (a) NODAS, (b) Chatoyant.

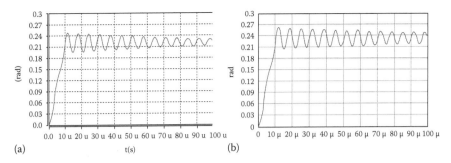

FIGURE 20.6
Transient response of a BEAM (a) NODAS, (b) Chatoyant.

The generalization of the previous case to an assembly of elements or mechanical structure is fairly straightforward. The general expression, seen in Equation 20.3, characterizes the whole structure defined by a set of nodes, from which every individual element shares a subset. The next step, similar to the previously considered electronic case, is merging the individual templates together, composing the general matrix representation for the composed structure. However, a common coordinate reference must be used for this characterization of mechanical structures since every template or element is characterized in a local reference system. The process of translation of these local templates to the global reference system can be described by [31]

$$[S] = [A]^T[\bar{S}][A] \tag{20.7}$$

where
  [A] represents the translation matrix from local displacements to global displacements (a function of the structure's geometry)
  [$\bar{S}$] represents the local template
  [S] is the corresponding global representation

The next step is the addition of these global representations into the general matrix form, using the matrices' nodal indexes as reference. Finally, the PWL solver can be used on the composed system's general matrix and simulated.

The use of a PWL general solver for mechanical simulation decreases the computational task and allows for a tradeoff between accuracy and speed. The additional advantage of using the same technique to characterize electrical and mechanical models allows us to easily merge both technologies in complex devices that interact in mixed domains.

20.2.6 Optical Propagation Models

Our optical propagation models are based on two techniques: Gaussian and diffractive scalar. Gaussian models give us fast, accurate results for macroscale systems and systems that exhibit limited diffraction. Slower diffractive scalar models must be used when diffraction effects dominate the system.

20.2.6.1 Gaussian Models

Our macroscale optical signal propagation technique is based on Gaussian beam analysis, allowing paraxial light to be modeled by scalar parameters, and components to be modeled by an optical "ABCD matrix," which describes how light is affected by a component [32]. The Gaussian beam is defined by the nine parameters seen in Table 20.1.

As the parameters in the table indicate, our optical propagation is actually a mixture of ray analysis and Gaussian analysis. We first find the position and direction of the center of the Gaussian beam, using ray propagation methods. We then "superimpose" the Gaussian beam over this ray to model its intensity, waist, phase, and depth of focus. The advantage of using this combination of ray and Gaussian beam analysis is its computational efficiency. The resulting simulation speed supports interactive system-level design.

The nine scalar parameters defining the optical signal are represented in Chatoyant's message class and passed between components. Each component "constructs" an internal model of the beam from these parameters, alters the beam according to the component function, and then "decomposes" the beam back into the propagation parameters, which are returned to the message class and passed to the next object. The internal model of the component might simply consist of beam steering, as below, or require optical power integration, imaging, or optoelectronic conversion.

Using Gaussian beam propagation, components are modeled with the previously mentioned ABCD matrix. For example, we examine the interaction between a Gaussian beam and a thin lens. To study the beam/lens interaction, we start with a definition of the Gaussian beam's q-parameter, which characterizes a Gaussian beam of known peak amplitude [32]:

$$q = z_{w0} + jz_0 \tag{20.8}$$

TABLE 20.1

Gaussian Beam Parameters

Parameter	Description
x, y	Central position of the Gaussian beam
Rho, theta	Directional cosines of the Gaussian beam
Intensity	Peak intensity of the Gaussian beam
z_0	Rayleigh range, depth of focus
z_{w0}	Distance to the next minimum waist
Lambda	Wavelength of the light
Phase	Phase of the central peak of the beam

where the real part is the distance to the minimum waist, and the imaginary is the Rayleigh range, from which the waist of the beam is determined. The new Gaussian beam is defined by the following:

$$q_2 = \frac{Aq_1 + B}{Cq_1 + D} \tag{20.9}$$

where A, B, C, D is the matrix that defines a component. In the case of a thin lens, $A = 1$, $B = 0$, $C = -1/f$, $D = 1$, where f is the focal length of the lens. Solving for q_2, and determining the real and imaginary parts, the new z_0' and zw_0' for the emerging Gaussian beam can be found:

$$z_0' = \frac{f^2 \cdot z_0}{(f - zw_0)^2 + z_0^2} \tag{20.10}$$

$$zw_0' = \frac{f(f \cdot zw_0 - zw_0^2 - z_0^2)}{(f - zw_0)^2 + z_0^2} \tag{20.11}$$

The position and direction of the beam is determined from common ray tracing techniques:

$$y_2 = Ay_1 + B\theta_1 \qquad \theta_2 = Cy_1 + D\theta_1 \tag{20.12}$$

However, as the systems that we wish to design continue to diminish in size, diffractive effects are a major concern. For example, in optical MEM design, the size of the components, apertures, and small structures bring diffractive effects into play, along with the use of diffractive elements such as Fresnel zone plates, binary lenses, gratings, and computer generated holograms (CGH) [33]. Therefore, new design methods are needed that utilize optical models that can provide accurate diffractive results with reasonable computational costs. In addition to diffractive effects, other characteristics of optical signals are important, such as polarization, scattering, phase, frequency (wavelength) dependence, and dispersion, this last being a requirement for modeling fiber optic components.

20.2.6.2 Scalar Diffractive Models

To identify which modeling technique is best suited for our needs, we need to analyze the MDSoCs that we wish to model and evaluate the available optical propagation techniques. Current optical MEM systems have component sizes of roughly tens to hundreds of microns and propagation distances in the hundreds of microns. With these sizes and distances on the order of ten to a thousand times the wavelength of light, optical diffractive models are required.

Figure 20.7 is a description of models of increasing abstraction that begins at the top with the fundamental vector wave equations, or Maxwell's equations, and branches through the different abstraction levels of scalar modeling techniques. Along the arrows, notes are added stating the limitations and approximations that are made for each formulation.

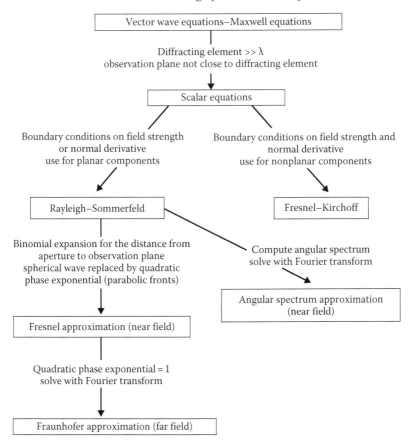

FIGURE 20.7
Scalar modeling techniques.

The size and scale of the optical components of MDSoC systems of at least 10 times greater than the wavelength of light lead to the use of scalar diffraction models. Scalar equations are directly derived from Maxwell's equations. Maxwell equations, with the absence of free charge, are [34–36]

$$\nabla \times \vec{E} = -\mu \frac{\partial \vec{H}}{\partial t} \qquad \nabla \times \vec{H} = -\varepsilon \frac{\partial \vec{E}}{\partial t} \qquad \nabla \cdot \varepsilon \vec{E} = 0 \qquad \nabla \cdot \mu \vec{H} = 0 \quad (20.13)$$

These equations can be recast into the following form:

$$\nabla^2 \vec{E} - \frac{n^2}{c^2} \frac{\partial^2 \vec{E}}{\partial t} = 0 \qquad \nabla^2 \vec{H} - \frac{n^2}{c^2} \frac{\partial^2 \vec{H}}{\partial t} = 0 \qquad (20.14)$$

If we assume that the dielectric medium is linear, isotropic, homogeneous, and nondispersive, all components in the electric and magnetic field can be summarized by the scalar wave equation:

$$\nabla^2 \vec{U} - \frac{n^2}{c^2}\frac{\partial^2 \vec{U}}{\partial t} = 0 \qquad (20.15)$$

For monochromatic light, $U(P,t)$ is the positional complex wave function, where P is the position of a point in space:

$$U(P,t) = a(P)e^{j\varphi(P)}e^{j2\pi v} \qquad (20.16)$$

By placing the positional complex wave function into the scalar wave equation, the result is the time-independent Helmholtz equation, which must be satisfied by the scalar wave:

$$(\nabla^2 + k^2)U(P) = 0 \qquad (20.17)$$

where $k = \frac{2\pi}{\lambda}$

The challenge is to determine the scalar wave function as it propagates through a diffractive element. One answer is based on the Huygens–Fresnel principle that states that every unobstructed point of a wave front at a given time serves as a source of spherical wavelets with the same frequency as the primary wave. The Huygens–Fresnel principle is mathematically described by the Rayleigh–Sommerfeld scalar diffraction formulation:

$$U_2(x,y) = \frac{z}{j\lambda}\iint U_1(\xi,\eta)\frac{e^{jkr_{12}}}{r_{12}^2}\partial\xi\partial\eta$$

where
ξ and η are the coordinates of the aperture plane
x and y are the coordinates of the observation plane

All scalar diffraction solutions are limited by two assumptions; the diffracting structures must be "large" compared to the wavelength of the light, and the observation screen cannot be "too close" to the diffracting structure. However, these dimensions are not clearly defined, questioning if scalar optical models are valid for all micro-optical systems. For some extremely small optical systems, our initial intuition of "adequate" scalar models might be invalid and full wave propagation models must be used.

When modeling scalar formulations, explicit integration of the wave front is performed at each interface, severely increasing the computation time. Using approximations to the scalar formulations, as seen in Figure 20.7, can reduce this time. For example, the Fraunhofer approximation is solved using a Fourier transform, where common FFT algorithms enable an efficient solution. However, the valid propagation ranges limit when these approximations can be used. Figure 20.8 shows where these different modeling techniques are valid with respect to the distance propagated past a diffracting element.

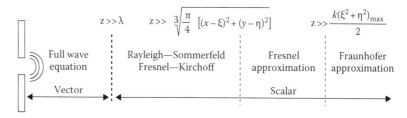

FIGURE 20.8
Valid propagation distances of scalar modeling techniques. (From Kurzweg, T.P. et al., *J. Model. Simul. Micro-Syst.*, 2, 21, 2001. With permission.)

Supporting diffractive propagation in Chatoyant requires additional parameters in the message class. Therefore, the class contains the user's requested optical propagation method (Gaussian, or scalar diffractive), along with the complex wave front of the beam as it propagates through the system. The wave front is gridded, defining the degree of accuracy of the model of the wave. As with the Gaussian propagation, it is the component model that alters the wave front as the component interacts with the light beam and returns the result in an outgoing message.

We implemented the Rayleigh–Sommerfeld diffractive formulation using a 96-point Gaussian quadrature method for our integration technique. In Figure 20.9, we show simulation results of an 850 nm plane wave

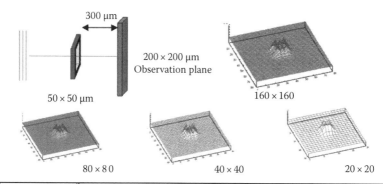

	Chatoyant		MathCAD	
	Time (min)	Error (%)*	Time (min)	Error (%)*
160 × 160	17.75	X	X	X
80 × 80	4.45	0.637	120	0
40 × 40	1.1	1.67	20	1.54
20 × 20	0.29	3.37	7	4.32

*RMS error of grid cells with respect to 80 × 80 MathCAD

FIGURE 20.9
Computation time versus accuracy using Chatoyant's scalar propagation models. (From Kurzweg, T.P. et al., *J. Model. Simul. Micro-Syst.*, 2, 21, 2001. With permission.)

propagating though a 50 μm aperture and striking an observation plane 300 μm away. We compare our simulations with an 80 × 80 "base case" from MathCAD, which uses a Romberg integration technique. The table in Figure 20.9 shows the computation time and relative error of the system (compared with the base case) for different grid spacing. Using our integration technique, we can decrease the computation time an order of magnitude and still remain within 2% accuracy.

20.2.6.3 Angular Spectrum Technique

As an alternative to direct integration over the surface of the wave front, the Rayleigh–Sommerfeld formulation can also be solved using a technique that is similar to solving linear, space invariant systems. In this case, the complex wave front is analyzed across its surface with a Fourier transform. By using the Fourier transform, the complex optical wave front is reduced into a set of simple linear exponential functions. This transform identifies the components of the angular spectrum, which are plane waves traveling in different directions away from the surface [35].

Examining the angular spectrum, we look at light propagating from an aperture plane at $z = 0$ to a parallel observation plane. The wave function $U(x,y,0)$ has a 2D Fourier transform, $A(v_x,v_y,0)$, in terms of angular frequencies, v_x and v_y:

$$A(v_x, v_y, 0) = \iint U(x, y, 0) \exp[-j2\pi(v_x x + v_y y)] \, \partial x \partial y$$

where $v_x = \sin\theta_x/\lambda$ and $v_y = \sin\theta_y/\lambda$.

$\sin(\theta_x)$ and $\sin(\theta_y)$ are the directional cosines of the plane wave propagating from the origin of the coordinate system, as seen in Figure 20.10. A is the complex amplitude of the plane wave decomposition defined by the specific angular frequencies.

To propagate the complex wave function to a parallel plane, a propagation phase term is used as a transfer function. The relationship of propagation in the frequency domain between $A(v_x,v_y,0)$ and $A(v_x,v_y,z)$ has been computed by satisfying the Helmhotz equation with the propagated complex wave function, $U(x,y,z)$ [35]:

$$A(v_x, v_y, z) = A(v_x, v_y, 0) \exp\left\{ jz2\pi\sqrt{\frac{1}{\lambda^2} - v_x^2 - v_y^2} \right\}$$

This describes the phase difference that each of the plane waves, differentiated by the angular, or spatial frequencies, experiences due to the propagation between the parallel planes. Therefore, the wave function after propagation can be transformed back into the spatial domain with the following inverse Fourier transform:

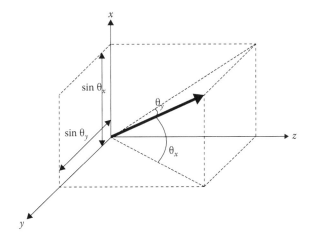

FIGURE 20.10
Angular spectrum frequencies.

$$U(x,y,z) = \iint A(v_x, v_y, 0) \exp\left\{ jz2\pi \sqrt{\frac{1}{\lambda^2} - v_x^2 - v_y^2} \right\}$$
$$\times \exp[j2\pi(v_x x + v_y y)]\, \partial v_x \partial v_y$$

It is interesting to note that the output complex wave function is simply the convolution of the input wave function and the propagation phase function. Using Fourier transform theory, the convolution in the spatial domain is performed by a multiplication in the frequency domain.

It is appropriate to discuss the physical effect of evanescent waves, which are defined in the case when $1/\lambda^2 - v_x^2 - v_y^2 < 0$. These waves carry very little power and die out in a couple of wavelengths of propagation [37]. In our simulations, we ignore these waves.

The angular spectrum method is restricted to propagation between parallel planes that share a common center. Removing these restrictions has been the goal of recent research. Tommoasi and Bianco have determined how to propagate to a plane that is tilted with respect to the initial plane [38]. Delen and Hooker have determined a way to allow offsets in the observation plane [39]. We summarize these two methods next.

For arbitrary angles between the aperture plane, $U(\xi, \eta, \zeta)$, and the observation plane, $U(x,y,z)$, a mapping of the spatial frequencies in each plane's coordinates system must occur. This mapping is possible due to the fact that the phase accumulation term does not change when the waves propagate to an observation plane that is not normal to the aperture plane. It can be found that the rotational matrix, M, relating (ξ, η, ζ) to (x,y,z), can be used to relate spatial frequencies in the two coordinate systems by [38].

$$(x, y, z)^t = M(\xi, \eta, \zeta)^t \qquad (v_\xi, v_\eta, v_\zeta)^t = M^t(v_x, v_y, v_z)^t$$

In the new tilted coordinate system, the incoming spatial frequencies are perceived as having a spatial frequency corresponding to the outgoing coordinate system. For example, the incoming aperture plane wave having spatial frequencies (0,0) correspond to angle $(-\phi, 0)$ in the observation plane with a ϕ tilt in the y direction. In all cases, even if the spatial frequencies are remapped, the amplitude of the plane wave remains constant.

For an observation plane, whose center is offset from the propagation axis of the aperture plane, the Fourier shifting theorem can be used to solve for the complex wave function [39]. With this relation between the offset of the coordinate systems, the function for free-space propagation between offset planes is

$$U(x, y, 0) = \iint A'(v_x, v_y, 0) \exp \left\{ jz2\pi \sqrt{\frac{1}{\lambda^2} - v_x^2 - v_y^2} \right\}$$

$$\exp[j2\pi(v_x x + v_y y)] \, \partial v_x \partial v_y$$

where $A'(v_x, v_y, 0) = A(v_x, v_y, 0) \exp[j2\pi(v_x(x' - x) - v_y(y' - y)]$

The angular spectrum technique for modeling propagation between the aperture and observation plane is summarized graphically in Figure 20.11. First, the forward Fourier transform is applied to the aperture surface, as seen in Figure 20.11A. In stage B, each plane wave is multiplied by the propagation phase term. If tilts are present, the remapping of spatial frequencies occurs, as denoted by C. If offsets between the planes occur, then the shifting theorem is applied, as shown in step D. Finally, stage E shows the inverse Fourier transform being applied, and the complex wave front on the surface of the observation plane is obtained.

The advantage of using the angular spectrum to model light propagation is that the method is based on Fourier transforms. In CAD tools, the

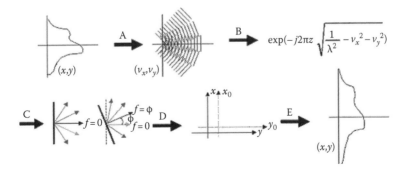

FIGURE 20.11
Angular spectrum frequencies.

Fourier transform can be implemented by one of the numerous fast Fourier transform (FFT) techniques. The computational order of the FFT for a 2D input is $O(N^2\log_2 N)$, obviously more efficient when compared to the direct integration method. We show this speed increase later through an example.

In continuous theory, the angular spectrum method is an exact solution of the Rayleigh–Sommerfeld formulation. However, when solving the algorithm on a digital computer, a discrete Fourier transform (DFT) must be used, resulting in the accuracy of the angular spectrum method being dependent on the resolution, or spacing, of the aperture and observation plane meshing. We call the physical size of the aperture and observation planes the "bounding box," defining the size of the optical wave front being propagated. Since the complex wave function is only nonzero for a finite space in the bounding box, the signal is not always bandwidth limited, and the Nyquist sampling theory does not always apply. It can be shown, however, that the resolution of the aperture and observation meshing must be $\lambda/2$ or smaller [39]. For many simulation systems without large degrees of tilt and hard diffractive apertures, the resolution can be coarser. In systems with high tilts, the resolution is most sensitive. With a mesh spacing of $\lambda/2$, the angular spectrum decomposition will model plane waves propagating from the aperture to the observation plane in a complete half circle, that is, between −90 and +90 degrees.

Other inaccuracies that can occur when using a DFT are aliasing and window truncation. Aliasing occurs when frequencies exist greater than the critical sampling frequency. In this case, these high frequencies are "folded over" into the sampled frequency range [40]. The effect of this is seen in our simulations as optical power "reflecting" off of the walls of the bounding box. If significant optical power reflects off the wall, interference between the propagating beam and these reflections can occur, resulting in inaccurate optical waveforms. The same effect can be seen when the bounding box truncates the signal. Truncation occurs when the waveform propagates into the edges of the bounding box. The simplest solution to ensure accurate results is having sufficient zero padding around the optical waveform, reducing the chance the waveform is aliased or truncated by walls of the bounding box.

In Chatoyant, the user can choose between using the Gaussian or scalar diffractive (angular spectrum) methods during simulation. The components in the optical library support both representations in the optical signal message class. Using these models we can simulate and analyze a variety of heterogeneous systems as presented in the next section.

20.2.7 Simulations and Analysis of Optical MEM Systems

In this section, we show how Chatoyant can model and simulate complete mixed-signal systems. The first system uses both electrical and optical signals to simulate a complete "4f" optoelectronic link which uses a four focal length image relaying optical system. The second example, building from the two signal 4f link, adds mechanical signals for simulation and analysis of an optical MEM system. This set of example systems is centered on an optical MEM scanning mirror. With this device we are able to simulate an optical scanning system and a self-aligning optical detection system. These systems show the ability to model a mixed system of mechanical MEMs, optics, and electronic feedback. The last example shows the power of the angular spectrum technique to model diffractive optical systems with the speed and accuracy required to perform system-level design.

20.2.7.1 Full Link Example

A complete optoelectronic simulation of a 4f optical communication link in Chatoyant is presented in Figure 20.12. The distance between the vertical cavity surface emitting laser (VCSEL) array and the first lens and the distance between the second lens and the detector array are both 1 mm. The distance between the lenses is 2 mm, with both lenses having a focal length of 1 mm, giving a 4f system. The top third of the figure shows the system as represented in Chatoyant. Each icon represents a component model, and each line represents a signal path (either optical or electrical) connecting the outputs of one component to the inputs of the next. Several of the icons, such as the VCSELs and receivers, model the optoelectronic components themselves, while others, such as the output graph, are used to monitor and display the behavior of the system. The input to the system is an electrical signal with speed varying from 300 MHz to 1.5 GHz. A Gaussian noise with variance of 0.5 V has been added to the multistage driver system to show the ability of our models to respond to arbitrary waveforms.

In the center of the figure, three snapshots (before the VCSEL, after the VCSEL, and after the detector) show the behavior of the CMOS drivers under

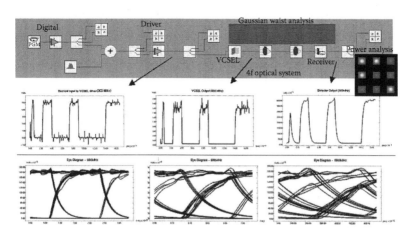

FIGURE 20.12
Chatoyant analysis of optoelectronic 4f communications link.

a 300 MHz noisy signal. In these snapshots, one can see the amplification of the system noise through the CMOS drivers, the clipping of subthreshold noise in the VCSEL, and the frequency response on the quality of the received signal. This last observation is better seen in the three eye diagrams, shown at the bottom of Figure 20.12, analyzed at 300 MHz, 900 MHz, and 1.5 GHz. For the component values chosen, the system operates with reasonable BER up to about 1 GHz.

For this 4f system, the VCSEL and driver circuits explicitly model the effects of bias current and temperature on the optoelectric conversion, L-I efficiency, of the lasers. Figure 20.13 shows the effects of temperature, T, and current bias, I_b, on the bit error rate (BER) of the link. Generally, the frequency response of the link is dominated by the design of the receiver circuit; however it is interesting to note that both the VCSEL temperature and bias have a significant effect on system performance, because of their impact in the power through the link. Perhaps most interesting is the fact that increasing bias current does not always correspond to better performance over the whole range of frequencies examined. Note that the curve for 1 mA bias offers the best performance below 600 MHz; however, the 0.5 mA bias (the nominal threshold of the VCSEL) crosses the curve for 1 mA and achieves the best performance at higher frequencies.

As an example of mechanical tolerancing, we analyze the system with varying-sized photodetectors (50, 30, and 20 μm). The detectors are displaced from ±10 μm to ±100 μm in detector position along the axis of optical propagation. This results in defocusing of the beam relative to the detector array. We calculate both the insertion loss and the worst case optical crosstalk as the detectors are displaced. The results are shown in Figure 20.14. Systems can be further analyzed for their sensitivity to mechanical tolerances using a Monte Carlo tolerancing method described in [8,9].

Two additional analyses are also shown in the Chatoyant representation in Figure 20.12. The first is the beam profile analysis, which graphically displays one beam's waist as it propagates between components, showing the possibility of clipping at the lenses. The second analysis shows the optical signals as they strike the detector array. This analysis also gives the user the amount of optical power captured on each of the detectors. From this analysis, optical crosstalk and system insertion loss can be calculated.

20.2.7.2   Optical Beam Steering/Alignment System

A torsion-scanning mirror is a micromachined 2D mirror built upon a micro-elevator by self assembly (MESA) structure [41,42]. The mirror and MESA structures are shown in Figure 20.15a and b, respectively. The scanning mirror can tilt along the torsion bars in both the x and y directions and is controlled electrostatically through four electrodes beneath the mirror, outlined in Figure 20.15a by the dashed boxes. For example, the mirror tilts in the positive x direction when voltage is applied to electrodes 1 and 2, and the

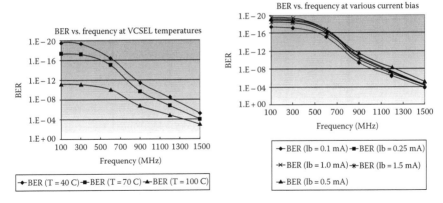

FIGURE 20.13
BER versus frequency at different VCEL temperatures and current biases.

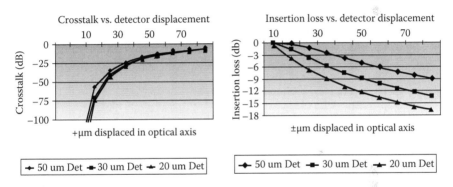

FIGURE 20.14
Insertion and crosstalk versus mechanical tolerancing. (From Kurzweg, T.P. et al., *J. Model. Simul. Micro-Syst.*, 2, 21, 2001. With permission.)

mirror tilts in the negative y direction when voltage is applied to electrodes 1 and 4.

The MESA structure is shown in Figure 20.15b. The mirror is elevated by four scratch drive actuator (SDA) sets pushing the support plates together, allowing for the scanning mirror to buckle and rise up off the substrate [43]. The MESA structure's height is required to be large enough such that the tilt of the mirror will not cause the mirror to hit the substrate. Post fabrication system alignment can also be performed by the MESA structure.

Figure 20.16 shows a drawing of the torsion-scanning mirror system. On the left one can see one VCSEL emitting light vertically through a lenslet, and a prism that reflects off a plane mirror. The light is then reflected off of the optical MEM scanning mirror, back to the plane mirror, and captured through a lenslet and prism onto detectors on the right. With the flexibility of the scanning mirror, this system could act as a switch, an optical scanner,

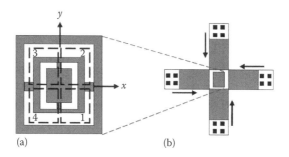

FIGURE 20.15
(a) Scanning torsion mirror, (b) MESA structure. (From Kurzweg, T.P. et al., CAD for optical MEMS, *Proceedings of the 36th IEEE/ACM Design Automation Conference (DAC'99)*, New Orleans, LA, June 20–25, 1999. With permission.)

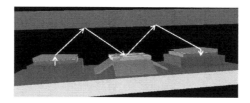

FIGURE 20.16
Scanning mirror system. (From Kurzweg, T.P. et al., CAD for optical MEMS, *Proceedings of the 36th IEEE/ACM Design Automation Conference (DAC'99)*, New Orleans, LA, June 20–25, 1999. With permission.)

or a reconfigurable optical interconnect. We have simulated systems using this scanning mirror configuration for switching and self-alignment through optical feedback. We first demonstrate an optical scanning system.

In this scanning system, we simulate a single source beam propagating through the 3 × 3 subsystem seen in Figure 20.16. With the appropriate voltage levels applied to the four electrodes, the scanning mirror tilts and directs the source to any of the nine detectors. This system, as represented in Chatoyant, is shown in Figure 20.17. The SDA arrays move the mirror to the correct height for alignment. We control the electrodes with a waveform generator, which applies the appropriate voltages on the four electrodes for the beam to scan or switch in a desired pattern.

As an example, we are able to scan a diamond pattern with the waveforms shown in Figure 20.18. The desired pattern is shown by the white arrow trace on the first output image. The other nine images show snapshots of the detector plane as the diamond pattern is scanned. Dashed lettered lines correspond to time intervals in the waveforms and in the snapshots. Mechanical alignment is critical in this system. For example, the lenslets in this simulation are only 100 μm in diameter. Therefore, when steering the

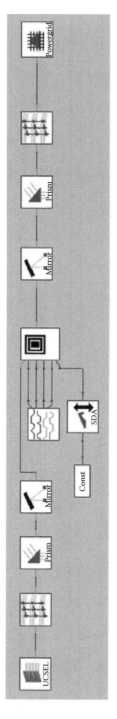

FIGURE 20.17

Scanning system as represented in Chatoyant. (From Kurzweg, T.P. et al., CAD for optical MEMS, *Proceedings of the 36th IEEE/ACM Design Automation Conference (DAC'99)*, New Orleans, LA, June 20–25, 1999. With permission.)

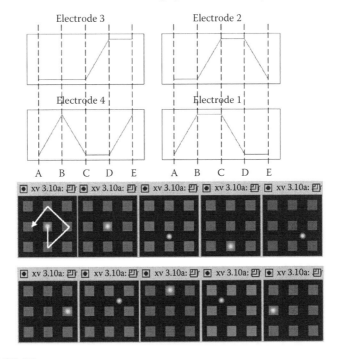

FIGURE 20.18
Scanning waveforms and scanned diamond pattern. (From Kurzweg, T.P. et al., CAD for optical MEMS, *Proceedings of the 36th IEEE/ACM Design Automation Conference (DAC'99)*, New Orleans, LA, June 20–25, 1999. With permission.)

beam, precision in the voltage waveforms is needed so that the light, bending through the prism, hits the desired detector's lenslet.

We next simulate a self-aligning system using optical feedback, using the same system setup as seen in Figure 20.16. Such a system could be used as a noise suppression system. The scanning mirror is used to actively align the system, with the electrodes now being controlled by a waveform generator with a programmed control algorithm. The waveform generator receives the power values detected on each of the detectors, determines where the beam is, and which electrodes to apply voltage to in order to steer the beam onto the center detector.

The system is considered aligned when the power detected on the center detector matches a threshold value set by the user. The user also specifies, in the control algorithm, the size of the voltage step that will be placed on the corresponding electrodes. With active feedback, the system will keep stepping enough voltage to the electrodes until the beam is steered onto the center detector and the system is aligned. The system, as displayed in Chatoyant, is shown in Figure 20.19.

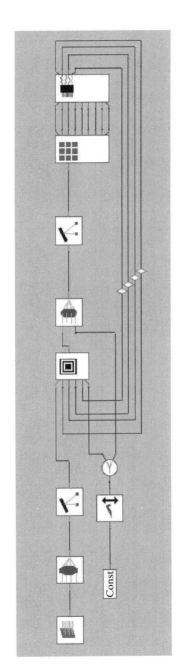

FIGURE 20.19

Self-aligning system using optical feedback. (From Kurzweg, T.P. et al., *J. Model. Simul. Micro-Syst.*, 2, 21, 2001. With permission.)

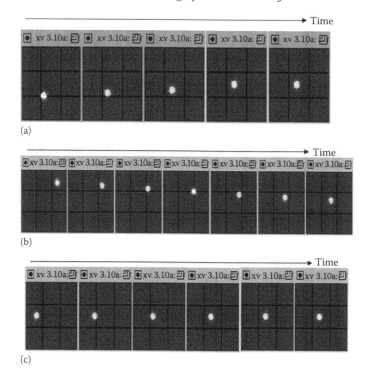

FIGURE 20.20
Self-alignment results. (From Kurzweg, T.P. et al., *J. Model. Simul. Micro-Syst.*, 2, 21, 2001. With permission.)

To simulate this self-aligning system, we introduced random offsets in the lenses and in the VCSEL position and observe as the beam moves toward focus on the center detector. Snapshots of the image at the detectors are given in Figure 20.20 for three cases. The first results, shown in Figure 20.20a, are when the second lens is offset 35 μm in the x-direction. Figure 20.20b shows the results of the second lenslet offset in both the $-x$- and y-direction by 35 μm. The final case has both lenses offset. The first is offset by 5 μm in the x-direction, and the second lens is offset by 35 μm in the $-x$-direction and 5 μm in the y-direction. The results are seen in Figure 20.20c. Notice that the beam on the final images is not exactly in the center of the middle detector. This is because of the power being detected at this point exceeding the power threshold (98.6%) we set for alignment.

20.2.7.3 Angular Spectrum Optical Simulation of the Grating Light Valve

In this section, we simulate and analyze a grating light valve (GLV) system in Chatoyant. This device has many display applications, including digital projection, HDTV, and vehicle displays. The GLV is simply a MEM (micro-electrical-mechanical) phase grating made from parallel rows of reflective ribbons. When all the ribbons are in the same plane, incident light that strikes normal to the surface reflects 180 degrees off the GLV. However, if alternating ribbons are moved down a quarter of a wavelength, a "square-well" diffraction pattern is created, and the light is reflected at an angle from that of the incident light. The angle of reflection depends on the width of the ribbons and the wavelength of the incident light. Figure 20.21 shows the ribbons, from both a top and side view, and also the reflection patterns for both positions of the ribbons.

The GLV component is fabricated using standard silicon VLSI technology, with ribbon dimensions approximately 3–5 μm wide and 20–100 μm long [44]. Each ribbon moves through electrostatic attraction between the ribbon and an electrode fabricated underneath the ribbon. This electrostatic attraction moves the ribbons only a few hundred nanometers, resulting in an approximate switching time of 20 ns. Since the GLV depends on a diffractive phenomenon to direct the light beam, a rigorous modeling technique is required for modeling the GLV system.

For the simulation of the GLV, we examine one optical pixel. A projected pixel is diffracted from a GLV composed of four ribbons, two stationary and two that are movable [44]. Each ribbon has a length of 20 μm and a width of 5 μm. Ideally, there is no gap between the ribbons, however, in reality, a gap is present and is a function of the feature size of the fabrication. Although this gap can be modeled in our tool, in these simulations, we provide an ideal GLV simulation with no gap.

The GLV is modeled as a phase grating, where the light that strikes the down ribbons propagates a half of a wavelength more than the light that strikes the up ribbons. In our model, light reflecting from the down ribbons is multiplied by a phase term. The phase term is similar to a propagation term through a medium: $U_{down\_ribbon} = U \exp(j2kd)$, where d is the distance that the ribbon is moved toward the substrate, typically $\lambda/4$ for the GLV.

Far-field diffraction theory states that the diffracted angle reflected from the square-well grating is [36]: $\theta = q\lambda/a$, where q is the diffraction mode

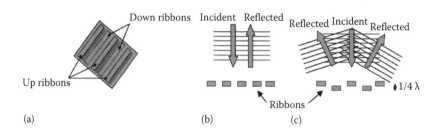

FIGURE 20.21
GLV device (a) top view and side view operation for, (b) up ribbons and, (c) down ribbons.

$(0, \pm 1, \pm 2, \pm 3, \dots)$, a is the period of the diffractive grating, and θ is in radians. In the special case of a square well, when light is diffracted by a grating with a displacement of $\lambda/4$ (a $\lambda/2$ optical path difference after reflection), all the optical power is diffracted from the even modes into the odd modes [45].

In the first simulation, the standard operation of the GLV is verified. We assume an incident plane wave of green light (λ_{green} 520 nm) striking the grating, with the square-well period defined by the ribbon width, and no gap. We simulate the GLV in both cases, that is, when all the ribbons are on the same plane and when the alternating ribbons are moved downward a distance of $\lambda/4$. In this example, the light is reflected off of the grating and propagated 1000 μm to an observation plane. A bounding box of 400 × 400 μm is used, with N equal to 2048. Intensity contours of the observation plane are presented in Figure 20.22a and b.

When the grating is moved into the down position, all of the optical power is not transferred into the expected odd far-field diffractive modes. This is seen in the center of Figure 20.22b, as small intensity clusters are scattered between the $\pm 1^{st}$ modes. This scattering is a near-field effect and demonstrates that in this system, light propagating 1000 μm, is not in the far field. If a designer used a tool propagating with the Fraunhofer far-field approximations, these scattering effects would not be detected. For example, when running the same simulation on LightPipes [46], a CAD tool using the Fraunhofer approximation for optical propagation, only the far-field pattern of light diffracted into the 1st and 3rd modes is seen, as presented in Figure 20.22c. When comparing this result to Figure 20.22b, it is shown that far-field approximation is not valid for this propagation distance. Through this example we have shown that using the angular frequency technique, we achieve the full Rayleigh–Sommerfeld accuracy, while obtaining the same computational speed of using the Fraunhofer approximation.

To show the advantage of the angular spectrum method, we compare the run time of the above simulation with the run time using the direct integration method. With $N = 2048$, the FFT simulation takes about 1.5 min.

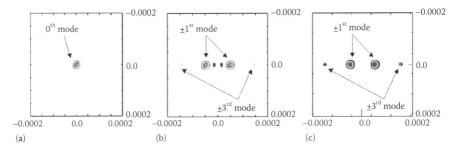

FIGURE 20.22
GLV operation (a) all ribbons up, (b) alternating ribbons down, (c) Fraunhofer approximation.

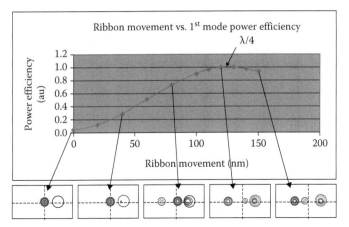

FIGURE 20.23
Transient analysis of ribbon movement and intensity contours.

The direct integration technique takes approximately 5.5 days to finish. If N is reduced to 1024, the simulation completes in approximately 25 s, whereas the direct integration simulation takes approximately 32 h. These simulations were run on a 1.7 GHz dual-processor PC running Linux, with 2 GB of main memory.

In the next simulation, we perform a transient sweep of the ribbon movement, from 0 to 150 nm. The rest of the system setup is exactly the same as before. However, this time, we simulate the normalized power efficiency captured in the 1st diffraction mode for different ribbon depths. To simulate this, a circular detector (radius = 12.5 μm) is placed on the positive 1st mode. Figure 20.23 is a graph that shows the simulated normalized power efficiency in this first mode. As the ribbons are moved downward, more optical power is diffracted into the nonzero modes. As the ribbons reach the $\lambda/4$ point, almost all the diffractive power is in the ± 1st mode. Figure 20.23 also includes intensity contours of selected wave fronts during the transient simulation, along with the markings of the system origin and circular detector position. From these wave fronts, interesting diffractive effects can be noted. As expected, when there is little or no ribbon movement, all the light is in the 0th mode. However, with a little ribbon movement, it is interesting to note that the 0th mode is "steered" at a slight angle from the origin. As the ribbons move downward about $\lambda/8$, the energy in the ± 1st modes are clearly defined. As the gratings move closer to the $\lambda/4$ point, the power is shifted from the 0th mode into the ± 1st modes, until there is a complete switch. As the ribbons move past the $\lambda/4$ point, optical power shifts back into the 0th mode.

In the final simulation, we present a full system-level example as we expand the system to show a complete end-to-end link used in a configuration of a color projection system. The system is shown in Figure 20.24.

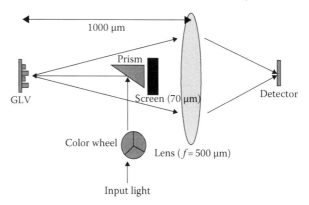

FIGURE 20.24
End-to-end GLV display link.

In this system, we model light, passing through a color wheel, striking a prism, reflecting off the GLV device, past a screen, focused by a lens, and striking a detector [44]. In this system, when the GLV ribbons are all up, the screen blocks the light's 0th mode and the pixel is not displayed. When the alternating ribbons are pulled down, the lens focuses the light found in the $\pm 1^{st}$ modes and converges them to the center of the system, displaying the pixel. Using a spinning color wheel to change the wavelength of the incident light, a frame-sequential GLV projection system uses red (680 nm), green (530 nm), and blue (470 nm) light on the same grating. Since the same grating is used for all wavelengths of light, the grating movement is tuned for the middle frequency: 130 nm ($\lambda_{green}/4$). During this simulation, we use a hybrid approach for the optical modeling. For the propagation through the color wheel and the prism, we use Gaussian propagation. Since propagating through these components does not diffract the beam, this Gaussian technique is not only efficient, but valid. However, as soon as the light propagates past the prism component, we switch the optical propagation technique to our full scalar method to accurately model the diffraction off the GLV device. The remainder of the simulation is propagated with the scalar technique.

We analyze the system by looking at the amount of optical power that is being received on a centered circular detector (radius 10 μm) for the different wavelengths of light, since we are using the same GLV that is tuned for the green wavelength for all wavelengths. A sweep of the distance between the focusing lens and the detector plane is simulated for 0–1500 μm, when the GLV ribbons are pulled down. The graph in Figure 20.25 shows the normalized power received on the circular detector for each wavelength along with selected intensity contours of the green wave front as the beam propagates past the lens. For clarity, the detector's size and position is added onto the intensity contours. For distances under 600 μm, the light remains in

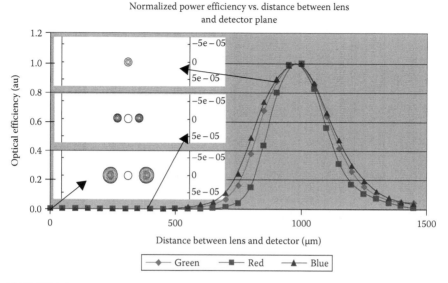

FIGURE 20.25

Wavelength power versus distance propagated.

its two positive and negative 1st modes, as the convergence of the beams has not occurred, resulting in zero power being received on the center detector. As expected, each of the wavelengths focuses at a different rate, as shown by each wavelength's specific curve in Figure 20.25. However, it is seen that all wavelengths focus and achieve detected maximum power at a distance past the lens of 1000 μm, or twice the lens' focal length. At this point, all three colors project on top of each other, creating a color pixel in the focal plane. With additional optics, this focal plane can be projected to a screen outside the projector. This simulation has shown that the grating, although tuned for the green wavelength, can be used for all three wavelengths.

Having shown the use of Chatoyant for modeling multi-domain ana-log systems, we now turn to the problem of co-simulation between the framework described above and a traditional HDL simulator. Co-simulation requires the solution of two problems at the interface between the simula-tors. First, a consistent model of time must be reached for when events occur. Second, a consistent model of signal values must be developed for signals crossing the interface. This is the subject of the next section.

20.3 HDL Co-Simulation Environment

The two levels of simulation discussed above, component and analog system that are supported by Chatoyant, have not been optimized to simulate designs that are specified in an HDL such as Verilog or VHDL. There are no components in the Chatoyant library that directly use HDL as an input language. On the other hand, there are many available commercial and research mixed-language HDL simulators. Mixed-language refers to the ability for a simulator to compile and execute VHDL, Verilog, and SystemC (or other C/C++ variants). In an earlier work we investigated the use of CoSim with Chatoyant models [47]. In this section, we explore an interface to a commercial system. Cadence, Mentor Graphics, Synopsys, and other EDA companies provide such simulators. One common feature among the more widely used simulators, such as ModelSim and NCSIM, is the ability to execute C-based shared object files embedded in HDL design objects. These simulators provide an application programmer's interface (API) to gain access to simulator data and control design components. ModelSim was chosen since it has a large set of C routines that allow access to simulator state as well as modifying design signals and runtime states. These functions and procedures are bundled in an extension package known as the foreign language interface (FLI) [48]. By creating a co-simulation environment between ModelSim and Chatoyant, a powerful MDSoC design and verification environment has been created. This environment is able to address the demand for a robust and efficient system architecture/ design space exploration and prototyping tool that can support the design of MDSoCs.

The rest of this chapter focuses on the development of the interface between Chatoyant and ModelSim and the performance of the resulting environment.

20.3.1 Architecture

The architecture of the co-simulation environment is kept simple to be as efficient and accurate as possible. There are two phases to the execution of the environment: a system generation phase and a runtime support environment. Each is a standalone process, but both are required for system simulation. Figure 20.26 illustrates this top-level structure.

20.3.1.1 System Generator

The System Generator allows the user to create the necessary files needed by both Chatoyant and ModelSim. For Chatoyant this includes a common header and object file used in both simulators as well as components (stars) used for the Chatoyant side of the interface. The same header and object file are used for ModelSim in addition to a shared object library file that is used for invoking the ModelSim FLI when ModelSim is loaded and elaborates a design.

The main input to this generator is the top-level or interface-specific VHDL file. This file contains the list of ports that represent the main conduit

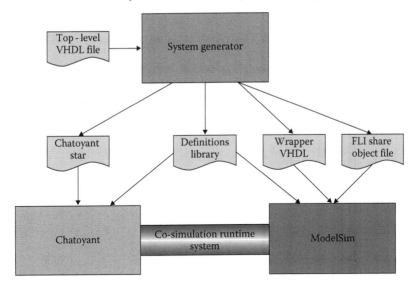

FIGURE 20.26
Co-simulation top-level structure.

between the digital domain running within ModelSim and the other domains handled in Chatoyant. When this file is loaded by the System Generator, the entity portion of the VHDL is parsed and a linked list of the ports is created. Each node in this linked list contains the port's name, its direction (in/out/bidirectional), and its width (1 bit for a signal and *n* bits for a bus).

Using a graphical user interface, the user can select which ports to include and the mapping for the analog voltage levels to be converted into and out of the MVL9 (Multi-Value Logic 9 signal representation standard) logic representation used by ModelSim. There are four fields for this including a high, a low, a cutoff for high, and a cutoff for low voltage values. The user also specifies a name for the system, used for code generation and library management. The outputs of the generator phase are the component star file for Chatoyant, the FLI source code for the ModelSim FLI, the header and source files for a common resource library for the system, a makefile for remaking the object files, a usage text file, and the first time compilation object files performed at the end of the generation.

With these files in place, the user can then proceed with the execution of the linked simulators.

20.3.1.2 *Runtime Environment: Application of Parallel Discrete Event Simulation*

The runtime system differentiates itself from other typical co-simulation environments in that there is no central simulation management system. Chatoyant and ModelSim are treated as two standalone processes and communicate only between themselves. This reduces the overhead of another application executing along with the two simulators as well as the additional message traffic produced by such an arbiter.

This philosophy is an application of a general parallel discrete event simulation (PDES) system. Since there are two standalone processes, each is treated as if it were its own DE processing node. Without a central arbiter, the two must (1) exchange event information by converting logic values into voltages and vice versa, and (2) synchronize their respective local simulation times. To exchange the event information, the system uses technology-specific lookup tables, created by the System Generator, that provide the conversion between a logic "1" and a logic "0" to a voltage in addition to determining what voltage level constitutes a logic "1" and "0."

The synchronization of the simulators is where the application of PDES methods enters [49]. The asynchronous DE simulation invokes both simulators to perform unique tasks on separate parts of a design in a nonsequential fashion. This is because of the fact that there is no master synchronization process as in [1]. For synchronization and scheduling there are two major approaches one can take, conservative or optimistic. We discuss our choice next.

20.3.1.3 Conservative versus Optimistic Synchronization

The conservative and optimistic approaches solve the parallel synchronization problem in two distinct ways. This problem is defined in [2] as the requirement for multiple processing elements to produce events of an equal timestamp in order to not violate the physical causality of the system. The conservative method solves this problem by constraining each processing node to remain in synchronicity with the others, never allowing one simulator's time to pass any other simulator. This can have the penalty of reducing the performance of a simulation by requiring extra overhead in the form of communication and deadlock avoidance.

The optimistic approach breaks the rule of maintaining strict causality by allowing each processing element to simulate without considering time in other processing element. This means that the simulators can run freely without having to synchronize, with the exception of communicating explicit event information. If, however, there is an event sent from one simulator to the other, and the second simulator has a local current time greater than the event's timestamp, then the receiving simulation process must stop and roll-back time to a known safe state that is before the timestamp of the incoming event. This approach requires state saving as well as rollback mechanisms. This can be costly in terms of memory usage and processing overhead for determining and recalling previous states, and thus increases the processing time of every event.

Both approaches are possible since ModelSim does have check-pointing and restoring methods available [48]. However, the conservative PDES method was chosen as the underlying philosophy for our co-simulation solution. Two factors went into this decision. The first consideration is that the co-simulation environment is executing as two processes on one workstation, so that exchanging timing information is not as costly as in a large physically distributed simulation environment. The second is that even with a dual-processor workstation, there is not an excess of computational or memory resources that is seen in a truly distributed PDES architecture, and therefore, a rollback would be too costly.

This was confirmed with a preliminary test of the fiber image guide system described below. For that system the amount of data required for a checkpoint file was on the order of 1 to 2MB. With an average of 10 checkpoint files needed to keep the two simulators within a common time horizon, rollback time took between 500 ms and 1.5 s.

On the other hand, the conservative approach gives a solution requiring significantly less memory at the expense of increased communication to ensure that both simulators are consistently synchronized. This becomes a matter of passing simple event time information between the two simulators. Thus, the only real design issue becomes the time synchronization method.

20.3.1.4 Conservative Synchronization Using UNIX IPC Mechanisms

As described in more detail below, the system was developed and tested on a Linux-based workstation. Therefore, UNIX-style IPC is used for the communication architecture. Event information is exchanged using shared memory, and synchronization is achieved by using named pipes in blocking mode. This is similar to the synchronized data transfer and blocking methodology described in [50]. With these two mechanisms, the conservative approach is implemented in the two algorithms seen in Figure 20.27.

The algorithm for the co-simulation is straightforward. Both simulators, running concurrently, reach a point in their respective execution paths where they enter the interface code in Figure 20.27. Both check to ensure that they are at the next synchronization point (next_sync), and if they are not, they exit this section of code and continue. If they are at the next synchronization point, defining the safe-point in terms of the conservative approach in PDES, then Chatoyant starts the exchange by checking for any change in its outputs to ModelSim. If there is any change in any bit of these ports, that port is marked dirty, and a change flag is set. When all the ports have been examined, Chatoyant sends ModelSim either a ModelSim_Bound event, if any port changed value, or a No_Change event.

Simultaneously, ModelSim waits for this event message from Chatoyant. Once received, it will update and schedule an event for those ports with dirty flags set, if any. It then jumps to check its own output ports, checking bit by bit for a change in each port's value. Once again, as in Chatoyant, if there is a difference, the dirty flag for that port is set, and the change flag in ModelSim is set true. Once this is done for every port, ModelSim will send a message to Chatoyant that there is either a change (Chatoyant_Bound) or No_Change.

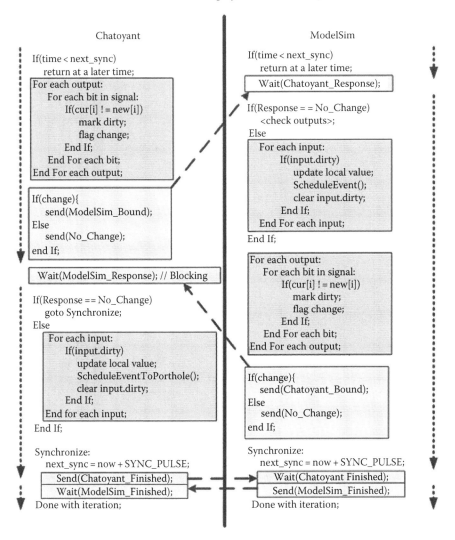

FIGURE 20.27
The synchronization in both simulators.

Chatoyant, waiting for this response, will receive it and take action similar to that of ModelSim in updating the inputs from ModelSim. Finally, the two will set their respective next synchronization times and handshake with one another to indicate it is safe to continue simulating. The No_Change messages are analogous to the null message passing scheme defined by Chandy and Misra [49], which has the benefit of avoiding simulation deadlock.

A key point is the concept of the next synchronization time (next_sync). This value is calculated based on a global parameter in the co-simulation environment known as the SYNC_PULSE. This parameter defines the resolution of how often synchronization occurs. This value ultimately defines the speed versus accuracy tradeoff ratio between the simulators. A higher resolution (smaller SYNC_PULSE value) means greater accuracy but slower runtime. Depending on a particular system, this could affect the quality of the simulation results.

20.3.2 Co-Simulation of Experimental Systems

To examine the effects of synchronization resolution on speed and accuracy, we simulate two example MDSoC systems. Both are large-scale systems, meaning there are many components in each domain, including multiple analog circuits, complex optics, and mixed wire and bus interconnects between the digital and analog domains.

20.3.2.1 Fiber Image Guide

The first of these systems is the fiber image guide, or FIG, system developed at the University of Pittsburgh [51]. FIG is a high-speed 64 × 64-bit opto-electronic crossbar switch built using an optical multi-chip module. FIG uses guided wave optics, analog amplification and filtering circuits, and digital control logic to create an 8 × 8, 8-bit bus crossbar switch. The switch is built as a multistage interconnection network (MIN) built with a shuffle-exchange architecture. The shuffle operations are performed by the wave guide, and the digital logic performs the exchange switching operation. Analog circuits amplify the digital signals and drive VCSEL arrays which in turn transmit light through the image guide. Photodetectors are used to convert the light back into an analog signal, which is amplified and fed back into the digital domain.

This system, illustrated in Figure 20.28, exercises the ability of the co-simulation environment to handle buses as well as the communications between domains without a synchronous clock. In other words, there is no clock signal traveling across the co-simulation interface, and thus the events occur in asynchronous fashion.

20.3.2.2 Smart Optical Pixel Transceiver

The smart optical pixel transceiver, or SPOT, was a development at the University of Delaware [52]. It provides a short-range free-space optical link between two custom-designed transceivers. Each transceiver either accepts or generates a parallel bus, in the digital domain. On the transmitter side, each bus is serialized into a double data rate data signal, along with a 4X clock (125 MHz clock doubled to 250 MHz in this test system). Serialization and de-serialization are handled in the digital domain. These serial data/clock streams are converted into analog signals that are amplified and used to drive VCSEL arrays, similar to FIG. Photodetectors convert the

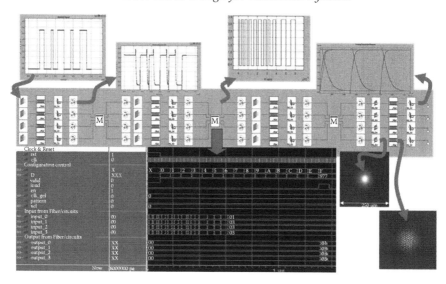

FIGURE 20.28
FIG test system block diagram: Areas in the digital domain are executed in ModelSim while areas in the analog and optical domains are executed in Chatoyant.

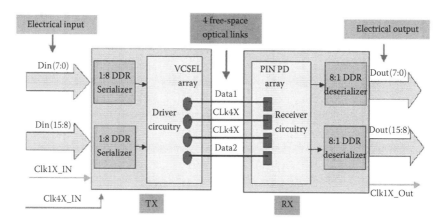

FIGURE 20.29
SPOT system block diagram showing the digital data entering in parallel to the UTSI transceiver chip, serialized transmitted over free-space optics, deserialized with clock-recovery, back into parallel data. (Courtesy of [52].)

propagated light back into analog signals at which point the analog circuits amplify and feed the de-serializing logic in the digital domain. Figure 20.29 shows the system block diagram.

SPOT tests the ability of the co-simulation environment to work with a global clock signal. This clock signal, generated by the digital domain, is transmitted with the data, and thus crosses the co-simulation interface. This means that there are a large number of periodic events occurring. This illustrates the simulation behavior of a synchronous system versus an asynchronous system in the co-simulation environment.

Given these two systems, we next show the results of runtimes and event traffic for different time resolutions of the SYNC_PULSE parameter. A total of four resolutions were tested, 1 ps, 10 ps, 100 ps, and 1 ns. These values were chosen since the systems run at relatively high frequencies, in the range of nanoseconds for both data bit rate and clock speed. All simulations were performed on a Dual 1.70 MHz Intel Xeon Processor Dell Precision with 3 GBs of RAM running Red Hat Linux 7.3, kernel version 2.4.18-3SMP.

20.3.2.3 FIG Runtimes

The following set of charts show the runtime and event traffic seen in each of the resolution steps. Figure 20.30 shows the runtime, in seconds versus the four different time resolutions. Figure 20.31 shows the event counts seen from the Chatoyant and ModelSim perspectives.

As seen in Figure 20.30, the runtimes decreased as the resolution became coarser. One thing to note is the logarithmic-like decay. This is most likely because of the total simulation time of the experiment rather than the time resolution. Since all simulations were performed for a simulation time of

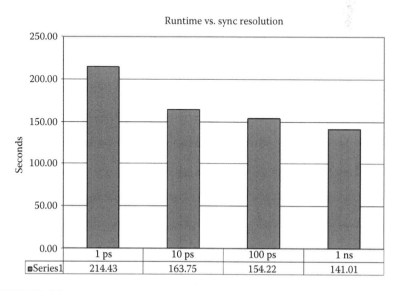

FIGURE 20.30
Runtime versus Sync resolution for FIG.

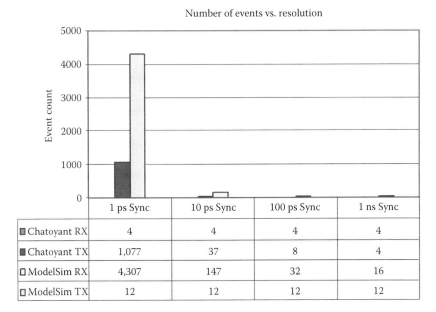

FIGURE 20.31
Number of events per simulation for FIG.

1.4 us, the closer the granularity of the resolution is to the magnitude of the end time, the smaller the difference will be in runtimes. This is explained by the notion that less event processing is performed since more events are ignored between synchronization points. Therefore, event processing overhead is reduced.

Also, the amount of event traffic decreases by two orders of magnitude between 1 ps and 1 ns resolutions. This is also related to the fact that more events are processed at higher resolutions.

20.3.2.4 SPOT Runtimes

The SPOT system yields a different perspective on the co-simulation system. Figure 20.32 shows the runtime results versus resolution and Figure 20.33 shows the event traffic at each resolution.

As seen in Figure 20.32, the runtimes do decrease, in general, with respect to an increasing granularity. The exception to this is the 10 ps resolution, which shows a slight increase in runtime compared to the 1 ps resolution. This may be because of the processing of more event changes given the periodicity of the clock signal in the system. Regardless of this outlier, there is still a general trend for decreasing runtimes as well as decreased event traffic with lower resolutions.

SPOT having a higher runtime versus FIG indicates the effect of the clock signal on performance. Since there is a clock having a consistent event

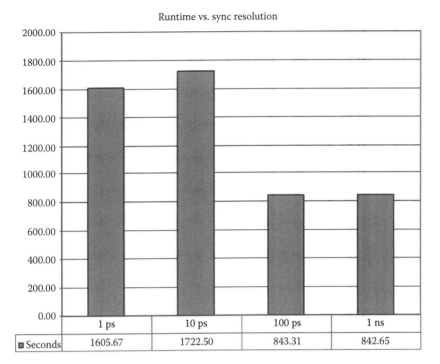

FIGURE 20.32
Runtime results versus resolution granularity.

change at a fixed frequency, the number of events per synchronization cycle increases. This is seen by the higher event counts in each SPOT simulation versus those for FIG. This amount, spread uniformly across the entire simulation in SPOT, versus FIG which has dense cluster of events separated by a large time gap, exemplifies the overhead associated with processing events.

20.4 Summary

In summary, we have presented a co-simulation environment for mixed-domain, mixed-signal simulation that spans the realms of HDL digital logic, analog electrical, optical, and mechanical systems. A variety of modeling techniques are used to develop analog component models that are evaluated using continuous time models. These component behaviors communicate via specific ports that pass complex messages between components. Those messages, and the corresponding execution of the component models, are coordinated by a DE simulation backbone. This backbone, built on Ptolemy, also runs in coordination with a commercial HDL simulator.

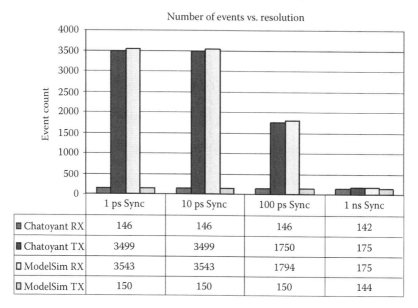

Number of events vs. resolution

	1 ps Sync	10 ps Sync	100 ps Sync	1 ns Sync
■ Chatoyant RX	146	146	146	142
■ Chatoyant TX	3499	3499	1750	175
◻ ModelSim RX	3543	3543	1794	175
◻ ModelSim TX	150	150	150	144

FIGURE 20.33
Event traffic versus resolution granularity.

This system, known as Chatoyant–ModelSim Co-Simulation Environment, provides an interface between the multi-domain analog realm handled by Chatoyant and the digital realm, handled by ModelSim.

As seen in the co-simulation experiments, there are a few factors that affect runtime performance. Asynchronous systems with more clustering of events within certain windows generally have a better runtime than synchronous systems that have a steady load of events. Also, as predicted, the resolution of synchronization, defined in the context of the PDES conservative approach and implemented using Unix IPC, has an effect on runtime performance by reducing the event processing overhead, at the cost of accuracy. This cost is assessed based on the system and requirements a particular user has for the simulation.

20.4.1 Conclusions

Multi-domain modeling and multi-rate simulation tools are required to support mixed-technology system design. This chapter has shown Chatoyant's support for simulating and analyzing optical MEM systems with models for optical, electrical, and mechanical models for components and signals. By supporting a variety of component and signal modeling techniques and multiple abstraction levels, Chatoyant has the ability to perform and analyze mixed-signal tradeoffs, which makes it valuable to multi-technology system designers. Keeping simulations, along with analysis techniques such as Monte Carlo for mechanical tolerancing, BER, crosstalk, and insertion loss within the Chatoyant framework allows for quick and efficient system-level design and analysis for the future development of next generation MDSoCs.

Acknowledgments

The work in this chapter was supported by DARPA, under Grant No. F3602-97-2-0122 and NSF, under Grant No. ECS-9616879 and CCR 88319. The University of Delaware is acknowledged for providing the SPOT system data for testing.

References

1. J. Buck, S. Ha, E. A. Lee, and D. Messerschmitt, Ptolemy: A framework for simulating and prototyping heterogeneous systems, *International Journal on Computer Simulation*, 4, 155–182, 1994.

2. K. Hines and G. Borriello, Dynamic communication models in embedded system co-simulation, *Proceedings of ACM SIGDA for the Design Automation Conference (DAC'97)*, Anaheim, CA, 1997.

3. C. Liem, F. Nacabal, C. Valderrama, P. Paulin, and A. Jerraya, System-on-a-chip cosimulation and compilation, *IEEE Design and Test of Computers*, 14(2), 16–25, April–June 1997.

4. S. P. Levitan, T. P. Kurzweg, P. J. Marchand, M. A. Rempel, D. M. Chiarulli, J. A. Martinez, J. M. Bridgen, C. Fan, and F. B. McCormick, Chatoyant: A computer-aided design tool for free-space optoelectronic systems, *Applied Optics*, 37(26), 6078–6092, September 1998.

5. E. A. Lee and H. Zheng, Modeling and simulation of mixed continuous and discrete systems, *Model-Based Design for Embedded Systems*, Taylor & Francis, Boca Raton, FL, 2009.

6. L. Gheorghe, G. Nicolescu, and H. Boucheneb, A generic methodology for the design of continuous/discrete co-simulation tools, *Model-Based Design for Embedded Systems*, Taylor & Francis, Boca Raton, FL, 2009.

7. S. P. Levitan, P. J. Marchand, T. P. Kurzweg, M. A. Rempel, D. M. Chiarulli, C. Fan, and F. B. McCormick, Computer-aided design of free-space opto-electronic systems, *Proceedings of the 1997 Design Automation Conference*, Anaheim, CA, June 1997, Best Paper Award, pp. 768–773.

8. T. P. Kurzweg, S. P. Levitan, P. J. Marchand, J. A. Martinez, K. R. Prough, and D. M. Chiarulli, CAD for optical MEMS, *Proceedings of the 36th IEEE/ACM Design Automation Conference (DAC'99)*, New Orleans, LA, June 20–25, 1999.

9. T. P. Kurzweg, S. P. Levitan, P. J. Marchand, J. A. Martinez, K. R. Prough, and D. M. Chiarulli., Modeling and simulating optical MEMS using Chatoyant, *Design, Test, and Microfabrication of MEMS/MOEMS*, Paris, France, March 30–April 1, 1999.

10. S. P. Levitan, J. A. Martinez, T. P. Kurzweg, A. J. Davare, M. Bails, M. Kahrs, and D. M. Chiarulli, System simulation of mixed-signal multi-domain microsystems with piecewise linear models, *IEEE Transactions on CAD of ICs and Systems*, 22(2), 139–154, February 2003.

11. A. Jose and M. S. Martinez, Piecewise linear simulation of optoelectronic devices with application to MEMS, Department of Electrical Engineering, University of Pittsburgh, Pittsburgh, PA, June 2000.

12. J. A. Martinez, S. P. Levitan, and D. M. Chiarulli, Nonlinear model order reduction using remainder functions, *IEEE Computer Society, Design, Automation and Test in Europe (DATE'06) ICM, IP2 Interactive presentations*, Paper No. 791, pp. 281–282, MESSE Munich, Germany, March 6–10, 2006.

13. J. Morikuni and S. Kang, *Computer-Aided Design of Optoelectronic Integrated Circuits and Systems*, Prentice-Hall, Inc., Upper Saddle River, NJ, Chapter 6, 1997.

14. E. M. Buturla, P. E. Cottrell, B. M. Grossman, and K. A. Salsburg, Finite element analysis of semiconductor devices: The FIELDAY program, *IBM Journal of Research and Development*, 25, 218–231, 1981.

15. M. R. Pinto, C. S. Rafferty, and R. W. Dutton, PISCES II—Poisson and continuity equation solver, Stanford Electronics Laboratory Technical Report, Stanford University, Stanford, CA, September 1984.

16. P. C. H. Chan and C. T. Sah, Exact equivalent circuit model for steady-state characterization of semiconductor devices with multiple energy-level recombination centers, *IEEE Transactions of Electronic Devices*, ED-26, 924–936, 1979.

17. R. Kielkowski, *SPICE, Overcoming the Obstacles of Circuit Simulation*, Chapter 4, McGraw-Hill, Inc., New York, 1994.

18. S. Kim, N. Gopal, and L. T. Pillage, Time-domain macromodels for VLSI interconnect analysis, *IEEE Transactions on CAD of Integrated Circuits and Systems*, 13(10), 1257–1270, October 1994.

19. A. Devgan and R. A. Rohrer, Adaptively controlled explicit simulation, *IEEE Transactions on CAD of Integrated Circuits and Systems*, 13, 746–762, June 1994.

20. P. Feldmann and R. W. Freund, Efficient linear circuit analysis by Padé approximation via the Lanczos process, *IEEE Transactions on Computer-Aided Design*, 14, 639–649, May 1995.

21. A. Odabasioglu, M. Celik, and L. T. Pileggi, PRIMA: Passive reduced-order interconnect macromodeling algorithm, *IEEE Transactions on Computer-Aided Design*, 17(8), 645–654, 1998.

22. L. T. Pillage and R. A. Rohrer, Asymptotic waveform evaluation for timing analysis, *IEEE Transactions on Computer-Aided Design*, 9(4), 352–366, April 1990.

23. L. W. Nagel, SPICE2, a computer program to simulate semiconductor circuits, Technical Report Memo UCB/ERL M520, University of California, Berkeley, CA, May 1975.

24. A. Salz and M. Horowitz, IRSIM: An incremental MOS switch-level simulator, *Proceedings of the 26th Design Automation Conference*, Las Vegas, NV, pp. 173–178, 1989.

25. R. Kao and M. Horowitz, Timing analysis for piecewise linear Rsim, *IEEE Transactions on Computer-Aided Design*, 13(12), 1498–1512, 1994.

26. J. Vlach and K. Singhai, *Computer Methods for Circuit Analysis and Design* (2nd edn.), John Wiley & Sons, Inc., New York, 1993.

27. J. Clark, N. Zhou, S. Brown, and K. S. J. Pister, Nodal analysis for MEMS simulation and design, *Proceedings of Modeling and Simulation of Microsystems Workshop*, Santa Clara, CA, April 6–8, 1998.

28. J. E. Vandemeer, Nodal design of actuators and sensors (NODAS), MS Thesis, Department of Electrical and Computer Engineering, Carnegie Mellon University, Pittsburgh, PA, 1997.

29. J. E. Vandemeer, M. S. Kranz, and G. K. Fedder, Hierarchical representation and simulation of micromachined inertial sensors, *Proceedings of Modeling and Simulation of Microsystems Workshop*, Santa Clara, CA, April 6–8, 1998.

30. W. J. Duncan, Reciprocation of triply partitioned matricies, *Journal Royal Aeronautical Society*, 60, 131–132, 1956.

31. J. S. Przemieniecki, *Theory of Matrix Structural Analysis*, McGraw-Hill, New York, 1968.

32. B. E. A. Saleh and M. C. Teich, *Fundamentals of Photonics*, Wiley-Interscience, New York, 1991.

33. S. J. Walker and D. J. Nagel, Optics and MEMS, Navel Research Lab, NRL/MR/6336-99-7975, May 15, 1999.

34. M. Born and E. Wolf, *Principles of Optics*, Pergamon Press, London, 1959.

35. J. W. Goodman, *Introduction to Fourier Optics* (2nd edn.), McGraw-Hill Companies, Inc., New York, 1996.

36. E. Hecht, *Optics* (2nd edn.), Addison-Wesley Publishing Company, Reading, MA, 1987.

37. M. W. Kowarz, Diffraction effects in the near field, PhD thesis, University of Rochester, Rochester, NY, 1995.

38. T. Tommasi and B. Bianco, Frequency analysis of light diffraction between rotated planes, *Optics Letters*, 17(8), 556–558, April 1992.

39. N. Delen and B. Hooker, Free-space beam propagation between arbitrarily oriented planes based on full diffraction theory: A fast Fourier transform approach, *JOSA*, 15(4), 857–867, April 1998.

40. W. L. Briggs and V. E. Henson, *The DFT: An Owner's Manual for the Discrete Fourier Transform*, SIAM, Philadelphia, PA, 1995.

41. M. C. Wu, Micromachining for optical and optoelectronic systems, *Proceedings of the IEEE*, 85(11), 1833–1856, November 1997.

42. W. Piyawattanametha, L. Fan, S. S. Lee, J. G. D. Su, and M. C. Wu, MEMS technology for optical crosslink for micro/nano satellites, *International Conference of Integrated Nano/Microtechnology for Space Applications (NANOSPACE98)*, NASA/Johnson Space Center, Houston, TX, November 1–6, 1998.

43. T. Akiyama, D. Collard, and H. Fujita, Scratch drive actuator with mechanical links for self-assembly of three-dimensional MEMS, *Journal of Microelectromechanical Systems*, 6(1), 10–17, March 1997.

44. D. M. Bloom, The grating light valve: Revolutionizing display technology, *Proceedings of SPIE*, 3013, Photonics West, *Projection Displays III*, 165–171, 1997.

45. O. Solgaard, Integrated semiconductor light modulators for fiber-optic and display applications, PhD Thesis, Stanford University, Stanford, CA, 1992.

46. Vdovin, G., LightPipes Manual, http://guernsey.et.tudelft.nl

47. L. Kriaa, W. Youssef, G. Nicolescu, S. Martinez, A. A. Jerraya, B. Courtois, S. Levitan, J. Martinez, and T. Kurzweg, System C-based cosimulation for global validation of MOEMS, *Design Test Integration and Packaging of MEMS/MOEMS (DTIP 2002)*, pp. 64–70, *SPIE Proceedings* Vol. 4755, Cannes-Mandelieu, France, May 6–8, 2002.

48. ModelSim SE Foreign Language Interface, Version 5.7g., 2003.

49. P. Banerjee, *Parallel Algorithms for VLSI Computer-Aided Design*, Prentice Hall, Englewood Cliffs, NJ, 1994.

50. H. Hübert, A Survey of HW/SW Cosimulation Techniques and Tools. Thesis Work, Royal Institute of Technology, Stockholm, Sweden, 1998.

51. Donald M. Chiarulli et al., Optics in Computing, OSA Technical Digest, Optical Society of America, Washington DC, 2001, pp. 125–127.

52. P. Gui, F. Kiamilev et al. Source synchronous double data rate (DDR) parallel optical interconnects, *Proceeding of InterPACK'03*, Maui, Hawaii, July 6–11, 2003.

21

Smart Sensors Modeling Using VHDL-AMS for Microinstrument Implementation with a Distributed Architecture

Carles Ferrer, Laura Barrachina-Saralegui, and Bibiana Lorente-Alvarez

CONTENTS

21.1 Introduction

The growing importance of microelectromechanical systems (MEMS) in a wide range of applications, which combine extreme sensitivity, accuracy, and compactness, has evidenced the need to simplify the design process in order to reduce the design time and cost. One of the possible solutions for MEMS design is the extension of the use of an integrated circuit design methodology to obtain a top-down design methodology that is possible thanks to the new available mixed-signal modeling languages, such as VHDL-AMS [1], analog and mixed-signal extension of VHDL language, which allows developing models that combine not only digital and analog signals, but also thermal, mechanical, and optical signals.

The main step in the process of designing for MEMS integration is to combine VHDL models with VHDL-AMS models to obtain a complete description of the multitechnological system. Considering a simplified design flow for mixed-signal models, the first step will be to define initial specifications including environment and technical characteristics. The next step is to define interfaces and partitioning into basic components (including sensors, actuators, analog, and digital circuitry) to design an abstract structure that meets the initial requirements. Each component modeling is realized with the most appropriate language and could be described at different representation levels (behavioral, structural, circuit, device, and physical) [2,3]. The modeling of each component becomes a complex task because of the different languages and abstraction levels required. Finally, after the fabrication and/or assembly of all the components, a test and qualification phase must be carried out in order to guaranty the expected quality levels for the target application. In one approach, the digital elements have been developed by using VHDL, while the nonelectronic (transducers) parts and the analog and mixed-signal circuitry models have been elaborated in VHDL-AMS.

The microinstrument that is modeled and to which this methodology is applied is an inertial measurement unit (IMU). An IMU is the main component of inertial guidance systems used in air-, space-, and watercraft. An IMU works by sensing motion including the type, rate, and direction of that motion, and it will be composed of three accelerometers and three gyroscopes with all these transducers based on MEMS technology. Additionally, the necessary processing circuitry and modules for digital communication have to be treated and modeled with the most suitable language depending on the nature of the element.

This chapter is structured in the following manner: Section 21.2 presents the distributed architecture; Section 21.3 deals with the design methodology for MEMS; Section 21.4 provides an example of an application case based on an IMU; Sections 21.5 and 21.6 present accelerometer and gyroscope sensor modeling, respectively; Section 21.7 describes the modeling of a complete smart sensor including the sensor and their associated electronic circuitry; Section 21.8 presents simulation and validation results; and Section 21.9 concludes the chapter.

21.2 Architecture

The associated electronic circuitry that measures a sensor must be considered, and it adds different and necessary functions, such as correcting offsets, temperature compensations, AD conversions, etc. All these functions

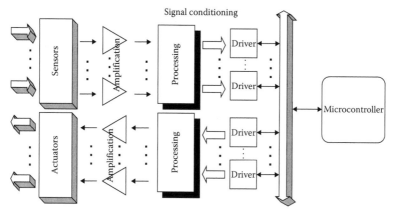

FIGURE 21.1
Distributed architecture.

have to be considered from early design phases of a smart sensor. Smart microsystems can not only process the signal coming from the sensor itself, they can also have the communication drivers related to these sensors [4].

All nodes or the set of smart sensors are connected to a host (computer or microcontroller) through a sensor bus called interconnection bus for integrated sensors (IBIS). IBIS was designed for a better serial connection between the smart sensors and the main controller. The communications between nodes is established through a designed protocol where each node is classified into master or slave and has its own logic address [5].

This bus is based on a distributed architecture (see Figure 21.1). The main advantage of the distributed architecture is that when it has to be extended because of the increasing number of smart sensors or actuators in a microsystem, additional smart sensors or actuators can easily be connected without the need to rearchitect. So, one of the solutions found was to specify a distributed architecture in which a bus sensor was implemented, and this way a specific interconnection was developed [6].

The distributed architecture introduces the advantage of modularity and interchangeability as it enables an easy communication applicable to different sets of microsystems. Its main characteristic is to own two buses. The sensor bus is used for relatively short distances, a few centimeters, and for connecting sensors and actuators on the same subsystem through a dedicated microcontroller. The use of miniature sensors in high numbers raises the problem of the size and mass of the interfacing cables and connectors, which are currently much higher than those of the sensors themselves. This increases the necessity to address the problem of whether it is possible to reduce, or even eliminate, the mass and volume of the interfacing devices.

An evolution of this second architecture can be seen in Figure 21.2. It is shown how the master is combined with several elements such as

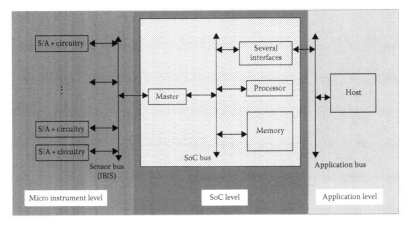

FIGURE 21.2
Enlarging the architecture.

memory, processor, and interfaces, forming a system on chip (SoC) that could be implemented with an ASIC or an FPGA, and how they can be integrated in the distributed architecture, combined with a higher lever (application level) and the lower level (sensor level) to built up the instrument.

As it is said, IBIS was designed for the better integration of the sensors in a system. We have decided to create a two-wire low rate synchronous monomaster bus. This bus can address up to 31 slaves, using the 32nd address to address all the slaves at the same time, when it is necessary to send a general reset, to do a self-test, or to initialize the sensors and actuators at the same time specially. However, any new necessary command that will affect the entire system could be implemented. The speed of an IBIS is about 1 MHz, and it has a bus-shaped topology.

21.3   Design Methodology for MEMS Design

The design of MEMS can often become a task more complex than designing an electronic circuit. This is because MEMS behavior cannot be considered a simple addition of separate mixed (fluidic, optical, thermal, etc.) and electrical behavior, but it is a simultaneous combination. This fact motivates the extension of the existent design methodology for integrated circuits to obtain a top-down design methodology for MEMS design [7]. It is based on a hierarchical design method with both abstract behavioral and functional models in device-analog-digital domain (see Figure 21.3). The development of a design hierarchy allows the designer to mix levels of abstraction to observe and evaluate interactions between interdependent subsystems.

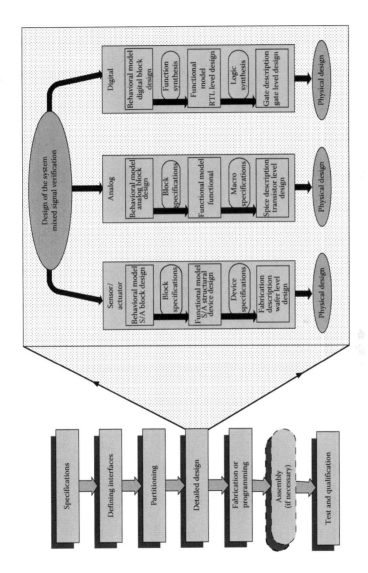

FIGURE 21.3
Top-down design flow for MEMS.

In this design flow, the functionality can be easily verified and analyzed at the top level. The specifications and characteristics of each stage (sensor/ actuator, analog, digital) can be fixed at a lower level, using specific simulators (such as SPICE for analog circuits) to optimize design at the lowest level blocks. This phase is very close to the convectional methods used in MEMS design. The following phases of the design include effects because of the extension of a top-down "digital" methodology: (a) The verification of the whole simulation system can be obtained, although it is not possible to directly synthesize the complete design with VHDL-AMS. (b) Rapid and easy design process from the top-down. We take physical characterization into consideration at the final design to validate abstract models developed in early design stages. (c) Evaluation of the entire system at any design stage. This is of greater importance in the analog and MEMS design, especially considering the multitechnolgical nature of MEMS.

In order to obtain the necessary models of the MEMS, there are several possible approaches. Generalized networks can be considered in MEMS modeling because many physical quantities are compared to flow or difference quantities and generalized Kirchhoff's laws can be applied. To obtain a generalized network, large systems can be interpreted as decompositions into basic network elements. This network concept is valid in many different domains, such as electrical, fluidic, mechanical, etc.

Another way to obtain the model is using order reduction. Modeling strategy with the real system can be described using partial differential equations for the entire system; producing reduced system matrices that simplify the simulation effort required comparing with the complexity of the equations that describes the MEMS device functionality. The last way is to obtain behavioral models as black-box models, derived from simulation results in the time or frequency domain. Once the model is developed and included in the complete system, the simulation process must consider a tuning phase through the optimization of parameter settings.

The available modeling tools as well as available multidomain libraries are usually incomplete. However, many system simulators can support standardized modeling languages, like VHDL-AMS, and therefore any others in the near future.

21.4 Application

The case example presented in this work focuses on the development of a complete behavioral model of an IMU in VHDL-AMS. This IMU combines MEMS sensing technology with analog and digital signal processing circuitry, all interconnected in a distributed architecture through the bus IBIS. The system is composed of three accelerometers and three gyroscopes, one for each direction of space, its conditioning signal circuitry, and the corresponding interfaces to connect them to the communication bus. The complete system model covers mixed-technology sensors and mixed-signal circuitry, so it has been developed in VHDL-AMS for the analog and mixed-technology modules, and VHDL for the digital ones. All these models have been cosimulated within the same environment following the design methodology design flow described in Figure 21.3. In this approach, a behavioral description of sensor/actuator and analog circuitry has been chosen in combination with a more structural description of the digital circuitry parts.

By employing an IMU, it is possible to know the position of an object in motion, so it has many applications in remote control systems and navigation systems. An IMU may have many architectures and designs, with different composing elements, depending on the technology and algorithm employed. In our case, the IMU is composed of three accelerometers and three gyroscopes, each one with its own processing signal circuitry, although it could be designed using a Kalman filter and gyroscopes. Both designs have been corrected from early conception to provide an adequate response for the analog-to-digital converter (ADC). They are basically composed of an amplifier stage and filtering modules to reduce the electrical noise. Finally, to connect each sensor to the communication bus, it has been necessary to develop an interface to synchronize and make the data types compatible. The IMU structure is shown in Figure 21.4.

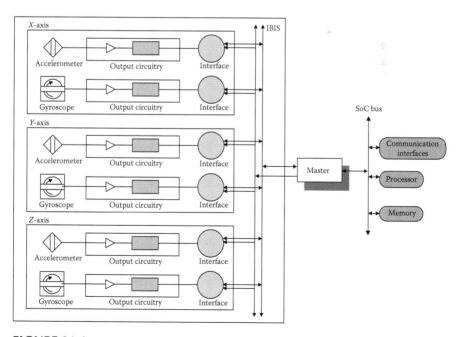

FIGURE 21.4
IMU structure.

Different models have been elaborated for each element of the system, trying to develop behavioral and functional models to be employed in large simulations, and structural models that allow descending to lower levels of abstraction. In this way, we obtain a complete description of the system, and it is thus possible to compare the results of both models in order to improve them and use the most suitable model in each simulation.

21.5 Accelerometer

The accelerometer that has been assembled as a part of the IMU can be considered a smart sensor, since it contains the necessary additional circuitry to connect it to a main controller digital system. The smart sensor is composed of the acceleration sensor, its output circuitry with an analog-to-digital conversion, and an interface to adapt the system to a digital device. Its internal structure is shown in Figure 21.5.

21.5.1 Description of the Accelerometer

The accelerometer is based on the piezoresistive effect, and was built into an SOI (silicon-on-insulator) wafer packaged on an MCM [8]. It is a Wheatstone bridge with four piezoresistances placed on a cantilever design, whose value will depend on the direction of the applied acceleration. We have developed three different representations of this accelerometer, which will be used depending on the level of abstraction and the features that we want to study. The three models are a behavioral model, a physical model, and a mathematical model.

- The behavioral model of the accelerometer is a simple model that shows a linear relation between the applied acceleration and the output voltage. The equation used has been obtained based on the experimental measurement. This simple model is useful to test other

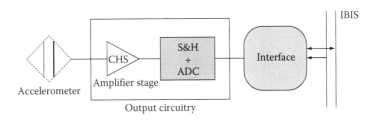

FIGURE 21.5
Complete system schematic.

elements of the sensor, and also to perform simulations that involve several accelerometers. These simulations could take much time and resources of the simulator. The model with its simulation can be seen in Figures 21.6 and 21.7.

- The physical model of the accelerometer takes care of the piezoresistive effect in the Wheatstone bridge. Given an acceleration, it first calculates all of the stress applied to the resistances because of this acceleration, depending on the direction that each resistance is placed. The relation between the acceleration and the stress is modeled as a linear relation, whose constants can be obtained through experimental measurement. Considering the stress applied to each resistance, the next step is calculating the change in the resistive value. This computation is performed by applying the known constants for the piezoresistive material of silicon for 100 wafers. With this data, the output voltage of the Wheatstone bridge is easily obtained by applying the characteristic equations of the system.

entity acc 2_5 v1 is

     generic

       (--------Default values for the generics are from D ZU-25g---------------------------

       ----

       ----------Static characteristics---

Z_in	:real	=	1888e3;	--[Input impedance] = Ohms
Z_out	:real	=	1882e3;	--[Output impedance] = Ohms
Offset	:real	=	0.00510;	--[Offset] = V/V
S_sens	:real	=	0.00046;	--[Sensitivity] = V/(V.G)
Vdd	:real	=	5.0;	--[Voltage supply] = V
N_lin	:real	=	037;	--[Non-linearity] = %PSO
Hyst	:real	=	0.12;	--[Hysteresis] = %PSO
Repet	:real	=	0.14;	--[Repetibility] = %PSO
C_sens_x:real		=	693;	--[X axis cross sensitivity] = %

FIGURE 21.6

Behavioral model for the accelerometer.

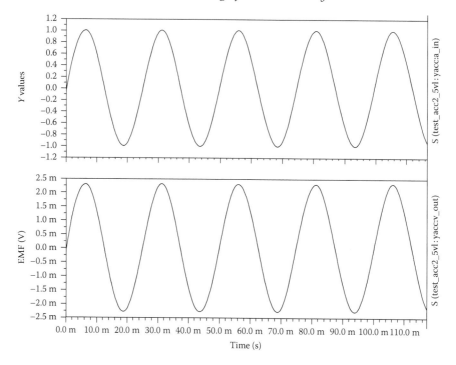

FIGURE 21.7
Simulation of the accelerometer behavioral model.

- The mathematical model describes the accelerometer as a mass-spring system. It applies the Newton's second law, which also deals with the damping force. Again, the constants used have been obtained through experimental measurement and known physical values. This model shows the possibilities of VHDL-AMS of solving differential equations.

21.5.2 Output Circuitry

The output circuitry of the accelerometer can be treated separately in two parts. The first one is the signal conditioning circuitry based on Chopper stabilization (CHS), so it amplifies the signal and eliminates the noise. The second part consists of a sample and hold (SH) device and an ADC, and makes possible the connection of the sensor to a digital device. The entire system scheme can be seen in Figure 21.8.

First, we address the signal conditioning circuitry. The operation principle in CHS is to avoid low level noise by moving the signal to higher frequencies and restoring it to its original frequency once amplified. The output voltage of the accelerometer reaches the first modulator, so it is transferred to the frequency imposed by the oscillator. The next step is to amplify the signal and pass through a band-pass filter, in order to eliminate undesirable signals.

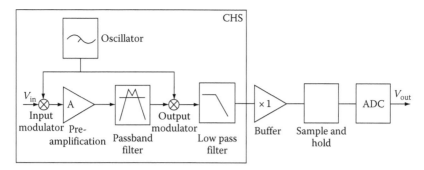

FIGURE 21.8
Accelerometer output circuitry.

Finally, the second modulator and the low-pass filter restore the original frequency signal.

The signal conditioning circuitry is composed of two modulators (input/output, which are controlled by an oscillator signal), a preamplification, a band-pass filter, an oscillator, a low-pass filter (40 dB/decade and 10 kHz Butterworth filter), and a frequency divider.

The oscillator is composed of an astable multivibrator (square-wave generator) and a comparator. It was designed to obtain the filter resonance frequency and oscillator frequency as closely as possible. The frequency of oscillation is defined by the relation between R_0 and R_1 (see Figure 21.9) and presents a value of 110 kHz. In spite of developing this model according to its schematic, its simulation has not been successful because it is impossible to define initial conditions on the simulator. For the global system simulation, a behavioral model of the oscillator has been used (Figure 21.10).

The modulator is composed of different logic gates (see Figure 21.11) and provides two square signals with the frequency of the oscillator, with a delay of 180° between them.

The preamplifier presents a gain of 100 dB, and it is composed of a differential amplifier with two amplification stages (see Figure 21.12).

The low-pass filter is a differential filter based on a 40 dB/decade Butterworth filter. The cutoff frequency of the filter is 10 kHz, and the model developed and the simulation results of the filter can be seen in Figures 21.9 and 21.10.

The band-pass filter is a differential filter based on a narrow band's band-pass filter (see Figure 21.13). The most important characteristics are the resonance frequency of 110 kHz, which is the frequency of the oscillator, and a 40 kHz bandwidth. These features can be seen in the results shown in Figure 21.14.

For this simulation, the behavioral description of the accelerometer has been used as excitation signal. As can be seen in Figure 21.15, the output signal has a linear relation with the input signal, and amplifies it in two orders of magnitude.

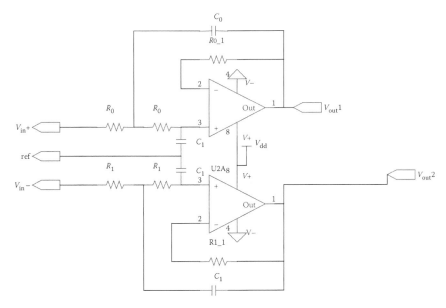

FIGURE 21.9
Schematic corresponding to the low-pass filter.

Next step is to model the second part of the output circuitry, which consists of an SH circuit and an ADC. Between these elements and the amplifier output circuitry simulated before, a simple buffer had to be added to make the offset levels compatible. The SH circuitry is based on a simple design with only one capacitor, where the sample is controlled by a switch and its clock signal. The converter operation is controlled also by a clock signal, which is the clock signal used by the IBIS bus. Its simulation can be seen in Figure 21.16, where the transformation of analog data to digital data is shown. The synchronization between the IBIS and these elements is achieved through an interface, which allows the connection between the digital and the analog parts of this smart sensor.

In Figure 21.16 the first signal shown is the excitation signal, followed by the same signal sampled, and finally intermediate signals that synchronize both elements. The command signal governs the beginning of the data capture, which finishes when the d_{out} signal takes a value.

21.5.3 IBIS Drivers

All the developed IBIS drivers have a similar inside structure. The bus codification is done in Manchester encoding, this allows a better transmission and its synchronization, but has the drawback of making the driver slightly more difficult to develop. The drivers are necessary to connect the slaves with the master of the bus, and they all are formed by different modules that

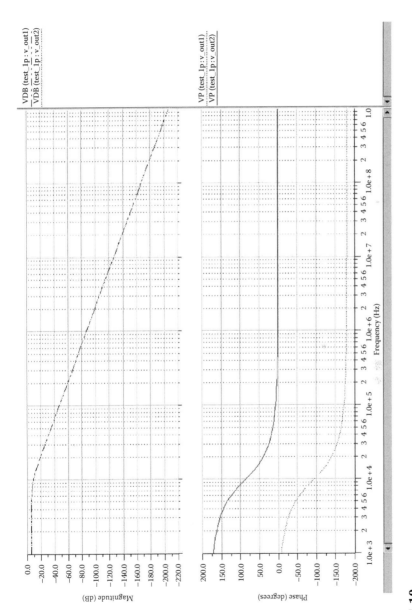

FIGURE 21.10
Results obtained for the low-pass filter.

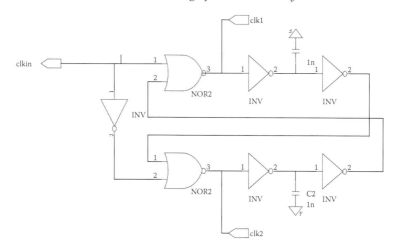

FIGURE 21.11
Schematic corresponding to the modulator.

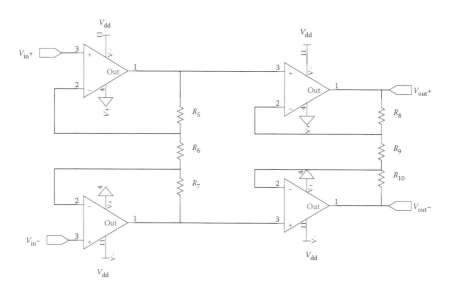

FIGURE 21.12
Schematic corresponding to the preamplifier.

comprise different functions, as for example, a fragmentation module, a Manchester encoding module, a Manchester decoding module, a module of frame formation, and all the involved mechanisms for the appropriate bus interaction. Examples of the later are the frequency divider or the intern buffers, to ensure that the data is not lost. Furthermore, the bus, because of its drivers, allows hot plugging and plug and play mode [9,10].

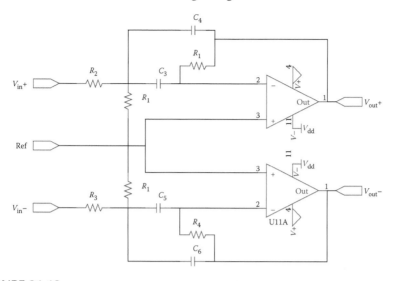

FIGURE 21.13
Schematic corresponding to the band-pass filter.

For easier design, all has been included in one single IP. Each device has a physical address programmed in it, and, as it has been said before, an IBIS can have up to 31 slaves.

In Figure 21.17 we can observe the structure of a connection between the driver and the master of the bus; the shaded part depicts the IBIS driver.

21.5.4 Interface of the Accelerometer

To enable the connection between the digital part IBIS and the sensor, it is necessary to introduce an interface. Its main function is to receive commands from the IBIS and to synchronize the data acquisition according to these commands [11].

While not receiving any command or when it receives the reset command, the state of the system does not change. However, when the IBIS sends a command signal, the interface sends signals to the SH and the ADC to begin the data acquisition. The SH holds the signal while the ADC converts it to digital data. When conversion is finished, the ADC informs the interface and sends the digital data. The interface transmits the data to the IBIS bus and proceeds to wait for new commands.

21.6 Gyroscope

Similar to the accelerometer, the assembled gyroscope itself can be considered a smart sensor, since it is prepared to be connected to a digital device

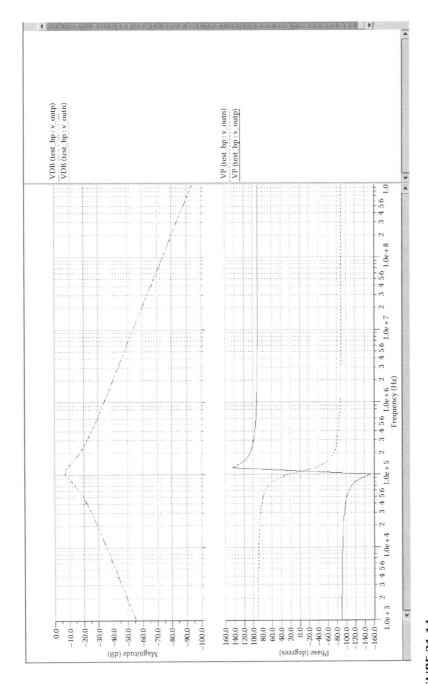

FIGURE 21.14
Results obtained for the band-pass filter.

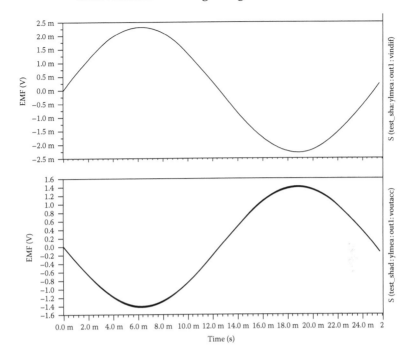

FIGURE 21.15
Simulation results obtained for the output circuitry.

without the need to solve compatibility issues. Its structure is similar to the accelerometer, except that instead of the CHS amplification, the signal processing circuitry recommended by the manufacturer has been utilized. The smart sensor schematic is shown in Figure 21.18.

The sensor employed is a gyroscope based on piezoelectric effects. It contains a ceramic bimorph vibrating unit, whose operating principle consists of the detection of the rotational motion from the generated Coriolis force. The ceramic material makes the bar vibrate, so when a rotational motion takes place, a Coriolis force is generated perpendicular to the original direction of vibration. This force is detected by other piezoelectric ceramics.

The conditioning signal circuitry contains a high-pass filter with a cutoff frequency of 0.3 Hz, and a low-pass filter with a cutoff frequency of approximately 1 kHz. To adapt this signal to the SH and ADC devices, a conditioning module has been added whose main objective is to amplify the signal to the range admitted by the ADC, and to make the offset levels compatible. This is achieved by means of a Zener diode and an operational amplifier. Its schematic can be seen in Figure 21.19.

In order to connect the gyroscope to the IBIS bus, the same interface developed for the accelerometers has been used. The only difference between the interfaces is the address of which the sensor begins the acquisition.

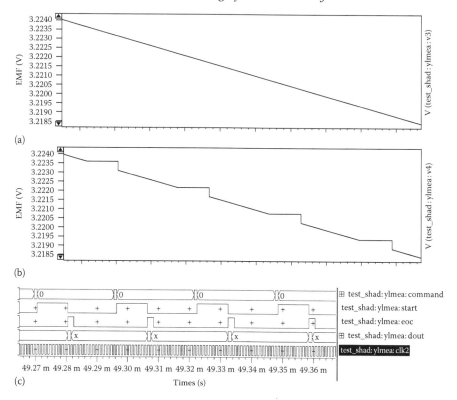

FIGURE 21.16
Simulation of the signal capture.

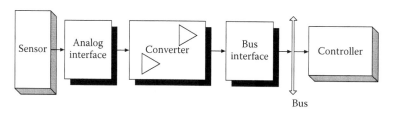

FIGURE 21.17
Structure: Sensor + IBIS driver + controller.

21.7 Smart Sensor Simulation

21.7.1 IBIS Drivers in Sensors

All the sensors and actuators need a specific driver to function properly, while the driver also implements the interface function with the dedicated

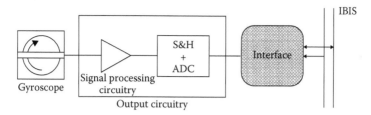

FIGURE 21.18
Gyroscope smart sensor schematic.

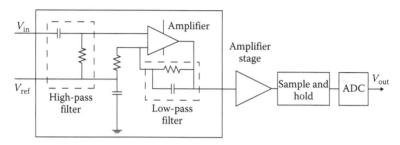

FIGURE 21.19
Gyroscope output circuitry.

bus sensors by containing the communication protocol. The implementation of the drivers has been developed in VHDL. The result has been validated in an FPGA, and the resulting IP model will be used in future developments based on the same bus sensor application. Two parts compose each driver: (a) the one for the specific communication with the sensor or actuator and its additional signal processing circuitry (this part is specific for each sensor, because it depends on the nature of it), and (b) the one that implements the communication bus protocol itself.

These drivers are necessary to connect the slaves with the master of the bus and they are composed of different modules such as a fragmentation module, a Manchester encoding module, a Manchester decoding module, a module of frame formation, and all the involved mechanisms for a good operation of the bus. As an example we can mention the frequency divider or the intern buffers, to ensure that the data is not lost. As a result, the bus, because of its drivers, allows hot plugging and plug and play mode.

21.7.2 Interface of the Gyroscope

The connection between the digital part of the IBIS bus and the sensor has been realized with the addition of an interface. Its main function is to receive commands from the IBIS and to synchronize the data acquisition according to these commands.

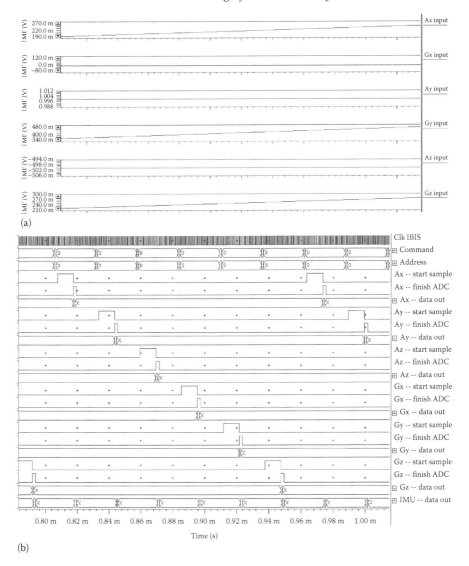

FIGURE 21.20
IMU simulation.

The behavior of the interface can be summarized as follows:

- The interface is always expecting data, so when no data are received the interface status does not change.
- When the IBIS sends a command, the interface sends signals to the SH and the ADC to start the data acquisition. The SH holds the signal while the ADC converts it to digital data. When the conversion is finished, the ADC informs the interface and sends the digital data.

The interface transmits the data to the IBIS bus, and waits for new commands.

21.8 Simulation and Validation

Once the models for the complete IMU system have been developed, it is possible to cosimulate the entire system. This simulation (see Figure 21.20) involves the three accelerometers and the three gyroscopes with their corresponding excitation signals, and the IBIS bus requesting and receiving data.

This simulation shows the complete process of acquiring the signal after the IBIS sends the command until the data is transmitted to the bus, with the main intermediate signals. Each sensor has been assigned to one address, so when the IBIS wants data from one sensor, it sends a frame composed of an address and a command. When the addressed sensor receives the command from the IBIS, the data capture takes place. When the process is finished, a signal is activated to indicate that the conversion has finished, and the final data is sent through the bus.

21.9 Conclusions

This chapter shows a VHDL-AMS approach to behavioral modeling of MEMS-based microinstrumentation using a design methodology with a distributed architecture. The development of behavioral models plays an important role in the design process, and thanks to the mixed-signal hardware description languages it is possible to cosimulate the entire system composed of elements of a different nature. These models allow reducing simulation time, and give the designer the possibility to mix different levels of abstraction to correct parameters and to evaluate interaction between modules. In future work, these results will help the design process allowing the simulation of electrical and nonelectrical features of microsystems applications since early phases of the design process.

Acknowledgments

The authors gratefully acknowledge the contributions and helpful comments of Guo Yi, Xavier Fitó, and Eleni Kanellou within the IMU modeling. The work presented in this chapter was supported by CICYT under ADDRESS project: TEC 2006-04123 and I3P scholarship program.

References

1. A. Vachoux, *Analog and Mixed-Signal Hardware Description Languages*, Kluwer Academic Publishers, Norwell, MA, July 1997.

2. A. Dewey, H. Dussault, J. Hanna, E. Christen, G. Fedder, B. Romanowicz, and M. Maher, Energy-based characterization of microelectromechanical systems(MEMS) and component modeling using VHDL-AMS, MSM99, *Technical Proceedings of the 1999 International Conference on Modelling and Simulation of Microsystems*, pp. 139–142, Puerto Rico, April 19–21, 1999.

3. B.J. Hosticka, *Circuit and System Design for Silicon Microsensors*, Fraunhofer Institute of Microelectronic Circuits and Systems, Duisburg, Germany, 1997.

4. B. Lorente, J. Oliver, and C. Ferrer, Towards a distributed architecture for MEMS integration, *Journal of Sensors & Actuators A: Physical*, 115(2–3), 2004, 470–475.

5. C. Ferrer and B. Lorente, Smart sensors development based on a distributed bus for microsystems applications, *Proceedings of the SPIE's First International Symposium on Microtechnologies for the New Millennium 2003: Smarts Sensors, Actuators and MEMS*, Maspalomas, Gran Canaria, Spain, May 19–21, 2003.

6. J. Hanna and R. Hillman, A common basis for mixed-technology microsystem modeling, *Technical Proceedings of the 1999 International Conference on Modeling and Simulation of Microsystems*, Puerto Rico, 1999.

7. A. Wada and K. Tan, Top-down design methodology of mixed signal with analog-HDL, *IEICE Transactions on Fundamentals*, E80-A(3), March 1997, 441–446.

8. A. Collado, J.A. Plaza, E. Cabruja, and J. Esteve, Adapting MCM-D technology to a piezoresistive accelerometer packaging, *Journal of Micromechanics and Microengineering*, Institute of Physics Publishing Ltd., Bristol, UK, June 41–44, 2003.

9. D. Potter, Smart plug and play sensors, *IEEE Instrumentation & Measurement Magazine*, 5(1), 28–30, 2002.

10. J. Zhang, Y. Ge, X. Wang, G. Song, and J. Jiang, P2-39: Design and application of the field-bus-based intelligent robot plug and play sensing system, *Sensors, 2002. Proceedings of IEEE*, Orlando, FL, 2002.

11. K. Lee, *Sensor Networking and Interface Standardization*, National Institute of Standards and Technology, Gaithersburg, MD, 2001.

Index